Basic Electrical E

S.N. Singh

Professor
Department of Electrical Engineering
Indian Institute of Technology Kanpur

PHI Learning Private Limited

Delhi-110092
2013

₹ 325.00

BASIC ELECTRICAL ENGINEERING
S.N. Singh

ISBN-978-81-203-4188-3

The export rights of this book are vested solely with the publisher.

Second Printing ... ... August, 2013

Published by Asoke K. Ghosh, PHI Learning Private Limited, Rimjhim House, 111, Patparganj Industrial Estate, Delhi-110092 and Printed by Baba Barkha Nath Printers, Bahadurgarh, Haryana-124507.

To
My Mother

Contents

Preface ....*xv*

1. INTRODUCTION 1–10

1.1   General   *1*
1.2   SI Units and Power of 10   *3*
1.3   Structure of Modern Electric Supply System   *5*
1.4   Energy Management System   *7*
1.5   Network Analysis Approaches   *9*
1.6   Outline of the Book   *9*

2. CIRCUIT ELEMENTS 11–27

2.1   Introduction   *11*
2.2   Energy Sources (Active Elements)   *12*
      2.2.1   Independent Sources   *12*
      2.2.2   Dependent or Controlled Sources   *13*
      2.2.3   Practical (Real) Energy Sources   *14*
2.3   Passive Elements   *15*
      2.3.1   Ideal Resistor   *15*
      2.3.2   Ideal Capacitor   *18*
      2.3.3   Ideal Inductor   *20*
2.4   Important Properties of Ideal Passive Elements   *23*
2.5   Network/Circuit Properties   *23*
      2.5.1   Linear and Non-linear Networks   *24*

2.5.2   Bilateral and Unilateral Networks   24
2.5.3   Lumped and Distributed Parameters Networks   24
2.6   Source Transformation   24
Problems   26

3. ANALYSIS OF DC CIRCUITS 28–69

3.1   Introduction   28
3.2   Circuit Laws   29
    3.2.1   Kirchhoff's Current Law (KCL)   29
    3.2.2   Kirchhoff's Voltage Law (KVL)   30
3.3   Applications of KCL and KVL   31
    3.3.1   Voltage and Current Dividers   31
    3.3.2   Series and Parallel Combinations   33
3.4   Star-Delta or Delta-Star Transformation   36
    3.4.1   Delta-Star Transformation   36
    3.4.2   Star-Delta Transformation   38
3.5   Mesh Current (Loop Current) Method   42
3.6   Node Voltage (or Node Pair Voltage) Method   47
3.7   Linearity and Superposition Theorem   51
    3.7.1   Superposition Theorem   51
3.8   Thevenin's Theorem   53
3.9   Norton's Theorem   58
3.10   Maximum Power Transfer Theorem   61
3.11   Reciprocity Theorem   63
3.12   Millman's Theorem   64
3.13   Two-Port (4-Terminal) Network   64
Problems   65

4. STEADY-STATE ANALYSIS OF AC CIRCUITS 70–112

4.1   Introduction   70
4.2   Sinusoidal Signal   70
    4.2.1   Effective (or RMS) Value   72
    4.2.2   Average Value   73
4.3   Phasor Representation   74
4.4   Applications of Kirchhoff's Laws   80
    4.4.1   Series and Parallel Combinations of Inductors and Capacitors   81
4.5   Definition of Impedance and Admittance   84
4.6   Star-Delta or Delta-Star Transformation   87
    4.6.1   Delta-Star Transformation   88
    4.6.2   Star-Delta Transformation   89
4.7   Network Theorems   90
    4.7.1   Mesh Current (or Loop Current) Method   90
    4.7.2   Node Voltage Method   91

      4.7.3   Thevenin's Theorem   *92*

      4.7.4   Norton's Theorem   *96*

      4.7.5   Maximum Power Transfer Theorem   *98*

  4.8   Power Calculation in AC Circuit   *100*

      4.8.1   Reactive Power   *102*

      4.8.2   Using Phasor Concept   *103*

      4.8.3   Apparent Power   *103*

  4.9   Superposition and Conservation of Power   *107*

  Problems   *108*

5. TRANSIENT ANALYSIS OF AC/DC CIRCUITS 113–141

  5.1   Introduction   *113*

  5.2   Transient Response of First-Order Circuits with DC Source   *113*

      5.2.1   Current through an Ideal Inductor   *113*

      5.2.2   Transient in *RL* Circuit   *114*

      5.2.3   Transient in Series *RC* Circuit   *117*

      5.2.4   Natural and Forced Response of First-Order Circuit   *120*

      5.2.5   Discharge Transients   *122*

      5.2.6   Decay Transients   *123*

  5.3   Transient Response of Second-Order Circuits   *126*

  5.4   Transients Response in Series *RLC* circuit   *127*

  5.5   Laplace Transform   *133*

  5.6   Transient Response in AC Circuits   *134*

  5.7   Initial and Steady-State Values   *136*

  Problems   *138*

6. ELECTRICAL MEASURING INSTRUMENTS AND MEASUREMENTS 142–163

  6.1   Introduction   *142*

  6.2   Types of Electrical Measuring Instruments   *143*

      6.2.1   AC and DC Instruments   *143*

      6.2.2   Absolute and Secondary Instruments   *143*

      6.2.3   Deflection and Null Type Instruments   *143*

      6.2.4   Analog and Digital Instruments   *143*

      6.2.5   Indicating, Recording and Integrating Instruments   *143*

      6.2.6   Direct Measuring and Comparison Instruments   *144*

  6.3   Main Components of Indicating Instruments   *144*

  6.4   Principle of Operation of Indicating Instruments   *145*

      6.4.1   Magnetic Effect   *145*

      6.4.2   Thermal Effect   *145*

      6.4.3   Electrostatic Effect   *145*

      6.4.4   Electromagnetic Effect   *146*

      6.4.5   Hall Effect   *146*

6.5   Controlling Mechanism   *146*
    6.5.1   Gravity Control   *146*
    6.5.2   Spring Control   *147*
6.6   Damping Mechanism   *147*
    6.6.1   Friction Damping   *148*
    6.6.2   Electromagnetic Damping   *148*
6.7   Permanent Magnet Moving Coil Instruments   *149*
6.8   Moving-Iron Instruments   *151*
    6.8.1   Attraction Type   *151*
    6.8.2   Repulsion Type   *151*
6.9   Electrodynamic or Dynamometer Instruments   *153*
6.10   Ammeter Design   *153*
6.11   Voltmeter Design   *155*
6.12   Wattmeter or Power Meter   *157*
    6.12.1   Measurement of Three-Phase Power with Single-Phase Wattmeters   *159*
    6.12.2   Three-Phase Wattmeters   *159*
6.13   Energy Meter   *159*
6.14   Performance of Indicating Instruments   *160*
    6.14.1   Errors   *160*
    6.14.2   Loading Effect   *161*
    6.14.3   Sensitivity and Efficiency   *161*
    6.14.4   Precautions   *162*
Problems   *162*

7.   THREE-PHASE AC CIRCUITS **164–190**
7.1   Introduction   *164*
7.2   Generation of Three-Phase Voltage   *165*
7.3   Star and Delta Connections   *166*
    7.3.1   Star Connection   *166*
    7.3.2   Delta Connection   *168*
7.4   Power in Three-Phase Circuit   *169*
7.5   Three-Phase Load Circuit   *171*
    7.5.1   Star Connected Three-Phase Load   *171*
    7.5.2   Delta-Connected Three-Phase Load   *173*
7.6   Measurement of Real Power   *175*
    7.6.1   One-Wattmeter Method   *175*
    7.6.2   Three-Wattmeter Method   *175*
    7.6.3   Two-Wattmeter Method   *177*
    7.6.4   Power Factor Determination for Three-Phase Balanced Load   *178*
7.7   Single-Phase Analysis of 3φ Circuit   *178*
    7.7.1   Star-Connected Balanced Load   *178*
    7.7.2   Delta-Connected Balanced Load   *179*

7.8   Unbalance Loads   *180*
    7.8.1   Delta-Connected Unbalanced Load   *181*
    7.8.2   Star-Connected Unbalanced Load   *181*
Problems   187

8. RESONANCE 191–202

8.1   Introduction   *191*
8.2   Series *RLC* Resonance   *191*
    8.2.1   Quality Factor   *193*
    8.2.2   Bandwidth and Selectivity   *194*
8.3   Parallel *RLC* Resonance   *196*
    8.3.1   Quality Factor   *198*
    8.3.2   Selectivity and Bandwidth   *199*
8.4   Practical Parallel Resonant Circuits   *200*
8.5   Applications of Resonant Circuits   *201*
Problems   202

9. MAGNETIC CIRCUIT 203–226

9.1   Introduction   *203*
9.2   Ampere's Law   *204*
9.3   Magnetic Resistance or Reluctance   *206*
9.4   Magnetic Circuit Analysis   *206*
9.5   Inductance   *211*
9.6   Mutual Inductance   *212*
    9.6.1   Sign Convention of Mutually Induced Voltages   *215*
    9.6.2   Induced Voltage Phasor   *215*
    9.6.3   Two-Port Representation   *216*
    9.6.4   Computation of Inductances   *219*
Problems   222

10. TRANSFORMERS 227–278

10.1   Introduction   *227*
10.2   Constructional Features   *228*
10.3   Principle of Transformer Action   *230*
    10.3.1   Ideal Transformer Case   *230*
    10.3.2   Induced *emf*–Flux Relationship   *232*
    10.3.3   Impedance Transformation   *233*
10.4   Transformer Rating   *236*
10.5   Losses in Transformer   *236*
    10.5.1   Hysteresis Loss   *237*
    10.5.2   Eddy Current Loss   *238*
    10.5.3   Core Loss   *239*
    10.5.4   Copper Loss   *240*

10.6   Practical Transformer   *240*
    10.6.1   Approximate Equivalent Circuit   *242*
    10.6.2   Exact Equivalent Circuit Analysis   *243*
10.7   Determination of Equivalent Circuit Parameters   *243*
    10.7.1   No Load Test or Open Circuit (OC) Test   *244*
    10.7.2   Short Circuit (SC) Test   *245*
10.8   Voltage Regulation   *248*
10.9   Efficiency of Transformer   *251*
10.10   All Day (Energy) Efficiency   *253*
10.11   Autotransformer   *254*
    10.11.1 Volt-Ampere Rating of Autotransformer   *256*
    10.11.2 Volt-Ampere Rating of Autotransformer   *257*
10.12   Three-Phase Transformers   *260*
    10.12.1 Single-Phase Equivalent   *261*
    10.12.2 Three-Phase Transformer on a Single Magnetic Core   *263*
10.13   Per Unit (PU) Representation   *264*
    10.13.1 Per Unit Representation of Transformers   *265*
    10.13.2 Advantages of Per Unit System   *267*
    10.13.3 Per Unit Impedance Diagram   *267*
Problems   *273*

11. ELECTROMECHANICAL ENERGY CONVERSION 279–289
11.1   Introduction   *279*
11.2   Force on Plunger (Linear Motion)   *280*
11.3   Coenergy   *283*
11.4   Rotational Motion   *284*
11.5   Motional Voltage/Rotational Voltage   *284*
11.6   Electromagnetic Force   *285*
Problems   *288*

12. DIRECT-CURRENT MACHINES 290–330
12.1   Introduction   *290*
12.2   DC Machine Construction   *291*
12.3   Induced Voltage and Torque Development   *292*
    12.3.1   Electro-Motive Force (*emf*) Equation   *293*
    12.3.2   Developed Electromagnetic Torque in DC Machine   *295*
12.4   Magnetization Curves   *296*
12.5   Types of DC Machines   *297*
12.6   DC Generator   *298*
    12.6.1   Separately Excited DC Generator   *298*
    12.6.2   DC Shunt Generator   *299*
    12.6.3   DC Series Generator   *304*
    12.6.4   Compound DC Generators   *304*

12.7   Armature Reaction   *305*

12.8   Voltage Regulation and Efficiency of DC Generators   *306*

    12.8.1   Losses in DC Generators   *307*

12.9   DC Motors   *311*

    12.9.1   DC Shunt Motor   *311*

    12.9.2   Speed Control of DC Shunt Motors   *313*

    12.9.3   Speed Control of Series DC Motors   *318*

12.10   Speed Control of DC Series Motors using Field Diversion Method   *320*

12.11   Ward Leonard Speed Control   *321*

12.12   DC Motor Starters   *322*

12.13   Efficiency   *322*

12.14   Mechanical Load   *326*

Problems   *327*

13. THREE-PHASE INDUCTION MACHINES 331–364

13.1   Introduction   *331*

13.2   Three-Phase Induction Motors   *331*

    13.2.1   Squirrel Cage Induction Motor   *332*

    13.2.2   Wound Rotor or Slip-Ring Induction Motor   *332*

13.3   Rotating Magnetic Field   *333*

13.4   Synchronous Speed   *335*

13.5   Pole Formation   *335*

13.6   Induced Voltages   *336*

    13.6.1   Standstill Condition   *336*

    13.6.2   Running Operation   *337*

13.7   Equivalent Circuit   *339*

    13.7.1   Stator Winding Equivalent   *339*

    13.7.2   Rotor Winding Equivalent   *340*

    13.7.3   Approximate Equivalent Circuit   *342*

    13.7.4   IEEE Recommended Equivalent Circuit   *343*

13.8   Determination of Equivalent Circuit Parameters   *344*

13.9   Torque Characteristic of Induction Motor   *344*

13.10   Performance of Induction Motors   *346*

13.11   Rotor and Stator Currents   *348*

13.12   Efficiency   *348*

13.13   Speed Control of Induction Motor (or Slip Control)   *354*

    13.13.1   Terminal Voltage Control   *354*

    13.13.2   Rotor Resistance Control   *355*

    13.13.3   Pole Changing Method   *356*

    13.13.4   Frequency Control   *356*

13.14   Starting of Induction Motors   *357*

    13.14.1   Auto-transformer Starters   *357*

    13.14.2   Primary Resistor Starters   *358*

13.14.3 Primary Reactor Starters   *358*
13.14.4 Star-Delta Starters   *358*
13.14.5 Solid State Starters   *359*
13.15   Increasing Starting Torque   *359*
Problems   *362*

14. THREE-PHASE SYNCHRONOUS MACHINES 365–388

14.1   Introduction   *365*
14.2   Constructional Details   *366*
14.3   Cylindrical Rotor Synchronous Generator   *367*
14.3.1 Equivalent Circuit   *370*
14.3.2 Power (Torque)-Angle Characteristic   *372*
14.4   Salient-Pole Synchronous Generator   *374*
14.5   Capability Curve of Synchronous Generator   *376*
14.5.1 Armature Heating Limit   *376*
14.5.2 Field Heating Limit   *377*
14.5.3 End Region Heating Limit   *378*
14.6   Parallel Operation of Synchronous Generator   *379*
14.7   Synchronous Motor   *380*
14.8   V-Curve of Synchronous Motor   *381*
14.9   Voltage Regulation and Efficiency   *384*
Problems   *386*

15. FRACTIONAL kW (HORSE-POWER) MOTORS 389–407

15.1   Introduction   *389*
15.2   Single-Phase Induction Motors   *390*
15.3   Equivalent Circuit of Single-Phase Induction Motor   *392*
15.4   Classification of Single-Phase Induction Motors   *394*
15.4.1 Split-Phase Induction Motors   *394*
15.4.2 Capacitor-Start Single-Phase Induction Motors   *395*
15.4.3 Capacitor-Run Single-Phase Induction Motor   *396*
15.4.4 Capacitor-Start Capacitor-Run Single-Phase Induction Motor   *397*
15.4.5 Shaded Pole Single-Phase Induction Motors   *397*
15.5   Stepper Motors   *398*
15.5.1 Variable Reluctance Stepper Motors   *399*
15.5.2 Permanent Magnet Stepper Motors   *402*
15.6   Single-Phase Series (or Universal) Motors   *402*
Problems   *406*

16. ELECTRIC POWER GENERATION TECHNOLOGIES 408–421

16.1   Introduction   *408*
16.2   Conventional Electric Power Plants   *408*
16.2.1 Thermal Power Plants   *408*

     16.2.2   Hydropower Plants   *409*

     16.2.3   Gas Power Plants   *410*

     16.2.4   Nuclear Power Plants   *412*

     16.2.5   Diesel Engines   *413*

16.3   Renewable and Alternate Energy Sources-based Power Generation   *413*

     16.3.1   Wind Power Generation   *413*

     16.3.2   Solar Power Generation   *414*

     16.3.3   Fuel Cells   *416*

     16.3.4   Tidal Power   *417*

     16.3.5   Geothermal Power   *418*

     16.3.6   Biomass Power   *418*

     16.3.7   Magneto Hydrodynamic (MHD) Generation   *419*

16.4   Some Other Energy Sources   *420*

     16.4.1   Cogeneration   *420*

     16.4.2   Combined Heat and Power (CHP)   *421*

     16.4.3   Micro-turbines   *421*

Appendix A *423–429*

Appendix B *430–432*

Appendix C *433*

Bibliography *435*

Index *437–443*

Preface

The impact of application of electric power on the quality of human life has been phenomenal. It has influenced the human life more than anything in the history of mankind. In the modern society, the electric power is a key element which is providing clean, efficient and convenient energy to use. The main purpose of this book is to outline and describe the fundamentals of electric engineering. It is a textbook designed for undergraduate students of almost all branches of engineering of Indian and foreign universities/institutions/colleges.

Basic electrical engineering course is taught to the first or second-year students in almost all branches of engineering education under different titles. This subject is one of the tough subjects to the engineering students. This book is an outcome of author's experience while teaching the course *Electrical Science* at Indian Institute of Technology, Rookree (formerly known as Roorkee University) and *Introduction to Electrical Engineering* at Indian Institute of Technology, Kanpur. While teaching the subjects, the author felt the need for an organized textbook on electrical engineering to provide the fundamental knowledge and basic concepts to the students. To have a clear concept is the prerequisite for basic and lucid knowledge which is a key to success for a teacher as well as a student. This subject is, of course, very important to the core electrical engineers as they have to go through the subjects which utilize these basic concepts. It was experienced that only problem-solving books would not be effective because problems can be formulated in several tricky ways.

Other than heavy mathematics and highly abstract concepts, this book aims at developing interest in the students to understand the subject matter. Most of the books in this area directly start with the problems while very small portion of theoretical background is provided. Furthermore, since this course is the first course for engineering students, it is important to deliver the clear concepts with minimum feed, as this area of course is very broad.

Starting with the evolution of electricity, the structure of electrical power system is described in Chapter 1 along with the correct usage of SI units and power of 10 to avoid common mistakes. Different types of circuit elements and source transformation are presented in Chapter 2. Analysis

xv

of dc circuit, utilizing various theorems is described in Chapter 3 while Chapter 4 elaborates the analysis of ac circuits along with the fundamentals of electric power and phasor notation that is useful to the students and also being used throughout this book. Chapter 5 presents the transient analysis due to switch operations in dc and ac circuits, whereas Chapter 6 is devoted to the measuring instruments and measurement techniques. Analysis of three-phase circuit with balanced and unbalanced loads and three-phase power measurement using wattmeters are explained in Chapter 7. Chapter 8 is devoted to the analysis of series and parallel RLC circuits.

The calculation of self-inductance and mutual inductance are developed using the concept of magnetic circuits in Chapter 9. Working principle of single-phase transformer, losses in transformer, and the equivalent electrical circuit are explained in Chapter 10, together with three-phase transformers and autotransformers. Chapter 11 explains the principle of energy conversion. DC machines (both generators and motors) are discussed in Chapter 12. Various types of dc machines and their performances are outlined. Three-phase induction machines and synchronous machines are explained in Chapters 13 and 14 respectively. Chapter 15 is devoted to single-phase motors used for domestic and some special purposes. A basic understanding of various types of electric power generation including the concept of distributed generations is given in Chapter 16.

Besides the theory and concept, the book provides a suitable number of examples to enhance the problem-solving capability of the students. Moreover, additional problems are also given at the end of chapter for further practice. Almost all information required now-a-days for the students to improve their knowledge and to be successful in study are provided in this text. This book would be rewarding, stimulating and will provide a considerable assistance to the students and engineers in the field of electrical engineering.

The author wishes to acknowledge the encouragement received from Prof. S.C. Srivastava, Electrical Engineering Department, IIT Kanpur, Dr. K.N. Sriavstava, ABB Sweden, Dr. G.K. Singh, IIT Roorkee, Mr. K.S. Verma, KNIT Sultanpur, Dr. K.G. Upadhaya, MMEC Gorakhpur, D. Saxena, IIET Bareily, Dr. J.G. Singh, AIT Bangkok, Dr. Bharat Singh Rajpurohit, IIT Mandi, Dr. Naran, M. Pindoriya, IIT Gandhinagar and Dr. B. Kalyan Kumar, IIT Madras. I appreciate the valuable help received from the students of IIT Kanpur, for their efforts in typing and drawing the diagrams. I thank the graduate and undergraduate students whose enthusiastic participation in the classroom discussions helped me to present many ideas and concepts, as discussed in this book, with greater clarity. I also thank the editorial and production team of PHI Learning, New Delhi, for their sincere cooperation.

Finally, no words are adequate to express my indebtedness to my family members for all their pains and sufferings they have undergone to bring me up to this stage. To them all, I bow in the deepest reverence. I deeply acknowledge the unfailing and unending support, constant encouragement, love and affection received from my wife Vandana and Sons Prashant, Praveen and Prakhar.

S. N. Singh

1

Introduction

1.1 GENERAL

Electricity is a very important form of energy for the socio-economical development. The electric power industry, over the past centuries, continues to shape and contribute to the welfare, progress, and technological advances for the human welfare. The growth of electric energy consumption in the world has been nothing but phenomenal. Per capita consumption of electricity is one of the indicating factors for the growth of any country.

Before the commercial use of electricity, several inventors contributed towards the development and understanding of this commodity (see Table 1.1).

TABLE 1.1 Inventors and their contributions

Name of Inventors	Country	Inventions	Remarks
William Gilbert (1544–1603)	English physician	Magnetic science	—
Charles A. Coulomb (1736–1806)	French engineer	Law of electrostatics	His name is used as unit of charge (Coulomb or C)
James Watt (1736–1819)	English inventor	Steam engine	His name is used to represent the unit of power (Watt or W)
Alessandro Volta (1745–1827)	Italian physicist	Electric piles	His name is used for the unit of potential or voltage (Volt or V)
André Marie Ampère (1775-1836)	French mathematician	Relation between electric current and magnetic field	The unit of current is named after him (Ampere or A)
Georg Simon Ohm (1789–1854)	German mathematician	Relation between voltage and current	His name is used to represent the unit of resistance or impedance (ohm or Ω)

(Contd.)

1

TABLE 1.1   Inventors and their contributions (*Contd.*)

Name of Inventors	Country	Inventions	Remarks
Michael Faraday (1791–1867)	English inventor	Electromagnetic induction, transformers, electromagnetic generators	His name is used to represent the unit of capacitance (Faraday or F)
Joseph Henry (1797–1878)	American physicist	Self-induction, telegraph	His name is used to represent the unit of inductance (Henry or H)
Carl Friedrich Gauss (1777–1855)	German mathematician	Measurement of the earth's magnetic field	Gauss is used as unit of magnetic strength
Wilhelm Eduard Weber (1804–1891)	German physicist	Electromagnetic telegraph	Weber (Wb) is used as unit of magnetic flux.
James Prescott Joule (1818–1889)	British inventor	Mechanical equivalent of head	Unit of energy is used as Joule.
James Clerk Maxwell (1831–1879)	Scottish physicist	Electromagnetic theory of light and law of electro-dynamics	
Ernst Werner Siemens (1816–1892)	German inventor	Invention and development of electrical machines	The unit of conductance (Siemens) is named after them.
Carl Wilhelm Siemens (1823–1883)	"	"	"
Gustav Robert Kirchhoff (1824–1887)	German scientist	Law of circuit analysis (voltage and current laws)	
Thomas Alva Edison (1847–1931)	US engineer	Lamp, motor, phonograph, dc power system	
Heinrich Rudolph Hertz (1857–1894)	German scientist	Nature of electromagnetic waves	His name is used to represent the unit of frequency (Hertz or Hz)
Nikola Tesla (1856–1943)	Croatian inventor	Poly-phase ac systems, induction motor	His name is used to represent the unit of magnetic flux density (Tesla or T)

The electricity, which is seen today, is contributed by several inventors for long time. The commercial use of electricity began in the late 1870s. The first electric power system consisting of generation, transmission and distribution system was established by Thomas Edison at the historic Pearl Station, New York in 1881, which began operational in September 1882. It should be noted that it was a direct current (dc) system. There were several parallel developments in increasing the voltage levels, feeding power to more customers, etc. DC motors invented by Frank Sprague in 1884 were added to the system for driving the mechanical load. The first practical ac distribution system was installed in USA by William Stanley at Great Barrington, Massachusetts, in 1886 for Westinghouse. The development of poly-phase system (three-phase) by N. Tesla increased the attraction of ac systems. In 1889, the first ac transmission line at 4 kV, single phase, 21 km was put into operation in Oregon, North America between Willamette Falls and Portland. Due to parallel developments in dc and ac systems, there was a great controversy in 1890s over the suitability and standardization of these systems. AC advocated by Westinghouse was preferred over dc system due to several advantages such as voltage transformation, possibility of high voltage transmission to increase power with less transmission loss, simpler and economical utilization, etc. In 1893, first three-phase line in Southern California, North America came into operation at 2.3 kV, which was12 km long.

There was phenomenal growth of electric generation over last few decades. In India, for example, the installed capacity of power generation has increased from 5654 MW during 1960–61 to 159398.49 MW as on 31.03.2010 consisting of 36863.4 MW hydro, 102453.98 MW thermal (including coal, gas and diesel), 4560 MW nuclear and 15521.11 MW renewable energy sources. It has gone up to 400 times in USA during last 30 years.

Early developments witnessed the different voltage levels in terms of generation and transmission, and different frequencies of operation. It was felt necessary to have one frequency in one region/country. Up to 1921, the ac system voltage used were 12 kV, 44 kV, and 60 kV (rms line-line), which rose to 165 kV in 1922, 220 kV in 1923, 287 kV in 1935, 330 kV in 1953, 500 kV in 1965, 735 kV in 1966, 765 kV in 1969 and 1100 kV in 1990. Every country has different standard ratings. In India, it is 132 kV, 220 kV for high voltage (HV), and 400 kV and 765 kV for extra high voltage (EHV). Generating unit sizes up to 1300 MW are in service, which was made operational in 1973 at Cumberland station of the Tennessee Valley Authority. The maximum generating voltage available in the world is 33 kV. In India, it is 21 kV and highest unit size is 500 MW till 2009. The growth of unit sizes is possible only after the interconnection and better cooling systems of generators.

With the advent of mercury valves in the early 1950s, High Voltage Direct Current (HVDC) transmission systems became economical for long distance transmission. With the development of new solid-state technology, HVDC has become even more attractive. Today numerous installations with voltages up to ± 800 kV dc have become operational in the world. The dc transmission may be advantageous to ac transmission for more than 500 km for overhead line and 50 km for underground cables. In India, ± 500 kV HVDC transmission from Rihand to Dadri is operational.

Due to limitations of HVDC systems such as cost, availability of land, etc., flexible ac transmission systems (FACTS) controllers which can be installed in the existing ac network are becoming popular because of several advantages. With the development of modern power systems, it becomes more and more important to control the power flow along the transmission corridor. The popular FACTS controllers include static VAr compensator (SVC), static compensators (STATCOM), thyristor controller series compensator (TCSC), thyristor controlled phase angle regulator (TCPAR), unified power flow controller (UPFC), etc.

At present, the most high-voltage transmission lines are operating below thermal ratings due to constraints such as voltage and transient stability limits. Power electronics-based FACTS technology can enhance transmission system control and increase line loading in some cases all the way up to thermal limits thereby without compromising the reliability. Based on these capabilities, bottlenecks can be eliminated, line capacity can be increased, and reliability can be improved. In India, the SVC is in operation at Kanpur station having rating of ±280 MVAr. Several transmission lines (400 kV) are now having TCSC and fixed capacitors.

1.2 SI Units and Power of 10

Modern engineering application uses international systems of units (known as *SI units,* from French Systemè International des Unités). SI units comprised of six quantities are shown in Table 1.2 and all the other units are derived in terms of these six fundamental units.

TABLE 1.2 SI units

Quantity	Unit	Symbol
Length	meter	m
Mass	gram	g
Time	second	s
Electric current	Ampere	A
Temperature	Kelvin	K
Luminous intensity	candela	cd

Many electrical calculations are performed using quantities that are multiplied by power of 10. In practice, it is common to omit writing powers of 10 and, instead, to attach certain prefixes to the units associated with the quantities. Each prefix represents a different power of 10. The most commonly used prefixes, their corresponding powers of 10 and their standard (SI) symbols are shown in Table 1.3. The prefixes can be thought of as multipliers.

TABLE 1.3 Standard prefixes

Prefix	Power of 10	Symbol
atto	10^{-18}	a
femto	10^{-15}	f
pico	10^{-12}	p
nano	10^{-9}	n
micro	10^{-6}	μ
milli	10^{-3}	m
centi	10^{-2}	c
deci	10^{-1}	d
deka	10^{1}	da
hecto	10^{2}	h
kilo	10^{3}	k
mega	10^{6}	M
giga	10^{9}	G
tera	10^{12}	T
peta	10^{15}	P
exa	10^{18}	E

It should be noted that these prefixes are never used alone and they are used with the quantities. Also the lower and upper cases must be used carefully. From table, small letter "m" and capital letter "M" have different powers. Similarly, small letter "k" is used for kilo and capital letter "K" is used for Kelvin as unit of temperature. The right notation of kilometer is "km" where letter "k" denotes the kilo and small letter "m" denotes meter. Thus, representing kilometer as Km or KM or kM is wrong. Similarly, the correct way to represent kilovolt is kV and not Kv or KV.

1.2 SI Units and Power of 10

Modern engineering application uses International systems of units (known as SI units). From French Systeme International des Units, SI units comprised of six quantities are shown in Table 1.2 and all the other units are derived in terms of these six fundamental units.

1.3 STRUCTURE OF MODERN ELECTRIC SUPPLY SYSTEM

Electric power is generated from the various generating plants using different fuels. A large share of electrical energy is generated by thermal, hydroelectric, nuclear and gas power stations, also called *conventional electric energy sources*. Thermal power plants normally use the coal and/or oil. Hydroelectric plants are operated with water which is passed through the turbine to rotate the shaft of electric generator. Nuclear power station uses nuclear energy to generate the steam, an energy transfer medium, for turbine to rotate the synchronous generators. Some amount of electricity is generated through non-conventional sources such as solar, wind, tidal, geo-thermal, etc. This power is also called *green power* as it emits less pollution. The details of various types of electric power generations are described in Chapter 16.

Generated electric power is transmitted to the load centres through power supply network, consisting of transmission lines, transformers and switchgears. Transmission networks are commonly classified into four categories, viz. transmission system, sub-transmission system, primary distribution system and secondary distribution system. The main propose of transmission system is to connect all major generating stations and load centres in the system without supplying any consumers en-route. The generating voltages are limited to 11 kV to 33 kV due to technical problems, such as heating and insulation problem, and are stepped up with the help of generating transformers those connect the generators and the transmission lines. The generating and transmission systems are often called *bulk supply system*. The interconnected transmission system of a state or a region is called the *grid* of state or region. State grids are interconnected with the help of tie lines and form the regional grid.

Generation, transmission and distribution of power, in most of the states of India, are owned and operated by different State Corporations/Boards/Nigams. In addition to these, a few private sector utilities operate in the metropolitan cities like Mumbai, Kolkata, Ahmedabad, etc. Five Regional Power Grids (RPGs), namely Northern RPG, Southern RPG, Western RPG, Eastern RPG and North-Eastern RPG exist to promote the integrated operation of the regional power systems. The responsibility of RPGs is to review the project progress, to plan integrated operation among the utilities in the region, to co-ordinate the maintenance schedules, to determine the availability of power for inter-state utilities transfer, to prescribe the generation schedule and to determine a suitable tariff for the inter-utility exchange of power. The names of states in each RPGs are given in Table 1.4 and Table 1.5 show the installed capacity and energy generation of RPGs.

Several central government organizations such as National Thermal Power Corporation (NTPC), National Hydro Power Corporation (NHPC), Nuclear Power Corporation (NPC), Damodar Valley Corporation (DVC), Bhakhara–Beas Management Board (BBMB) etc. are involved in building and operating large thermal, hydro and nuclear power plants and supply bulk power to the other electric utilities for transmission and distribution. Power Grid Corporation is responsible for bulk power transmission through Extra High Voltage (EHV) transmission lines besides the state networks. Northern RPG and Western RPG have 60% share in the total installed capacity of India.

TABLE 1.4   Organization structure of regional electricity boards

NRPG	WRPG	SRPG	ERPG	NERPG
Haryana	Gujarat	Karnataka	Bihar	Assam
Himanchal Pradesh	Madhya Pradesh	Andhra Pradesh	Andman and Nicobar	Arunachal Pradesh
Delhi	Maharastra	Kerala	West Bengal	Meghalaya
Punjab	Goa	Tamilnadu	Sikkim	Nagaland
Rajasthan	Daman and Diu	Puducherry	Orissa	Tripura
Jammu and Kasmir	Dadar and Nagar Haveli	Lakhadweep	Jharkhand	Manipur
Chandigarh	Chhattisgarh	—	—	Mizoram
Uttar Pradesh	—			—
Uttaranchal	—			

TABLE 1.5   Region-wise installed capacity MW as on 31-03-2010

Region	Hydro	Thermal			Nuclear	RES[1] (MNRE)	Grand Total
		Coal	Gas	Diesel			
NREB	13310.75	21275.00	3536.26	12.99	1620.00	2407.33	42189.33
WREB	7447.50	28145.50	8143.81	17.48	1840.00	4630.74	50225.03
SREB	11107.03	17822.50	4392.78	939.32	1100.00	7938.87	43300.50
EREB	3882.12	16895.38	190.00	17.20	0.00	334.76	21319.46
NEREB	1116.00\	60.0.00	766.00	142.74	0.00	204.16	2288.90
Islands	0.00	0.00	0.00	70.02	0.00	5.25	75.27
All India	36863.40	84198.38	17055.85	1199.75	4560.00	15521.11	159398.49
% share	23.13	52.82	10.70	0.75	2.86	9.74	100

Captive generating capacity connected to the grid (MW) = 19509
[1]RES -Renewable Energy Sources includes Small Hydro Project (SHP), Biomass Gas (BG), Biomass Power (BP), Urban and Industrial waste power (U and I), and Wind Energy.

Source: Central Electricity Authority, New Delhi.

Large electric consumers sometimes get supply from the low voltage transmission lines (132 kV and 66 kV). It is very difficult to distinguish sub-transmission system from the main transmission system. In India, 132 kV systems come under transmission system, whereas 66 kV system is sub-transmission system. In early days when the transmission voltage was not high, lower voltages were listed under transmission. Due to system expansion and increase in voltage level of transmission system, the lower voltage transmission systems are termed as subtransmission systems.

Utilization of electrical power, such as heat, light and mechanical energy, is restricted to low voltage only. Before feeding the powers to consumers, the transmission voltages are stepped down and power is transmitted over distribution lines. Distribution systems are further divided into two categories—primary and secondary distribution systems. Sub-transmission systems form the link between the main receiving station and the secondary substation. At the secondary substations, the voltage is stepped down and power is fed into the primary distribution system, which feeds power to medium large consumers at distribution voltage of 33 kV, 25 kV, 11 kV and 6.6 kV in India. The secondary feeders supply residential and commercial customers at 400 V. Distribution systems have the largest share in power system network.

Figure 1.1 shows the basic elements of modern power system. Every power system network need not necessarily has all the components. In some cases, there is one level of transmission and secondary transmission does not exist.

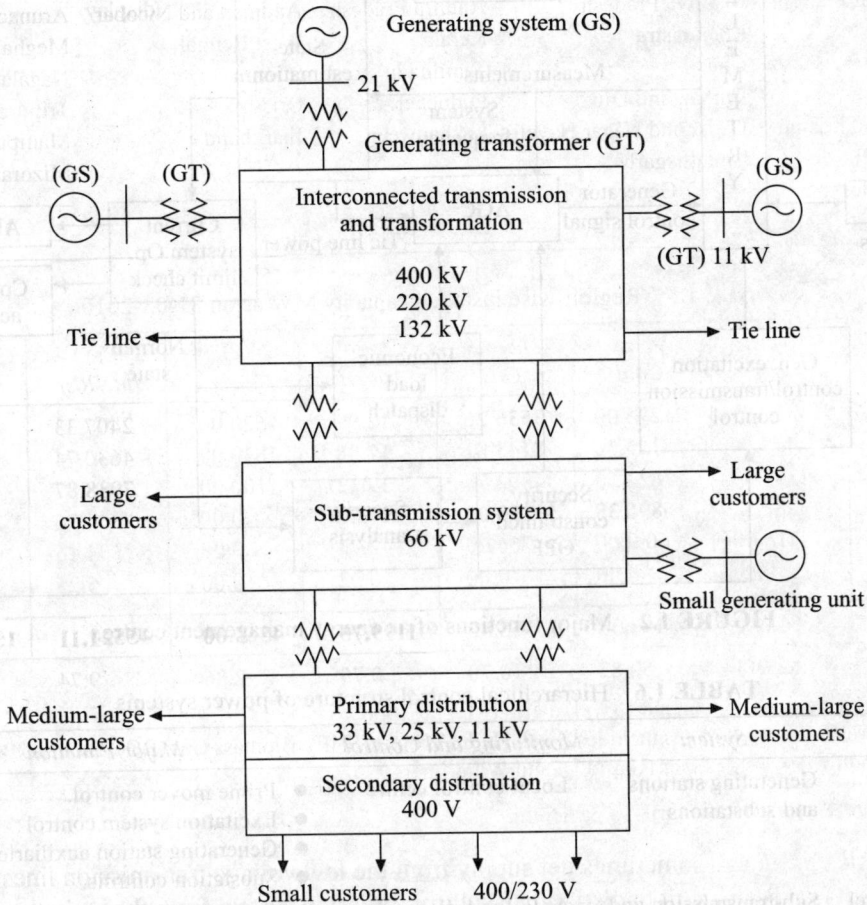

FIGURE 1.1   Structure of power system.

1.4 ENERGY MANAGEMENT SYSTEM

Energy management system (EMS), heart of control centre, extends the scope of the supervisory control and data acquisition (SCADA) systems by the provision of power application software to assist the operator in the monitoring and control of power system. Its importance is further increased due to increase in unit sizes, growth in interconnections and the need to maintain the system security, stability and reliability. Figure 1.2 shows the block diagram of an energy management system. Power systems are characterized by strong hierarchical control structure as given in Table 1.6.

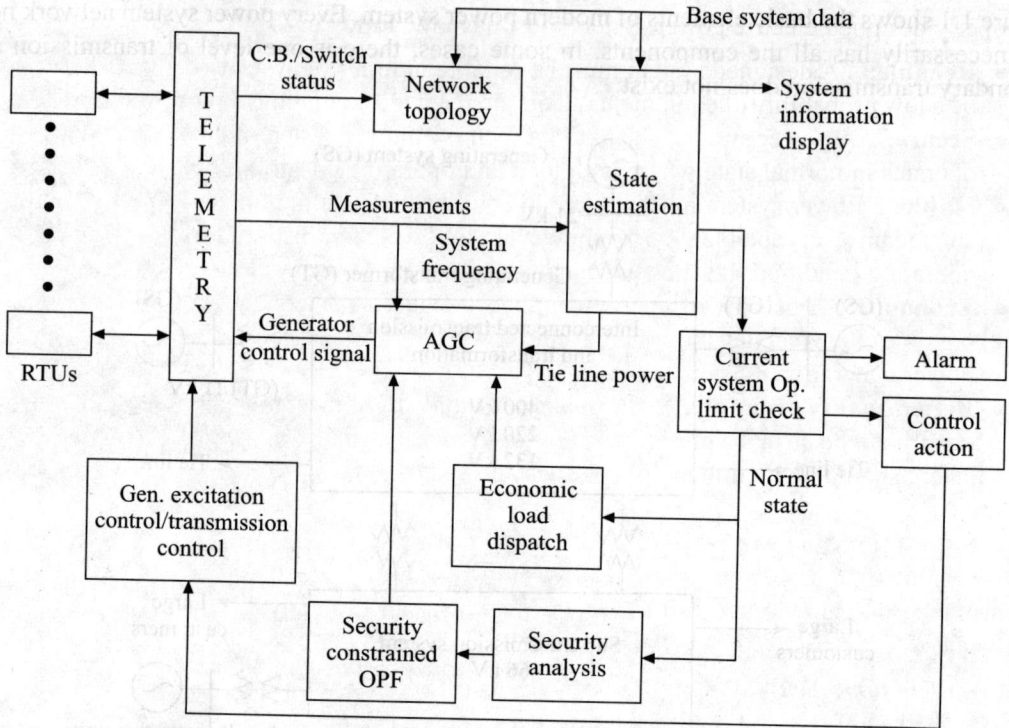

FIGURE 1.2 Major functions of an energy management centre.

TABLE 1.6 Hierarchical control structure of power systems

Level	System	Monitoring and Control	Major Functions
First level	Generating stations and substations	Local control centre	• Prime mover control. • Excitation system control. • Generating station auxiliaries controls. • Substation controls.
Second level	Sub-transmission and transmission networks	Area load dispatch centre	• Generation and load control. • Control as per instruction of third level.
Third level	Transmission system	State load dispatch centre	• System generation and load monitoring and control. • State-wise monitoring and control. • Load shedding and load restoration. • Planning and monitoring of system operations.
Fourth level	Interconnected power system	Regional load centre	• Integrated operation of state load dispatch centres. • Operation and maintenance schedule of generating units/ transmission lines, etc. • Monitor and control of inter-state/ inter-regional power transactions.

A properly designed electric power system must have the following features:

- **Reliable:** A designed system must be reliable in terms of system outages. There should be a low probability of equipment outages to meet the continuous supply of the consumers.
- **Secure:** A power system is said to be secure if power system, following a contingency, operates in normal state where both load and operating constraints are satisfied.
- **Stable:** Power system must operate in equilibrium state in normal operating conditions and regain an acceptable state of equilibrium following any disturbance occurring in normal operating condition. It states the dynamic behaviour of the system.
- **Economic:** The cost of energy supplied to the consumers should be as minimum as possible while meeting the system continuity, reliability and security. For this, the economic power generation must be adopted.
- **Better quality of supply:** The quality of supply must meet the minimum standards. The supply should be in the prescribed range of frequency, voltage and total harmonic distortion. There should be minimum voltage sags and swells.

1.5 NETWORK ANALYSIS APPROACHES

Performance or behaviour of an electrical network can be analyzed by the following two approaches:

(a) *Field approach:*

- It uses electric field or magnetic field concept.
- Analysis is worked out in terms of flux (Φ), field (E, B), *mmf.*
- This is very close to physical phenomenon.
- It is, in general, very cumbersome.

(b) *Circuit approach:*

- Circuit theory is an approximation of field theory.
- Analysis is worked out using voltage (V) and current (I).
- It is very simple and widely used for analyzing the network.
- In this approach, Kirchhoff's voltage law (KVL) and current law (KCL) are used which are a modification of Faraday's law of induction and Amperes circuital law, respectively.

1.6 OUTLINE OF THE BOOK

This book is divided into sixteen chapters. There are three appendices addressing the matrix algebra, trigonometric identities and Laplace transforms of commonly used functions.

Chapter 1 starts with a list of various inventions along with their inventors and their contributions to the present day electricity. The structure of modern electric power system is described along with the energy management system. The correct usage of SI units and power of 10 are presented to avoid common mistakes.

In *Chapter 2*, different electric circuit elements including the active and passive elements are described. The network properties and source transformation are explained. Fundamental concepts of various network theorems are explained in the *Chapter 3* along with the proofs and demonstrated using numerical examples of dc circuits to establish the clear understanding. The theorems discussed in *Chapter 3* are utilized to apply for ac network in *Chapter 4*. The fundamentals of electric power and phasor notation are described that is useful to the students and also being used throughout this book. A detailed description along with the numerical examples and problems is presented in this chapter.

The transient analysis due to closing and opening of switches in dc and ac circuits are elaborated in *Chapter 5*. Laplace transform is introduced to solve the responses of ac and dc circuits. *Chapter 6* is devoted to measuring instruments and measurement. Different types of measuring instruments such as voltmeter, ammeter, wattmeter, energy meter are described in brief. Analysis of three-phase circuit with balanced and unbalanced loads is explained in *Chapter 7*. Basics of three-phase power measurement using wattmeter are discussed with numerical examples. *Chapter 8* is devoted to the analysis of series and parallel RLC circuits. Applications of practical resonant circuits are highlighted.

Fundamentals of magnetic circuits used in various chapters of this book are established in *Chapters 9*. Calculations of self-inductance and mutual inductance are demonstrated with the help of numerical examples. Analogy between the electrical and magnetic circuit for easy understanding of the subject matter is established. The impact of mutual inductance in circuit analysis is explained using the sign (dot-rule) convention.

Starting with the construction details of transformers and various losses in it, the equivalent electrical circuit is developed in *Chapter 10*. The principle of transformer action is elaborated for ideal and practical transformers. Open-circuit tests and short-circuit tests are discussed to determine the parameters of equivalent circuit of transformers. The performance of transformers in terms of voltage regulation and efficiency is demonstrated with examples. Three-phase transformers and autotransformers are also explained in this chapter. *Chapter 11* is devoted to explain the principle of energy conversion.

DC machines (both generators and motors) are discussed in *Chapter 12* including constructional details, induced voltage and torque development, magnetization characteristics, etc. Various types of dc machines and their performances are outlined. The speed control of dc motors along with starters and their efficiencies are described with examples.

Chapter 13 and *Chapter 14* describe the two types of three-phase ac machines: induction machines and synchronous machines, respectively. However, *Chapter 15* is devoted to single-phase motors used for domestic and some special purposes. In *Chapter 13*, the main focus is on three-phase induction motor, whereas three-phase synchronous generator is discussed in *Chapter 14*. *Chapter 13* also elaborates the types of three-phase induction machines, rotating magnetic field and equivalent circuit developments, determination of equivalent circuit parameters, performance analysis, starting and speed control of three-phase induction motors. In *Chapter 14*, salient pole and cylindrical rotor synchronous machines equivalent circuits are derived and the performances of three-phase synchronous generator and motor are analyzed with examples. To derive the equivalent circuit of single-phase induction motor, double revolving theory is used and explained in *Chapter 15*.

Chapter 16 describes the different types of energy sources such as wind power, solar power, fuel cells, tidal power, biogas power, etc. The other sources of electric power generation such as cogeneration, combined heat and power, have also been introduced.

2

Circuit Elements

2.1 INTRODUCTION

All the circuit elements can, broadly, be categorized into the passive and active elements. They are connected by wires to form the closed path to flow the charge, known as *electric circuit*. Current is measured in terms of the rate at which charge is passed through a predetermined area and defined as

$$i = \frac{\Delta q}{\Delta t} = \frac{dq}{dt} \tag{2.1}$$

where Δq is unit of charge flowing through a cross-sectional area in Δt units of time. The smallest amount of charge that exists, is the charge carried by an electron, equal to -1.602×10^{-19} coulomb. Please note that the possible charge on any object will be integer multiple of an electron charge. The magnitude of electron charge is also equal to the proton charge. One coulomb of negative charge is the total charge carried by 6.242×10^{18} electrons. In a closed path, the electrons flow from the negative terminal to the positive terminal of the source and thus, this current flowing from – to +, is often called *electrons current*. The flow of electron is opposite to the proton flow and thus, conventional current is assumed as flow of positive charge and to flow from positive terminal (coming out) to negative terminal (entering into) of source.

To establish a flow of charge in any closed path, it is necessary to exert a kind of force on the electrons that carries the charge. This force is called *electromotive force* (*emf*). The total work per unit charge associated with the motion of charge between any two points, is called *voltage* or *potential difference* between those points. A source of electromotive force may be voltage source or current source. Any electrical apparatus/device can be represented by active and passive elements. With help of electrical circuit representations, the performance and behaviour of any electrical/electronic device can be studied in simple, accurate and efficient manners.

2.2 ENERGY SOURCES (ACTIVE ELEMENTS)

Active element is also known as *energy source* which can be voltage source or current source. An energy source is called as *ideal source* if its output is unaffected by the circuit conditions.

In an ideal voltage source, the terminal voltage of the source is constant, whereas current drawn from the source can vary depending upon the connected load. Similarly, an ideal current source delivers constant current into the circuit, whereas the voltage across the ideal current source varies depending on the circuit parameters. However, practically it is not possible to construct an ideal source.

Depending on the dependency on other elements, energy sources (ideal or practical) can be defined as

- ❑ Independent voltage/current sources
- ❑ Dependent or controlled sources: it can be further classified as
  - • Voltage dependent current source
  - • Voltage dependent voltage source
  - • Current dependent current source
  - • Current dependent voltage source

2.2.1 Independent Sources

An independent source can either be a voltage source or a current source. The output voltage of an independent voltage source does not depend on any circuit element, whereas the output current of an independent current source never depends on any circuit element. An ideal voltage source has zero internal resistance and can never be short circuited. Figure 2.1 shows the independent ideal voltage sources. The output characteristic of an ideal independent voltage source is represented in Figure 2.2. The voltage V in Figure 2.2 shows the voltage magnitude of dc voltage source and root mean square (rms) value of voltage in ac voltage source. For dc quantities (current and voltage), the magnitude is the same as *rms* value. Thus, we can say that quantities in Figure 2.2 are in *rms* values. Direct current (dc) generators and batteries are the examples of dc voltage source, whereas synchronous alternators (generators) are the examples of ac voltage sources.

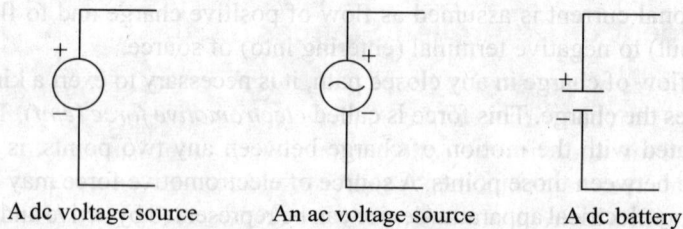

A dc voltage source   An ac voltage source   A dc battery

FIGURE 2.1 Ideal voltage sources.

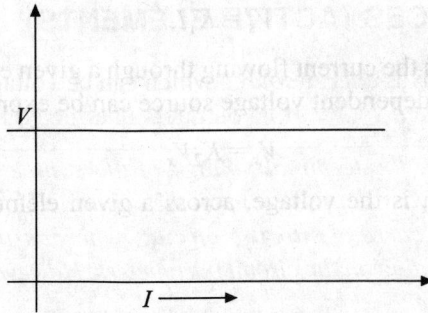

FIGURE 2.2 Voltage and current characteristic of ideal voltage source.

Independent ideal current source has infinite internal resistance and can never be open circuited. It is always connected to the network as shown in Figure 2.3(a). Figure 2.3(b) shows the voltage-current characteristic of an ideal current source. From Figure 2.3(b), the internal impedance can be calculated as

$$\text{Internal impedance} = \frac{\Delta v}{\Delta i} = \infty \text{ ohm}$$

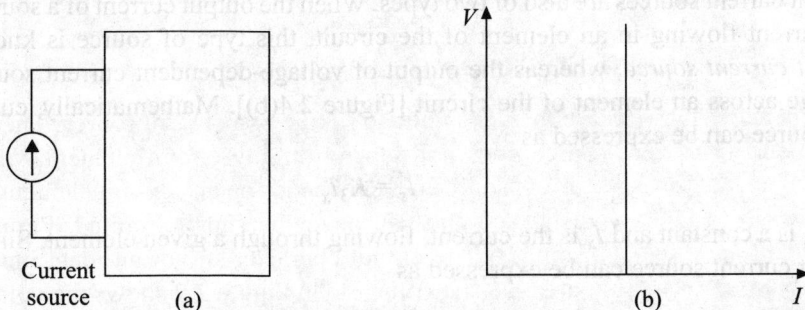

FIGURE 2.3 Ideal current source.

In practice, there is no current source but it can be realized by some circuits. A voltage source with very high internal resistance can be considered as current source as current flows from this source is normally constant.

2.2.2 Dependent or Controlled Sources

Dependent sources are also known as *controlled sources* whose outputs depend on some other variable (voltage or current) in a circuit. A common example of a dependent source is the equivalent current source used for modelling the collector junction in a transistor. Typically, this is modelled as a current-dependent current source, in which collector current is taken to be directly dependent on emitter current. In a current dependent voltage source, the output voltage of the source is dependent on the current, flowing in an element of the circuit, whereas output of a voltage-dependent voltage source depends on the voltage across an element of the circuit. Mathematically, current-dependent voltage source can be expressed as

$$V_s = K_1 I_x \tag{2.2}$$

where K_1 is a constant and I_x is the current flowing through a given element exiting somewhere in the circuit. Similarly, voltage-dependent voltage source can be expressed as

$$V_s = K_2 V_x \tag{2.3}$$

where K_2 is a constant and V_x is the voltage, across a given element. Figure 2.4(a) shows the dependent voltage source.

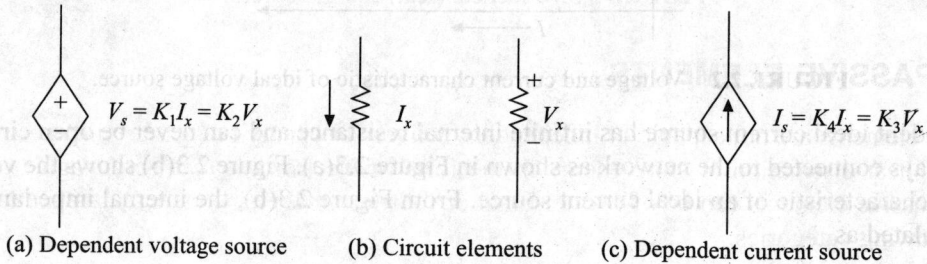

$V_s = K_1 I_x = K_2 V_x$ I_x V_x $I_s = K_4 I_x = K_3 V_x$

(a) Dependent voltage source   (b) Circuit elements   (c) Dependent current source

FIGURE 2.4   Dependent sources.

Dependent current sources are also of two types. When the output current of a source is dependent on the current flowing in an element of the circuit, this type of source is known as *current-dependent current source*, whereas the output of voltage-dependent current source depends on the voltage across an element of the circuit [Figure 2.4(b)]. Mathematically, current-dependent current source can be expressed as

$$I_s = K_3 I_x \tag{2.4}$$

where K_3 is a constant and I_x is the current, flowing through a given element. Similarly, voltage-dependent current source can be expressed as

$$I_s = K_4 V_x \tag{2.5}$$

where K_4 is a constant and V_x is the voltage across a given element. Figure 2.4(c) shows the dependent current source.

In a single source circuit, the direction of positive current flows out of positive terminal of a voltage source. However, in a multiple source circuit, this condition may not be valid. The direction of current in the current source is always in the directed (tip of arrow) direction.

2.2.3   Practical (Real) Energy Sources

It is impossible to have an ideal voltage or current source because all the real (practical) voltage and current sources have internal impedances. Due to the internal impedance (resistance in the case of dc source), a voltage source output is not constant when current is drawn from the source. Similarly, a current source cannot deliver constant current into the circuit due to internal impedance. Practical voltage and current sources are shown in Figure 2.5. Normally, the internal resistance of a voltage source is very small (zero for ideal voltage source), whereas for current source, the internal resistance is very high (infinite for ideal current source).

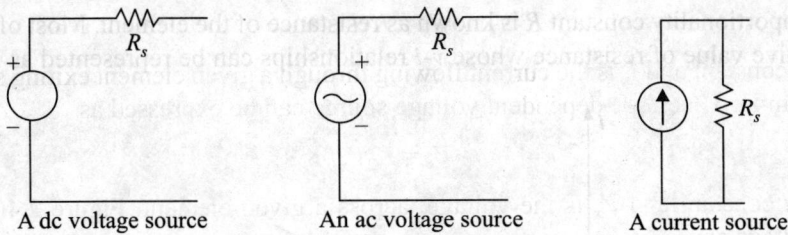

FIGURE 2.5 Practical (real) energy sources.

2.3 PASSIVE ELEMENTS

Passive elements are not capable of independent delivery of energy but can handle energy in the form of heat dissipation or store energy in the form of electrostatic or electromagnetic fields. Basic passive elements can be of two-terminal or four-terminal elements and are classified into the following categories:

(a) Resistor ⎫
(b) Inductor ⎬ two terminal elements
(c) Capacitor ⎭
(d) Coupled coils or mutual inductances (four terminal elements)

These basic elements are used for modelling any practical electrical device. The behaviour of such elements can be defined in form of their v-i characteristic which is important for circuit analysis. The following notations are used.

- Quantities of ac
  $i \rightarrow$ instantaneous current (amp)
  $v \rightarrow$ instantaneous voltage (volt)
  $I \rightarrow$ rms current (amp)
  $V \rightarrow$ rms voltage (volt)

- Quantities of dc
  $I \rightarrow$ current (amp)
  $V \rightarrow$ voltage (volt)

2.3.1 Ideal Resistor

A resistor resists the flow of current. The v-i characteristic is derived from Ohm's law. Georg Simon Ohm was a German physicist and gave the following law in the early nineteenth century. This law holds good for both ac and dc circuits and is valid for only linear resistors.

Ohm's Law: *Voltage across a resistor is directly proportional to the amount of current, flows through it.*

Mathematically, Ohm's law can be expressed as

$$v \propto i \quad \Rightarrow \quad v = Ri \tag{2.6}$$

where, the proportionality constant R is known as resistance of the element. Most of the elements have the positive value of resistance whose v-i relationships can be represented as Figure 2.6.

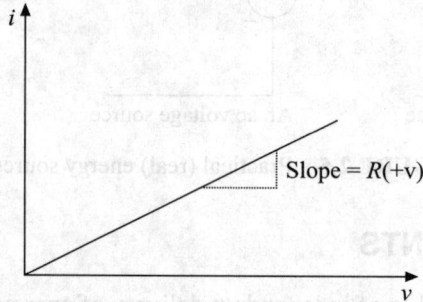

FIGURE 2.6 v-i relationship of an ideal resistor.

The basic units of resistance are Ω (ohm), k Ω (kilo-ohm) and M Ω (Mega-ohm). The value of resistance R (in Ω) can be given by

$$R = \rho \frac{l}{a} \tag{2.7}$$

where l is the length of the element (in metre), a is the area of the element (in metre^2) and ρ is the resistivity of the element (in Ω-metre). The inverse of the resistivity is called *conductivity* and is denoted by σ.

Due to the property of a resistor, the voltage is developed across a resistor in such a way that it opposes the flow of current through it. This voltage developed is also known as *voltage drop across the resistor* as shown in Figure 2.7.

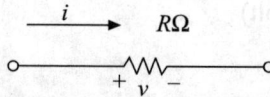

FIGURE 2.7   Voltage drop across resistor.

Certain elements have negative resistance characteristics. Figure 2.8 shows a thyristor characteristic where, in certain portion, resistance is negative. Several elements offer non-linear resistance characteristic. Electronic elements are the best examples of this. Figures 2.8 and 2.9 are the examples of non-linear resistances.

The v-i characteristic of an element can also be expressed as

$$i = \frac{1}{R} v = Gv \tag{2.8}$$

where the reciprocal of resistance $\left(\dfrac{1}{R}\right)$ is known as *conductance* (G). The unit of conductance is "Siemens" (S) which is named after German inventor Carl Wilhelm Siemens. The other unit of conductance is "mho" ($\mho$).

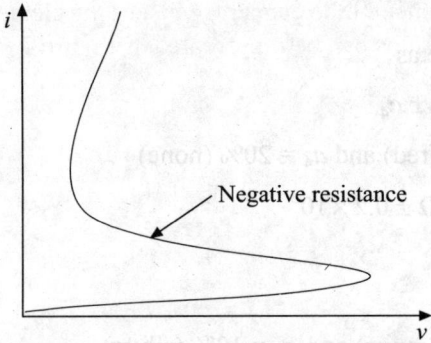

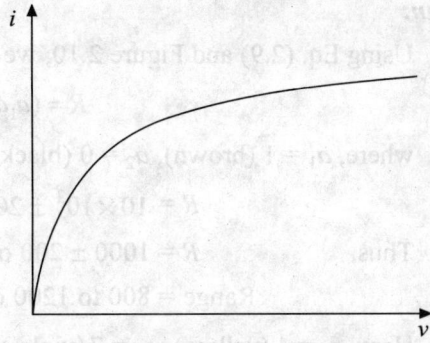

FIGURE 2.8 *v-i* characteristic of a thyristor.   **FIGURE 2.9** Non-linear characteristic of an element.

Physical resistors are made of carbon and metal films. Resistors made of cylindrical section of carbon are very common and available in a wide range of values for several power ratings. The resistance value of many resistors can be determined by reading a series of coloured bands imprinted on the resistor body. Each colour represents a different decimal digit. The common combinations of the digits are the sequence of four, $a_1a_2a_3a_4$, (sometimes three, $a_1a_2a_3$) colour bands. The value of resistor will be

$$R = (a_1a_2) \times 10^{a_3} \pm a_4 \text{ ohm} \qquad (2.9)$$

Figure 2.10 shows resistor colour code with associated digits. The last digit (fourth band), a_4 is called a *tolerance* and represents the percentage tolerance in the actual value. Sometimes, this is not represented and the resistance value will be

$$R = (a_1a_2) \times 10^{a_3} \text{ ohm} \qquad (2.10)$$

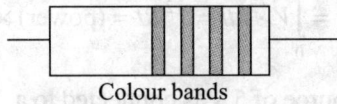

Colour bands

Black	0	Blue	6
Brown	1	Violet	7
Red	2	Gray	8
Orange	3	White	9
Yellow	4	Gold	5%
Green	5	Silver	10%
		None	20%

FIGURE 2.10 Resistor colour code.

EXAMPLE 2.1 Find the nominal resistance and the possible range corresponding to the following colour codes.

(a) Brown, black, red
(b) Yellow, violet, orange, silver

Solution:

(a) Using Eq. (2.9) and Figure 2.10, we can write as

$$R = (a_1 a_2) \times 10^{a_3} \pm a_4$$

where, $a_1 = 1$ (brown), $a_2 = 0$ (black), $a_3 = 2$ (red) and $a_4 = 20\%$ (none)

$$R = 10 \times 10^2 \pm 20\% = 10^3\,\Omega \pm 0.2 \times 10^3$$

Thus, $\qquad R = 1000 \pm 200$ ohm

$$\text{Range} = 800 \text{ to } 1200 \text{ ohm}$$

(b) Here, $a_1 = 4$ (yellow), $a_2 = 7$ (violet), $a_3 = 3$ (orange) and $a_4 = 10\%$ (silver)

$$R = 47 \times 10^3 \pm 10\% = 47\,k\Omega \pm 0.1 \times 47\,k\Omega$$

Thus, $\qquad R = 47 \pm 4.7\,k\Omega$

$$\text{Range} = 42.3 \text{ to } 51.7\,k\Omega$$

Power and energy in a resistor

Power (P) consumed by the resistor is given by

$$P = V \times I = \frac{V^2}{R} = I^2 R \qquad (2.11)$$

From above equation, it can be seen that whatever may be the voltage polarity or the direction of current, the power consumed by a resistor is always positive which is dissipated in the form of heat.

The energy consumed can be expressed as

$$\text{Energy} = \int_0^t V \cdot I dt = I^2 R t = (\text{power}) \times (\text{time}) \qquad (2.12)$$

EXAMPLE 2.2 A dc voltage source of 5 V is connected to a resistor and an unknown circuit element as shown in Figure 2.11. Find the power supplied to or by the dc source,

(a) If current $I = 2$ A
(b) If current $I = -1$ A

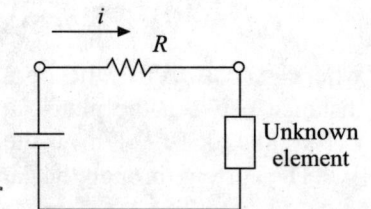

Solution:

(a) Power supplied by the source $= VI = 5 \times 2 = 10$ Watt
(b) Power supplied by the source $= VI = 5 \times -1 = -5$ Watt
Negative shows that power is supplied to the source.

FIGURE 2.11 Example 2.2.

2.3.2 Ideal Capacitor

Generally, a capacitor is made of two plates with a dielectric medium between them. When it is connected to a battery, the plate on the positive side gets positively charged and one on the negative side gets negatively charged. The charge on the plate is proportional to the applied voltage. The steady-state current through the capacitor is zero when it is connected with the dc source. The current flows only till the charging of the plates of capacitor. Figure 2.12(a) shows the charging of capacitor connected with a voltage source V whereas Figure 2.12(b) shows the q-V characteristic of an ideal capacitor which has a linear characteristic.

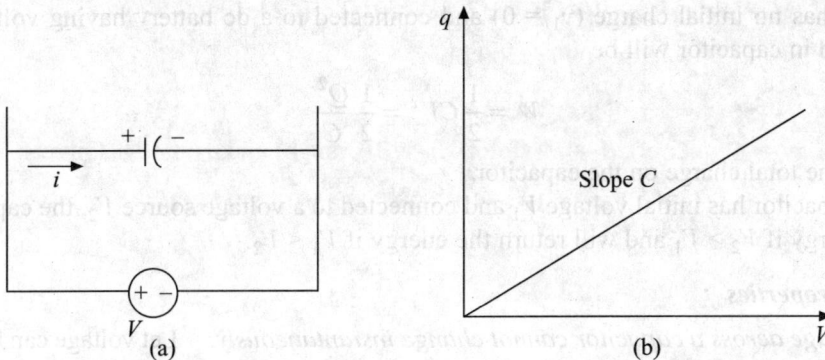

FIGURE 2.12   Charging of a capacitor.

Mathematically, the charge on the capacitor can be expressed as

$$q \propto V \implies q = CV \tag{2.13}$$

where C is a proportionality constant and known as *capacitance* of the capacitor. The unit of capacitance is Farad (F) or coulomb/volt (q/V). Normally, the value of capacitance is very small and therefore, practical unit is micro-farad (μF). Capacitance is a passive circuit element that is related to the geometry and material used between the plates. The capacitance of a capacitor is defined as

$$C = \frac{\varepsilon A}{d} \tag{2.14}$$

where ε ($= \varepsilon_0 \varepsilon_r$) is the permittivity of the material, A is the area of plate and d is the distance between the plates. ε_r is the relative permittivity of the dielectric material whereas ε_0 ($= 0.8842 \times 10^{-11}$ F/m) is the permittivity of air. Its unit is Farad per meter.

The current through the capacitor can be given by

$$i(t) = \frac{dq(t)}{dt} = C \frac{dv(t)}{dt} \tag{2.15}$$

Thus, the voltage across capacitor will be

$$v(t) = v(0) + \frac{1}{C} \int_0^t i(t)\, dt \tag{2.16}$$

where, $v(0)$ is the initial voltage on the capacitor. For sinusoidal voltage, a capacitor offers a linear characteristic.

Energy stored in a capacitor

The energy stored in a capacitor is given by

$$W = \int_v^t v * i\,dt = \int_{V_1}^{V_2} Cv\,dv = \frac{1}{2}C(v_2^2 - v_1^2)$$ (2.17)

If capacitor has no initial charge ($v_1 = 0$) and connected to a dc battery having voltage V, the energy stored in capacitor will be

$$W = \frac{1}{2}CV^2 = \frac{1}{2}\frac{Q^2}{C}$$ (2.18)

where Q is the total charge on the capacitor.

If the capacitor has initial voltage V_1 and connected to a voltage source V_2, the capacitor will store the energy if $V_2 > V_1$ and will return the energy if $V_1 < V_2$.

Important properties

(a) *Voltage across a capacitor cannot change instantaneously:*   Let voltage can be changed instantaneously ($\Delta t = 0$) at any time t as shown in Figure 2.13. The current through the capacitor will be

$$i = C\frac{dV}{dt} = \infty$$ (2.19)

The infinite current is not possible (not practical). Thus, the voltage across capacitor cannot be changed instantaneously.

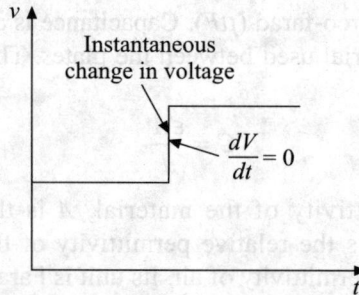

FIGURE 2.13   Instantaneous voltage change across a capacitor.

(b) *Acts as open circuit to dc source in steady state:*   When the capacitor is fully charged, the change in voltage or charge will be zero.

Thus

$$i = C\frac{dV}{dt} = C\frac{dQ}{dt} = 0$$

2.3.3   Ideal Inductor

When a conducting coil is wound on a magnetic core, it behaves as an inductor. If flux produced by the current in the inductor having N turns is ϕ, the flux linkage (λ) will be

$$\lambda = N\phi$$ (2.20)

The flux linkage is proportional to current through the inductor. Mathematically, it can be written as

$$\lambda \propto i \quad \text{or} \quad \lambda = L\,i \tag{2.21}$$

Thus,

$$L = \frac{\lambda}{i} = \frac{N\phi}{i} \tag{2.22}$$

The proportionality constant L is known as *self-inductance* or *inductance* of the coil as shown in Figure 2.14. Figure 2.15 shows the λ-i characteristic of an ideal inductor.

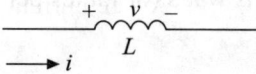

FIGURE 2.14 An ideal inductor. **FIGURE 2.15** λ-i characteristic of an ideal inductor.

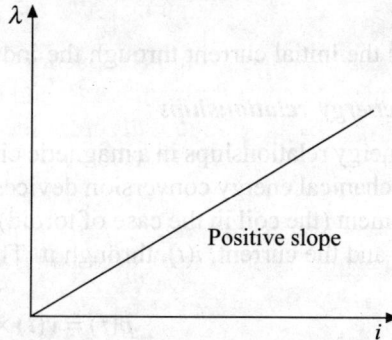

Inductance is a passive circuit element that is related to the geometry and material properties of the structure. From this point of view, inductance is a ratio of total flux linkages in the structure to the current by which flux is produced. The inductance L is related to the reluctance ($\mathfrak{R}$) of the magnetic structure as

$$L = \frac{N^2}{\mathfrak{R}} \tag{2.23}$$

where reluctance is defined in terms of area (A), length of coil (l), relative permeability (μ_r) and permeability of air ($\mu_0 = 4\pi \times 10^{-7}$ Henry per meter), as

$$\mathfrak{R} = \frac{l}{\mu_0 \mu_r A} \tag{2.24}$$

There is no single definition of inductance where the medium is not linear.

A change in a magnetic field establishes an electric field that is manifested as an induced voltage. This basic fact is due to Faraday's experiments and is known as Faraday's Law of Electromagnetic Induction. The electromotive force (*EMF*), or induce voltage (v), is thus equal to the rate of change of flux linkages in the structure.

$$v = \frac{d\lambda}{dt} = N\frac{d\phi}{dt} \tag{2.25}$$

Using relation $\lambda = Li$, above equation can be rewritten as

$$v = \frac{dLi}{dt} = L\frac{di}{dt} + i\frac{dL}{dt} \tag{2.26}$$

In electromechanical energy conversion devices, the reluctance varies with time and thus L also varies with time. Note that if L is constant, we get

$$v = L\frac{di}{dt} \tag{2.27}$$

If the time varying voltage is applied across the inductor, the current through it will be obtained by the following expression.

$$i(t) = i(0) + \frac{1}{L}\int_0^t v(t)\,dt \tag{2.28}$$

where $i(0)$ is the initial current through the inductor.

Power and energy relationships

Power and energy relationships in a magnetic circuit are important for evaluating the performance of electromechanical energy conversion devices. Starting with the fundamental definition, power $p(t)$ in an element (the coil in the case of toroid) is given as the product of the voltage, $v(t)$, across its terminals and the current, $i(t)$, through it. The unit of power is watts (or joules per second).

$$p(t) = v(t) \times i(t) = i(t)\frac{d\lambda}{dt} \tag{2.29}$$

Let us recall the basic relation stating that power $p(t)$ is the rate of change of energy $W(t)$. Thus,

$$dW = \int p(t)\,dt \tag{2.30}$$

or

$$dW = \int p(t)\,dt = \int i(t)\,d\lambda = \int \frac{\lambda}{L}\,d\lambda \tag{2.31}$$

Using above equation, the change in energy

$$\Delta W = \int_{\lambda_1}^{\lambda_2} i(t)\,d\lambda = \frac{1}{2L}(\lambda_2^2 - \lambda_1^2) \tag{2.32}$$

In terms of current change, the change in stored energy will be

$$\Delta W = W_2 - W_1 = \int_{\lambda_1}^{\lambda_2} i(t)\,d\lambda = \int_{i_1}^{i_2} Li(t)\,di(t) = \frac{L}{2}(I_2^2 - I_1^2) \tag{2.32}$$

If I_2 is more than I_1, inductor stores the energy in the form of magnetic energy, whereas in the case of $I_2 < I_1$, inductor releases magnetic energy. If the initial current in the inductor is zero ($I_1 = 0$), the energy stored in the inductor will be

$$W = \frac{L}{2}I^2 \tag{2.33}$$

Basic properties

(a) *Current in the inductor cannot change instantaneously*:   Whether current can be changed instantaneously at any time, t is shown in Figure 2.16. The voltage induced in the inductor will be

$$v = L\frac{di}{dt} = \infty \tag{2.34}$$

The infinite voltage is not possible (not practical). Thus, the current in the inductor cannot be changed instantaneously.

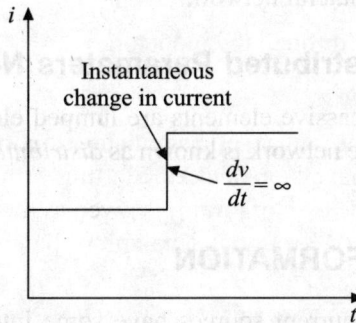

FIGURE 2.16   Instantaneous current change across an inductor.

(a) *Inductor acts as short circuit to dc source*:   When the dc source ($di = 0$) is applied, the induced voltage across the inductor will be zero and thus behaves as a short circuit.

2.4   IMPORTANT PROPERTIES OF IDEAL PASSIVE ELEMENTS

The main properties of ideal passive elements are as follows:

1. Values are constant and independent of the current flowing through them or voltage across them.
2. They are linear and invariant.
3. They are *bilateral* in the sense that if voltage polarity reverses, direction of the current also reverses and vice versa.
4. They are lumped elements which mean that circuit elements carrying no information about their spatial dimensions. There are several elements whose values depend on the length of the element. A transmission line has resistance, inductance and capacitance where the values are directly proportional to the length of line. It is distributed element.

2.5   NETWORK/CIRCUIT PROPERTIES

Electrical networks or circuits are made of several passive and active elements. Normally, an electric network is called a circuit if it has closed paths. The network and circuits are used in the several books, invariably, for the same meaning. A network can be categorized in several ways.

2.5.1   Linear and Non-linear Networks

A network is said to be linear if all the connected passive elements are linear elements. If one or more non-linear elements is (are) connected, the network is known as *non-linear network*. A circuit having electronic element is non-linear network.

2.5.2   Bilateral and Unilateral Networks

If all the passive elements of the network are bilateral, the network is known as *bilateral network*. If one or more unilateral elements is (are) connected, the network is known as *unilateral network*. A circuit having diode is a unilateral network.

2.5.3   Lumped and Distributed Parameters Networks

In a lumped network, all the passive elements are lumped elements. If one or more distributed elements is (are) connected, the network is known as *distributed network*. A transmission line is a distributed parameter network.

2.6   SOURCE TRANSFORMATION

All the practical voltage and current sources have some internal resistance. A voltage source can be converted into a current source and vice versa. This is true for both ac and dc sources. Figure 2.17 shows the conversion of a voltage source, having voltage $\bar{V}$ with an internal impedance $\bar{Z}$, to a current source. The internal impedance of the current source will be the same as $\bar{Z}$ whereas value of current source, $\bar{I}$, will be given by*

$$\bar{I} = \frac{\bar{V}}{\bar{Z}} \quad \text{(for ac energy source)} \tag{2.35}$$

and for dc source

$$I = \frac{V}{R} \tag{2.36}$$

where R is the internal resistance of the voltage source.

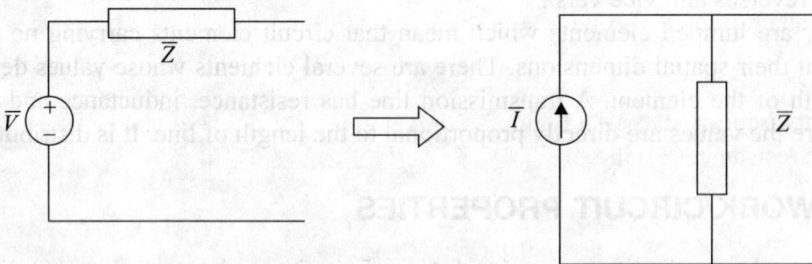

FIGURE 2.17   Conversion of ac voltage source to ac current source.

*The variables with bar are used to denote complex qualities of ac system. This will be discussed in Chapter 4.

Similarly, a current source $\overline{I}$ having internal impedance $\overline{Z}$ can be converted to a voltage source as shown in Figure 2.18 where the internal impedance will be $\overline{Z}$ and voltage will be

$$\overline{V} = \overline{I}\,\overline{Z} \tag{2.37}$$

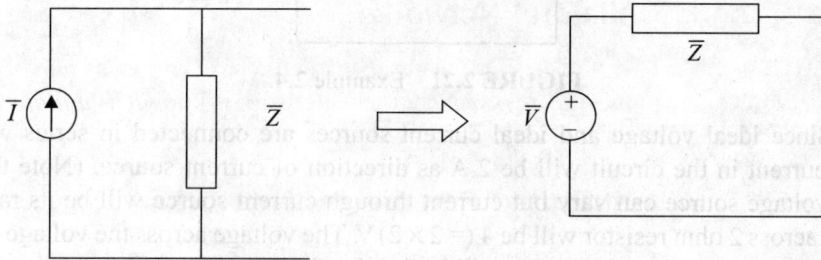

FIGURE 2.18   Conversion of current source to voltage source.

For dc system, the impedance Z is replaced with residence R and complex variables are replaced with real variables.

EXAMPLE 2.3   A current source having internal resistance of 10 ohm is connected to a resistance $R = 5$ ohm as shown in Figure 2.19. Find the current in the resistor R using source transformation.

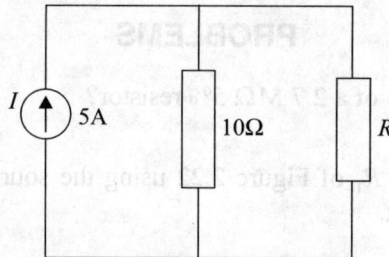

FIGURE 2.19   Example 2.3.

Solution: Changing the voltage source from current source, the value of voltage source will be

$$V = 5 \times 10 = 50 \text{ V}$$

The equivalent circuit is shown in Figure 2.20.
The current in resistance R will be

$$I = \frac{V}{R_s + R} = \frac{50}{10 + 5} = 3.33 \text{ A}$$

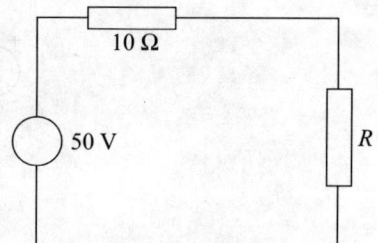

FIGURE 2.20   Equivalent circuit of Example 2.3.

EXAMPLE 2.4   Find the voltage across 2 ohm resistor (Figure 2.21). Also find the power consumed/generated by each element.

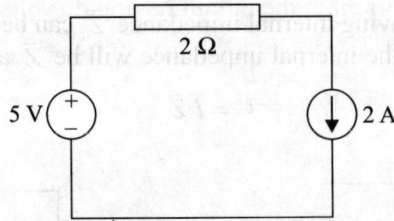

FIGURE 2.21 Example 2.4.

Solution: Since ideal voltage and ideal current sources are connected in series with 2 ohm resistor, the current in the circuit will be 2 A as direction of current source. (Note that current through the voltage source can vary but current through current source will be its rated value). Thus voltage across 2 ohm resistor will be 4 (= 2 × 2) V. The voltage across the voltage source will be 5 V and voltage across current source will be 1(= 5 – 4)V.

Power in various elements will be as

(a) In resistor = $I^2R = 2^2 \times 2 = 8$ W (consuming)
(b) In voltage source = $VI = 10$ W (generating as current is coming out from the positive terminal of the voltage source)
(c) In current source = $VI = 1 \times 2 = 2$ W (consuming as voltage across the source is opposite to the direction of current, i.e. current is coming out from negative terminal).

PROBLEMS

2.1 What is the colour code of a 2.7 MΩ 5% resistor?

[Ans. red, violet, green, gold**]**

2.2 Find the current in the R_1 of Figure 2.22 using the source transformation. Given that $R_1 = 2.5$ Ω and $R_2 = 5$ Ω.

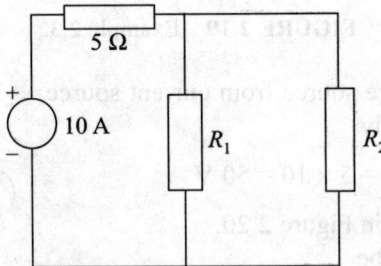

FIGURE 2.22 Problem 2.2.

[Ans. 20/3 A**]**

2.3 A resistive circuit is having a voltage source of 10 V and consuming the 10 W power. What is the resistance seen by the source?

[Ans. 10 ohm**]**

2.4 Find the steady-state current in the circuit having dc voltage source of 10 V as shown in Figure 2.23.

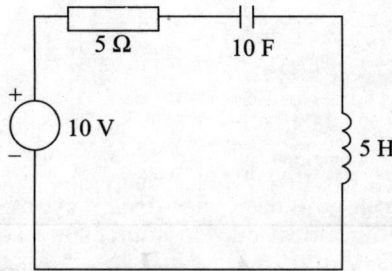

FIGURE 2.23   Problem 2.4.

[**Ans.** 0 A]

2.5 Find the voltage across inductance (1 H) connected in series with a resistance ($R = 5\ \Omega$) and a dc voltage source (10 V). What will be the voltage across the resistor?

[**Ans.** 0 V, 10 V̇]

2.6 One ideal voltage source of 2 V, one ideal current source of 5 A and one resistor of 5 Ω are connected in series. Assume the direction of current is coming out the voltage source. Find the voltage across the current source. Also find the power delivered/consumed by each element in the circuit.

[**Ans.** 23 V, 10 W, 115 W, 125 W]

3

Analysis of DC Circuits

3.1 INTRODUCTION

The performance or behaviour of an electric network can be analyzed by two approaches—field approach and circuit approach. Field approach utilizes the field concepts and the analysis is carried out in terms of flux, field strength, magneto-motive force (MMF), flux density and other magnetic concepts. Though, much close to the physical phenomenon, it is in general, cumbersome. The circuit approach analyzes in terms of voltage across and current through various elements which are represented as a combination of resistors, inductors and capacitors. The circuit approach is much simpler and is widely used for analyzing the behaviour of the system or network.

There are several terms, which are explained below, used in the circuit approach.

(a) **Node:** A node in a network is an equipotential surface at which two or more circuit elements are joined. It is a connection point of various terminals together. The identification of proper nodes are very important in the analysis of electrical network. In Figure 3.1, *a, b, c, d* and *e* are the nodes.

(b) **Junction:** Junction is a node where three or more circuit elements are joined. (Sometimes, it is also referred to as *independent node*). Nodes *b* and *e* of Figure 3.1 are the junctions.

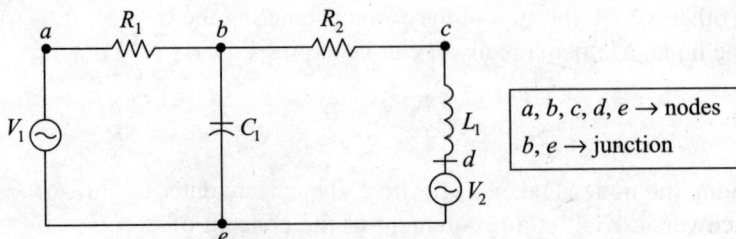

FIGURE 3.1 Nodes and junctions.

28

(c) **Branch:**   A branch is that part of network which lies between two junction points. A branch can have one or more circuit elements. Figure 3.1 has three branches: one consisting of R_1 and V_1, second is R_2, L_1, V_2 and third one is C_1.

(d) **Loop:**   Loop is any closed path of the network. It begins at one terminal of a branch and ends at opposite terminal. Figure 3.1 has three loops: *abea*, *bcdeb* and *abcdea*.

(e) **Mesh:**   It is a minimum number of loop which covers all the elements of the network. It must be so defined that it cannot be further devided into other loop. The network shown in Figure 3.1 has two meshes: *abea* and *bcdeb* whereas *abcdea* is not a mesh. It must be identified very carefully. Mathematically, it can be calculated as

Number of mesh (m) = Number of branches (l) – [number of junctions (j) – 1]   (3.1)

Thus, for Figure 3.1, the number of meshes (independent loops or basic loops) will be

$$m = 3 - (2 - 1) = 2$$

3.2 CIRCUIT LAWS

A German Professor Gustav Robert Kirchhoff (1848) was the first to provide systematic formulation of the principles governing the behaviour of electric circuit. He formulated two laws: a current law and a voltage law which together are popularly known as Kirchhoff's Laws. In fact, electric circuit theory is mainly dependent on these two basic laws.

The electric circuit law/theory deals with two primitive quantities—voltage and current. Current is the actual flow of charged carriers, while difference in potential (voltage) is the force that causes the current flow. Voltage is a single-valued function that may uniquely be defined over the nodes of a network. Current, on the other hand, flows through the branches of the network. Network topology is determined by the interconnection of its elements. The elementary network elements affect voltages and currents in one of the three ways:

- Voltage sources constrain the voltage at node to be of some fixed value (the value of the source).
- Current sources constrain the current through the branch to be of some fixed value (the value of source).
- All other elements impose some sort of relationship, either linear or non-linear, between voltage across and current through the branch.

3.2.1 Kirchhoff's Current Law (KCL)

This law states that the algebraic sum of the currents (ac or dc) into a node (junction) at any instant is zero. In other words, the sum of the currents entering the node is equal to the sum of the current leaving the node. Mathematically, it can be expressed as

$$\sum_{\text{node}} I_k = 0 \qquad (3.2)$$

If the current leaving the node is taken as positive, the current entering the node should be taken as negative or vice versa. KCL is a restatement of the concept of total charge flow. Figure 3.2 shows the KCL concept.

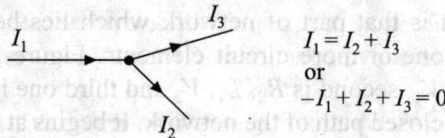

$$I_1 = I_2 + I_3$$
or
$$-I_1 + I_2 + I_3 = 0$$

FIGURE 3.2 Kirchhoff's current law.

3.2.2 Kirchhoff's Voltage Law (KVL)

Kirchhoff's voltage law states that at any time (instant), the algebraic sum of the voltages (ac or dc), taken in the same direction, around a loop (closed path) is zero. In other words, the net voltage around a closed circuit is zero.

$$\sum_{\text{loop}} V_k = 0 \tag{3.3}$$

This law is valid regardless of the direction in which loop is considered. It should also be noted that, in a circuit, the direction of a loop does not have to be the same as the direction of current. However, the sign convention is very important. If you are travelling with from negative to positive, the positive voltage will rise. In Figure 3.3, the sum of voltage across loop *abcdea* (i.e. $a \to b \to c \to d \to e \to a$) is zero.

$$-V_1 - V_2 - V_3 - E_2 + E_1 = 0 \tag{3.4}$$

or voltage across loop (opposite to the current direction) *aedcba* (i.e. $a \to e \to d \to c \to b \to a$) is zero

$$-E_1 + V_1 + V_2 + V_3 + E_2 = 0 \tag{3.5}$$

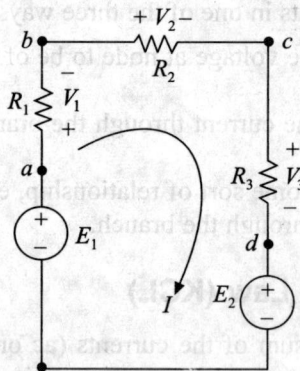

FIGURE 3.3 Loop voltage.

The polarities of impedances are decided by assuming the direction of current. The developed potential difference always opposes the flow of current. However, the polarities of voltage sources remain the same. Normally, moving from the positive terminal to negative terminal in an element, is taken as negative, whereas opposite is also true. It can be seen from Eq. (3.4) while moving $a \to b \to c \to d \to e \to a$. In Eq. (3.4), voltage E_2 has been taken as negative because equation is written by travelling in the direction of current I. Please note that the actual direction of current will be determined solving the above equations. Kirchhoff's law is applicable to both ac and dc circuits.

In fact, KVL is a restatement of law of energy conversion. When current leaves the positive terminal of voltage source, the source supplies energy otherwise it absorbs the energy. If current enters the resistance positive terminal, it absorbs energy. Please note that a resistor always absorbs the energy. If voltage across the current source is positive at the leaving terminal of current, then it generates the power otherwise it absorbs. From Figure 3.3, the energy is supplied by source E_1 whereas source E_2 absorbs the energy. Thus, using energy conservation law,

$$W = W_1 + W_2 + W_3 + W_4 \tag{3.6}$$

where, W is the energy supplied by the source E_1 and W_1, W_2 and W_3 are the energy dissipated in the resistances R_1, R_2, and R_3 respectively. W_4 is the energy absorbed in the source E_2.

If Q charge flows in the loop,

$$\frac{W}{Q} = \frac{W_1}{Q} + \frac{W_2}{Q} + \frac{W_3}{Q} + \frac{W_4}{Q} \tag{3.7}$$

Thus,

$$E_1 = V_1 + V_2 + V_3 + E_2 \quad \text{or} \quad E_1 - V_1 - V_2 - V_3 - E_2 = 0$$

which is similar to Eq. (3.4).

3.3 APPLICATIONS OF KCL AND KVL

3.3.1 Voltage and Current Dividers

Figure 3.4 may be used as an example to show the use of KVL. It has only one loop and three nodes. Around the loop *abca*, KVL is:

$$E - V_1 - V_2 = 0 \tag{3.8}$$

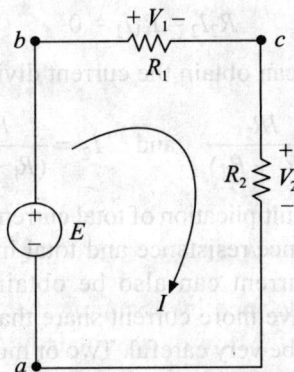

FIGURE 3.4 Voltage divider.

The constitutive relations imposed by the resistances are:

$$V_1 = R_1 I \quad \text{and} \quad V_2 = R_2 I \tag{3.9}$$

From Eqs. (3.8) and (3.9), we have

$$E = (R_1 + R_2)\, I \tag{3.10}$$

Eliminating current, I, from Eq. (3.9), we can get so called *voltage divider* relationship as

$$V_1 = \frac{ER_1}{(R_1 + R_2)} \quad \text{and} \quad V_2 = \frac{ER_2}{(R_1 + R_2)} \tag{3.11}$$

From Eq. (3.11), the voltage across each resistor in a series circuit is directly proportional to the ratio of its resistances to the total resistance of the circuit.

A second example is illustrated by Figure 3.5. KCL at node a will be

$$I - I_1 - I_2 = 0 \tag{3.12}$$

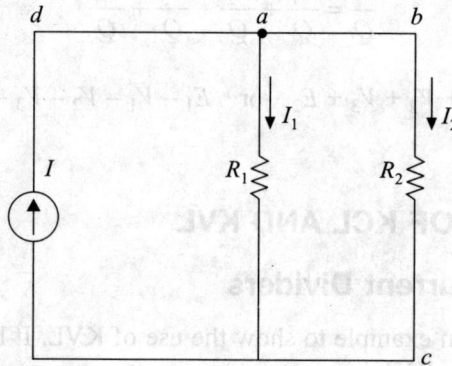

FIGURE 3.5   Current divider.

Using KVL relationship in loop *abca*, we can obtain

$$R_2 I_2 - R_1 I_1 = 0 \tag{3.13}$$

Solving Eqs. (3.12) and (3.13), we can obtain the current divider relationship as

$$I_1 = \frac{IR_2}{(R_1 + R_2)} \quad \text{and} \quad I_2 = \frac{IR_1}{(R_1 + R_2)} \tag{3.14}$$

Thus, the current in branch 1 is the multiplication of total current (going into branch 1 and branch 2) to the ratio of the branch 2 impedance/resistance and total impedance/resistance (branch 1 and branch 2). Similarly, branch 2 current can also be obtained. The smaller resistance path (total resistance of the path) will have more current share than the larger resistance path. While deciding the parallel path, one must be very careful. Two or more circuit elements are said to be in parallel if the same voltage appears across each of the elements [see Eq. (3.13)]. It should be noted that if any path has an active element, this rule is not applicable.

EXAMPLE 3.1   Find the current in 5 ohm resistor of Figure 3.6.

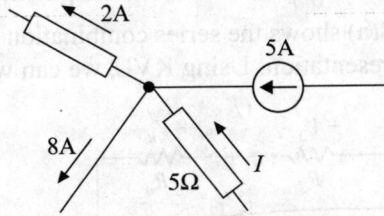

FIGURE 3.6   Example 3.1.

Solution: Let current I into the junction through 5 ohm resistance as shown in Figure 3.6. Using the KCL (assuming current entering into the node is positive and leaving the node is negative). we get

$$I + 5 = 2 + 8$$

or

$$I = 5 \text{ A}$$

EXAMPLE 3.2   Find the current in 5 ohm resistor of Figure 3.7(a) and Figure 3.7(b).

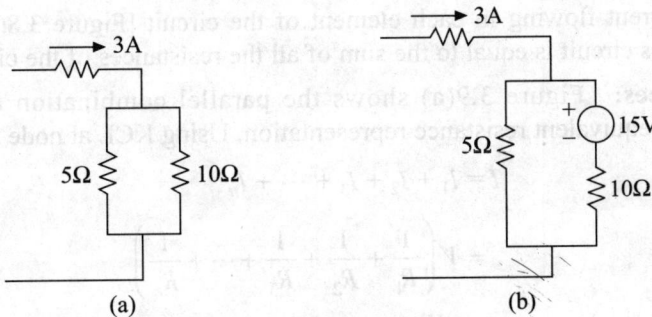

FIGURE 3.7   Example 3.2.

Solution: The current in 5 ohm resistor (using Eq. (3.14)) of Figure 3.7(a) will be

$$I_{5-\Omega} = \frac{3 \times 10}{(5 + 10)} = 2\text{A}$$

In Figure 3.7(b), the current in 5 ohm resistor cannot be calculated using current divider rule. Let current in 5 ohm resistor be I_1. Thus, current in 10 ohm resistor branch will be $(3 - I_1)$. Thus using KVL, we can write as

$$5 \times I_1 = 15 + 10 \times (3 - I_1)$$

Thus, $I_1 = 3$ A. This shows that there is no current in the 10 ohm resistor.

3.3.2   Series and Parallel Combinations

Two or more elements of a circuit are said to be in series if the same current flows through each of the elements. Similarly, two or more circuit elements are said to be in parallel if they are connected to the same nodes and the same voltage is appearing across each of the elements. Series passive elements of one type can be reduced to one equivalent element of the same type. The parallel passive elements of one type can also be reduced to one equivalent element of the same type.

Series resistances: Figure 3.8(a) shows the series combination of *n*-resistors and Figure 3.8(b) is the equivalent resistance representation. Using KVL, we can write

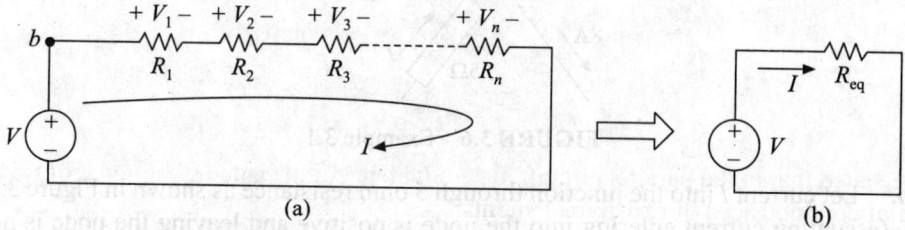

FIGURE 3.8   Series combination of resistances.

$$V = V_1 + V_2 + V_3 + \cdots + V_n$$
$$= I(R_1 + R_2 + R_2 + \cdots + R_n) \tag{3.15}$$
$$= IR_{eq}$$

Since the same current flowing in each element of the circuit [Figure 3.8(b)], the equivalent resistance of a series circuit is equal to the sum of all the resistances of the circuit.

Parallel resistances: Figure 3.9(a) shows the parallel combination of *n*-resistors and Figure 3.9(b) is the equivalent resistance representation. Using KCL at node *a*, we can write

$$I = I_1 + I_2 + I_3 + \cdots + I_n$$
$$= V\left(\frac{1}{R_1} + \frac{1}{R_2} + \frac{1}{R_3} + \cdots + \frac{1}{R_n}\right) \tag{3.16}$$
$$= \frac{V}{R_{eq}}$$

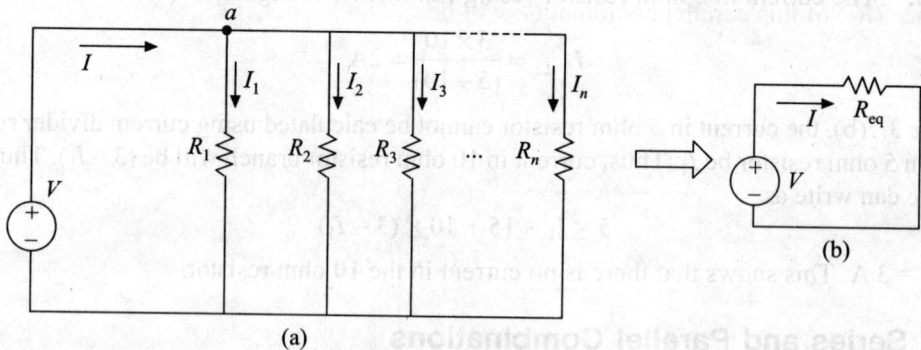

FIGURE 3.9   Parallel combination of resistances.

From Figure 3.9(b), we can also write as

$$I = \frac{V}{R_{eq}}$$

Thus,

$$\frac{1}{R_{eq}} = \left(\frac{1}{R_1} + \frac{1}{R_2} + \frac{1}{R_3} + \cdots + \frac{1}{R_n} \right) \text{ohm} \qquad (3.17)$$

or

$$G_{eq} = G_1 + G_2 + G_3 + \cdots + G_n \text{ Siemen or Mho} \qquad (3.18)$$

It can be stated that in the parallel circuit, the equivalent conductance (reciprocal of resistance) is the sum of conductances of resistive elements.

The series and parallel combinations of capacitors and inductors are discussed in Chapter 4.

EXAMPLE 3.3   Find the equivalent resistances of the series-parallel combination as shown in Figure 3.10.

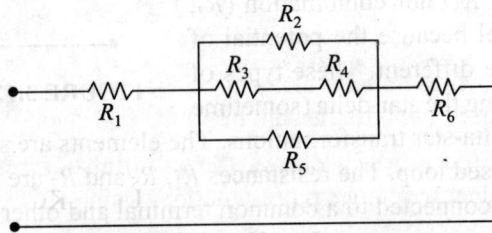

FIGURE 3.10   Example 3.3.

Solution: To solve the circuit, first see which elements are strictly in series or in parallel. In Figure 3.10, resistances R_3 and R_4 are in series because the current in R_3 and R_4 will be the same. Now the equivalent circuit will be as Figure 3.11(a). In Figure 3.11(a), resistance R_{34} will be the sum of R_3 and R_4 ($R_{34} = R_3 + R_4$). The resistances R_2, R_{34} and R_5 are in parallel as they are connected between the same nodes (potential across them will be the same). The equivalent resistance (R_7) of this parallel combination will be

$$\frac{1}{R_7} = \frac{1}{R_2} + \frac{1}{R_{34}} + \frac{1}{R_5}$$

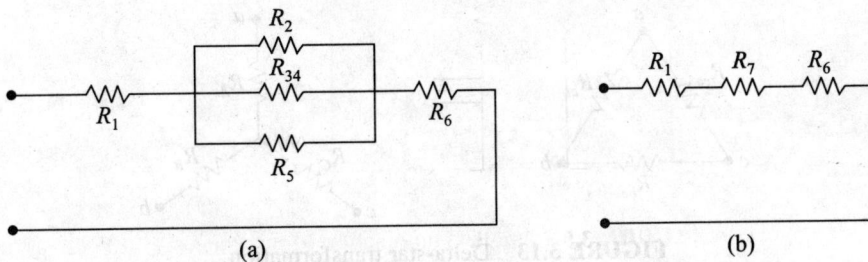

FIGURE 3.11   Equivalent circuits of Example 3.3.

After simplifying the parallel combination, the equivalent circuit can be represented as Figure 3.11(b) in which resistances R_1, R_7 and R_6 are in series as for any applied voltage, the current in these resistances will be the same. Thus, equivalent resistance of the circuit will be

$$R_{eq} = R_1 + R_7 + R_6$$

3.4 STAR-DELTA OR DELTA-STAR TRANSFORMATION

Sometimes network/circuit reduction is not possible using series-parallel combinations. Figure 3.12 shows a typical circuit which cannot be solved using series and parallel combination.

In Figure 3.12, neither resistances R_1 and R_2 nor resistances R_3 and R_4 are in parallel because the voltage at nodes b and c may be different. Similarly neither combination (R_1, R_5 or R_1, R_3) nor combination (R_2, R_5 or R_2, R_4) are in parallel because the potential of nodes a, b, c and d may be different. These types of circuit can be simplified using the star-delta (sometime

FIGURE 3.12 Typical delta-star circuit.

called *wye-delta*) or/and delta-star transformations. The elements are said to be connected in the delta (Δ), if they form a closed loop. The resistances R_1, R_5 and R_2 are in a delta form. If one end of the three resistances are connected to a common terminal and other ends to other points, they are said to be connected in star or wye (Y). The resistances R_1, R_5 and R_3 are in a star form because one end of these resistances is connected at a common point b. The analysis of Figure 3.12 becomes easy if one of the deltas can be transformed into equivalent star. The main concepts behind the star-delta or delta-star transformation are the following.

- The terminals (nodes) are unaffected.
- The equivalent resistance between any two terminals in either of the formation is the same.

3.4.1 Delta-Star Transformation

Using above concepts, delta connection of Figure 3.13 can be transformed into star connection. The resistance values can be computed as follows:

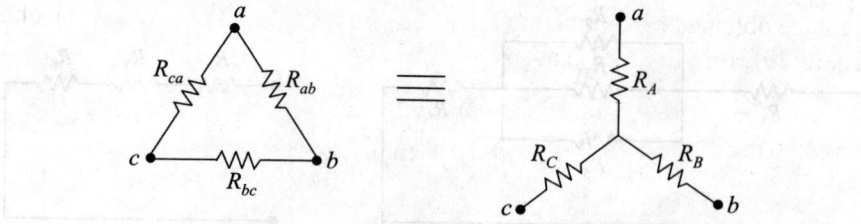

FIGURE 3.13 Delta-star transformation.

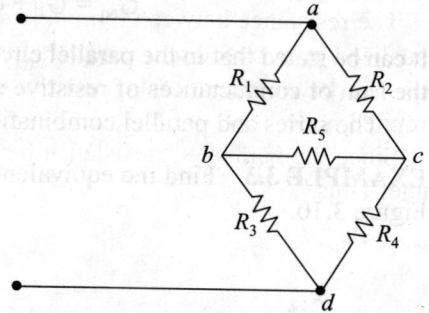

Resistance between terminals a and b of delta connection will be equal to the resistance between terminals a and b of star connection. The resistance between terminals a and b of delta connection will be

$$R_{ab} \| (R_{bc} + R_{ca}) = \frac{R_{ab}(R_{ca} + R_{bc})}{(R_{ab} + R_{ca} + R_{bc})} \qquad (3.19)$$

where, sign $\|$ represents the parallel combination which can be visualized by connecting a voltage source between terminals a and b. Current will be divided between two paths: acb (having resistances R_{ca} and R_{bc}) and ab (having resistance R_{ab}).

The resistance between terminals a and b of star connection will be $(R_A + R_B)$ which can be visualized by connecting a voltage source between terminals a and b of star connection. Current will flow between terminals a and b through resistances R_A and R_B. Terminal c will be as open-circuited node.

Since the equivalent resistance between any two terminals in either of the formation is the same. Thus,

$$R_A + R_B = \frac{R_{ab}(R_{ca} + R_{bc})}{(R_{ab} + R_{ca} + R_{bc})} \qquad (3.20)$$

Similarly, resistance between terminals b and c can be equated as

$$R_B + R_C = \frac{R_{bc}(R_{ab} + R_{ca})}{(R_{ab} + R_{ca} + R_{bc})} \qquad (3.21)$$

and resistance between terminals c and a can be equated as

$$R_C + R_A = \frac{R_{ca}(R_{ab} + R_{bc})}{(R_{ab} + R_{ca} + R_{bc})} \qquad (3.22)$$

Adding Eqs. (3.20–3.22) and simplifying, we get

$$R_A + R_B + R_C = \frac{R_{ab}R_{bc} + R_{bc}R_{ca} + R_{ca}R_{ab}}{(R_{ab} + R_{ca} + R_{bc})} \qquad (3.23)$$

Resistance R_A can be obtained by subtracting Eq. (3.21) from Eq. (3.23) as

$$R_A = \frac{R_{ab}R_{ca}}{(R_{ab} + R_{ca} + R_{bc})} \qquad (3.24)$$

Similarly, R_B can be obtained by subtracting Eq. (3.22) from Eq. (3.23) and R_C can be obtained by subtracting Eq. (3.20) from Eq. (3.27) as

$$R_B = \frac{R_{ab}R_{bc}}{(R_{ab} + R_{ca} + R_{bc})} \qquad (3.25)$$

$$R_C = \frac{R_{bc}R_{ca}}{(R_{ab} + R_{ca} + R_{bc})} \qquad (3.26)$$

Rule: *Any arm of the equivalent Y from delta connection is found by the product of two adjacent arms of the delta connection and dividing the sum of the resistances of the delta connection.*

3.4.2 Star-Delta Transformation

Approach 1

There are several ways to obtain the equivalent resistances of delta connection from star connection. This can be easily proved by applying current source as shown in Figure 3.14.

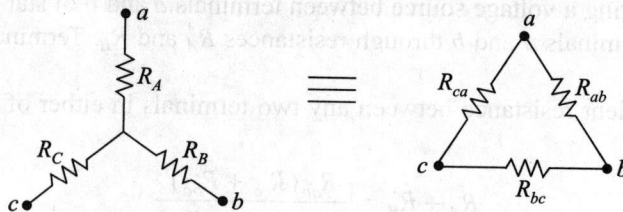

FIGURE 3.14   Star-delta transformation.

Shorting the terminals a and b and measuring the resistance between terminals a and c in both the connections, the equivalent resistances will be equal in either connection. The equivalent resistance [between terminals a and c after shorting the terminals a and b as shown in Figure 3.15(a)] in star connection will be

$$R_C + \frac{R_A R_B}{R_A + R_B} = \frac{R_A R_B + R_B R_C + R_C R_A}{R_A + R_B} \tag{3.27}$$

and in delta connection [as shown in Figure 3.15(b)], it will be

$$\frac{R_{ca} R_{bc}}{R_{ca} + R_{bc}} = 1 \bigg/ \left(\frac{1}{R_{ca}} + \frac{1}{R_{bc}} \right) \tag{3.28}$$

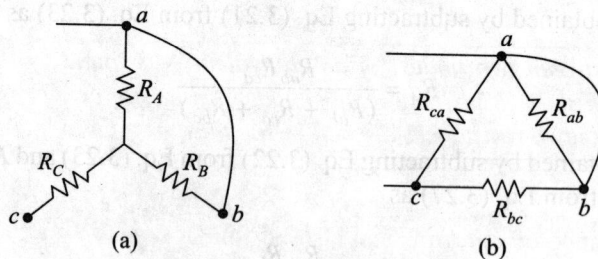

(a)   (b)

FIGURE 3.15   Shorting the terminal.

Since both the values [Eq. (3.27) and Eq. (3.28)] are equal, thus

$$R_C + \frac{R_A R_B}{R_A + R_B} = \frac{R_{ca} R_{bc}}{R_{ca} + R_{bc}}$$

or

$$\left(\frac{1}{R_{ca}} + \frac{1}{R_{bc}} \right) = \frac{R_A + R_B}{R_A R_B + R_B R_C + R_C R_A} \qquad (3.29)$$

Similarly, the following equation can be obtained by shorting terminals b and c, and measuring the resistance between terminals a and b, as

$$\left(\frac{1}{R_{ab}} + \frac{1}{R_{ca}} \right) = \frac{R_B + R_C}{R_A R_B + R_B R_C + R_C R_A} \qquad (3.30)$$

and the following equation can be obtained by shorting terminals c and a, and measuring the resistance between terminals a and b, as

$$\left(\frac{1}{R_{bc}} + \frac{1}{R_{ca}} \right) = \frac{R_C + R_A}{R_A R_B + R_B R_C + R_C R_A} \qquad (3.31)$$

Subtracting Eq. (3.29) from the addition of Eqs. (3.29) to (3.31), we get

$$\left(\frac{1}{R_{ab}} \right) = \frac{R_C}{R_A R_B + R_B R_C + R_C R_A} \qquad (3.32)$$

or

$$R_{ab} = R_A + R_B + \frac{R_A R_B}{R_C} \qquad (3.33)$$

Similarly, subtracting Eq. (3.30) from the addition of Eqs. (3.29) to (3.31) and solving, we have

$$R_{bc} = R_B + R_C + \frac{R_B R_C}{R_A} \qquad (3.34)$$

and subtracting Eq. (3.31) from the addition of Eqs. (3.29) to (3.31) and solving, we have

$$R_{ca} = R_C + R_A + \frac{R_C R_A}{R_B} \qquad (3.35)$$

> **Rule:** *Any arm (between two nodes) of an equivalent Δ is found by the taking sum of the individual resistances of the arms (between two nodes) of the star connection and the term equal to the product of these two resistances divided by the third arm resistance of star connection.*

Approach 2

The equivalent resistance of star arms can also be found by solving Eqs. (3.24) to (3.28). The ratio of Eq. (3.24) and Eq. (3.25) can be written as

$$\frac{R_A}{R_B} = \frac{R_{ca}}{R_{bc}} \qquad (3.36)$$

or

$$R_{ca} = \frac{R_A}{R_B} R_{bc} \qquad (3.37)$$

Similarly, dividing Eq. (3.24) to Eq. (3.25) and solving, we have

$$\frac{R_A}{R_C} = \frac{R_{ab}}{R_{bc}} \qquad (3.38)$$

or

$$R_{ab} = \frac{R_A}{R_C} R_{bc} \qquad (3.39)$$

Putting the values of Eqs. (3.37) and (3.39) in Eq. (3.24), we get

$$R_A = \frac{R_{ab} R_{ca}}{(R_{ab} + R_{ca} + R_{bc})} = \frac{\dfrac{R_A}{R_C} R_{bc} \dfrac{R_A}{R_B} R_{bc}}{\left(\dfrac{R_A}{R_C} R_{bc} + \dfrac{R_A}{R_B} R_{bc} + R_{bc}\right)} \qquad (3.40)$$

or

$$R_A = \frac{R_A^2 R_{bc}}{\left(\dfrac{R_A}{R_C} + \dfrac{R_A}{R_B} + 1\right) R_B R_C} = \frac{R_A^2 R_{bc}}{(R_A R_B + R_B R_C + R_C R_A)} \qquad (3.41)$$

Thus,

$$R_{bc} = \frac{(R_A R_B + R_B R_C + R_C R_A)}{R_A} = R_B + R_C + \frac{R_B R_C}{R_A} \qquad (3.42)$$

Equation (3.42) is the same as Eq. (3.34). Similarly, other delta resistances can be obtained. Equation (3.42) can also be written as the following rule.

> **Rule:** *Any arm of the equivalent Δ connection is found by the taking sum of the products of the arm resistances, taken two at a time, and dividing it by the resistance of the arm of star connection to a terminal to which the delta arm in the consideration is not connected.*

EXAMPLE 3.4 Using the star-delta transformation, find the current in the branch *bc* of Figure 3.16 where all the values are in ohm.

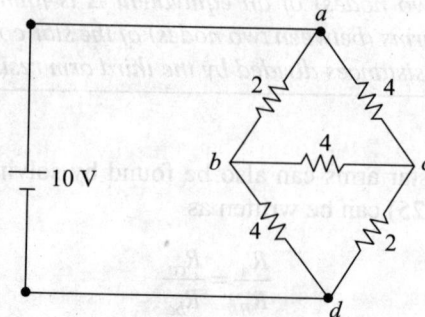

FIGURE 3.16 Example 3.4.

Solution: Converting delta-connection *abc* into star connection *abc* as shown in Figure 3.17.

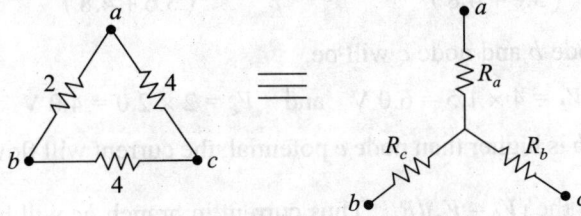

FIGURE 3.17   Star conversion.

The values of R_a, R_b and R_c can be obtained from Eqs. (3.24), (3.25) and (3.26) respectively. Thus,

$$R_a = \frac{2 \times 4}{(2 + 4 + 4)} = 0.8 \text{ ohm}$$

$$R_b = \frac{2 \times 4}{(2 + 4 + 4)} = 0.8 \text{ ohm}$$

$$R_c = \frac{4 \times 4}{(2 + 4 + 4)} = 1.6 \text{ ohm}$$

Now the equivalent circuit will be as shown in Figure 3.18.

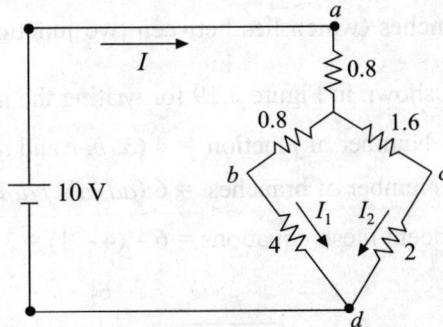

FIGURE 3.18   Circuit after delta to star transformation.

Now series resistances 0.8 and 4 ohms are in parallel with the series resistances 1.6 and 2 ohms.

The equivalent will be $\dfrac{4.8 \times 3.6}{4.8 + 3.6} = 2.057$ ohm. The current (*I*) from source to node *a* will be

$3.5 \left(= \dfrac{10}{0.8 + 2.057} \right)$ A.

To calculate the current in the branch element *bc*, there is no need to go back to the original network. The potential of nodes *b* and *c* can be obtained with respect to node *d*. Using, the current divider approach, the I_1 and I_2 will be

$$I_1 = I\left(\frac{3.6}{3.6+4.8}\right) = 1.5 \quad \text{and} \quad I_2 = I\left(\frac{4.8}{3.6+4.8}\right) = 2.0 \text{ A}$$

Thus, the voltages at node b and node c will be

$$V_b = 4 \times 1.5 = 6.0 \text{ V} \quad \text{and} \quad V_c = 2 \times 2.0 = 4.0 \text{ V}$$

Since potential of node b is higher than node c potential, the current will flow from node b to node c which will be equal to the $(V_b - V_c)/R_{bc}$. Thus current in branch bc will be $\left(\frac{6-4}{4}\right) = 0.5$ A.

Remark: *This transformation will not be valid if some of the elements are with the dependent sources, and is only valid for passive elements.*

3.5 MESH CURRENT (LOOP CURRENT) METHOD

In mesh analysis method, KVL is used. Mesh current flows around the perimeter of a mesh. For convention and for generalization, the clockwise direction of current is normally taken in the loop. The branch current is the actual current flows in a branch and is the same as the mesh current of that branch if the branch is not common to other loops. Otherwise, the branch current is the algebraic sum of the mesh currents flowing in that branch. In any network, the number of independent loop equations (m) will be

$$m = l - (j - 1) \tag{3.43}$$

where, l is the number of branches (which lies between two junctions) and j is the number of junctions (independent nodes).

Let us consider the circuit shown in Figure 3.19 for writing the mesh equations. It has

$$\text{Number of junction} = 4 \ (a, b, c \text{ and } d)$$

$$\text{Number of branches} = 6 \ (ad, ab, bd, bc, cd, \text{ and } ac)$$

$$\text{Number of independent mesh equations} = 6 - (4 - 1) = 3$$

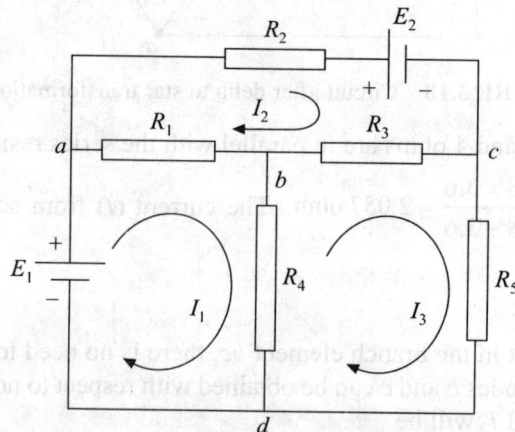

FIGURE 3.19 Mesh equations.

Writing the voltage equation for different mesh as

Mesh 1 (abda and having current I_1):

$$E_1 - R_1 (I_1 - I_2) - R_4 (I_1 - I_3) = 0 \qquad (3.44)$$

Normally, the mesh equations are written along with the mesh current direction whereas the branch currents are considered in writing the voltage equations. The current in branch *ab* (flowing from node *a* to node *b*) is $(I_1 - I_2)$. The voltage drop in the branch will be $R_1(I_1 - I_2)$ which will be opposing the flow of current $(I_1 - I_2)$. It should be noted that the choice of current direction will not affect the *mesh equation*. Consider the current in the branch current is flowing from node *b* to node *a*. The current will be now $(I_2 - I_1)$. The voltage drop will be $R_1(I_2 - I_1)$ and will be opposing the flow of current $(I_2 - I_1)$. The sign of voltage is reverse of the previous case. Thus, voltage drop is the same. Similarly, the voltage drop in element *bd* will be $R_4(I_1 - I_4)$. Equation (3.44) can be rearranged as

$$(R_1 - R_4)I_1 - R_1I_2 - R_4I_3 = E_1 \qquad (3.45)$$

Similarly the mesh equations can be written for mesh 1 and mesh 2 as

Mesh 2 (acba and having current I_2):

$$-E_2 - R_2I_2 - R_3(I_2 - I_3) - R_1(I_2 - I_1) = 0$$

Please note that in mesh 2 equation, term $R_1(I_2 - I_1)$ is taken as positive drop. After rearranging, we have

$$-R_1I_1 + (R_1 + R_2 + R_3)I_2 - R_3I_3 = -E_2 \qquad (3.46)$$

Mesh 3 (bcdb and having current I_3):

$$R_3(I_3 - I_2) + R_5I_3 + R_4 (I_3 - I_1) = 0$$

or

$$-R_4I_1 - R_3I_2 + (R_3 + R_4 + R_5)I_3 = 0 \qquad (3.47)$$

Equations (3.45) to (3.47) can be arranged in a matrix form as

$$\begin{bmatrix} (R_1 + R_4) & -R_1 & -R_4 \\ -R_1 & (R_1 + R_2 + R_3) & -R_3 \\ -R_4 & -R_4 & (R_3 + R_4 + R_5) \end{bmatrix} \begin{bmatrix} I_1 \\ I_2 \\ I_3 \end{bmatrix} = \begin{bmatrix} E_1 \\ -E_2 \\ 0 \end{bmatrix} \qquad (3.48)$$

The above equation can be written in the compact form as

$$[R_{\text{loop}}] \, [I_{\text{loop}}] = [V_{\text{loop}}] \qquad (3.49)$$

In general form, Eq. (3.51) will be written as

$$\begin{bmatrix} R_{11} & R_{12} & R_{13} \\ R_{12} & R_{22} & R_{23} \\ R_{13} & R_{23} & R_{33} \end{bmatrix} \begin{bmatrix} I_1 \\ I_2 \\ I_3 \end{bmatrix} = \begin{bmatrix} E_1 \\ E_2 \\ E_3 \end{bmatrix} \qquad (3.50)$$

Comparing Eq. (3.48) and looking at the network, it can be seen that the diagonal elements, also known as *self-elements*, of the matrix is the sum of the resistances of the mesh, whereas the off-diagonal elements are the negative of the sum of the resistance common to the loop. Thus,

R_{ii} = the sum of the resistances of loop i

$$R_{ij} = \begin{cases} -\sum (\text{Resistance common to the loop } i \text{ and loop } j, \text{ if } I_i \text{ and } I_j \text{ are} \\ \quad \text{in opposite direction in common resistances}) \\ \qquad\qquad\qquad\qquad\qquad \text{or} \\ +\sum (\text{Resistance common to the loop } i \text{ and loop } j, \text{ if } I_i \text{ and } I_j \text{ are} \\ \quad \text{in the same direction in common resistances}) \end{cases} \quad (3.51)$$

The above equation is only true when all the mesh currents are taken in clockwise direction. The sign of voltage vector is decided by the considered current direction. If the mesh current is entering into the positive terminal of the voltage source, the direction of voltage vector element will be negative otherwise it will be positive.

Thus, if the network is not having any dependent sources, the mesh equations as Eq. (3.50) can be written directly without writing the individual mesh equation. The mesh current or loop current can be solved by taking inverse of matrix $[R_{\text{loop}}]$ and pre-multiplied with the voltage vector $[V_{\text{loop}}]$. Mathematically, we can write as

$$[I_{\text{loop}}] = [R_{\text{loop}}]^{-1} [V_{\text{loop}}]$$

Equation (3.50) can also be solved using Cramer's rule as

$$\Delta_1 = \begin{bmatrix} V_1 & R_{12} & R_{13} \\ V_2 & R_{22} & R_{23} \\ V_3 & R_{32} & R_{33} \end{bmatrix}; \quad \Delta_2 = \begin{bmatrix} R_{11} & V_1 & R_{13} \\ R_{21} & V_2 & R_{23} \\ R_{31} & V_3 & R_{33} \end{bmatrix}; \quad \Delta_3 = \begin{bmatrix} R_{11} & R_{12} & V_1 \\ R_{21} & R_{22} & V_2 \\ R_{31} & R_{32} & V_3 \end{bmatrix}$$

$$\Delta = \begin{bmatrix} R_{11} & R_{12} & R_{13} \\ R_{21} & R_{22} & R_{23} \\ R_{31} & R_{32} & R_{33} \end{bmatrix} \quad I_1 = \frac{\Delta_1}{\Delta}; \quad I_2 = \frac{\Delta_2}{\Delta}; \quad I_3 = \frac{\Delta_3}{\Delta} \quad (3.52)$$

Main steps for solving the network using Mesh equation method
1. *Find the independent loops (or mesh) and identify them.*
2. *Associate a current with each mesh going clockwise direction.*
3. *Write voltage equations in terms of branch current for each mesh.*
4. *Solve the equation.*

If there is no dependent source, only two steps are required as

1. Form R_{loop} matrix as Eq. (3.52) and voltage vector.
2. Solve the equations.

Note: *If any mesh or loop is having a current source, that current source should be transformed into voltage source and then mesh equation is derived and solved. If it is not possible to convert current source (ideal) to voltage source, this current will be one mesh current and the remaining mesh current equation should be solved.*

EXAMPLE 3.5 Find the currents in each branch of Figure 3.20 using Mesh current method.

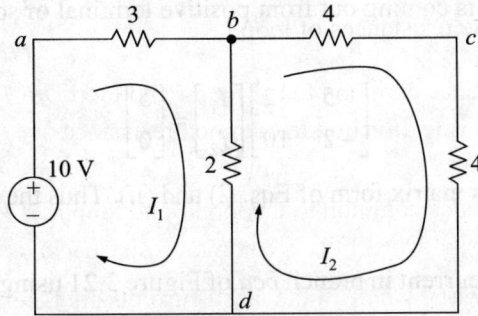

FIGURE 3.20 Example 3.5.

Solution: Figure 3.20 has 2 junctions (b and d) and 3 branches (bad, bd and bcd). Thus the number of independent mesh will be $2 (= 3 - (2 - 1))$. The maximum possible loops will be $abda$, $abcda$, and $bcdb$ but only two loops will be independent. Here, loops $abda$ and $bcdb$ are considered having mesh currents I_1 and I_2, respectively, however, any two loops can be considered and it will not affect the solution.

Method 1:

Voltage equation for mesh 1 will be

$$10 - 3I_1 - 2(I_1 - I_2) = 0$$

or

$$5I_1 - 2I_2 = 10 \qquad \text{(i)}$$

Similarly, voltage equation of mesh 2 will be

$$-4I_2 - 4I_2 - 2(I_2 - I_1) = 0$$

or

$$-2I_1 + 10I_2 = 0 \qquad \text{(ii)}$$

Solving Eqs. (i) and (ii), we get

$$I_1 = 1.85 \text{ A} \quad \text{and} \quad I_2 = 0.43 \text{ A}$$

The current in branch dab will be 1.85 A (as the direction of I_1) whereas current in branch bcd will be 0.43 A (as the direction of I_2). The current in branch bd will be $1.42 (= 1.85 - 0.43)$ A flowing from node b to node d.

Method 2:

Since network has no dependent sources, the matrices of Eq. (3.50) can be formed as follows. Please note that order of matrix **[R]** will be 2×2 as there are only two independent loops. Thus,

R_{11} = sum of resistance of loop 1 = $3 + 2 = 5$ ohm

R_{12} = – (common resistance between loop 1 and loop 2) = -2 ohm

R_{22} = sum of resistance of loop 2 = $4 + 4 + 2 = 10$ ohm

$R_{21} = R_{12} = -2$ ohm

$E_1 = 5$ V (current I_1 is coming out from positive terminal of source)

$E_2 = 0$

$$\begin{bmatrix} 5 & -2 \\ -2 & 10 \end{bmatrix} \begin{bmatrix} I_1 \\ I_2 \end{bmatrix} = \begin{bmatrix} 5 \\ 0 \end{bmatrix} \tag{iii}$$

It can be seen that Eq. (iii) is matrix form of Eqs. (i) and (ii). Thus the solution of Eq. (iii) will be the same.

EXAMPLE 3.6   Find the current in branch *bcd* of Figure 3.21 using the Mesh current method.

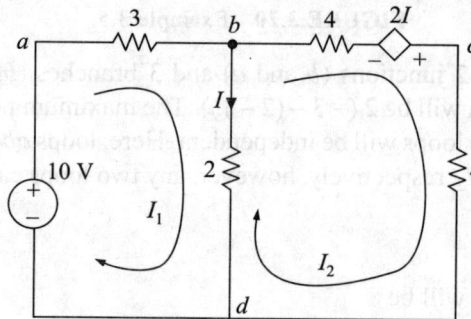

FIGURE 3.21   Example 3.6.

Solution: This figure has also two independent loops as shown.

Voltage equation for mesh 1 will be

$$10 - 3I_1 - 2(I_1 - I_2) = 0$$

or

$$5I_1 - 2I_2 = 10 \tag{i}$$

Similarly voltage equation of mesh 2 will be

$$-4I_2 - 4I_2 - 2(I_2 - I_1) + 2I = 0 \tag{ii}$$

The current *I* which depends on the current in branch *bd* will be $I_1 - I_2$. Equation (3.59) will be

$$-4I_2 - 4I_2 - 2(I_2 - I_1) + 2(I_1 - I_2) = 0$$

or

$$-4I_1 + 12I_2 = 0 \tag{iii}$$

Solving Eqs. (i) and (iii), we get

$$I_1 = 2.31 \text{ A} \quad \text{and} \quad I_2 = 0.77 \text{ A}$$

The currents in branch *bcd* will be 0.77 A (as the direction of I_2).

EXAMPLE 3.7   Find the current in branch *bcd* of Figure 3.22 using Mesh current method.

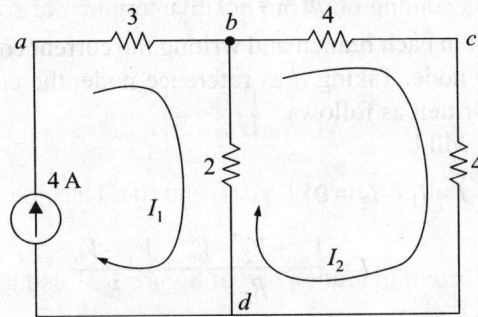

FIGURE 3.22   Example 3.7.

Solution: Figure 3.22 has an ideal current source between node *a* and node *d*. The conversion from current source to the voltage source is not possible. The mesh current I_1 (= 4 A) will be the same as current source. Only mesh equation for loop 2 is to be solved. Voltage equation of mesh 2 will be

$$-4I_2 - 4I_2 - 2(I_2 - I_1) = 0 \tag{i}$$

Putting the vale of I_1 in Eq. (i), we get

$$I_2 = 0.8 \text{ A}$$

The currents in branches *bcd* will be 0.8 A (as the direction of I_2).

3.6 NODE VOLTAGE (OR NODE PAIR VOLTAGE) METHOD

This method uses KCL for analysis. Using this method, one of the nodes (or junctions) is taken as reference node and voltage of other nodes are expressed with respect to it. This is also logical as one point in the electrical circuit is invariably grounded or the voltages of nodes are expressed with respect to the grounded node. Number of independent nodes equations (*n*) will be

$$n = \text{number of junctions } (j) - 1 \tag{3.53}$$

If the numbers of node equations (*n*) are less than the numbers of mesh equations (*m*), the node voltage method is more attractive. Consider the circuit shown in Figure 3.23.

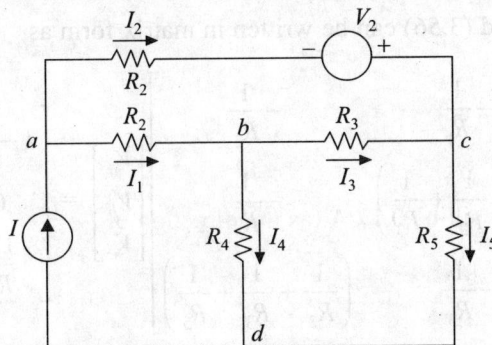

FIGURE 3.23   Node voltage method.

The number of independent nodes will be

$$m = 4 - 1 = 3$$

Assigning the branch current in each branch and writing the current equation using KCL for each junction except at reference node. Taking d as reference node, the current equations at node a, node b and node d will be written as follows.

At node a, the equation will be

$$I - I_1 - I_2 = 0$$

$$I = \frac{V_a + V_2 - V_c}{R_2} + \frac{V_a - V_b}{R_1}$$

or

$$V_a\left(\frac{1}{R_1} + \frac{1}{R_2}\right) - \frac{V_b}{R_1} - \frac{V_c}{R_2} = I - \frac{V_2}{R_2} \qquad (3.54)$$

where V_a, V_b and V_c are the node voltages at node a, node b and node c respectively.

At node b, it will be

$$I_1 - I_3 - I_4 = 0$$

$$\frac{V_a - V_b}{R_1} - \frac{V_b - V_c}{R_3} - \frac{V_b}{R_4} = 0$$

or

$$-\frac{V_a}{R_1} + V_b\left(\frac{1}{R_1} + \frac{1}{R_3} + \frac{1}{R_4}\right) - \frac{V_c}{R_3} = 0 \qquad (3.55)$$

At node c, the current equation will be

$$I_1 - I_3 - I_4 = 0$$

$$\frac{V_a + V_2 - V_c}{R_2} + \frac{V_b - V_c}{R_3} = \frac{V_c}{R_5}$$

or

$$-\frac{V_a}{R_2} - \frac{V_b}{R_3} + V_c\left(\frac{1}{R_2} + \frac{1}{R_3} + \frac{1}{R_5}\right) = \frac{V_2}{R_2} \qquad (3.56)$$

Equations (3.54), (3.55) and (3.56) can be written in matrix form as

$$\begin{bmatrix} \left(\dfrac{1}{R_1} + \dfrac{1}{R_2}\right) & -\dfrac{1}{R_1} & -\dfrac{1}{R_2} \\ -\dfrac{1}{R_1} & \left(\dfrac{1}{R_1} + \dfrac{1}{R_2} + \dfrac{1}{R_3}\right) & -\dfrac{1}{R_3} \\ -\dfrac{1}{R_2} & -\dfrac{1}{R_3} & \left(\dfrac{1}{R_2} + \dfrac{1}{R_3} + \dfrac{1}{R_5}\right) \end{bmatrix} \begin{bmatrix} V_a \\ V_b \\ V_c \end{bmatrix} = \begin{bmatrix} I - \dfrac{V_2}{R_2} \\ 0 \\ \dfrac{V_2}{R_2} \end{bmatrix} \qquad (3.57)$$

or

$$\begin{bmatrix} (G_1 + G_2) & -G_1 & -G_2 \\ -G_1 & (G_1 + G_2 + G_3) & -G_3 \\ -G_2 & -G_3 & (G_2 + G_3 + G_5) \end{bmatrix} \begin{bmatrix} V_a \\ V_b \\ V_c \end{bmatrix} = \begin{bmatrix} I - \dfrac{V_2}{R_2} \\ 0 \\ \dfrac{V_2}{R_2} \end{bmatrix} \qquad (3.58)$$

In compact form, we can write the node voltage equations as

$$[Y_{bus}] \ [V_{bus}] = [I_{bus}] \qquad\qquad (3.59)$$

where vector $[V_{bus}]$ are the node voltage vector, matrix $[I_{bus}]$ is the vector of sum of the current entering to the node. If there are no dependent sources, matrix $[Y_{bus}]$ can be formed by inspection as its element will be

Y_{ii} = the sum of the conductances at node i.
Y_{ij} = negative of conductance between node i and node j.

The node voltage of Eq. (3.59) can be solved by Cramer's Rule or by taking the inverse of matrix $[Y_{bus}]$ and pre-multiplying it to the current vector as.

$$[V_{bus}] = [Y_{bus}]^{-1} \ [I_{bus}] \qquad\qquad (3.60)$$

Main steps of node voltage method:

1. *Chose a reference node (ground node) and label all the nodes except reference node.*
2. *Assign the direction of currents in every branch.*
3. *Write the KCL at all the nodes except the reference node.*
4. *Express the equations of step 3 in terms of node voltages (unknowns).*
5. *Solve the resulting equations.*

EXAMPLE 3.8   Find the currents in each branch of Figure 3.24 using node voltage method.

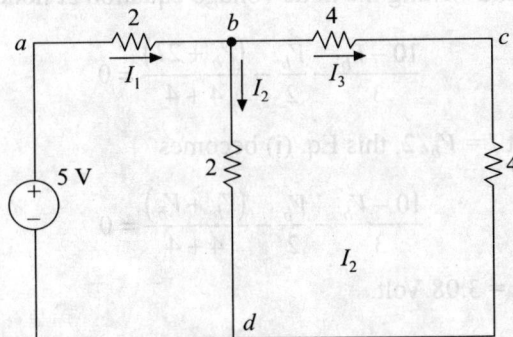

FIGURE 3.24   Example 3.8.

Solution: Figure 3.24 has only 2 junctions (b and d) and thus, the number of independent node voltage equation will be one ($= 2 - 1$). Node-d is taken as reference node ($V_d = 0$). The current direction assigned is shown in Figure 3.24. The KCL at node-b can be written as

$$I_1 - I_2 - I_3 = 0 \qquad\qquad\qquad\qquad (i)$$

The values of currents will be

$$I_1 = \frac{V_a - V_b}{2} = \frac{5 - V_b}{2}; \quad I_2 = \frac{V_b - V_d}{2} = \frac{V_b}{2}; \quad I_3 = \frac{V_b - V_d}{(4+4)} = \frac{V_b}{8} \tag{ii}$$

Putting the values of currents in Eq. (i), we have

$$\frac{5 - V_b}{2} - \frac{V_b}{2} - \frac{V_b}{8} = 0$$

Solving above equation, we get

$$V_b = 2.22 \text{ Volt.}$$

Thus, the current in branches will be

$$I_1 = 1.39 \text{ A}, \quad I_2 = 1.11 \text{ A} \quad \text{and} \quad I_3 = 0.275 \text{ A}$$

EXAMPLE 3.9   Find the current in branches of Figure 3.25 using the node voltage method.

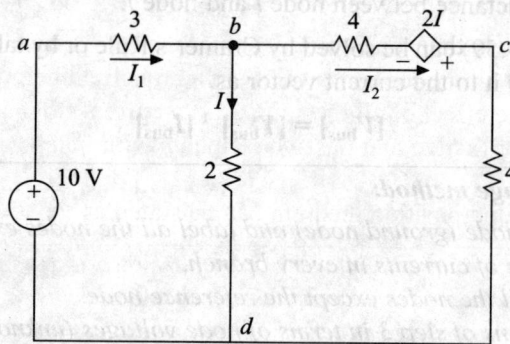

FIGURE 3.25   Example 3.9.

Solution: This figure has also two independent loops but only one independent node. Taking node d is reference node and writing the node voltage equation at node 3, we get

$$\frac{10 - V_b}{3} - \frac{V_b}{2} - \frac{(V_b + 2I)}{4+4} = 0 \tag{i}$$

Since, the value of current $I = V_b/2$, this Eq. (i) becomes

$$\frac{10 - V_b}{3} - \frac{V_b}{2} - \frac{(V_b + V_b)}{4+4} = 0$$

Solving it, the value of $V_b = 3.08$ Volt.
 Thus, the currents

$$I = 1.54 \text{ A}; \quad I_1 = 2.31 \text{ A} \quad \text{and} \quad I_2 = 0.77 \text{ A}$$

The result matches with the Example 3.6.

3.7 LINEARITY AND SUPERPOSITION THEOREM

An extraordinarily powerful notion of network theory is linearity. This property has two essential elements, stated as follow:

1. For any single input x yielding output y, the response to an input kx is ky for any value of k.
2. If, in a multi-input network, the input x_1 by itself yields output y_1 and a second input x_2 by itself yields output y_2, then the combination of inputs x_1 and x_2 yields the output $y = y_1 + y_2$.

Superposition is a very powerful technique to analyze multi-source electric circuit which holds linearity property. Using the superposition principle, the effect of several energy sources acting, simultaneously, on a system can be obtained by analyzing the effect of each source acting alone, and then combining (superimposing) those effects.

3.7.1 Superposition Theorem

If a network contains more than one source, sometimes, it is easier to find out the effect of considering one source at a time for current flow in a branch. Superposition theorem is useful in finding out the cumulative effect of all the sources in a current flow or voltage in a branch. This theorem is valid for linear (active and passive elements with independent sources) and bilateral networks.

In other words, the superposition theorem can be stated as

> **Statement:** *The total current or voltage in any part of a linear, bilateral network having more than one source is equal to the algebraic sum of the currents or voltages with the sources acting individually while other sources are replaced by their internal resistances (short circuiting the voltage sources and open circuiting the current sources).*

"If cause and effect are linearly related, the total effect of several causes acting simultaneously is equal to the sum of effects of individual causes acting one at a time."

> **Procedure for circuit analysis using Superposition theorem:**
>
> 1. *Select any one source in the circuit and remove all other sources (replace voltage source by short circuit and current source by open circuit keeping the internal resistances).*
> 2. *Calculate the desired voltage or current in an element with only one source selected in the step 1.*
> 3. *Repeat the step 1 and step 2 for all the sources, one by one.*
> 4. *Add all the computed values of element obtained with each source acting alone. The sum is the actual voltage or current when all the sources are present and acting simultaneously. The polarity of voltage and direction of current must be taken carefully while adding the quantities.*

Consider a network as shown in Figure 3.26 having two sources. The current in the resistance R_4 will be the sum of the currents I_1 and I_2 of Figure 3.27(a) and Figure 3.27(b), respectively.

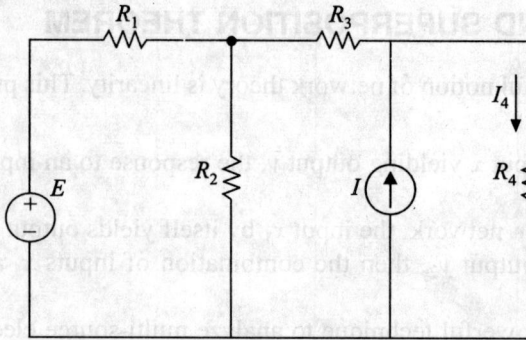

FIGURE 3.26 Superposition theorem example.

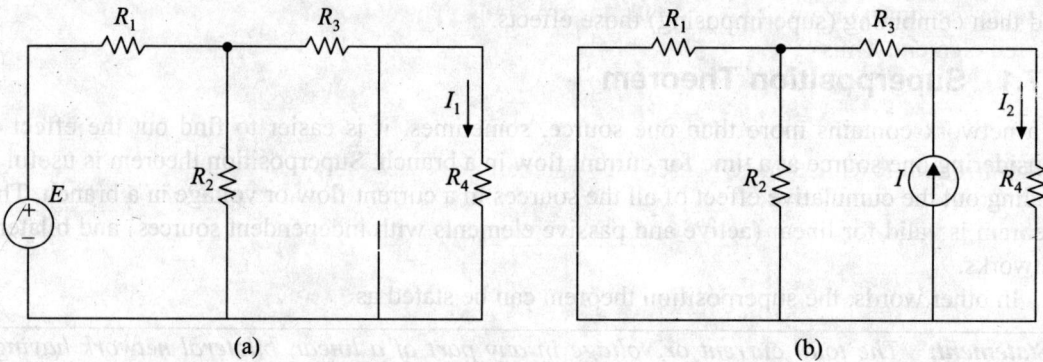

(a)   (b)

FIGURE 3.27 Responses on individual sources.

This shows that Figure 3.26 is equivalent to Figure 3.27(a) plus Figure 3.27(b).

EXAMPLE 3.10 Find the current in branch *ab* of Figure 3.28 using superposition theorem. All the values of resistances are in ohm.

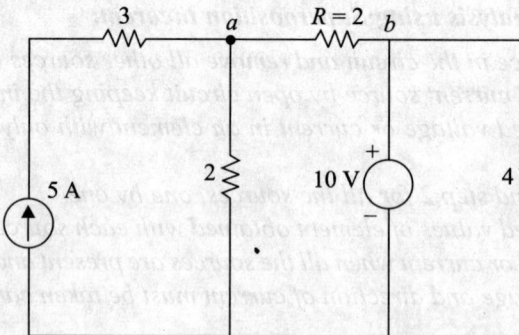

FIGURE 3.28 Example 3.10.

Solution: Let us find the current in *R* with current source only. The equivalent circuit is shown in Figure 3.29.

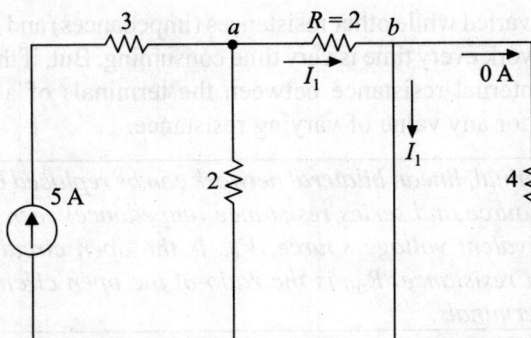

FIGURE 3.29 Response of current source of Example 3.10.

It can be seen that current of 5A is divided in two paths. The current I_1 will flow through R and the shorted element. Thus

$$I_1 = 5 \times \frac{2}{(2+2)} = 2.5 \text{ A}$$

Now current in R with voltage source only can be found by circuit shown in Figure 3.30.

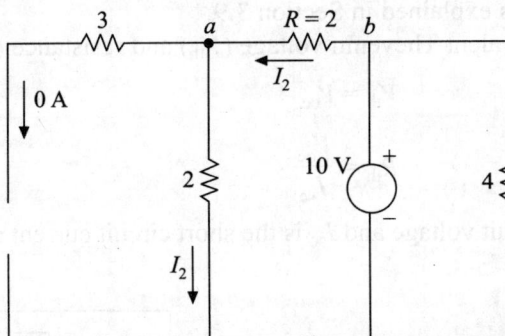

FIGURE 3.30 Response of voltage source of Example 3.10.

The current (I_2), which is assumed to flow in opposite direction of current I, due to voltage source, will be easily calculated as

$$I_2 = \frac{10}{4} = 2.5 \text{ A}$$

The total current due to both the source taken at a time will be

$$I = I_1 - I_2 = 0 \text{ A}$$

3.8 THEVENIN'S THEOREM

Thevenin's theorem and Norton's theorem (discussed in Section 3.9) are very important in solving the network problems having many active and passive (linear, bilateral) elements. Using these theorems, the network is reduced to an equivalent single active source and single passive element. Network reduction, which has been discussed in previous sections so far, involved only passive elements. Sometimes, it is desirable to find out a particular branch current in a network when the resistance of that branch is varied while other resistances (impedances) and sources remain constant. In this case solving the network every time is very time consuming. But, if the network is represented as a single source with internal resistance between the terminals of a varying resistance, the calculation becomes easy for any value of varying resistance.

> **Theorem:** *Any two-terminal, linear, bilateral network can be replaced by an equivalent circuit consisting of a voltage source and series resistance (impedance) seen into the network from that terminals. The equivalent voltage source, V_{th}, is the open circuit voltage between the terminals and equivalent resistance, R_{th}, is the ratio of the open circuit voltage to the short circuit current at these terminals.*

Figure 3.31(a) shows a network having active and passive elements and a load resistance R_L whereas Figure 3.31(b) is the equivalent network representation, having an equivalent voltage source in series with an equivalent resistance, of network shown in Figure 3.31(a). This representation is known as *Thevenin equivalent*. The values of equivalent voltage source and equivalent resistance are obtained by the Thevenin theorem. The network of Figure 3.31(a) can also be represented in terms of an equivalent current source in parallel with an equivalent resistance [Figure 3.31(c)]. Recall the source transformation. This representation is known as Norton's equivalent. The values of equivalent current source and equivalent resistance are obtained by the Norton's theorem which is explained in Section 3.9.

Mathematically, equivalent Thevenin voltage (V_{th}) and resistance (R_{th}) will be written as

$$V_{th} = V_{oc} \tag{3.61}$$

$$R_{th} = \frac{V_{oc}}{I_{sc}} \tag{3.62}$$

where V_{oc} is the open circuit voltage and I_{sc} is the short circuit current at terminals of Thevenin's equivalent representation.

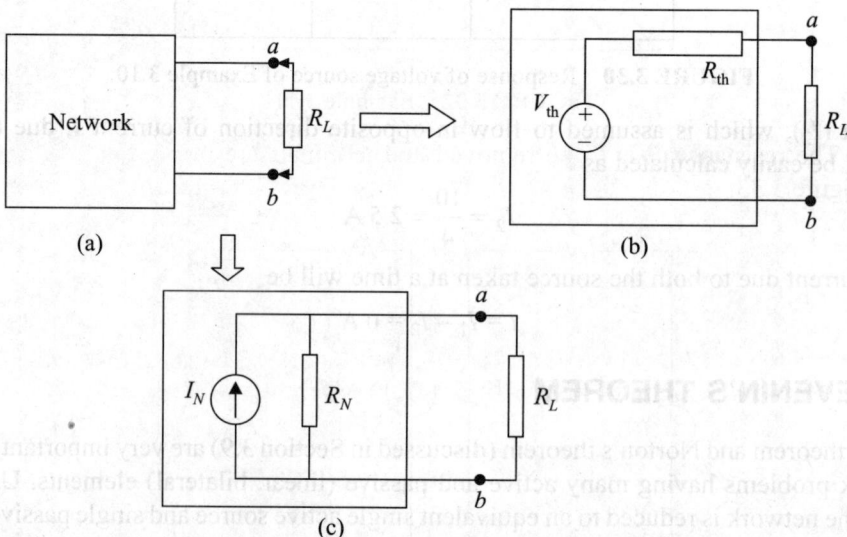

FIGURE 3.31 Network equivalents.

Alternatively, if network has no dependent sources, R_{th} is the equivalent resistance looking into the network from the load terminal (between terminals a and b) when ideal voltage sources are replaced by a short circuit and ideal current sources are replaced by open circuit. In other words, sources are made inactive and replaced by their internal resistances. If network has any dependent source, the R_{th} is only determined by Eq. (3.62).

Main steps to find Thevenin equivalent:

1. *Mark the terminals (say, a and b) and remove the resistance if any in which current is to be obtained.*
2. *Find the open-circuit voltage across terminals.*
3. *Find the Thevenin resistance between the terminals, if there is no dependent source, by looking into the terminals when all the sources are replaced by their internal resistance. Otherwise, find the current flowing between terminals if these are short circuited. The Thevenin resistance will be calculated using Eq. (3.62).*
4. *Replace the network with the Thevenin voltage source in series with Thevenin resistance and the resistance in which current is to be obtained, if any.*

EXAMPLE 3.11   Find the current in resistance R of Figure 3.32.

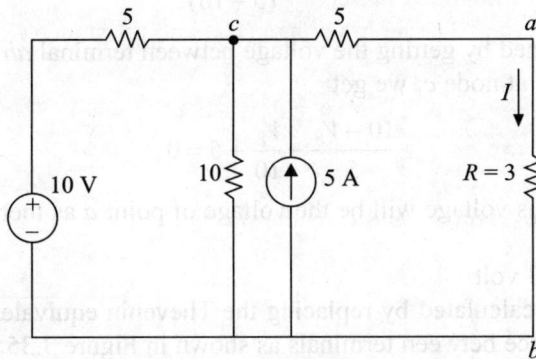

FIGURE 3.32   Example 3.11.

Solution: The resistance R is to be removed and terminals are marked as a and b which is shown in Figure 3.33.

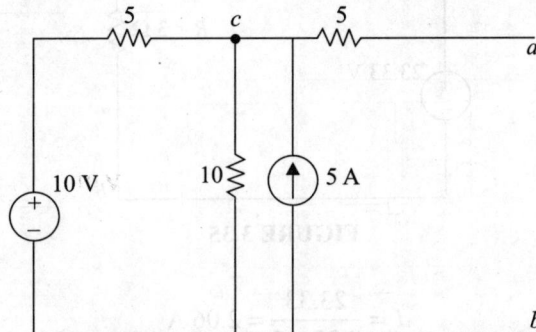

FIGURE 3.33

Since the network of Figure 3.33 does not have any dependent source, the Thevenin resistance can be obtained by looking back into the network from terminals a and b. Voltage source is replaced by the short circuit and current source is replaced by open circuit as shown in Figure 3.34.

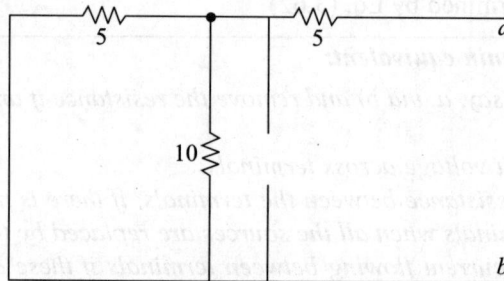

FIGURE 3.34

Since resistances 5 ohm and 10 ohm are in parallel, the resistance between terminal ab will be

$$R_{ab} = 5 + (5 \| 10) = 5 + \frac{5 \times 10}{(5 + 10)} = 8.33\ \Omega = R_{th}$$

Thevenin voltage is obtained by getting the voltage between terminal ab of Figure 3.32. Writing the node voltage equation at node c, we get

$$\frac{10 - V_c}{5} - \frac{V_c}{10} + 5 = 0$$

Thus, $V_c = 23.33$ Volt. This voltage will be the voltage of point a as there is no current between node c and node a.

Therefore, $V_{th} = 23.33$ volt

Thus, the current will be calculated by replacing the Thevenin equivalent between terminal ab and correcting the resistance between terminals as shown in Figure 3.35.

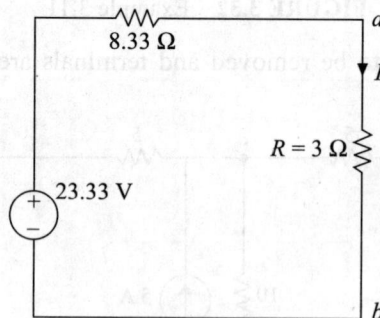

FIGURE 3.35

$$I = \frac{23.33}{8.33 + 3} = 2.06\ \text{A}$$

EXAMPLE 3.12 Find the Thevenin equivalent between terminals *a* and *b* of Figure 3.36.

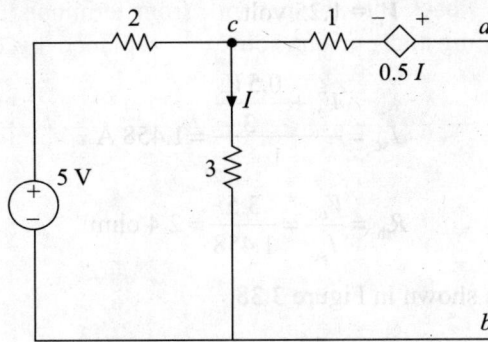

FIGURE 3.36 Example 3.12.

Solution: To calculate the open circuit voltage between terminals *a* and *b*, find the current *I* as

$$I = \frac{5}{3+2} = 1 \text{ A}$$

Thus, voltage $V_c = 3I = 3$ volt. The open circuit voltage will be

$$V_{oc} = V_{ab} = V_c + 0.5\ I = 3.5 \text{ volts}$$

To calculate the short circuit current between terminal *a* and *b*, the terminal will be short-circuited as shown Figure 3.37.

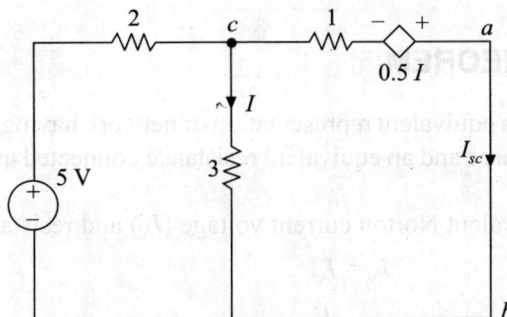

FIGURE 3.37

Use node voltage method at node *c*,

$$\frac{5 - V_c}{2} - I - \frac{0.5\ I + V_c}{1} = 0 \tag{i}$$

and

$$I = \frac{V_c}{3} \tag{ii}$$

Putting Eq. (ii) in Eq. (i) and solving we get

$$V_c = 1.25 \text{ volt.}$$

Thus

$$I_{sc} = \frac{V_c + \dfrac{0.5\,V_c}{3}}{1} = 1.458 \text{ A}$$

Therefore

$$R_{th} = \frac{V_{oc}}{I_{sc}} = \frac{3.5}{1.458} = 2.4 \text{ ohm}$$

The Thevenin equivalent is shown in Figure 3.38.

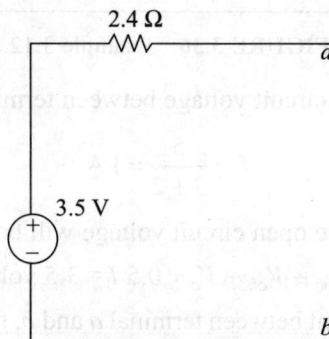

FIGURE 3.38

3.9   NORTON'S THEOREM

The Norton's theorem is an equivalent representation of network having several active and passive elements into a current source and an equivalent resistance connected across the current source as shown in Figure 3.31(c).

   Mathematically, equivalent Norton current voltage (I_N) and resistance (R_N) will be

$$I_N = I_{sc} \tag{3.63}$$

$$R_N = \frac{V_{oc}}{I_{sc}} \tag{3.64}$$

where V_{oc} is the open circuit voltage and I_{sc} is the short circuit current at terminals of Norton's equivalent representation.

> **Theorem:**   *Any two-terminal, linear, bilateral network can be replaced by an equivalent circuit consisting of a current source parallel with the resistance (impedance) seen from that terminals. The equivalent current source, I_N, is the short circuit current between the terminals and equivalent resistance, R_N, is the ratio of the open circuit voltage to the short circuit current at these terminals.*

From Eqs. (3.62) and (3.64), we can get

$$R_N = R_{th} \tag{3.65}$$

Thus, R_N can be obtained as in the case of Thevenin's equivalent resistance. It can be noted that source transformation as shown in Figure 3.39 is a byproduct of Thevenin's and Norton's theorems. Since,

$$V_{th} = V_{oc} = \frac{V_{oc}}{I_{sc}} I_{sc} = R_N I_N \qquad (3.66)$$

or

$$I_N = \frac{V_{th}}{R_{th}} \qquad (3.67)$$

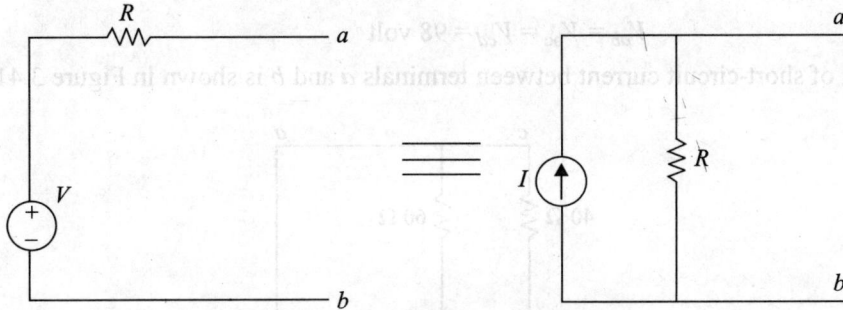

FIGURE 3.39   Source transformation.

Main steps to find Norton's equivalent

1. *Mark the terminals (say, a and b) and remove the resistance, if any, in which current or voltage across is to be obtained.*
2. *Find the short-circuit current between the terminals.*
3. *Find the equivalent resistance between the terminals, if there is no dependent source, by looking into the terminals when all the resources are replaced by their internal resistance. Otherwise find the open current voltage between terminals. The Norton's resistance will be calculated using Eq. (3.64).*
4. *Replace the network with the Norton's current source in parallel with the equivalent resistance and the resistance in which current is to be obtained, if any.*

EXAMPLE 3.13   Find the current in resistance R_L having 25 ohm of network shown in Figure 3.40 using Thevenin's theorem and Norton's theorem. Take $E_1 = 120$ V and $E_2 = 65$ V.

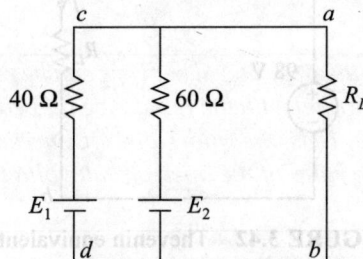

FIGURE 3.40   Example 3.13.

Solution: To find the Thevenin equivalent, remove the resistance between the terminals a and b and calculate the open circuit voltage between these terminals. After removing the resistance R_L, the current will flow between resistances 40 ohm and 60 ohm only and will be equal to

$$\frac{120 - 65}{60 + 40} = 0.55 \text{ A}$$

The voltage between terminals c and d will be

$$V_{cd} = 120 - 40 \times 0.55 \text{ (or } 65 + 60 \times 0.55)$$

Thus,

$$V_{ab} = V_{oc} = V_{cd} = 98 \text{ volt}$$

Calculation of short-circuit current between terminals a and b is shown in Figure 3.41.

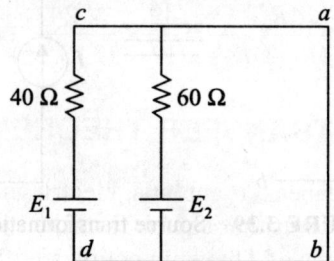

FIGURE 3.41

$$I_{sc} = \frac{120}{40} + \frac{65}{60} = \frac{49}{12} \text{ A}$$

(The effect of source E_1 and E_2 can be taken separately.)

Therefore,

$$R_{th} = R_N = \frac{V_{oc}}{I_{sc}} = \frac{98 \times 12}{49} = 24 \text{ ohm}$$

Thus, Thevenin equivalent is shown in Figure 3.42. The load current $I_L = \dfrac{98}{24 + 25} = 2$ A.

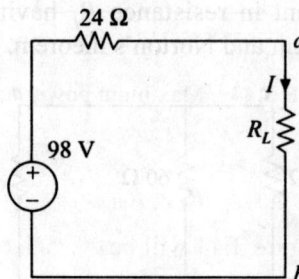

FIGURE 3.42 Thevenin equivalent.

The Norton's equivalent is shown in Figure 3.43.

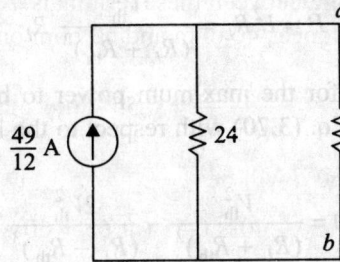

FIGURE 3.43 Norton's equivalent.

The load current will be

$$I_L = I_N \frac{R_N}{(R_N + R_L)} = \frac{49}{12} \times \frac{24}{(24 + 25)} = 2 \text{ A}$$

3.10 MAXIMUM POWER TRANSFER THEOREM

In many applications, such as loudspeaker, antenna, electric motor and other useful devices, it is required to find the maximum power that can be furnished to a load. Normally, the network other than load is expressed in terms of Thevenin equivalent. For maximum power transfer in load R_L as shown in Figure 3.44, the load resistance must be equal to the equivalent Thevenin resistance (R_{th}).

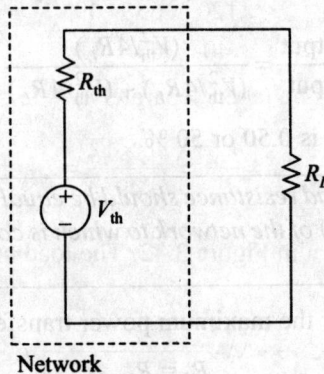

Network

FIGURE 3.44 Maximum power transfer.

Mathematically, it can be written as

$$R_L = R_{th} \tag{3.68}$$

The load current in the circuit of Figure 3.44 will be

$$I = \frac{V_{th}}{R_{th} + R_L} \tag{3.69}$$

The power consumed by the load will be

$$P = I^2 R_L = \frac{V_{th}^2}{(R_L + R_{th})^2} R_L \tag{3.70}$$

The value of load resistance R_L for the maximum power to be consumed by the load can be obtained by taking derivative of Eq. (3.70) with respect to the load resistance and equating it to zero as

$$\frac{\partial P}{\partial R_L} = 0 = \frac{V_{th}^2}{(R_L + R_{th})^2} - \frac{2V_{th}^2}{(R_L + R_{th})^3} R_L = 0 \tag{3.71}$$

Solving Eq. (3.71), we get

$$(R_L + R_{th}) - 2R_L = 0 \quad \Rightarrow \quad \boxed{R_L = R_{th}}$$

Thus, for maximum power transfer, load resistance should be equal to the resistance of the Thevenin equivalent of the network. The maximum power which will flow to the load will be

$$P_{max} = \frac{V_{th}^2}{(R_L + R_L)} \times R_L = \frac{V_{th}^2}{4R_L} \tag{3.72}$$

The power loss in the internal resistance of the network will be

$$P_{loss} = \frac{V_{th}^2}{(R_L + R_L)} \times R_{th} = \frac{V_{th}^2}{4R_L} \tag{3.73}$$

At the maximum power transfer, the efficiency (η) of the system will be

$$\eta = \frac{\text{output}}{\text{input}} = \frac{(V_{th}^2/4R_L)}{(V_{th}^2/4R_L) + (V_{th}^2/4R_L)} = 0.50 \tag{3.74}$$

Thus, the efficiency of the system is 0.50 or 50 %.

> *For maximum power transfer, load resistance should be equal to the resistance of the Thevenin equivalent (or Norton equivalent) of the network to which is connected. However, the efficiency of the system is 50%.*

Using the Norton's equivalent, for the maximum power transfer

$$R_L = R_N \tag{3.75}$$

and maximum power transferred will be

$$P_{max} = \frac{I_N^2 R_L}{4} \tag{3.76}$$

For ac circuit, for maximum power transfer, the load current will be in phase with the source voltage which will be shown in Chapter 4.

EXAMPLE 3.14 In Example 3.13, what is the maximum power that can be absorbed by the load if it is varied? Find the efficiency of the system. What is the efficiency at $R_L = 25$ ohm?

Solution: Since the Thevenin equivalent of Example 3.13 is 24 ohm, for maximum power transfer, the load resistance should be equal to the Thevenin resistance. Thus, load resistance will be

$$R_L = 24 \text{ ohm}$$

And maximum power consumed will be

$$P_{max} = \frac{98^2}{4 \times 24} = 100.04 \text{ watt}$$

The efficiency at maximum power transfer

$$\eta = \frac{\text{Output}}{\text{Input}} = \frac{100.04}{2 \times 100.04} = 0.50 \text{ (or 50\%)}$$

The power consumed by the load when $R_L = 25$ ohm,

$$P_{max} = \frac{98^2}{(24 + 25)^2} \times 25 = 100 \text{ watt}$$

Power consumed in the internal resistance of the network

$$P_{max} = \frac{98^2}{(24 + 25)^2} \times 24 = 96 \text{ watt}$$

Thus, efficiency at this load will be

$$\eta = \frac{\text{Output}}{\text{Input}} = \frac{100}{(100 + 96)} = 0.51 \text{ (or 51\%)}$$

Figure 3.45 shows the maximum power transfer and efficiency (%) for variation of load resistance.

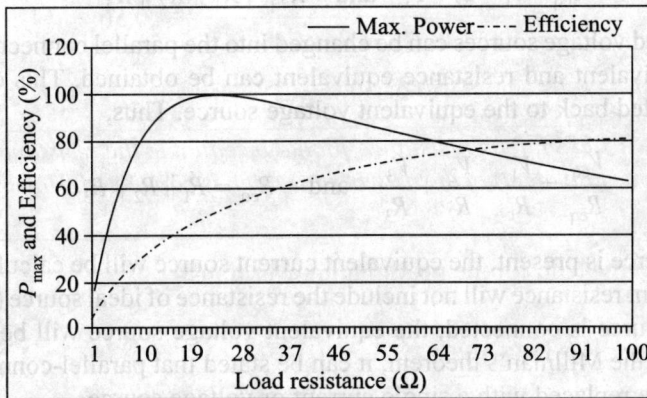

FIGURE 3.45 Maximum power and efficiency with load variation.

3.11 RECIPROCITY THEOREM

In any passive linear, bilateral network, if a voltage V in branch 1 causes a current I to flow in branch 2, then, voltage V applied in branch 2 will cause current I to flow in branch 1.

3.12   MILLMAN'S THEOREM

The Millman's theorem states that parallel connected current sources can be replaced by a single current source as shown in Figure 3.46. It can also be stated as parallel-connected voltage sources can be replaced by a single equivalent voltage source as shown in Figure 3.47.

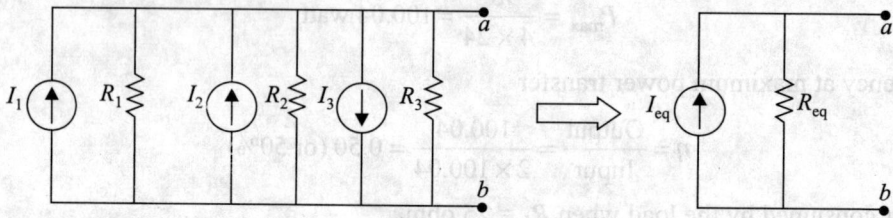

FIGURE 3.46   Equivalent current source.

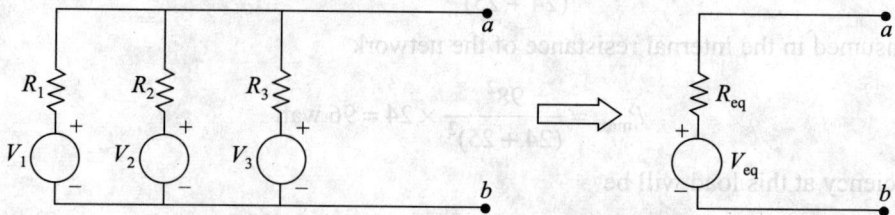

FIGURE 3.47   Equivalent voltage source.

The equivalent current (I_{eq}) and equivalent resistance (R_{eq}) in Figure 3.46 will be

$$I_{eq} = I_1 + I_2 - I_3 \quad \text{and} \quad R_{eq} = R_1 \parallel R_2 \parallel R_3 \tag{3.77}$$

The parallel-connected voltage sources can be changed into the parallel connected current sources and then current equivalent and resistance equivalent can be obtained. This equivalent current source can be converted back to the equivalent voltage source. Thus,

$$\frac{V_{eq}}{R_{eq}} = \frac{V_1}{R_1} - \frac{V_2}{R_2} + \frac{V_3}{R_3} \quad \text{and} \quad R_{eq} = R_1 \parallel R_2 \parallel R_3 \tag{3.78}$$

If an ideal current source is present, the equivalent current source will be calculated as Eq. (3.77) however, the equivalent resistance will not include the resistance of ideal source (which is infinite). If an ideal voltage source is connected, the equivalent voltage source will be the ideal voltage source. To generalize the Millman's theorem, it can be stated that parallel-connected voltage and current sources can be replaced with a single current or voltage source.

3.13   TWO-PORT (4-TERMINAL) NETWORK

So far, networks considered are single-terminal-pair or two terminals or one port circuits. Networks consisting of two pairs of terminals or two-port entries to which either sources or loads can be connected are coupling circuit, filter circuit, etc. One of the terminal pairs may be called as input and the other as output terminal as shown in Figure 3.48. There are four terminal variables and out of these four, two can be independent. This is discussed in detail in Chapter 4.

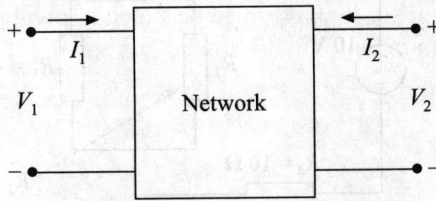

FIGURE 3.48   Two-port network.

PROBLEMS

3.1 Using the KVL, find the voltage across terminal a and b as shown in Figure 3.49.

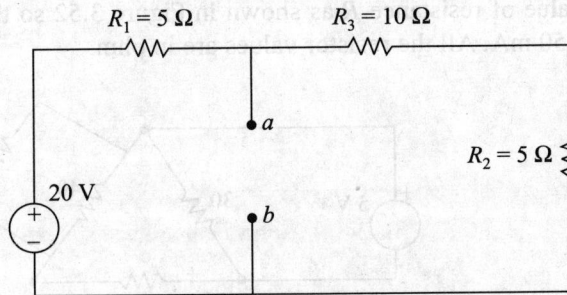

FIGURE 3.49   Problem 3.1.

[**Ans.** 15 V]

3.2 Find the unknown currents in the different section of the circuit shown in Figure 3.50 using KCL. Also find the unknown voltages in each resistor and source.

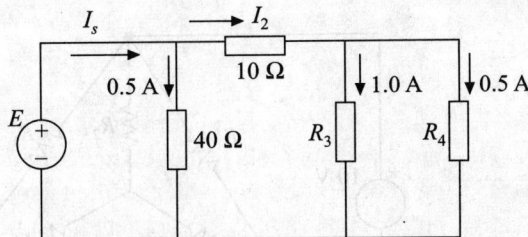

FIGURE 3.50   Problem 3.2.

[**Ans.** $I_2 = 1.5$ A, $I_s = 2$ A, $E = 20$ V, $V_{10-\Omega} = 15$ V, $V_3 = V_4 = 5$ V]

3.3 In Figure 3.51, the measured voltage across resistor R_3 is 5 V. Find all possible combinations of defects that would account for measurement.

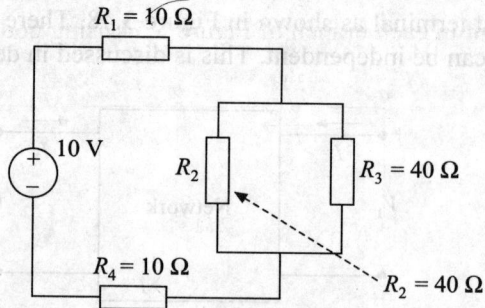

FIGURE 3.51   Problem 3.3.

[Ans. None]

3.4 When a 100 ohm resistor is connected across a 20 A current source, it is found that only 18 A current is flowing in the load. Find the value of internal resistance.

[Ans. 900 ohm]

3.5 Find the value of resistance R as shown in Figure 3.52 so that current drawn from the source is 250 mA. All the resistor values are in ohm.

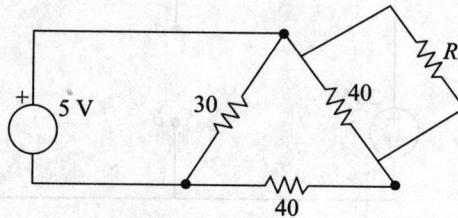

FIGURE 3.52   Problem 3.5.

[Ans. 40 ohm]

3.6 Find the current drawn from the source and each resistor of Figure 3.53 using star-delta transformation. Take $R_1 = 300\ \Omega$ and $R_2 = 100\ \Omega$.

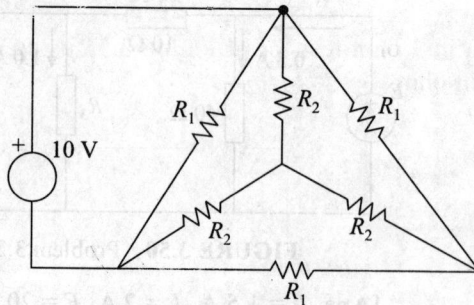

FIGURE 3.53   Problem 3.6.

[Ans. 0.1 A]

3.7 Find the current in each branch of Figure 3.54 using nodal analysis method.

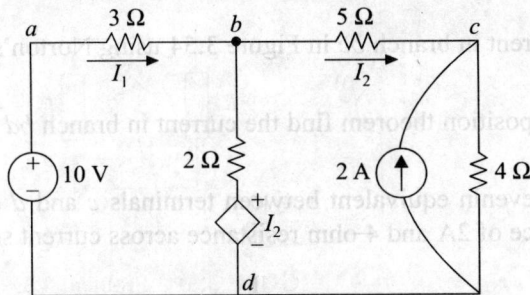

FIGURE 3.54   Problem 3.7.

[**Ans.** 23/12 A, –5/12 A]

3.8 Find the current in each branch of Problem 3.7 using mesh current method.

[**Ans.** 23/12 A, –5/12 A]

3.9 Find the current in 2 ohm resistor of Figure 3.55 using Thevenin's theorem.

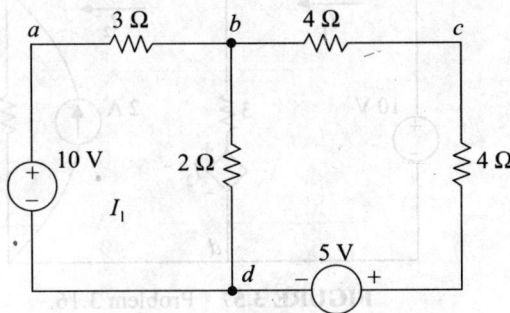

FIGURE 3.55   Problem 3.9.

[**Ans.** 95/46 A]

3.10 Find the voltage across each branch in Figure 3.56 using mesh current method.

[**Ans.** $V_{ab} = 0.2632$, $V_{ac} = 2.894$, $V_{cd} = 0.2632$, $V_{bd} = 2.8948$]

3.11 Find the current in 3 ohm resistor of Figure 3.56 using node equation method. All the resistances are in ohm.

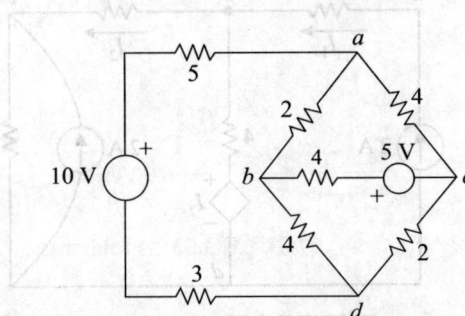

FIGURE 3.56   Problem 3.11.

[**Ans.** 0.855 A]

3.12 Find the current in 5 ohm resistor in Figure 3.54 using Thevenin's theorem.

[**Ans.** −5/12 A]

3.13 Find the current in branch *bc* in Figure 3.54 using Norton's theorem.

[**Ans.** −5/12 A]

3.14 Using superposition theorem find the current in branch *bd* in Figure 3.55.

[**Ans.** 95/46A]

3.15 Find the Thevenin equivalent between terminals *c* and *d* of Figure 3.54 removing the current source of 2A and 4 ohm resistance across current source.

[**Ans.** 4 V, 28/5 Ω]

3.16 In Figure 3.57, all the values of resistances are in ohms. Find the following,
- (a) Current I_1 using mesh current method
- (b) Find the Thevenin equivalent between terminal *c* and *d* removing the current source of 2A and 5 ohm resistance across current source.

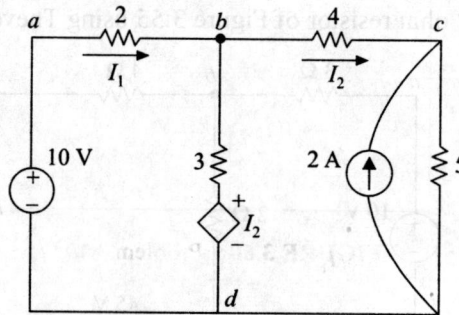

FIGURE 3.57   Problem 3.16.

[**Ans.** (a) 90/49 A; (b) 6 V, 24/5 Ω]

3.17 In Figure 3.58, all the values of resistances are in ohms. Find the following,
- (a) Current in branch *bd* using mesh current method.
- (b) Find Thevenin equivalent across terminal *a* and *b* without removing 2 ohm resistance.

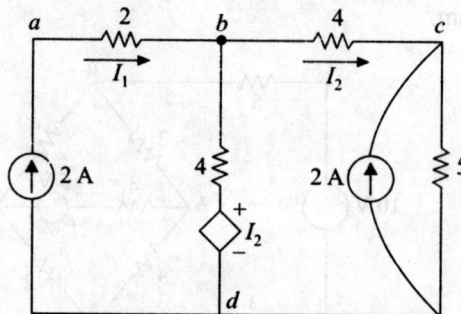

FIGURE 3.58   Problem 3.17.

[**Ans.** (a) 13/6 A; (b) 4 V, 2 Ω]

3.18 Find I (leaving node B) in Figure 3.59 using nodal method.

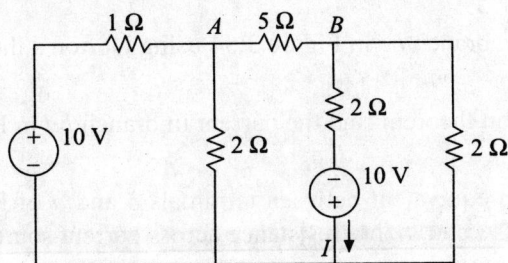

FIGURE 3.59   Problem 3.18.

[**Ans.** 3.375 A]

3.19 Find the Thevenin equivalent between terminals a and b as shown in Figure 3.60.

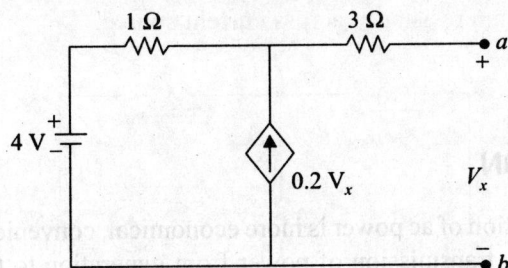

FIGURE 3.60   Problem 3.19.

[**Ans.** 5 V, 5 Ω]

4

Steady-State Analysis of AC Circuits

4.1 INTRODUCTION

The generation and utilization of ac power is more economical, convenient and efficient compared to the dc power, however, transmission of power from generation to the load centres is mostly over long distance ac transmission lines. Thus, it is important to understand the ac power and ac circuits.

4.2 SINUSOIDAL SIGNAL

Sinusoidal signals are the periodic varying signals. Time expression of such signals can either be written in terms of sine function or cosine function. AC voltage and current are sinusoidal signals which are alternating in nature. Sinusoids are used in power generation and transmission of electric power since the derivatives and integrals of sinusoids are themselves sinusoidal and a sinusoidal source always produces sinusoidal response in any linear circuit. A sinusoidal signal can be expressed as

$$a = A_m \sin (\omega t + \alpha) \tag{4.1}$$

or

$$a = A_m \cos (\omega t + \alpha) \tag{4.2}$$

where α is known as *phase angle* (in radians), $\omega (= 2\pi f)$ is the angular frequency (radian/sec), A_m is the peak or maximum value of the periodic function and t is the time in sec. f is the frequency (Hz). The unit of frequency is given by Heinrich Hertz, a German physicist, who demonstrated the existence of the radio waves. The graphical representation of signals of Eqs. (4.1) and (4.2) are shown in Figure 4.1 and Figure 4.2, respectively.

One cycle of a sinusoidal signal corresponds to the portion of the wave between two consecutive zero crossings of curves at positive slope. In other words, the minimum portion between two

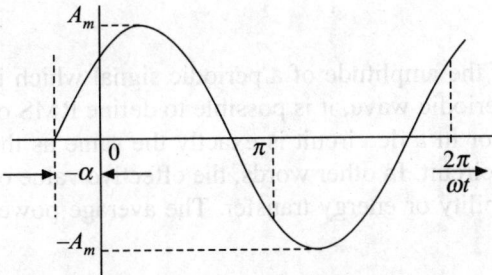

FIGURE 4.1 Sine function.

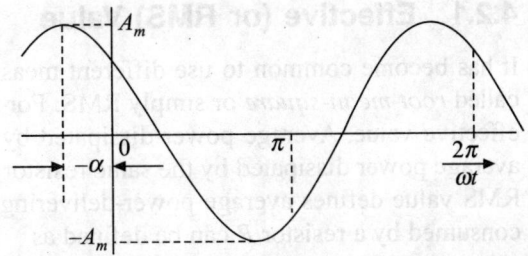

FIGURE 4.2 Cosine function.

points of the curve is having the same magnitudes with same slope. The time period (T), in second, is defined as the reciprocal of the frequency at which the signal is alternating. Mathematically, it can be written as

$$f = \frac{1}{T} \tag{4.3}$$

or

$$T = \frac{2\pi}{\omega} \tag{4.4}$$

Periodic signals have instantaneous value varying between minimum to the maximum. A signal is said to be periodic if it satisfies the following relation.

$$f(t + nT) = f(t) \qquad \forall \text{ integer } n \tag{4.5}$$

Some other periodic signals or waves are shown in Figure 4.3.

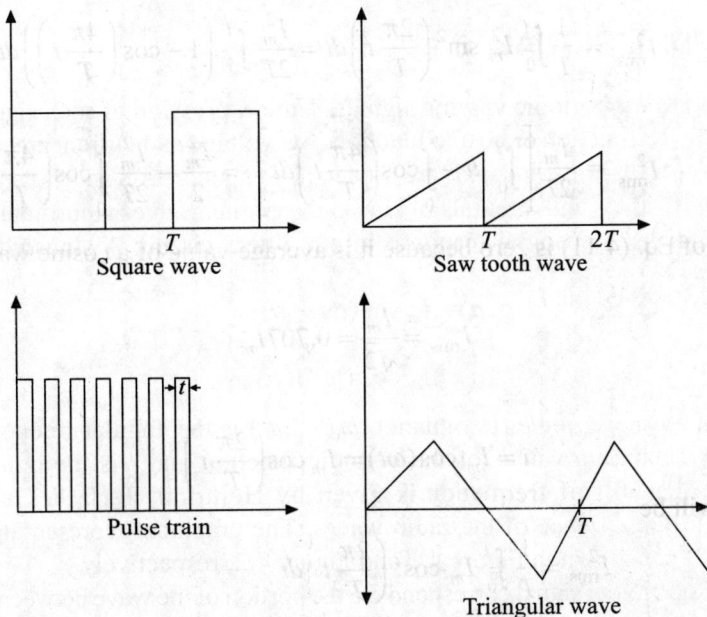

Square wave

Saw tooth wave

Pulse train

Triangular wave

FIGURE 4.3 Periodic signals.

4.2.1   Effective (or RMS) Value

It has become common to use different measure of the amplitude of a periodic signal which is called *root-mean-square* or simply RMS. For any periodic wave, it is possible to define RMS or effective value. Average power dissipated by resistor in a dc circuit is exactly the same as the average power dissipated by the same resistor in ac circuit. In other words, the effective value or RMS value defines average power-delivering capability or energy transfer. The average power consumed by a resistor R can be defined as

$$P_{av} = \frac{1}{T} \int_0^T p \, dt \qquad (4.6)$$

The power $p = i^2 R$ is dissipated in the resistor R. Thus the average power will be

$$P_{av} = \frac{1}{T} \int_0^T i^2 R \, dt = I_{eff}^2 \, R = I_{DC}^2 R \qquad (4.7)$$

or

$$I_{eff} = \sqrt{\left(\frac{1}{T} \int_0^T i^2 \, dt \right)} = I_{rms} \qquad (4.8)$$

This equation is applicable to any periodic wave.

RMS value of sinusoid wave

For sine wave:

$$i = I_m \sin(\omega t) = I_m \sin\left(\frac{2\pi}{T} t \right) \qquad (4.9)$$

The RMS value will be

$$I_{rms}^2 = \frac{1}{T} \int_0^T I_m^2 \sin^2\left(\frac{2\pi}{T} t \right) dt = \frac{I_m^2}{2T} \int_0^T \left(1 - \cos\left(\frac{4\pi}{T} t \right) \right) dt \qquad (4.10)$$

or

$$I_{rms}^2 = \frac{I_m^2}{2T} \left[\int_0^T dt - \int_0^T \cos\left(\frac{4\pi}{T} t \right) dt \right] = \frac{I_m^2}{2} - \frac{I_m^2}{2T} \int_0^T \cos\left(\frac{4\pi}{T} t \right) dt \qquad (4.11)$$

The second term of Eq. (4.11) is zero because it is average value of a cosine wave having period 4π. Thus,

$$I_{rms} = \frac{I_m}{\sqrt{2}} = 0.707 I_m \qquad (4.12)$$

For cosine wave:

$$i = I_m \cos(\omega t) = I_m \cos\left(\frac{2\pi}{T} t \right) \qquad (4.13)$$

The RMS value will be

$$I_{rms}^2 = \frac{1}{T} \int_0^T I_m^2 \cos^2\left(\frac{2\pi}{T} t \right) dt \qquad (4.14)$$

$$= \frac{I_m^2}{2T} \int_0^T \left(1 + \cos\left(\frac{4\pi}{T} t \right) \right) dt = \frac{I_m^2}{2}$$

Thus,

$$I_{\text{rms}} = \frac{I_m}{\sqrt{2}} = 0.707 I_m \qquad (4.15)$$

For the sinusoidal functions, the effective or RMS value is $\frac{1}{\sqrt{2}}$ times the maximum (peak) value.

It can also be seen from the Eqs. (4.12) and (4.15) that the RMS values of sine and cosine wave are the same. Therefore, either sine or cosine function can be taken for ac voltage or current.

4.2.2 Average Value

The average value of any periodic function (wave) is defined as the integral value of function over the time period T divided by the T. Mathematically, it can be expressed as

$$F_{\text{av}} = \frac{1}{T} \int_0^T f(t)\,dt \qquad (4.16)$$

For alternating voltage and current, the average value will be zero. It is obvious from the waveforms (Figures 4.1 and 4.2) which, in their first cycle has equal positive and negative area. Mathematically it can also be proved as

$$I_{\text{av}} = \frac{1}{T} \int_0^T I_m \sin\left(\frac{2\pi}{T}\right) dt = \frac{1}{2\pi} \int_0^{2\pi} I_m \sin\theta\,d\theta \qquad (4.17)$$

In Eq. (4.17), the first and second integrals are the same where $\theta = \omega t$. Thus,

$$I_{\text{av}} = \frac{I_m}{2\pi} \left[-\cos\theta\right]_0^{2\pi} = 0 \qquad (4.18)$$

EXAMPLE 4.1 Find the effective (RMS) value and average value of a full-wave rectifier voltage as shown in Figure 4.4.

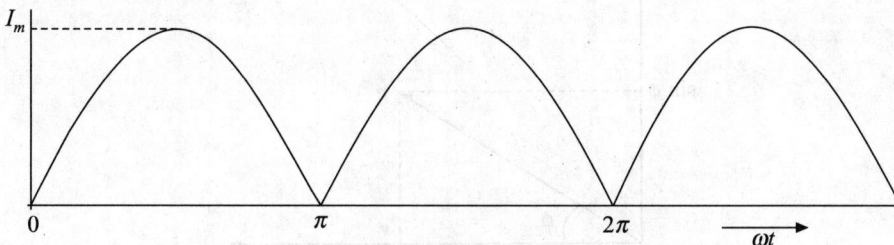

FIGURE 4.4 Example 4.1.

Solution: It can be seen that time period of the function is π (as it is repeating after every π degree). Between zero and π, the function is a perfect sine wave. Thus,

$$I_{\text{rms}}^2 = \frac{1}{\pi} \int_0^\pi (I_m \sin\theta)^2 d\theta = \frac{I_m^2}{2\pi} \int_0^\pi (1 - \sin 2\theta)\,d\theta \qquad (i)$$

or

$$I_{rms}^2 = \frac{I_m^2}{2\pi} \times \pi - \frac{I_m^2}{2\pi} \int_0^\pi (\sin 2\theta)\, d\theta = \frac{I_m^2}{2} - 0 \tag{ii}$$

Thus,
$$I_{rms} = \frac{I_m}{\sqrt{2}}$$

Average value will be computed as

$$I_{av} = \frac{1}{\pi} \int_0^\pi I_m \sin\theta\, d\theta = \frac{I_m}{\pi} \left[-\cos\theta\right]_0^\pi = -\frac{I_m}{\pi} \left[(-1) - 1\right]$$

Thus,
$$I_{av} = \frac{2I_m}{\pi}$$

4.3   PHASOR REPRESENTATION

Using the Euler's formula, the complex exponential expression can be written as

$$e^{j\phi} = \cos\phi + j\sin\phi = 1\angle\phi \tag{4.19}$$

Equation (4.19) has a representation as shown in Figure 4.5. It is straightforward to show that

$$\cos\phi = \frac{e^{j\phi} + e^{-j\phi}}{2} \tag{4.20}$$

$$\sin\phi = \frac{e^{j\phi} - e^{-j\phi}}{2j} \tag{4.21}$$

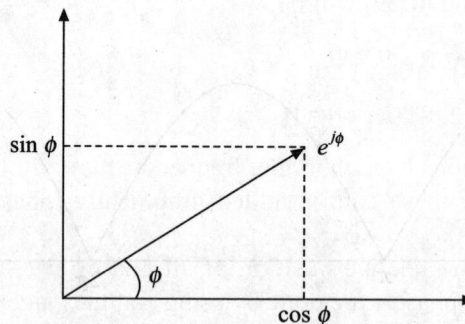

FIGURE 4.5   Representation of $e^{j\phi}$.

The complex exponential is a useful function as the product of two numbers, expressed as exponential, is the same as the exponential of the sums of the two exponents. Also the reciprocal of a number in the exponential notation is just the exponential of the negative exponent. Mathematically, these can be expressed as

$$e^a \times e^b = e^{a+b} \tag{4.22}$$

$$\frac{1}{e^a} = e^{-a} \qquad (4.23)$$

A voltage sinusoidal function can be represented by

$$v_1 = \sqrt{2}\, V_1 \cos(\omega t + \alpha) \qquad (4.24)$$

where v_1 is the instantaneous voltage and $V_1 \left(= \dfrac{V_m}{\sqrt{2}} \right)$ is the *rms* value. Using the Eq. (4.19), Eq. (4.24) can be rewritten as*

$$v_1 = \mathrm{Re}\{\sqrt{2}\, V_1\, e^{j(\omega t + \alpha)}\} = \mathrm{Re}\{V_1\, e^{j\alpha} \times \sqrt{2}\, e^{j\omega t}\} \qquad (4.25)$$

It is possible to write Eq. (4.25) as

$$v_1 = \mathrm{Re}\{\overline{V}_1 \times \sqrt{2}\, e^{j\omega t}\} \qquad (4.26)$$

where

$$\overline{V}_1 = V_1\, e^{j\alpha} \qquad (4.27)$$

The term $\sqrt{2}\, e^{j\omega t}$ in Eq. (4.26) has the frequency exponent which is common to all the functions having the same frequency, hence, it is dropped. This term does not have any other information.

Thus, the phasor representation of voltage function v_1 can be expressed as

$$\overline{V}_1 = V_1\, e^{j\alpha} \quad or \quad V_1 \angle \alpha \qquad (4.28)$$

Let another voltage function (with the same frequency) as written below

$$v_2 = \sqrt{2}\, V_2 \cos(\omega t) \qquad (4.29)$$

The voltage can be written in phasor form as

$$\overline{V}_2 = V_2 \angle 0^\circ \qquad (4.30)$$

The addition of Eqs. (4.24) and (4.29) will be

$$v_1 + v_2 = \mathrm{Re}\{\sqrt{2}\, V_1\, e^{j(\omega t + \alpha)} + \sqrt{2}\, V_2\, e^{j\omega t}\} = \mathrm{Re}\{\sqrt{2}\, e^{j\omega t}[V_1\, e^{j\alpha} + V_2]\}$$
$$= \mathrm{Re}\{\sqrt{2}\, e^{j\omega t}[\overline{V}_1 + \overline{V}_2]\} \qquad (4.31)$$

Since, both the voltage functions have the same frequency, they will be stationary to each other in the complex plane. Thus, these can be added graphically (phasor diagram) as shown in Figure 4.6.

Phasor diagram shows the relative position of all phasor at $t = 0$. Addition, subtraction, multiplication and division of phasors are done by using complex algebra. Phasors can be treated as vectors as they have both magnitude and phase angle. Advantage of phasor is that these can be added/subtracted using phasor diagram.

> *Please note that in above equations, cosine function has been taken as the reference. A sine function can also be represented in the same manner. It will not affect the solution.*

---

*Re {.} is used as Real of function{.}

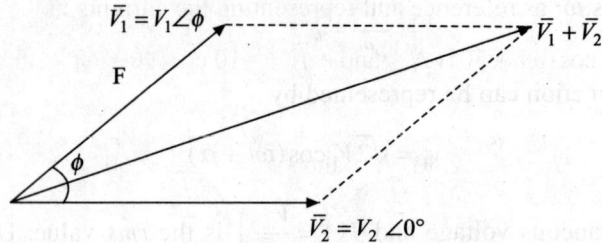

FIGURE 4.6 Addition of two voltage phasors.

Any instantaneous function can be presented in term of phasor as shown in Table 4.1 (taking cosine function as reference).

TABLE 4.1 Phasors

Instantaneous Function	Phasor Representation
$I_m \cos(\omega t)$	$\dfrac{I_m}{\sqrt{2}} \angle 0°$
$I_m \sin(\omega t)$	$\dfrac{I_m}{\sqrt{2}} \angle -\dfrac{\pi}{2}$
$I_m \cos(\omega t + 30°)$	$\dfrac{I_m}{\sqrt{2}} \angle 30°$

EXAMPLE 4.2 Find the sum of two instantaneous currents. Also draw the phasor diagram.

$$i_1 = 5\cos(\omega t + 30°)\,\text{A} \quad \text{and} \quad i_2 = 10\sin(\omega t - 30°)\,\text{A}$$

Solution: The total instantaneous current i will be

$$i = i_1 + i_2 = 5\cos(\omega t + 30°) + 10\sin(\omega t - 30°)\,\text{A}$$

The current i can be obtained by several ways as

Method 1: Using the trigonometric expansion, we have

$$i = 5(\cos \omega t \times \cos 30° - \sin \omega t \times \sin 30°) + 10(\sin \omega t \times \cos 30° - \cos \omega t \times \sin 30°)$$

Putting the values of sin 30° and cos 30° in above equation and simplifying we get,

$$i = 5\left(\frac{\sqrt{3}}{2} - 1\right)\cos \omega t + 5(\sqrt{3} - 0.5)\sin \omega t = -0.67\cos \omega t + 6.16\sin \omega t$$

Let $6.16 = A\cos\theta$ and $0.67 = A\sin\theta$, then $A = \sqrt{6.16^2 + 0.67^2} = 6.20$ and $\theta = \tan^{-1}(0.67/6.16)$ $= 6.31°$

Thus,

$$i = 6.20 \sin(\omega t - \theta)\,\text{A} \tag{i}$$

Method 2: Taking $\cos \omega t$ as reference and representing the currents in the cosine form forms as

$$i_1 = 5 \cos(\omega t + 30°) \text{ A} \quad \text{and} \quad i_2 = -10 \cos(90 + \omega t - 30°) \text{ A}$$

The phasor form will be

$$\overline{I}_1 = \frac{5}{\sqrt{2}} \angle 30° \text{ A} \quad \text{and} \quad \overline{I}_2 = -\frac{10}{\sqrt{2}} \angle 60° \text{ A}$$

Thus

$$\overline{I} = \overline{I}_1 + \overline{I}_2 = \frac{5}{\sqrt{2}} (\cos 30° + j \sin 30°) - \frac{10}{\sqrt{2}} (\cos 60° + j \sin 60°)$$

or

$$\overline{I} = \left(\frac{5\sqrt{3}}{2\sqrt{2}} - \frac{5}{\sqrt{2}} \right) + j \left(\frac{5}{2\sqrt{2}} - \frac{10\sqrt{3}}{2\sqrt{2}} \right) = (-0.48 - j4.35) = -4.38 \angle 83.71° \text{ A}$$

Therefore

$$i = -4.38\sqrt{2} \cos(\omega t + 83.71°) = -6.19 \cos(\omega t + 83.71°) \text{ A} \tag{ii}$$

The phasor diagram shown is Figure 4.7.

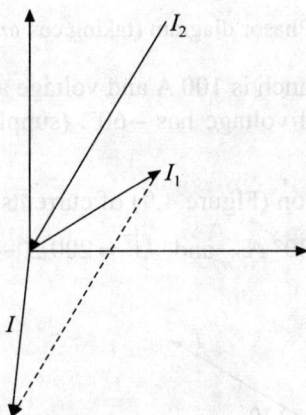

FIGURE 4.7 Phasor diagram (taking $\cos \omega t$ as reference).

Method 3: Taking $\sin \omega t$ as reference and representing the currents in the sine forms as

$$i_1 = 5 \sin(90° - \omega t - 30°) = -5 \sin(\omega t - 60°) \text{ A} \quad \text{and} \quad i_2 = 10 \sin(\omega t - 30°) \text{ A}$$

The phasor form of above current will be

$$\overline{I}_1 = -\frac{5}{\sqrt{2}} \angle (-60°) \text{ A} \quad \text{and} \quad \overline{I}_2 = \frac{10}{\sqrt{2}} \angle (-30°) \text{ A}$$

Thus,

$$\overline{I} = \overline{I}_1 + \overline{I}_2 = -\frac{5}{\sqrt{2}} (\cos 60° - j \sin 60°) + \frac{10}{\sqrt{2}} (\cos 30° - j \sin 30°) \text{ A}$$

or

$$\overline{I} = \left(-\frac{5}{2\sqrt{2}} + \frac{10\sqrt{3}}{2\sqrt{2}}\right) + j\left(\frac{5\sqrt{3}}{2\sqrt{2}} - \frac{5}{\sqrt{2}}\right) = (4.35 - j\,0.48) = 4.38 \angle -6.30° \text{ A}$$

Therefore,

$$i = 4.38\sqrt{2}\,\sin(\omega t - 6.30°) = 6.19\sin(\omega t - 6.30°) \text{ A} \tag{iii}$$

Phasor diagram is shown in Figure 4.8.

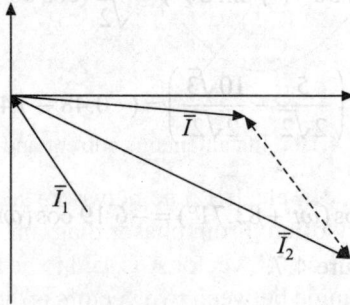

FIGURE 4.8   Phasor diagram (taking cos ωt as reference).

EXAMPLE 4.3   If current in a branch is 100 A and voltage across the branch is 200 volts. The current has phase angle of 30° and voltage has −60°. (supply frequency is 50 Hz). Draw the phasor diagram.

Solution: The phasor representation (Figure 4.9) of currents and voltage will be

$$\overline{I} = 100 \angle 30° \text{ A} \quad \text{and} \quad \overline{V} = 200 \angle -60° \text{ volts}$$

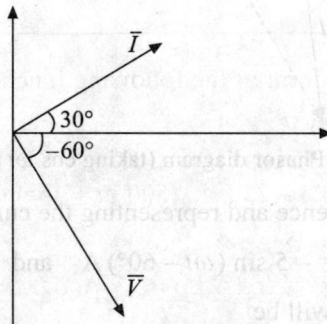

FIGURE 4.9   Phasor diagram of Example 4.3.

Please note that if nothing is mentioned about the magnitude of phasor, it is normally taken as the *rms* value. The current and voltage can be represented in terms of the instantaneous values (taking any reference). If reference is cos ωt, then instantaneous value of current and voltage will be

$$i = 100\sqrt{2}\,\cos(\omega t + 30°) = 141\cos(100\,\pi t + 30°) \text{ A}$$

$$v = 200\sqrt{2}\,\cos(\omega t - 60°) = 282\cos(100\,\pi t - 60°) \text{ V}$$

Current and voltage can be shown in terms of time varying quantities (instantaneous) as given in Figure 4.10.

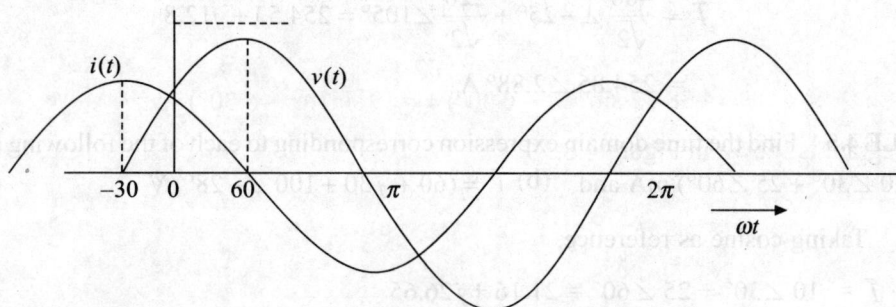

FIGURE 4.10   Instantaneous current and voltage.

Lagging and Leading Concept: In electrical ac network, the quantities (voltage and current) use leading and lagging terms very often. From phasor diagram, if a vector **A** is ahead of another vector (taking anti-clockwise direction) **B**, vector **A** is said to be leading to vector **B** or vector **B** is lagging behind the vector **A**. The angle between two vectors is the value by which they are leading or lagging. From Figure 4.9, it is clear that current is leading to voltage by 90° or voltage is lagging behind current by 90°.

From time domain variation, leading and lagging concept can also be understood. If the peak of a wave is ahead of the peak of second wave, that first wave is lagging behind the second wave. The lagging degree will be the minimum angle between to maximum peaks of both the waves. From Figure 4.10, it can be seen that voltage wave is lagging behind the current wave as peak of voltage wave is ahead of current wave.

> *If current (voltage) is leading to the voltage (current), the voltage (current) is lagging behind the current (voltage).*

EXAMPLE 4.4   Find the phasor form of the following functions (take cosine as reference).

   (a) $v(t) = 311 \cos (377\,t - 25°)$ V,
   (b) $i(t) = 10 \cos (10t + 60°) + 15 \cos (10t - 30°)$ A
   (c) $i(t) = 460 \cos (500\,\pi t - 25°) - 220 \sin (500\,\pi t + 15°)$ A

Solution:

   (a) $$\bar{V} = \frac{311}{\sqrt{2}} \angle -25° = 219.9 \angle -25° \text{ V}$$

   (b) Phasor current will be

$$\bar{I} = \frac{10}{\sqrt{2}} \angle 60° + \frac{15}{\sqrt{2}} \angle -30° = 3.54 + j6.12 + 9.18 - j5.30$$

$$= 12.75 \angle 3.69° \text{ A}$$

   (c) The sine function can be converted into cosine function as

$$i(t) = 460 \cos (500\,\pi t - 25°) + 220 \cos (500\,\pi t + 15° + 90°) \text{ A}$$

The above function can be written into phasor form as

$$\bar{I} = \frac{460}{\sqrt{2}} \angle -25° + \frac{220}{\sqrt{2}} \angle 105° = 254.53 + j12.8$$

$$= 254.85 \angle 2.88° \text{ A}$$

EXAMPLE 4.5   Find the time domain expression corresponding to each of the following phasors.
(a) $\bar{I} = (10 \angle 30° + 25 \angle 60°)$ mA and   (b) $\bar{V} = (60 + j30 + 100 \angle -28°)$ V

Solution: Taking cosine as reference,

(a) $\bar{I} = 10 \angle 30° + 25 \angle 60° = 21.16 + j26.65$

$= 34.03 \angle 51.55°$ mA

or $i = 48.18 \cos(\omega t + 51.55°)$ mA

(b) $\bar{V} = (60 + j30 + 100 \angle -28°)$ V

$= 149.26 \angle -6.52°$

$v = 211.1 \cos(\omega t - 6.52)°$ V

4.4   APPLICATIONS OF KIRCHHOFF'S LAWS

Kirchhoff's laws are also applicable for the ac network having the ac active and passive elements.
It is shown in the following examples.

EXAMPLE 4.6   Find the voltage equation of a series *RLC* circuit as shown in Figure 4.11.

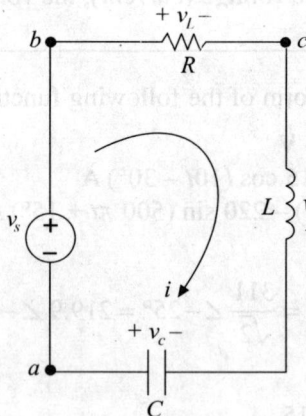

FIGURE 4.11   Series *RLC* circuit.

Solution: Using KVL, we can have

$$v_s = v_R + v_L + v_C \qquad \text{(i)}$$

Putting the values of voltages of passive elements in Eq. (i), the following equation is obtained.

$$v_s = iR + L\frac{di}{dt} + \frac{1}{C}\int i\,dt \qquad\qquad \text{(ii)}$$

If capacitor has initial voltage (charge) v_{co}, the voltage v_c can be written as

$$v_c = \frac{1}{C}\int i\,dt + v_{co} \qquad\qquad \text{(iii)}$$

Thus, Eq. (ii) can be rewritten as

$$v_s = iR + L\frac{di}{dt} + \frac{1}{C}\int i\,dt + v_{co} \qquad\qquad \text{(iv)}$$

EXAMPLE 4.7   Find the current equation of a parallel *RLC* circuit as shown in Figure 4.12.

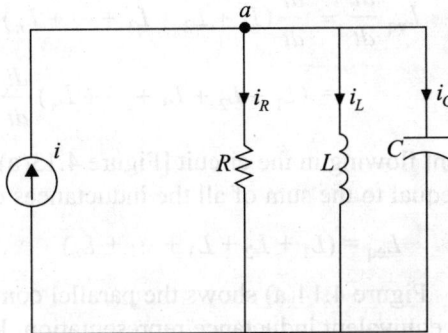

FIGURE 4.12   Parallel *RLC* circuit.

Solution: Applying KCL at node *a*, we get

$$i = i_R + i_L + i_C \qquad\qquad \text{(i)}$$

The currents in each branch can be written as

$$i_R = \frac{v_a}{R}, \quad i_c = C\frac{dv_a}{dt} \quad \text{and} \quad i_L = i_{L0} + \frac{1}{L}\int v_a\,dt \qquad\qquad \text{(ii)}$$

where i_{L0} is the initial current in the inductor and v_a is the node *a* voltage. If there is no initial current in the inductor, i_{L0} will be zero.

Thus, Eq. (i) can be written as

$$i = \frac{v_a}{R} + C\frac{dv_a}{dt} + i_{L0} + \frac{1}{L}\int v_a\,dt \qquad\qquad \text{(iii)}$$

4.4.1   Series and Parallel Combinations of Inductors and Capacitors

Series combination of inductors can be represented as one equivalent inductor whose value is the sum of all the inductances, whereas the parallel combination of inductors can be reduced to one equivalent inductance given as follows:

(a) Series inductors: Figure 4.13(a) shows the series combination of n-inductors and Figure 4.13(b) is the equivalent inductor representation. Using KVL, we can write

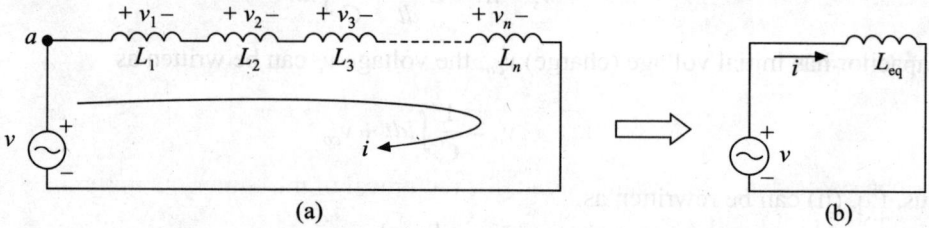

FIGURE 4.13 Series combination of inductors.

$$v = v_1 + v_2 + v_3 + \cdots + v_n$$

$$L_{eq} \frac{di}{dt} = \frac{di}{dt}(L_1 + L_2 + L_3 + \cdots + L_n) \tag{4.32}$$

$$= (L_1 + L_2 + L_3 + \cdots + L_n)\frac{di}{dt}$$

Since the same current flowing in the circuit [Figure 4.13(a)], the equivalent inductance of a series circuit is equal to the sum of all the inductances of the circuit.

$$L_{eq} = (L_1 + L_2 + L_3 + \cdots + L_n) \tag{4.33}$$

(b) Parallel inductors: Figure 4.14(a) shows the parallel combination of n-inductors and Figure 4.14(b) is the equivalent inductance representation. Using KCL, we can write

$$i = i_1 + i_2 + i_3 + \cdots + i_n$$

$$= \int\left(\frac{v}{L_1} + \frac{v}{L_2} + \frac{v}{L_3} + \cdots + \frac{v}{L_n}\right)dt \tag{4.34}$$

$$= \frac{1}{L_{eq}}\int v\,dt$$

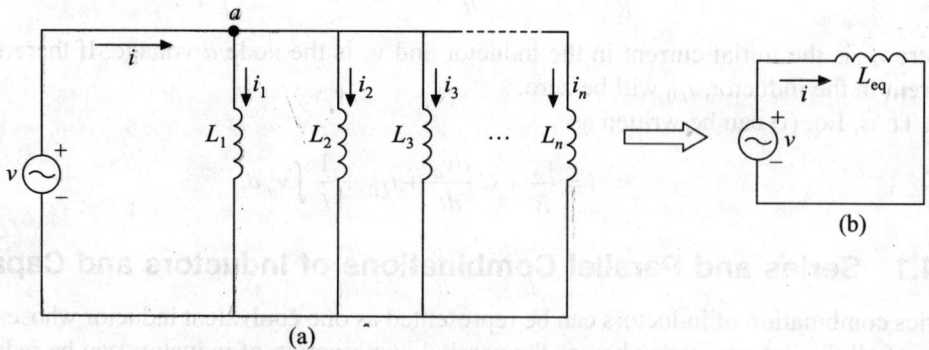

FIGURE 4.14 Parallel combination of resistances.

From Figure 4.14(b), we can also write as

$$i = \frac{1}{L_{eq}} \int v dt \qquad (4.35)$$

Thus,

$$\frac{1}{L_{eq}} = \left(\frac{1}{L_1} + \frac{1}{L_2} + \frac{1}{L_3} + \cdots + \frac{1}{L_n} \right) \qquad (4.36)$$

It can be stated that, in a parallel circuit, the reciprocal of the equivalent inductance is the sum of the reciprocal of the inductances.

(a) **Series capacitors:** Figure 4.15(a) shows the series combination of n-capacitors and Figure 4.15(b) is the equivalent capacitor representation. Using KVL, we can write

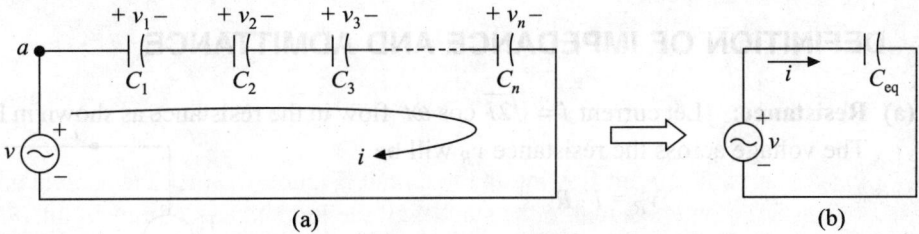

(a)   (b)

FIGURE 4.15   Series combination of capacitors.

$$v = v_1 + v_2 + v_3 + \cdots + v_n$$

$$\frac{1}{C_{eq}} \int i dt = \left(\frac{1}{C_1} \int i dt + \frac{1}{C_2} \int i dt + \frac{1}{C_3} \int i dt + \cdots + \frac{1}{C_n} \int i dt \right) \qquad (4.37)$$

$$= \left(\frac{1}{C_1} + \frac{1}{C_2} + \frac{1}{C_3} + \cdots + \frac{1}{C_n} \right) \int i dt$$

Thus, the equivalent capacitance of a series circuit can be written as

$$\frac{1}{C_{eq}} = \left(\frac{1}{C_1} + \frac{1}{C_2} + \frac{1}{C_3} + \cdots + \frac{1}{C_n} \right) \qquad (4.38)$$

(b) **Parallel capacitors:** Figure 4.16(a) shows the parallel combination of n-capacitors and Figure 4.16(b) is the equivalent capacitance representation. Using KCL, we can write

$$i = i_1 + i_2 + i_3 + \cdots + i_n$$

$$C_{eq} \frac{dv}{dt} = \frac{dv}{dt} \left(C_1 \frac{dv}{dt} + C_2 \frac{dv}{dt} + C_3 \frac{dv}{dt} + \cdots + C_n \frac{dv}{dt} \right) \qquad (4.39)$$

$$= \left(C_1 + C_2 + C_3 + \cdots + C_n \right) \frac{dv}{dt}$$

or

$$C_{eq} = \left(C_1 + C_2 + C_3 + \cdots + C_n \right) \qquad (4.40)$$

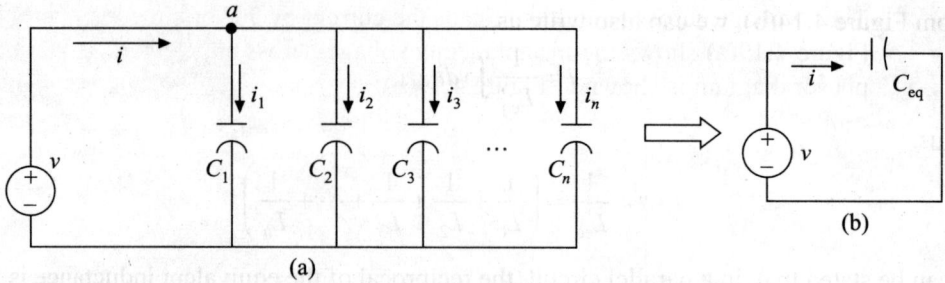

FIGURE 4.16   Parallel combination of capacitors.

It can be stated that in the parallel circuit, the equivalent capacitance is the sum of the series capacitances.

4.5 DEFINITION OF IMPEDANCE AND ADMITTANCE

(a) Resistance: Let current $i = \sqrt{2}I \cos \omega t$ flow in the resistance as shown in Figure 4.17. The voltage across the resistance v_R will be

$$v_R = i \cdot R$$

$$= \sqrt{2} \, IR \cos \omega t = \sqrt{2} V_R \cos \omega t$$

The phasor notation of voltage and current will be

$$\overline{I} = I \angle 0°$$

$$\overline{V}_R = V_R \angle 0°$$

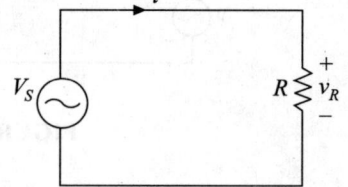

FIGURE 4.17   Flows in resistance.

This impedance will be

$$\overline{Z} = \frac{V_R \angle 0°}{I \angle 0°} = \frac{R \cdot I}{I} = R \angle 0°$$

Thus the voltage across the resistor is in phase with current.

(b) Inductance: Let current $i = \sqrt{2}I \cos \omega t$ flows in the inductor as shown in Figure 4.18. The voltage across the inductor v_L will be

$$v_L = L \frac{di}{dt} = -L \sqrt{2} I \omega \sin (\omega t)$$

$$= I \omega L \sqrt{2} \cos (\omega t + 90°)$$

The phasor notation of voltage and current will be

$$\overline{I} = I \angle 0°$$

$$\overline{V}_L = \omega L I \angle 90°$$

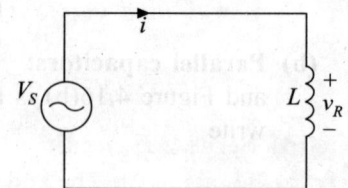

FIGURE 4.18   Flows in inductor.

Thus, impedance will be

$$\overline{Z} = \frac{\overline{V}_L}{\overline{I}} = \frac{I \omega L \angle 90°}{I \angle 0°} = \omega L \angle 90° = j \omega L$$

Thus, voltage across an inductor leads the current by 90° or current lags voltage by 90°. Figure 4.19(a) shows the instantaneous voltage across and current in the inductor. The phasor diagram is shown in Figure 4.19(b).

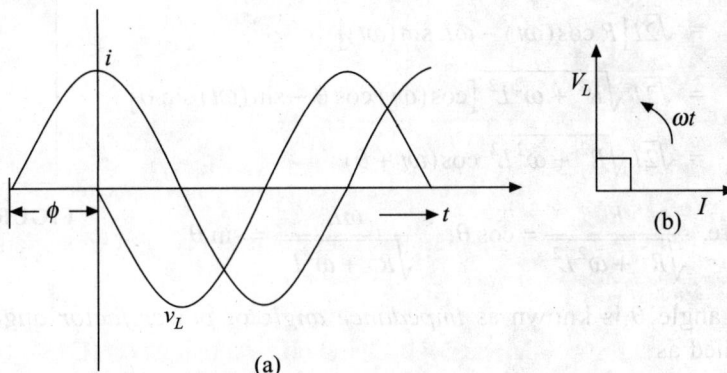

FIGURE 4.19 (a) Instantaneous current and voltage across the inductor, (b) phasor diagram.

(c) **Capacitor:** Let voltage ($v_C = \sqrt{2}V \cos \omega t$) across the capacitor as shown in Figure 4.20. The current in the capacitor v_C will be

$$i_C = C\frac{dv_C}{dt} = \sqrt{2}V\omega C \cos(\omega t + 90°)$$

or phasor will be

$$\bar{I}_C = V\omega C \angle 90°$$

The impedance will be

$$\bar{Z} = \frac{\bar{V}_C}{\bar{I}_C} = \frac{V\angle 0°}{V\omega C\angle 90°} = \frac{1}{\omega C}\angle -90° = \frac{-j}{\omega C} = \frac{1}{j\omega C}$$

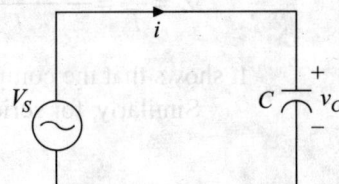

FIGURE 4.20 Current in capacitor.

This shows that voltage across a capacitor lags the current by 90° or current leads voltage by 90°.

Thus,
- *Voltage across a resistor is in phase with current*
- *Voltage across an inductor leads current by 90°*
- *Voltage across a capacitor lags current by 90°*

(d) **Series *RL* circuit:** Let current $i = \sqrt{2}I \cos(\omega t)$ flow in the *RL* circuit as shown in Figure 4.21. The voltage across the resistor (v_R) and across the inductor (v_L) will be

$$v_R = \sqrt{2}I\,R \cos(\omega t)$$

$$v_L = L\frac{di}{dt} = \sqrt{2}I\,\omega L \cos(\omega t + 90°)$$

Thus,

$$v_S = v_R + v_L$$

$$= \sqrt{2}I\left[R\cos(\omega t) + \omega L\cos(\omega t + 90°)\right]$$

$$= \sqrt{2}I\left[R\cos(\omega t) - \omega L\sin(\omega t)\right]$$

$$= \sqrt{2}I\sqrt{R^2 + \omega^2 L^2}\left[\cos(\omega t)\cos\theta - \sin(\omega t)\sin\theta\right]$$

$$= \sqrt{2}I\sqrt{R^2 + \omega^2 L^2}\cos(\omega t + \theta)$$

where, $\dfrac{R}{\sqrt{R^2 + \omega^2 L^2}} = \cos\theta;$ $\dfrac{\omega L}{\sqrt{R^2 + \omega^2 L^2}} = \sin\theta$

FIGURE 4.21   *RL* circuit.

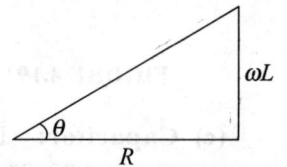

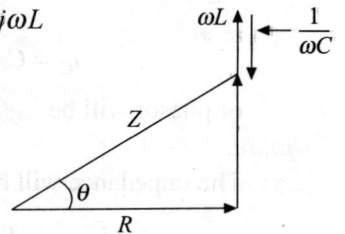

The angle θ is known as *impedance angle* or *power factor angle* and can also be defined as

$$\theta = \tan^{-1}\frac{\omega L}{R}$$

Therefore, the impedance will be

$$\bar{Z} = \frac{\bar{V}_s}{\bar{I}} = \frac{I\sqrt{R^2 + \omega^2 L^2}\angle\theta}{I\angle 0°} = \sqrt{R^2 + \omega^2 L^2}\angle\theta = R + j\omega L$$

It shows that the complex impedances are added.

Similarly, for series *RLC* circuit, we can write

$$\bar{Z} = R + j\omega L - \frac{j}{\omega C}$$

If there are several complex impedances are connected in series, the total impedance ($\bar{Z}_T$) will be the summation of all the impedances as

$$\bar{Z}_T = \bar{Z}_1 + \bar{Z}_2 + \cdots$$

or in the admittance form

$$\frac{1}{\bar{Y}_T} = \frac{1}{\bar{Y}_1} + \frac{1}{\bar{Y}_2} + \cdots$$

Similarly, it can be seen that total impedance and admittance ($\bar{Y}_T$) of the parallel connected impedances will be

$$\bar{Y}_T = \bar{Y}_1 + \bar{Y}_2 + \cdots$$

$$\frac{1}{\bar{Z}_T} = \frac{1}{\bar{Z}_1} + \frac{1}{\bar{Z}_2} + \cdots$$

Please note that

$$\bar{Y} = \frac{1}{\bar{Z}} = G + jB$$

$G \rightarrow$ Conductance

$B \rightarrow$ Susceptance

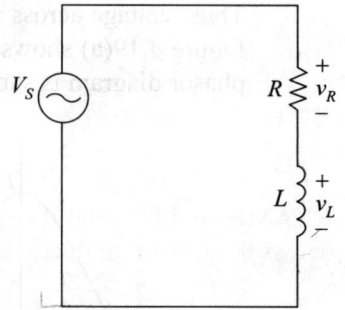

and

$$\overline{Z} = Z \angle \theta$$

$$\overline{Y} = \frac{1}{Z \angle \theta} = \frac{1}{Z} \angle - \theta = Y \angle - \theta$$

EXAMPLE 4.8 Find the supply voltage magnitude in Figure 4.22, if the voltage magnitudes across the resistor, inductor and capacitor are 3, 10 and 6 volts, respectively.

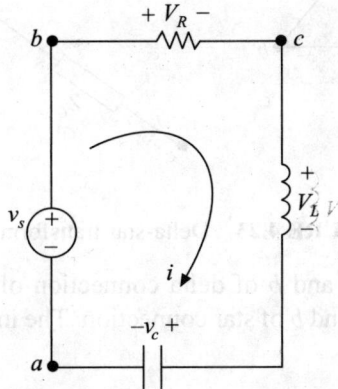

FIGURE 4.22 Example 4.8.

Solution: Taking the current as reference ($\overline{I} = I \angle 0°$), the phasor voltages across the resistor, inductor and capacitor will be

$$\overline{V}_R = 3 \angle 0°, \overline{V}_L = 10 \angle 90° \quad \text{and} \quad \overline{V}_C = 6 \angle - 90° \text{ V}$$

Thus

$$\overline{V}_s = \overline{V}_R + \overline{V}_L + \overline{V}_C = 3 \angle 0 + 10 \angle 90° + 6 \angle - 90°$$

or

$$\overline{V}_s = 3 + j10 - j6 = 3 + j4 = 5 \angle 53.13°$$

> *Please note that, in ac circuit, the voltages are treated as the phasors and not as the magnitude only as in the dc circuit.*

Thus, supply voltage magnitude will be 5 volts.

4.6 STAR-DELTA OR DELTA-STAR TRANSFORMATION

The star-delta or delta-star transformation in ac circuit is similar to the dc circuit as explained in Section 3.4 of Chapter 3. In ac circuit, in place of resistances, impedances are used which is a complex quantity. The major concepts behind the Y-Δ or Δ-Y transformations are

- The terminals are unaffected.
- The equivalent impedance between any two terminals in either formation is the same.

4.6.1 Delta-Star Transformation

Using the concept, the star impedances can be obtained as follows:

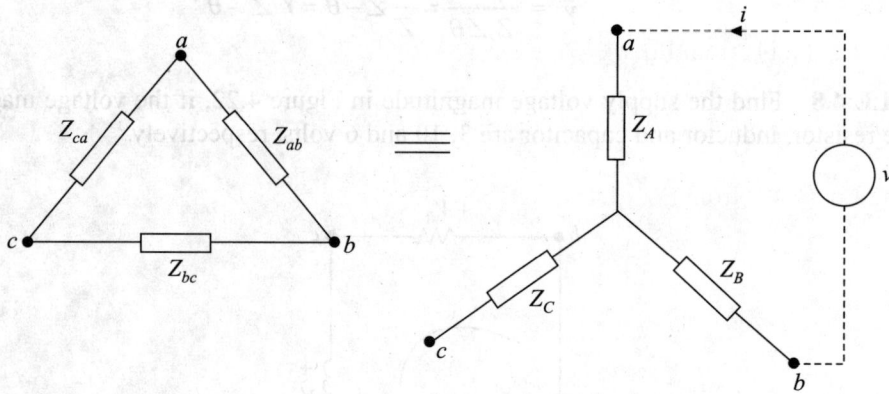

FIGURE 4.23   Delta-star transformation.

Impedance between terminals a and b of delta connection of Figure 4.23 will be equal to the impedance between terminals a and b of star connection. The impedance between terminals a and b of delta connection will be

$$\bar{Z}_{ab} \| (\bar{Z}_{ca} + \bar{Z}_{bc}) = \frac{\bar{Z}_{ab}(\bar{Z}_{ca} + \bar{Z}_{bc})}{\bar{Z}_{ab} + \bar{Z}_{bc} + \bar{Z}_{ca}} \tag{4.41}$$

where, $\|$ represents the parallel combination which can be visualized by connecting a voltage source between terminals a and b. The resistance between terminals a and b of star connection will be $(\bar{Z}_A + \bar{Z}_B)$ which can be visualized by connecting a voltage source between terminals a and b of star connection. Thus,

$$\bar{Z}_A + \bar{Z}_B = \frac{\bar{Z}_{ab}(\bar{Z}_{ca} + \bar{Z}_{bc})}{\bar{Z}_{ab} + \bar{Z}_{bc} + \bar{Z}_{ca}} \tag{4.42}$$

Similar equation can be written for terminals b and c as

$$\bar{Z}_B + \bar{Z}c = \frac{\bar{Z}_{bc}(\bar{Z}_{ab} + \bar{Z}_{ca})}{\bar{Z}_{ab} + \bar{Z}_{bc} + \bar{Z}_{ca}} \tag{4.43}$$

and for terminals a and c as

$$\bar{Z}_A + \bar{Z}_C = \frac{\bar{Z}_{ca}(\bar{Z}_{ab} + \bar{Z}_{bc})}{\bar{Z}_{ab} + \bar{Z}_{bc} + \bar{Z}_{ca}} \tag{4.44}$$

Adding Eqs. (4.42), (4.43) and (4.44), and simplifying, we get

$$(\bar{Z}_A + \bar{Z}_B + \bar{Z}_C) = \frac{(\bar{Z}_{ab}\bar{Z}_{bc} + \bar{Z}_{bc}\bar{Z}_{ca} + \bar{Z}_{ca}\bar{Z}_{ab})}{\bar{Z}_{ab} + \bar{Z}_{bc} + \bar{Z}_{ca}} \tag{4.45}$$

Subtracting Eq. (4.43) from Eq. (4.45), we get

$$\bar{Z}_A = \frac{\bar{Z}_{ca}\bar{Z}_{ab}}{\bar{Z}_{ab} + \bar{Z}_{bc} + \bar{Z}_{ca}} \tag{4.46}$$

Subtracting Eq. (4.44) from Eq. (4.45), we get

$$\bar{Z}_B = \frac{\bar{Z}_{bc}\bar{Z}_{ab}}{\bar{Z}_{ab} + \bar{Z}_{bc} + \bar{Z}_{ca}} \tag{4.47}$$

Subtracting Eq. (4.42) from Eq. (4.45), we get

$$\bar{Z}_C = \frac{\bar{Z}_{bc}\bar{Z}_{ca}}{\bar{Z}_{ab} + \bar{Z}_{bc} + \bar{Z}_{ca}} \tag{4.48}$$

Rule: *The impedance of any arm of the equivalent Y(star) is found by taking the product of the impedances of two adjoined arm of the Δ network and dividing them by the sum of the impedances of Δ network arms.*

4.6.2 Star-Delta Transformation

From Eq. (4.46) and (4.47), we have the following relation,

$$\frac{\bar{Z}_A}{\bar{Z}_B} = \frac{\bar{Z}_{ca}}{\bar{Z}_{bc}} \tag{4.49}$$

and from Eqs. (4.46) and (4.48), we have

$$\frac{\bar{Z}_A}{\bar{Z}_C} = \frac{\bar{Z}_{ab}}{\bar{Z}_{bc}} \tag{4.50}$$

From Eq. (4.46), (4.49) and (4.50), we get

$$\bar{Z}_A = \frac{\bar{Z}_{ca}\bar{Z}_{ab}}{\bar{Z}_{ab} + \bar{Z}_{bc} + \bar{Z}_{ca}} = \frac{\dfrac{\bar{Z}_A}{\bar{Z}_B}\bar{Z}_{bc}\dfrac{\bar{Z}_A}{\bar{Z}_C}\bar{Z}_{bc}}{\dfrac{\bar{Z}_A}{\bar{Z}_B}\bar{Z}_{bc} + \bar{Z}_{bc} + \dfrac{\bar{Z}_A}{\bar{Z}_C}\bar{Z}_{bc}} = \frac{\bar{Z}_A{}^2\bar{Z}_{bc}{}^2}{\left(\dfrac{\bar{Z}_A}{\bar{Z}_B} + 1 + \dfrac{\bar{Z}_A}{\bar{Z}_C}\right)\bar{Z}_{bc} \cdot \bar{Z}_B\bar{Z}_C}$$

$$= \frac{\bar{Z}_A{}^2\bar{Z}_{bc}}{(\bar{Z}_A\bar{Z}_C + \bar{Z}_B\bar{Z}_C + \bar{Z}_A\bar{Z}_B)} \tag{4.51}$$

Therefore,

$$\bar{Z}_{bc} = \frac{\bar{Z}_A\bar{Z}_C + \bar{Z}_B\bar{Z}_C + \bar{Z}_A\bar{Z}_B}{\bar{Z}_A}$$

or

$$\bar{Z}_{bc} = \bar{Z}_B + \bar{Z}_C + \frac{\bar{Z}_B\bar{Z}_C}{\bar{Z}_A} \tag{4.52}$$

Similarly, we can get following relations.

$$\overline{Z}_{ab} = \overline{Z}_A + \overline{Z}_B + \frac{\overline{Z}_A \overline{Z}_B}{\overline{Z}_C} \tag{4.53}$$

$$\overline{Z}_{ca} = \overline{Z}_A + \overline{Z}_C + \frac{\overline{Z}_A \overline{Z}_C}{\overline{Z}_B} \tag{4.54}$$

4.7 NETWORK THEOREMS

4.7.1 Mesh Current (or Loop Current) Method

This method is the same as explained in Section 3.5. In this analysis method, KVL is used. In any network, the number of independent loop equations (m) will be

$$m = b - (j - 1) \tag{4.55}$$

where, b is the number of branch (which lies between two junctions) and j is the number of junctions (independent nodes). The following example explains the same.

EXAMPLE 4.9   Find the number of independent mesh equations of circuit shown in Figure 4.24 and write the mesh equations.

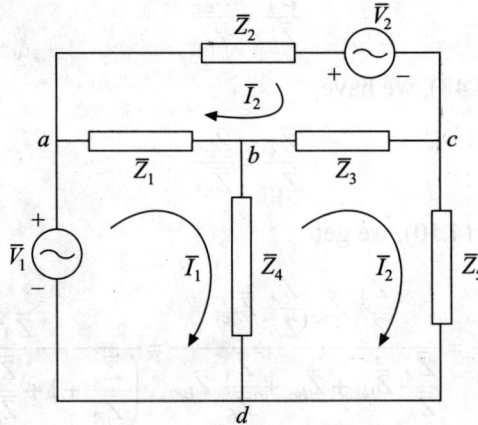

FIGURE 4.24   Example 4.9.

Solution: In Figure 4.24,

  Number of junction = 4 (a, b, c, d)
  Number of branches = 6 (ad, ab, bc, cd, ac, bd)
  Number of independent loop (mesh) equations = $6 - (4 - 1) = 3$
  Three loops to be considered are shown in Figure 4.24.

Mesh 1 Equation:

$$\overline{V}_1 - \overline{Z}_1(\overline{I}_1 - \overline{I}_2) - \overline{Z}_4(\overline{I}_1 - \overline{I}_3) = 0$$

or

$$(\bar{Z}_1 - \bar{Z}_4)\bar{I}_1 - \bar{Z}_1\bar{I}_2 - \bar{Z}_4\bar{I}_3 = \bar{V}_1 \tag{i}$$

Mesh 2 Equation:

$$\bar{V}_2 + \bar{Z}_2\bar{I}_2 + \bar{Z}_3(\bar{I}_2 - \bar{I}_3) + \bar{Z}_1(\bar{I}_2 - \bar{I}_1) = 0$$

or

$$-\bar{Z}_1\bar{I}_1 + (\bar{Z}_1 + \bar{Z}_2 + \bar{Z}_3)\,\bar{I}_2 - \bar{Z}_3\bar{I}_3 = -\bar{V}_2 \tag{ii}$$

Mesh 3 Equation:

$$\bar{Z}_3(\bar{I}_3 - \bar{I}_2) + \bar{Z}_5\bar{I}_3 + \bar{Z}_4(\bar{I}_3 - \bar{I}_1) = 0$$

or

$$-\bar{Z}_4\bar{I}_1 - \bar{Z}_3\bar{I}_2 + (\bar{Z}_3 + \bar{Z}_4 + \bar{Z}_5)\,\bar{I}_3 = 0 \tag{iii}$$

In matrix form, it will be

$$\begin{bmatrix} (\bar{Z}_1 + \bar{Z}_4) & -\bar{Z}_1 & -\bar{Z}_4 \\ -\bar{Z}_1 & (\bar{Z}_1 + \bar{Z}_2 + \bar{Z}_3) & -\bar{Z}_3 \\ -\bar{Z}_4 & -\bar{Z}_4 & (\bar{Z}_3 + \bar{Z}_4 + \bar{Z}_5) \end{bmatrix} \begin{bmatrix} \bar{I}_1 \\ \bar{I}_2 \\ \bar{I}_3 \end{bmatrix} = \begin{bmatrix} \bar{V}_1 \\ -\bar{V}_2 \\ 0 \end{bmatrix} \tag{iv}$$

Above equation (iv) can be solved, using any method.

4.7.2 Node Voltage Method

The node voltage equations can be written similar to the equations written for dc circuit as explained in Section 3.6. The quantities are taken into the phasor forms. Example 4.10 explains the node voltage method.

EXAMPLE 4.10 Find the number of independent node equations of circuit shown in Figure 4.24 and write the node voltage equations.

Solution: In Figure 4.24,
  Number of junction = 4
  Number of node voltage equations = (4 – 1) = 3
Taking the junction d as reference, the independent nodes will be a, b and c. The node voltage at node a is known (which is $\bar{V}_1$) and thus there is no need to write the equation for that node. The node equations for nodes b and c will be as

At node b

$$\frac{\bar{V}_a - \bar{V}_b}{\bar{Z}_1} - \frac{\bar{V}_b - \bar{V}_c}{\bar{Z}_3} - \frac{\bar{V}_b}{\bar{Z}_4} = 0$$

or

$$\bar{V}_b\left(\frac{1}{\bar{Z}_1} + \frac{1}{\bar{Z}_3} + \frac{1}{\bar{Z}_4}\right) - \frac{\bar{V}_c}{\bar{Z}_3} - \frac{\bar{V}_a}{\bar{Z}_1} = \frac{\bar{V}_1}{\bar{Z}_1} \tag{i}$$

where $\bar{V}_a, \bar{V}_b$ and $\bar{V}_c$ are the node voltages at node a, node b and node c, respectively.

At node c

$$\frac{\bar{V}_a + \bar{V}_2 - \bar{V}_c}{\bar{Z}_2} + \frac{\bar{V}_b - \bar{V}_c}{\bar{Z}_3} = \frac{\bar{V}_c}{\bar{Z}_5}$$

or

$$-\frac{\bar{V}_b}{\bar{Z}_3} + \bar{V}_c\left(\frac{1}{\bar{Z}_2} + \frac{1}{\bar{Z}_3} + \frac{1}{\bar{Z}_5}\right) = \frac{\bar{V}_2}{\bar{Z}_2} + \frac{\bar{V}_a}{\bar{Z}_2} \qquad\qquad\text{(ii)}$$

Equations (i) and (ii) can be written in matrix form as

$$\begin{bmatrix} \left(\dfrac{1}{\bar{Z}_1} + \dfrac{1}{\bar{Z}_3} + \dfrac{1}{\bar{Z}_4}\right) & -\dfrac{1}{\bar{Z}_3} \\[2ex] -\dfrac{1}{\bar{Z}_3} & \left(\dfrac{1}{\bar{Z}_2} + \dfrac{1}{\bar{Z}_3} + \dfrac{1}{\bar{Z}_5}\right) \end{bmatrix}\begin{bmatrix} \bar{V}_b \\[1ex] \bar{V}_c \end{bmatrix} = \begin{bmatrix} \dfrac{\bar{V}_a}{\bar{Z}_1} \\[2ex] \left(\dfrac{\bar{V}_a}{\bar{Z}_2} + \dfrac{\bar{V}_2}{\bar{Z}_2}\right) \end{bmatrix} \qquad\text{(iii)}$$

Equation (iii) can be solved for the unknown node voltages.

4.7.3 Thevenin's Theorem

Any two-terminal, linear, bilateral network can be replaced by an equivalent circuit consisting of a voltage source and series impedance seen from that terminals. The equivalent voltage source, $\bar{V}_{th}$, is the open circuit voltage between the terminals and equivalent impedance, $\bar{Z}_{th}$, is the ratio of the open-circuit voltage to the short-circuit current at these terminals. Figure 4.25 shows the Thevenin's equivalent of the network between terminals *a* and *b*. This theorem has been explained in detail in Chapter 3.

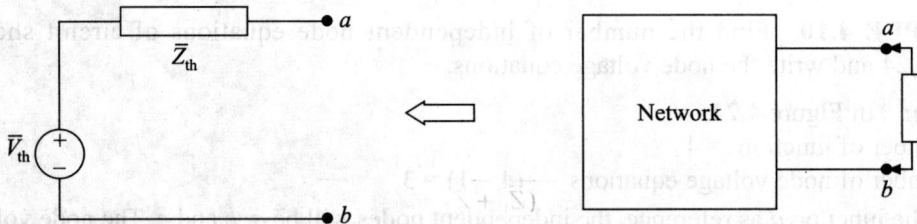

FIGURE 4.25   Thevenin's equivalent.

Mathematically, equivalent Thevenin voltage $(\bar{V}_{th})$ and resistance $(\bar{Z}_{th})$ will be

$$\bar{V}_{th} = \bar{V}_{oc} \qquad\qquad\qquad (4.56)$$

$$\bar{Z}_{th} = \frac{\bar{V}_{oc}}{\bar{I}_{sc}} \qquad\qquad\qquad (4.57)$$

where $\bar{V}_{oc}$ is the open-circuit voltage and $\bar{I}_{sc}$ is the short-circuit current at terminals.

EXAMPLE 4.11   Find the current $\overline{I}$ in Figure 4.26 using the Thevenin's theorem.

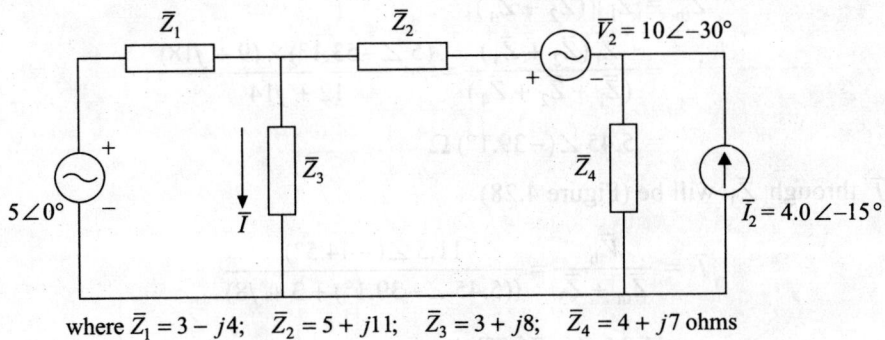

where $\overline{Z}_1 = 3 - j4$; $\overline{Z}_2 = 5 + j11$; $\overline{Z}_3 = 3 + j8$; $\overline{Z}_4 = 4 + j7$ ohms

FIGURE 4.26

Solution: The impedance $\overline{Z}_3$ is removed and terminals are marked as a and b which is shown in Figure 4.27. The current source is transformed into the voltage source.

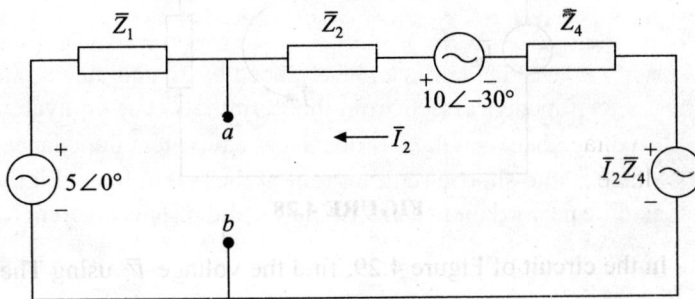

FIGURE 4.27

The current $\overline{I}_2$ will be obtained from the circuit of Figure 4.27 as follows:

$$\overline{I}_2 = \frac{\overline{I}_1 \overline{Z}_4 + 10 \angle(-30°) - 5 \angle 0°}{(\overline{Z}_1 + \overline{Z}_2 + \overline{Z}_3)}$$

$$= \frac{32.25 \angle 45.25° + 10 \angle(-30°) - 5 \angle 0°}{12 + j14}$$

$$= \frac{26.36 + j17.90}{12 + j14} = 1.73 \angle(-15.2°) \text{ A}$$

Thus Thevenin voltage (voltage between terminals a and b) will be

$$\overline{V}_{ab} = \overline{V}_{th} = 5 \angle 0° + \overline{I}_2 \overline{Z}_1$$

$$= 5 \angle 0° + (1.73 \angle -15.2°) \times (3 - j4)$$

$$= 11.5 \angle(-44.5°) \text{ volt}$$

Thevenin impedance will be

$$\bar{Z}_{th} = \bar{Z}_1 \| (\bar{Z}_2 + \bar{Z}_4)$$

$$= \frac{\bar{Z}_1(\bar{Z}_2 + \bar{Z}_4)}{(\bar{Z}_1 + \bar{Z}_2 + \bar{Z}_4)} = \frac{(5\angle -53.1°) \times (9 + j18)}{12 + j14}$$

$$= 5.45 \angle(-39.1°)\ \Omega$$

Current $\bar{I}$ through $\bar{Z}_3$ will be (Figure 4.28)

$$\bar{I} = \frac{\bar{V}_{th}}{\bar{Z}_{th} + \bar{Z}_3} = \frac{11.5\angle(-44.5°)}{((5.45\angle -39.1°) + 3 + j8)}$$

$$= 1.34 \angle(-76.7°)\ A$$

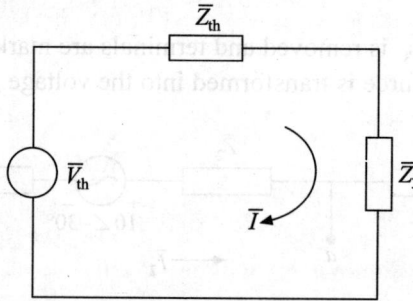

FIGURE 4.28

EXAMPLE 4.12   In the circuit of Figure 4.29, find the voltage $\bar{V}$ using Thevenin theorem.

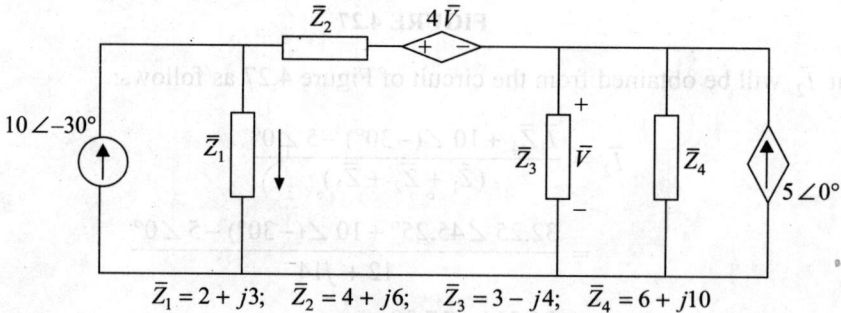

$$\bar{Z}_1 = 2 + j3; \quad \bar{Z}_2 = 4 + j6; \quad \bar{Z}_3 = 3 - j4; \quad \bar{Z}_4 = 6 + j10$$

FIGURE 4.29

Solution: Converting current sources to voltage source and removing the impedance $\bar{Z}_3$ to obtain open circuit voltage $\bar{V}_{oc}$ between terminals *a* and *b*, Figure 4.30 can be obtained.

The current in the circuit will be

$$\bar{I} = \frac{36.1\angle 26.31° - 4\bar{V}_{ab} - 58.3\angle 59.04°}{\bar{Z}_1 + \bar{Z}_2 + \bar{Z}_4}$$

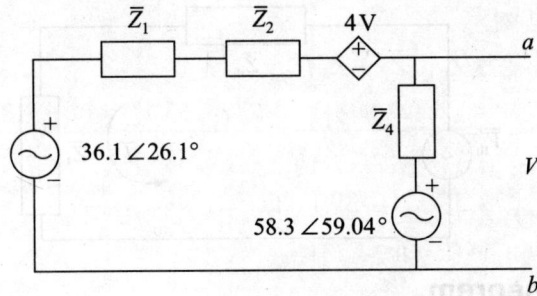

FIGURE 4.30

$$\overline{V}_{ab} = \overline{V}_{oc} = 58.3 \angle 59.04° + \overline{IZ}_4$$

$$\overline{V}_{oc} = 58.3 \angle 59.04° + \frac{11.66 \angle 59.04°}{12 + j19} [36.1 \angle 26.31° - 4\overline{V}_{oc} - 58. \angle 359.04°]$$

$$- \frac{11.66 \angle 59.04°}{12 + j19} \overline{V}_{oc}$$

$$\overline{V}_{oc} = \frac{45.46 \angle 45.92°}{3.078 \angle 0.89°} = 14.77 \angle 45.03° = \overline{V}_{th}$$

The short-circuit current, $\overline{I}_{sc}$ (Figure 4.31) will be $(\overline{V} = 0)$

$$\overline{I}_{sc} = \frac{36.1 \angle 26.31°}{\overline{Z}_1 + \overline{Z}_2} + \frac{58.3 \angle 59.04°}{\overline{Z}_4} = \frac{36.1 \angle 26.31°}{6 + j9} + 5 \angle 0° = 8.06 \angle -11.95° \text{ A}$$

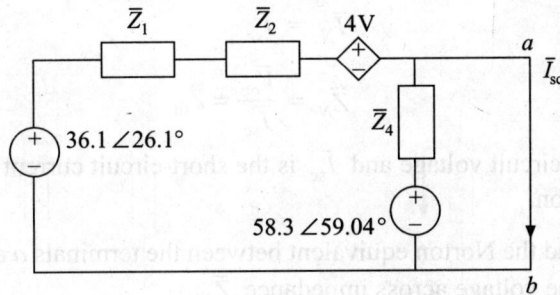

FIGURE 4.31

$\therefore$ Thevenin impedance $\overline{Z}_{th} = \dfrac{\overline{V}_{oc}}{\overline{I}_{sc}} = \dfrac{14.77 \angle 45.03°}{8.06 \angle -11.95°} = 1.83 \angle 56.98° \ \Omega$

Voltage across $\overline{Z}_3$ will be

$$\overline{V} = \frac{\overline{V}_{th} \cdot \overline{Z}_3}{\overline{Z}_3 + \overline{Z}_{th}} = \frac{14.77 \angle 45.03° \times 5 \angle -53.13°}{(1.83 \angle 56.98° + 5 \angle -53.13°)} = 15.71 \angle 23.56° \text{ Volt}$$

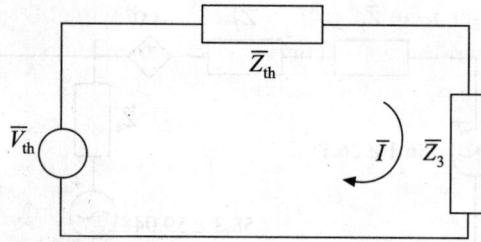

4.7.4 Norton's Theorem

Any two-terminal, linear, bilateral network can be replaced by an equivalent circuit consisting of a current source parallel with the impedance seen from that terminals. The equivalent current source, $\overline{I}_N$, is the short-circuit current between the terminals and equivalent impedance, $\overline{Z}_N$, which is the ratio of the open-circuit voltage to the short circuit current at these terminals as shown in Figure 4.32.

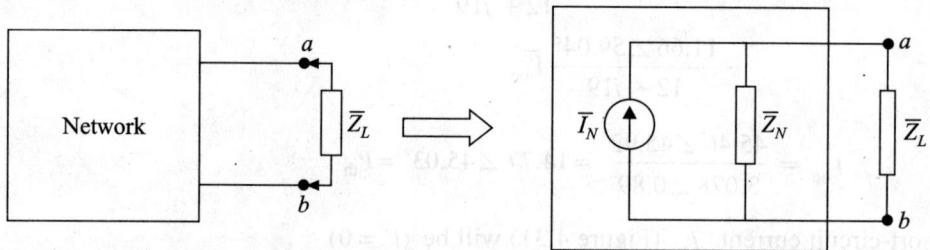

FIGURE 4.32   Norton's equivalent.

Mathematically, equivalent Norton current voltage $(\overline{I}_N)$ and resistance $(\overline{Z}_N)$ will be

$$\overline{I}_N = \overline{I}_{sc} \tag{4.58}$$

$$\overline{Z}_N = \frac{\overline{V}_{oc}}{\overline{I}_{sc}} = \overline{Z}_{th} \tag{4.59}$$

where $\overline{V}_{oc}$ is the open-circuit voltage and $\overline{I}_{sc}$ is the short-circuit current at terminals of Norton's equivalent representation.

EXAMPLE 4.13   Find the Norton equivalent between the terminals a and b of circuit shown in Figure 4.33 and find the voltage across impedance $\overline{Z}_4$.

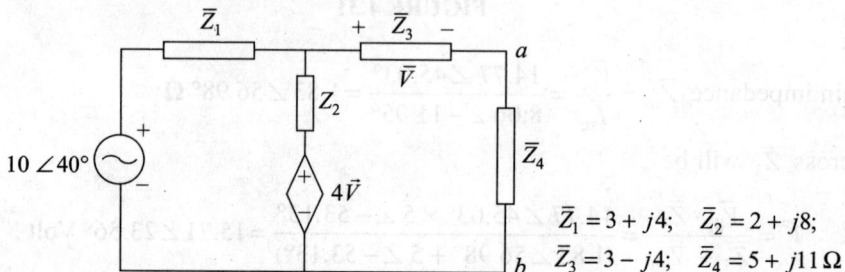

$$\overline{Z}_1 = 3 + j4; \quad \overline{Z}_2 = 2 + j8;$$
$$\overline{Z}_3 = 3 - j4; \quad \overline{Z}_4 = 5 + j11\,\Omega$$

FIGURE 4.33

Solution: After opening of the terminal a and b, the current will not flow through the impedance $\bar{Z}_3$ and thus the voltage $\bar{V}$ will be zero. The value of dependent source will also be zero. The current through the impedances $\bar{Z}_1$ and $\bar{Z}_2$ will be

$$\bar{I}_1 = \left(\frac{10(\cos 40° + j\sin 40°)}{5 + j12}\right) = 0.6831 - j.0354 \text{ A}$$

The open-circuit voltage will be

$$\bar{V}_{oc} = \bar{I}_1(2 + j8) = 4.196 + j4.7519 \text{ V}$$

To obtain the short-circuit current between the terminals a and b, the impedance $\bar{Z}_4$ will be removed and the terminals a and b will be shorted as shown in Figure 4.34.

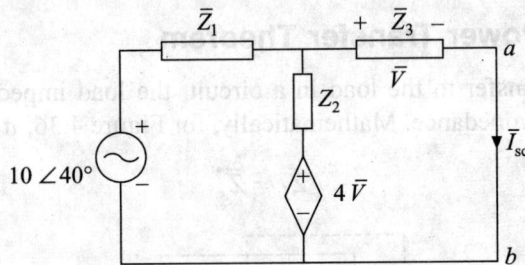

FIGURE 4.34

Writing the nodal equation, we get

$$\frac{10 \angle 40° - \bar{V}}{3 + j4} + \frac{4\bar{V} - \bar{V}}{2 + j8} = \frac{\bar{V}}{3 - j4}$$

or

$$\bar{V} = 0.9170 - j5.124 = 5.206 \angle - 79.85° \text{ V}$$

The short-circuit current will be

$$\bar{I}_{sc} = \frac{\bar{V}}{3 - j4} = 0.93 - j0.468 = 1.04 \angle 26.72° \text{ A}$$

Thus the Norton's equivalent impedance will be

$$\bar{Z}_{eq} = \frac{\bar{V}_{oc}}{\bar{I}_{sc}} = 1.54 + j5.89 = \bar{Z}_N$$

Thus the current in impedance $\bar{Z}_4$ will be (Figure 4.35)

$$\bar{I}_{z4} = \bar{I}_{sc}\left(\frac{\bar{Z}_N}{\bar{Z}_N + \bar{Z}_4}\right) = 0.328 - j0.121 = 0.35 \angle - 20.19° \text{ A}$$

and voltage across $\overline{Z}_4$ will be

$$\overline{V}_{ab} = \overline{I}z_4 \times \overline{Z}_4 = 2.97 + j3.0 = 4.22 \angle 45.35° \text{ V}$$

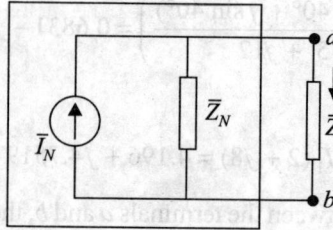

FIGURE 4.35

4.7.5 Maximum Power Transfer Theorem

For maximum power transfer to the load in a circuit, the load impedance should be complex conjugate of the source impedance. Mathematically, for Figure 4.36, it can be written as

$$\overline{Z}_L = \overline{Z}_s^* \tag{4.60}$$

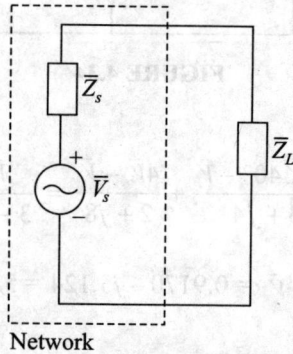

FIGURE 4.36   Maximum power transfer theorem.

The power transfer to the load will be the maximum when the circuit is purely resistive. In other words, the current should be in phase with the source voltage (or total reactive impedance should be zero), i.e*

$$\text{Im}(\overline{Z}_L) + \text{Im}(\overline{Z}_s) = 0$$

$$\text{Im}(\overline{Z}_L) = -\text{Im}(\overline{Z}_s) \tag{4.61}$$

For resistive circuit, we have proved in Section 3.10 that load resistance should be equal to the source resistance, i.e.,

$$R_L = R_s$$

---

*Im(.) is used for imaginary value of function (.).

or

$$\text{Re}(\bar{Z}_L) = \text{Re}(\bar{Z}_s) \tag{4.62}$$

Adding Eqs. (4.61) and (4.62), we get

$$\text{Re}(\bar{Z}_L) + j\,\text{Im}(\bar{Z}_L) = \text{Re}(\bar{Z}_s) - j\,\text{Im}(\bar{Z}_s) \tag{4.63}$$

or

$$\bar{Z}_L = \bar{Z}_s^* \tag{4.64}$$

EXAMPLE 4.14 Find current $\bar{I}_L$ using the Thevenin Theorem. Find the value of extra impedance to be connected in series with the impedance $\bar{Z}_1$ of Figure 4.37 for maximum power to be transferred through $\bar{Z}_1$.

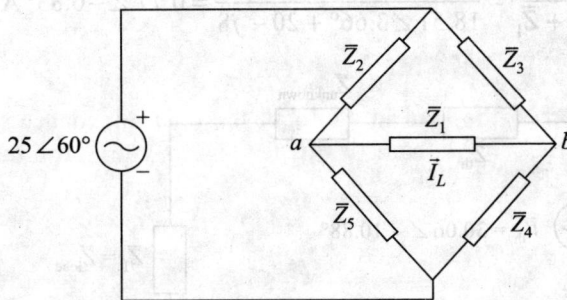

$$\bar{Z}_1 = 20 - j8; \quad \bar{Z}_2 = 2 + j9; \quad \bar{Z}_3 = 8 + j4; \quad \bar{Z}_4 = 3 - j4; \quad \bar{Z}_5 = 4 - j9 \ \Omega$$

FIGURE 4.37 Example 4.14.

Solution: Removing the impedance between the terminals a and b, and keeping them open, the voltage across the terminal will be

$$\bar{V}_{ab} = \bar{V}_{oc} = \frac{\bar{V} \cdot \bar{Z}_5}{\bar{Z}_5 + \bar{Z}_2} - \frac{\bar{V} \cdot \bar{Z}_4}{\bar{Z}_4 + \bar{Z}_3} = \bar{V}_{th}$$

where $\bar{V}$ is the source voltage. Putting the values, we get

$$\bar{V}_{th} = 9.53 - j5.67 = 30.06 \ \angle -10.88° \ \text{V}$$

Since there is no dependent source, the Thevenin impedance between open circuit terminals can be simply obtained by replacing the voltage source by the short circuit. Thus,

$$\bar{Z}_{th} = (\bar{Z}_2 \| \bar{Z}_5) + (\bar{Z}_3 \| \bar{Z}_4) = \frac{\bar{Z}_2 \cdot \bar{Z}_5}{\bar{Z}_2 + \bar{Z}_5} + \frac{\bar{Z}_3 \cdot \bar{Z}_4}{\bar{Z}_3 + \bar{Z}_4}$$

$$= \frac{(2 + j9)(4 - j9)}{6} + \frac{(8 + j4)(3 - j4)}{11} \ \Omega$$

Therefore,

$$\bar{Z}_{th} = 18.47 + j1.18 = 18.51 \ \angle 3.66° \ \Omega$$

The equivalent circuit is shown below:

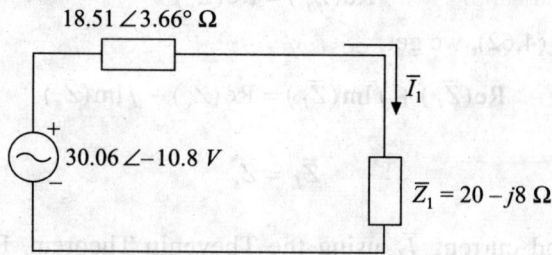

The current $\bar{I}_1$ will be

$$\bar{I}_1 = \frac{\bar{V}_{th}}{\bar{Z}_{th} + \bar{Z}_1} = \frac{30.06 \angle -10.88°}{18.51 \angle 3.66° + 20 - j8} = 0.77 \angle -0.83° \, A$$

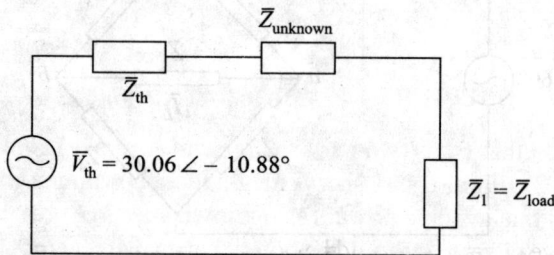

For maximum power transfer,

$$\bar{Z}_1^* = \bar{Z}_{th} + \bar{Z}_{unknown}$$

or

$$(20 - j8)^* = 18.47 + j1.18 + \bar{Z}_{unknown}$$

Thus,

$$\bar{Z}_{unknown} = 1.53 + j6.81 \, ohm$$

4.8   POWER CALCULATION IN AC CIRCUIT

Consider the circuit shown in Figure 4.38 having a voltage source, $v = \sqrt{2} V \cos \omega t$. The current in the circuit (assuming inductive) will be lagging and can be written as

$$i = \sqrt{2} I \cos(\omega t - \theta) \tag{4.65}$$

where θ is the angle between voltage and current phasor and also known as *power factor* angle*. V and I are the rms values of voltage and current magnitudes, respectively.

---

* It should be noted that angle θ = difference of voltage angle and current angle. If $v = \sqrt{2} V \cos(\omega t + \theta_1)$ and $i = \sqrt{2} I \cos(\omega t + \theta_2)$, then angle $\theta = (\theta_1 - \theta_2)$.

The instantaneous power in an element is defined as the product of instantaneous voltage and instantaneous current, and can be written as

$$p = vi = \sqrt{2}\,V\cos\omega t \cdot \sqrt{2}\,I\cos(\omega t - \theta) = VI\left[2\cos\omega t \cdot \cos(\omega t - \theta)\right]$$

$$= VI\left[\cos\theta + \cos(2\omega t - \theta)\right] \qquad (4.66)$$

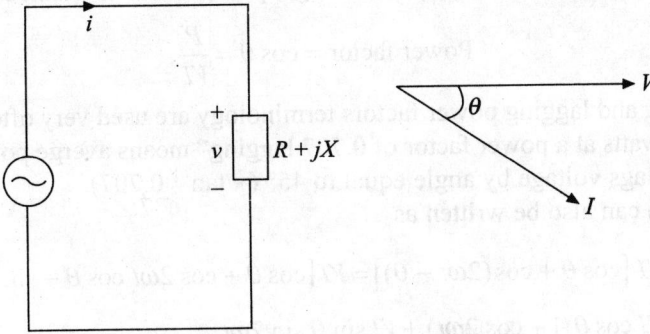

FIGURE 4.38   An inductive circuit.

From Eq. (4.66), it is seen that p has two parts: a constant (average) component and a sinusoidal component of frequency 2ω. Figure 4.39 shows the relationship among variables p, v, and i. From this figure, it is observed that average power (also known as *active* or *real power*) is positive, but it can be positive, negative or zero. It should be noted that an element (active or passive) consumes real power, if the current enters through the positive terminal of the element. If current leaves the positive terminal of the element, it produces real power. It should also be noted that a linear resistor always consumes real power, however, average power is zero in pure inductance or pure capacitance.

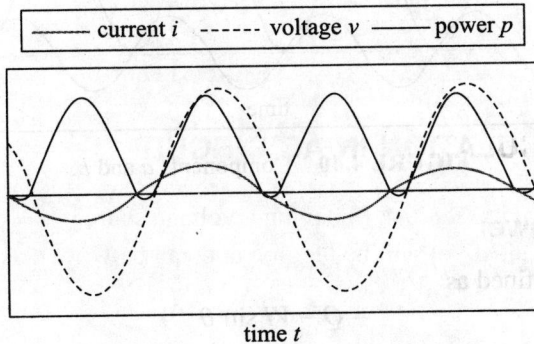

time t

FIGURE 4.39   Instantaneous power.

If power generated is taken as positive, the power consumed will be considered as negative and vice versa. The average power P over one time-period, $T = 2\pi/\omega$, can be defined as

$$P_{av} = \frac{1}{T}\int_0^T p\,dt = VI\cos\theta = P \qquad (4.67)$$

The real power (P) is the average power (P_{av}) of the circuit. In all applications, generally, P is used rather than p. If a circuit has resistance R, the real power can also be written as

$$P = VI \cos \theta = I^2 R \tag{4.68}$$

The power factor (p.f.) is defined as the cosine of angle between current and voltage waves (or phasors). In terms of power, current and voltage, power factor is defined as

$$\text{Power factor} = \cos \theta = \frac{P}{VI} \tag{4.69}$$

In ac power, leading and lagging power factors terminology are used very often. In a description, "a load draws 100 watts at a power factor of 0.707 lagging" means averge power consumption is 100 W and current lags voltage by angle equal to 45° ($= \tan^{-1} 0.707$).

Equation (4.66) can also be written as

$$p = VI \left[\cos \theta + \cos (2\omega t - \theta) \right] = VI \left[\cos \theta + \cos 2\omega t \cos \theta + \sin 2\omega t \sin \theta \right] \tag{4.70}$$

$$= \underbrace{VI \cos \theta (1 + \cos 2\omega t)}_{a} + \underbrace{VI \sin \theta \sin 2\omega t}_{b}$$

The instantaneous power p can be decomposed into two components. First component (marked as a in Eq. (4.70)) pulsates around the same average value as before but never goes negative. The second term (as marked b in Eq. (4.70)) has a zero average value. This can be seen in Figure 4.40.

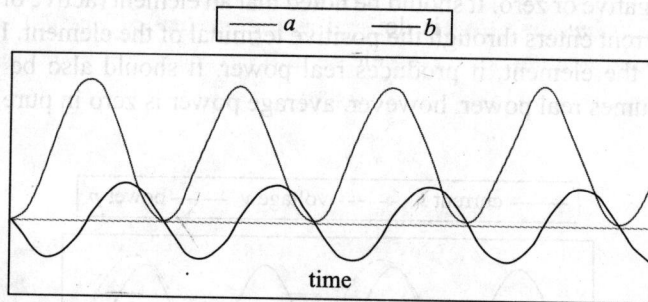

FIGURE 4.40   Components a and b.

4.8.1 Reactive Power

Reactive power (Q) is defined as

$$Q = VI \sin \theta \tag{4.71}$$

Equation (4.70) can be written in terms of real and reactive powers as

$$p = P(1 - \cos 2\omega t) - Q \sin 2\omega t \tag{4.72}$$

From above equation, the reactive power Q is equal to the peak value of that power components, which travels back and forth on the line, resulting in zero average power and therefore, capable of doing no useful work. The physical dimension of both P and Q is watt. But the symbolized unit of P is watt (W) and of Q is var (VAR). More practical and bigger units of real power are kW, MW and for Q, it is kVAR and MVAR.

4.8.2 Using Phasor Concept

Voltage and current phasors (for the circuit shown in Figure 4.38) can be represented as

$$\bar{V} = V \angle 0 \quad \text{and} \quad \bar{I} = I \angle -\theta$$

Thus, the impedance will be

$$\bar{Z} = \frac{\bar{V}}{\bar{I}} = \frac{V}{I} \angle \theta = Z \angle \theta \tag{4.73}$$

Using the analogy to dc circuit (except R is replaced by $\bar{Z}$), the complex power

$$\bar{S} = \left|\bar{I}^2\right|\bar{Z} = \bar{I}^*\bar{I}\,\bar{Z} = \bar{V}\bar{I}^* = I^2 R + jI^2 X = P + jQ$$

or

$$\bar{S} = \bar{V}\bar{I}^* = V\angle 0 \cdot I\angle\theta = \bar{V}\bar{I}\angle\theta = VI\cos\theta + jVI\sin\theta = P + jQ \tag{4.74}$$

Using Eq. (4.74), the real power (P) can also be written as

$$P = \mathrm{Re}(\bar{V}\bar{I})^* \tag{4.75}$$

and reactive power will be

$$Q = \mathrm{Im}(\bar{V}\bar{I})^* \tag{4.76}$$

4.8.3 Apparent Power

Apparent power or *volt-ampere* (VA) is defined as the product of *rms* values of current flows through the element and voltage across the element. It is very useful in defining the rating of an apparatus. Mathematically, it can be expressed as

$$S = \left|\bar{S}\right| = VI = \sqrt{P^2 + Q^2} \tag{4.77}$$

kVA and MVA are the practical units of apparent power.

Vector representation of complex power (*R-L* circuit) is shown in Figure 4.41.

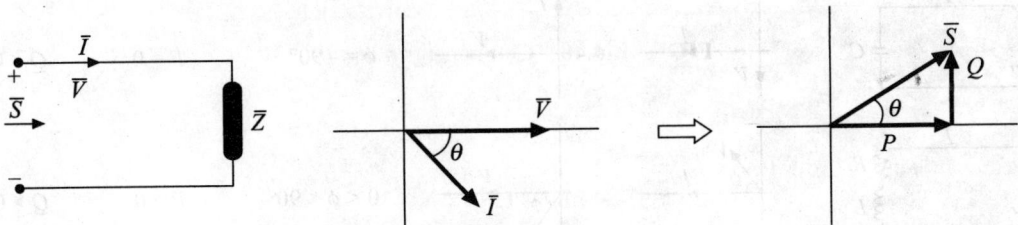

FIGURE 4.41   Complex power.

The real, reactive and apparent powers are equal in magnitude to the three sides of a right triangle known as *power triangle*. If voltage leads current, which is the same as current lags voltage, both P and Q will be positive. For voltage lags current or current leads voltage, the reactive power Q will be negative. Complex power can also be written in terms of voltage and current.

The power factor angle in terms of powers can be expressed as

$$\text{p.f. angle } (\theta) = \tan^{-1}\frac{P}{Q} = \cos^{-1}\frac{P}{S} = \sin^{-1}\frac{Q}{S} \tag{4.78}$$

Thus, power factor is a function of the apparent power which is actually being utilized or part of energy which is being transferred into other forms. In an inductive circuit, the power factor will be lagging, whereas in a capacitive circuit, it is leading. In a pure resistive circuit, the power factor is always unity ($\theta = 0$).

For a network supplied by several independent sources, all at the same frequency, the sum of the complex power supplied by the independent sources is equal to the sum of the complex power received by all the other branches of the network. This is known as *Theorem of conservation of complex power.*

Table 4.2 shows the phasor relationship for different types of loads. Different terminology and units of power are presented in Table 4.3.

TABLE 4.2   Phasor relationship for different types of loads

Type of Load	Phasor Diagram		Phase Angle	Power Absorbed by Load	
	Taking Current as Reference	Taking Voltage as Reference		Real Power	Reactive Power
R			$\phi = 0$	$P > 0$	$Q = 0$
L			$\phi = 90°$	$P = 0$	$Q > 0$
C			$\phi = -90°$	$P = 0$	$Q < 0$
R, L			$0 < \phi < 90°$	$P > 0$	$Q > 0$
C, R			$-90° < \phi < 0$	$P > 0$	$Q < 0$

TABLE 4.3   Notations and units of power

Notation	Terminology	Descriptive Units
p	Instantaneous power	Watts: W, kW, MW
$\overline{S}$	Complex power	Voltamperes: VA, kVA, MVA
S	Apparent power	Voltamperes: VA, kVA, MVA
P	Real or active or average power	Watts: W, kW, MW
Q	Reactive power	Voltamperes reactive: VAR, kVAR, MVAR

EXAMPLE 4.15   For $\overline{Z} = R + j\,\omega L$, consider the current is $i = \sqrt{2}I\sin(\omega t + \theta)$:

(a) Calculate P and Q
(b) Calculate the instantaneous power.

Solution:

(a) $\overline{S} = \overline{V}\,\overline{I}^{*} = \overline{Z}I^{2} = (R + j\omega L)I^{2}$
Therefore, the real power $P = RI^{2}$ and reactive power, $Q = \omega L I^{2}$

(b) Instantaneous voltage v can be written as

$$v = Ri + L\frac{di}{dt} = R\sqrt{2}I\sin(\omega t + \theta) + L\sqrt{2}I\omega\cos(\omega t + \theta)$$

Instantaneous power can be calculated as

$$p = vi = 2RI^{2}\sin^{2}(\omega t + \theta) + 2\omega LI^{2}\sin(\omega t + \theta)\cos(\omega t + \theta)$$

$$= 2P\sin^{2}(\omega t + \theta) + Q\sin 2(\omega t + \theta) = P(1 - \cos 2(\omega t + \theta)) + Q\sin 2(\omega t + \theta)$$

EXAMPLE 4.16   Compute the powers in each elements of the circuit shown in Figure 4.42. Take supply voltage of 220 volt.

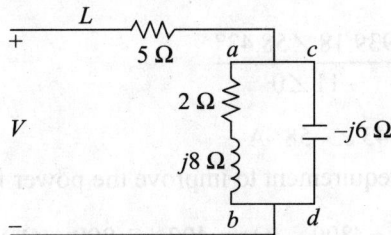

FIGURE 4.42   Example 4.16.

Solution: Let us find the equivalent impedance of parallel branches.

$$\overline{Z}_{eq} = \frac{(2 + j8)\times(-j6)}{(2 + j8) + (-j6)} = 9.0 - j15.0\ \Omega$$

$$= 17.50\ \angle -59.04°\text{ ohm}$$

The current $\bar{I}$ will be

$$\bar{I} = \frac{220\angle 0°}{5\angle 0° + 17.5\angle -59.04°} = 7.31 + j7.84 \text{ A}$$

The power in 5 ohm resister $= I^2R = 574.39$ watts.

If current in branch ab is $\bar{I}_1$ and current in branch cd is $\bar{I}_2$, these can be calculated as,

$$\bar{I}_1 = \frac{\bar{I}\,\bar{Z}_{cd}}{\bar{Z}_{ab} + \bar{Z}_{cd}} = \frac{(7.31 + j7.84)\times(-j6)}{(2 + j8) + (-j6)} = 22.81\angle -88.03° \text{ A}$$

$$\bar{I}_2 = \frac{\bar{I}\,\bar{Z}_{ab}}{\bar{Z}_{ab} + \bar{Z}_{cd}} = \frac{(7.31 + j7.84)\times(2 + j8)}{(2 + j8) + (-j6)} = 31.26\angle 77.94° \text{ A}$$

The power loss in resister of branch $ab = I_1^2R = 1040.59$ W.

Power loss in inductor $= I_1^2 X_L = 4162.37$ VAR.

Power loss in capacitor $= I_1^2 X_c = -5863.13$ VAR

EXAMPLE 4.17   An induction furnace load requires 400 kW and 800 kVAR at 11 kV (single phase), 50Hz. Find

    (a) power factor and p.f. angle,

    (b) line current,

    (c) value of C required across load to improve the p.f. 0.9 lagging.

Solution: Complex power will be

$$\bar{S} = P + jQ = 400 + j800 \text{ kVA} = 939.18\angle 58.43° \text{ VA}$$

    (a) p.f. $= \cos(58.43) = 0.523$ (lagging)

        p.f. angle $= -58.43°$

    (b) Line current $\bar{I}^* = \dfrac{\bar{S}}{\bar{V}} = \dfrac{939.18\angle 58.43°}{11\angle 0}$

$$\bar{I} = 85.3843\angle -58° \text{ A}$$

    (c) Let the reactive power requirement to improve the power factor is Qc

$$\bar{S} = 400 + j800 - jQc = 400 + j(800 - Qc)$$

$$\cos\theta = \frac{400}{\sqrt{400^2 + (800 - Qc)^2}} = 0.9$$

$$Qc = 561.95 \text{ kVAR}; \quad Qc = \frac{V^2}{Xc} = V^2\omega C \Rightarrow C = 14.7\ \mu\text{F}$$

4.9 SUPERPOSITION AND CONSERVATION OF POWER

Let v and i be the terminal voltage and current into a two-terminal sub-circuit in a linear circuit having two independent sources. The superposition principle always holds to the current and voltage in any linear circuit, thus

$$i = i_1 + i_2 \quad \text{and} \quad v = v_1 + v_2 \tag{4.79}$$

The subscripts in above equation represent the responses due to each source separately taking one at a time. The instantaneous power consumed by the network will be

$$p = vi = (v_1 + v_2)(i_1 + i_2) = v_1 i_1 + v_2 i_2 + (v_2 i_1 + v_1 i_2)$$

or

$$p = p_1 + p_2 + (v_2 i_1 + v_1 i_2) \tag{4.80}$$

where, p_1 and p_2 are the instantaneous powers due to each source separately. From Eq. (4.80), it can be seen that the total power is different than the sum of individual instantaneous power*. Thus, *superposition does not in general hold for instantaneous power***. Similarly, it can be proved for ac circuits that

  1. *Superposition does not in general hold for complex power.*
  2. *Superposition does not in general hold for real power.*
  3. *Superposition does not in general hold for real power.*

Consider an ac circuit having n-node. Applying KCL at node j, we get

$$\sum_{k=1}^{n} \overline{I}_{jk} = 0 \tag{4.81}$$

where, current flows from node k to node j is denoted at $\overline{I}_{jk}$. Taking the complex conjugate of KCL equation and pre-multiplying by the complex node voltage $\overline{V}_j$, we get

$$\sum_{k=1}^{n} \overline{V}_j \overline{I}_{jk}^* = 0 \tag{4.82}$$

There is one such equation for each node in the circuit. Adding all such equations provides,

$$\sum_{j=1}^{n} \left[\sum_{k=1}^{n} \overline{V}_j \overline{I}_{jk}^* \right] = 0 \tag{4.83}$$

The current between every pair of node appears twice in the first double sum as $\overline{I}_{jk} = -\overline{I}_{kj}$. It should be noted that $\overline{I}_{jj} = \overline{I}_{kk} = 0$. From Eq. (4.83), we can write as

$$\sum_{j=1}^{n} \left[\sum_{k=j+1}^{n} (\overline{V}_j - \overline{V}_k) \overline{I}_{jk}^* \right] = 0 \tag{4.84}$$

---

* It should be noted that extra term in Eq. (4.80) may not always be positive and it all depends on the connection of the active sources.

** But superposition holds for average power P, reactive power Q and complex power S when the sources are of different frequencies.

Each term in Eq. (4.84) is the complex power $\overline{S}_{jk}$ absorbed by sub-circuit or element connected between node pair (j, k) and each node pair in the circuit will appear only once. Thus, Eq. (4.84) can be simplified as

$$\sum_{\text{node pair } (j,k)} \overline{S}_{jk} = 0 \qquad (4.85)$$

Thus, it can be stated that *the sum of the complex power absorbed by all the elements in a circuit equals zero**. By putting the negative and positive terms of power absorption, the conservation principle can be restated as *the net power absorbed by all elements in a circuit equals the net power supply by all the sources*.

This conservation principle holds equally for complex, real (average) and reactive power.

PROBLEMS

4.1   Find the phasor form (taking cosine as reference) of the following functions.

     (a) $v(t) = 311 \cos (377t - 25°)$ V

     (b) $i(t) = 10 \cos (10t + 60°) + 15 \cos (10t - 30°)$ A.

     (c) $i(t) = 460 \cos (500 \pi t - 25°) - 220 \sin (500 \pi t + 15°)$ A

               [Ans. (a) 219.9 $\angle -25°$ V, (b) 12.746 $\angle 3.69°$ A, (c) 254.85 $\angle 2.88°$ A **]**

4.2   Find the time-domain expression corresponding to each of the following phasors

     (a) $\overline{I} = (10 \angle 30° + 25 \angle 60°)$ mA

     (b) $\overline{V} = (60 + j30 + 100 \angle -28°)$ V

               [Ans. (a) 48.12 $\cos (\omega t + 51.55°)$ mA, (b) 211.1 $\cos (\omega t - 6.52°)$ V**]**

4.3   A series *RLC* circuit is composed of 10 Ω resistance, one 0.1 H inductance and one 50.0 μF capacitance. A voltage $v(t) = 141.4 \cos (100 \pi t)$ V is impressed upon the circuit.

     (a)   Find the phasor current in the circuit.

     (b)   Find the expression for instantaneous current.

     (c)   Calculate voltage drops V_R, V_L and V_C across resistor, inductor and capacitor, respectively.

     (d)   Draw the phasor diagram showing these voltage relations.

               [Ans. (a) 2.958 $\angle 72.79°$ A, (b) 4.183 $\cos (100 \pi t + 72.79)$ A,

               (c) 29.58 $\angle 72.79°$ V, 92.99 $\angle 162.79°$ V, 188.4 $\angle -17.2°$ V**]**

4.4   A source (377 rad/s) is connected to a load Z_L as shown in Figure 4.43. Find the value of capacitance for load to be completely resistive. What is the actual impedance that the source sees with this value of capacitor?

               [Ans. 510.10 μF, 26 Ω**]**

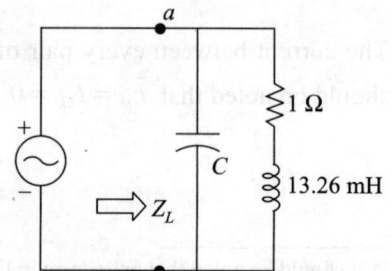

FIGURE 4.43   Problem 4.4.

---

* Net power supplied by the active sources is taken as negative absorption.

4.5 Find Z_{ab} in Figure 4.44, if $\omega = 10^5$ rad/s.

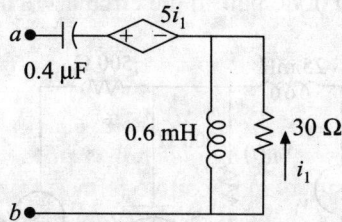

FIGURE 4.44   Problem 4.5.

[**Ans.** 25 $\angle -36.87°$ Ω]

4.6 Find the value of ω, if the circuit in Figure 4.45, is operating in steady state and
$v_s = 40 \cos (\omega t - 15°)$ V
$i_0 = 40 \sin (\omega t + 21.87°)$ mA

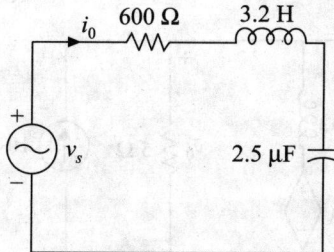

FIGURE 4.45   Problem 4.6.

[**Ans.** 500 rad/sec]

4.7 The current $\bar{I}_a$ in the circuit shown in Figure 4.46 is 2 $\angle 0°$ A.

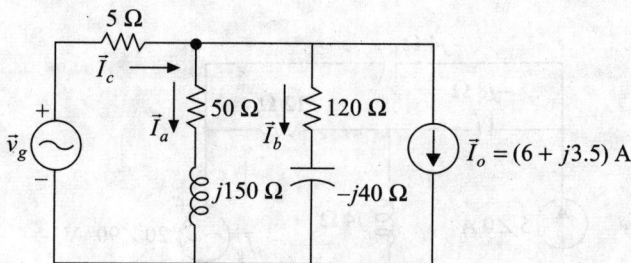

FIGURE 4.46   Problem 4.7.

(a) Find $\bar{I}_b, \bar{I}_c$ and $\bar{V}_g$

(b) Draw the phasor diagram showing $\bar{I}_a, \bar{I}_b, \bar{I}_c, \bar{I}_o, \bar{V}_a, \bar{V}_g$

[**Ans.** (a) 2.5 $\angle 90°$ A, 10 $\angle 36.87°$ A, 358.47 $\angle 67°$ V]

4.8 Find the steady state expression for $v_0(t)$ in the circuit shown in Figure 4.47 by using the technique of source transformations. The source voltages are

$$v_1 = 500 \sin(8000t + 126.87°) \text{ V}$$
$$v_2 = 1000 \cos(8000t - 90°) \text{ V}$$

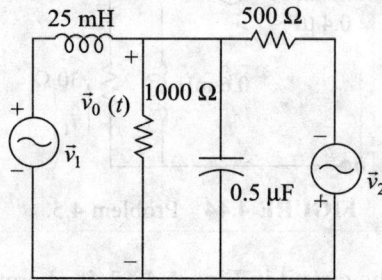

FIGURE 4.47   Problem 4.8.

[**Ans.** 474.38 cos (8000t + 18.44°) V]

4.9 Using the node voltage method, find the phasor voltage $\bar{V}_0$ in Figure 4.48.

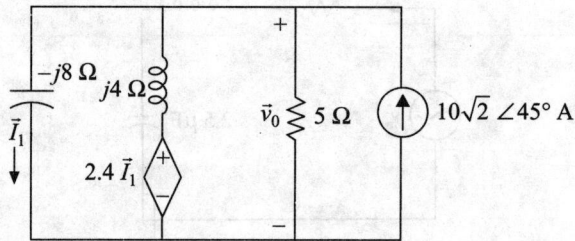

FIGURE 4.48   Problem 4.9.

[**Ans.** 80 ∠90° V]

4.10 Using the mesh current method, find the phasor current $\bar{I}_g$ in Figure 4.49.

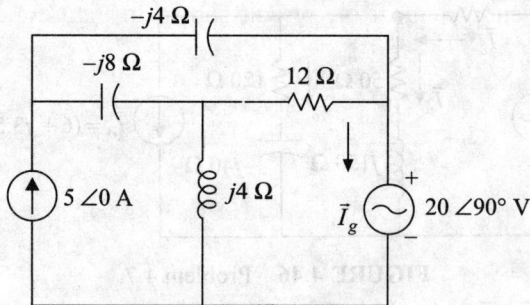

FIGURE 4.49   Problem 4.10.

[**Ans.** 4.47∠−26.56° A]

4.11 Find the current $\bar{I}$ in Figure 4.50 using Thevenin theorem.

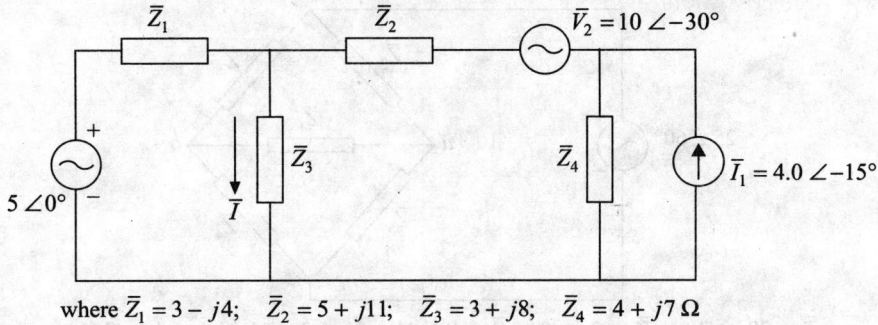

where $\bar{Z}_1 = 3 - j4; \quad \bar{Z}_2 = 5 + j11; \quad \bar{Z}_3 = 3 + j8; \quad \bar{Z}_4 = 4 + j7\,\Omega$

FIGURE 4.50   Problem 4.11.

[**Ans.** $1.34\ \angle -76.7°$ A]

4.12 In the circuit of Figure 4.51, find the voltage $\bar{V}$ using Thevenin theorem.

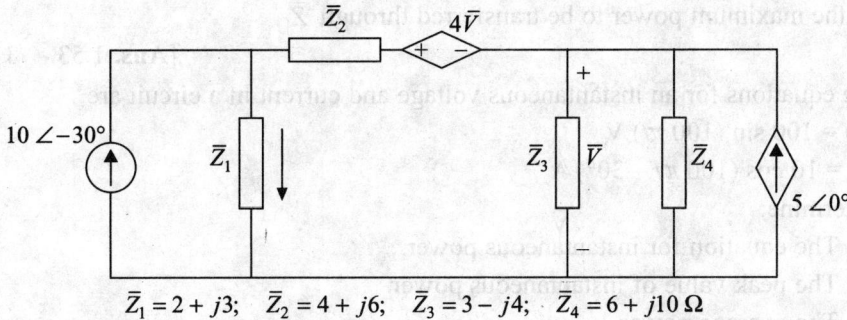

$\bar{Z}_1 = 2 + j3; \quad \bar{Z}_2 = 4 + j6; \quad \bar{Z}_3 = 3 - j4; \quad \bar{Z}_4 = 6 + j10\,\Omega$

FIGURE 4.51   Problem 4.12.

[**Ans.** $15.75\ \angle 23.66°$ A]

4.13 Find the Norton equivalent of the circuit shown in Figure 4.52, and find the voltage across impedance Z_4.

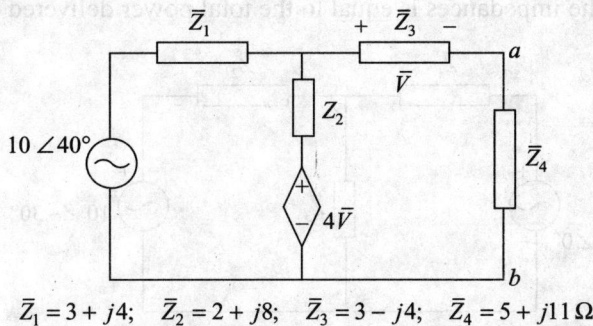

$\bar{Z}_1 = 3 + j4; \quad \bar{Z}_2 = 2 + j8; \quad \bar{Z}_3 = 3 - j4; \quad \bar{Z}_4 = 5 + j11\,\Omega$

FIGURE 4.52   Problem 4.13.

[**Ans.** $\bar{V}_{th} = 6.34\ \angle 48.58°$ V, $\bar{Z}_{th} = 6.13\ \angle -285.26°\ \Omega, V_{ab} = 4.23\ \angle 45.36°$ V]

4.14 Find current $\vec{I}_L$ in circuit shown in Figure 4.53 using the Thevenin theorem.

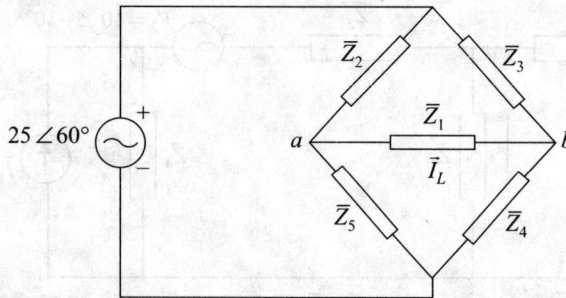

$\bar{Z}_1 = 20 - j30; \quad \bar{Z}_2 = 2 + j9; \quad \bar{Z}_3 = 8 + j4; \quad \bar{Z}_4 = 3 - j4; \quad \bar{Z}_5 = 4 - j9 \, \Omega$

FIGURE 4.53   Problem 4.14.

[**Ans.** 0.61 ∠−55.9° A]

4.15 Find the value of extra impedance to be connected in series across *a* and *b* of Figure 4.53 for the maximum power to be transferred through $\bar{Z}_1$.

[**Ans.** 1.53 − *j*31.18 ohm]

4.16 The equations for an instantaneous voltage and current in a circuit are

$v(t) = 100 \sin (100 \, \pi t)$ V

$i(t) = 10 \cos (100 \, \pi t - 30°)$ A

Determine:

(a) The equation for instantaneous power.

(b) The peak value of instantaneous power.

(c) The average power.

(d) Complex power, reactive power and apparent power.

[**Ans.** (a) 500[0.5 + sin (200 πt − 30°)], (b) 750 W, (c) 250 W,

(d) 250 − *j*433.01, 433.01 VAR, 500 VA]

4.17 Find the power consumed in each impedance of Figure 4.54. Show that total power consumed by the impedances is equal to the total power delivered by the sources.

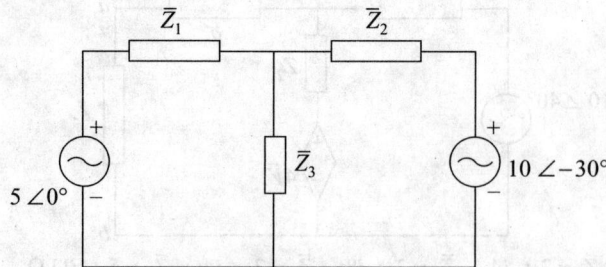

where $\bar{Z}_1 = 3 + j4; \quad Z_2 = 2 + j8; \quad Z_3 = 3 - j4 \Omega$

FIGURE 4.54   Problem 4.17.

[**Ans.** $\bar{S}_{Z_1} = 4.25 + j5.66, \ \bar{S}_{Z_2} = 5.47 - j5.29, \ \bar{S}_{Z_3} = 0.80 + j3.20$ VA]

5

Transient Analysis of AC/DC Circuits

5.1 INTRODUCTION

Transients which are generated due to abrupt changes in operating conditions in the presence of energy storing elements, such as capacitors and inductors, occur in ac as well as dc circuits. The word transient means temporary, or short-lived. When a circuit is on/off (abrupt change in the condition), the circuit condition reaches the steady state and the period from abrupt change and steady state condition is known as *transient period*. During the transient period, the phasor relation does not hold good. The degree of complexity in transient analysis depends on the number of energy storage elements in the circuit. A circuit having single energy storage element (an inductor or a capacitor) and regardless of number of resistors, is known as a *first-order circuit*. In the higher-order circuit, there will be several energy storage elements.

5.2 TRANSIENT RESPONSE OF FIRST-ORDER CIRCUITS WITH DC SOURCE

5.2.1 Current through an Ideal Inductor

An ideal inductor having inductance L is excited by a dc voltage source V as shown in Figure 5.1. When switch S is closed at time $t = 0$, assuming that there is no initial current flowing through the ideal indictor (i.e. $i(0) = 0$), the Kirchhoff's voltage law (KVL) equation can be written as

$$V = v_L(t)$$

$$V = L \frac{di(t)}{dt} \tag{5.1}$$

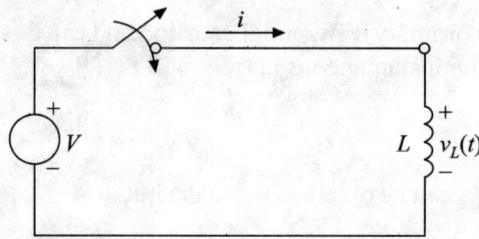

FIGURE 5.1   Ideal inductor connected to a dc voltage source.

Equation (5.1) can be solved as

$$\int_0^{i(t)} di(t) = \frac{V}{L} \int_0^t dt \quad \Rightarrow \quad i(t) - i(0) = \frac{V}{L} t \tag{5.2}$$

Thus,

$$i(t) = \frac{V}{L} t \tag{5.3}$$

Equation (5.3) implies that current in the inductor increases with time and becomes infinite at $t \to \infty$ but a practical inductor will always have resistance. The current response for that case is discussed in the following section.

5.2.2   Transient in *RL* Circuit

A series *RL* circuit connected through a switch and a dc voltage source is shown in Figure 5.2. Due to closing the switch, current in the inductor will try to change, so a voltage $v_L(t)$ is induced across its terminals in opposition to the applied voltage V. At the instant the switch closes, the rate of change of current is maximum and the induced voltage is nearly equal to the applied voltage. The rate of change of current decreases with time which allows current to reach a maximum value that can be determined using Ohm's law.

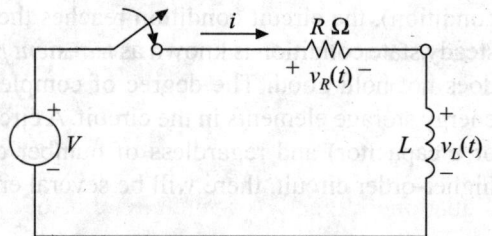

FIGURE 5.2   An elementary *RL* circuit.

Using KVL, we can write,

$$V = v_R(t) + v_L(t) \tag{5.4}$$

Using the relationships (voltage and current) for the ideal resistor and ideal inductor, Eq. (5.4) can be rewritten as

$$V = Ri(t) + L \frac{di(t)}{dt}$$

or

$$\frac{di(t)}{dt} + \frac{R}{L} i(t) = \frac{V}{L} \tag{5.5}$$

Equation (5.5) is a first order, ordinary differential equation and can be solved by several methods. Using the calculus identity, the instantaneous current will be

$$i(t) = k_1 e^{-\frac{R}{L}t} + k_2 \qquad (5.6)$$

The value of constant k_1 and k_2 can be obtained using the initial ($t = 0$) and final ($t = \infty$) conditions. Thus,

$$i(0) = k_1 e^{-\frac{R}{L} \times 0} + k_2 = k_1 + k_2$$

Since before the closing of switch, there was no current in the circuit, therefore $i(0) = 0$.

$$k_1 + k_2 = 0 \qquad (5.7)$$

At final condition (also known as *steady-state condition*), the voltage across inductor will be zero and the current is governed by the resistor only which is V/R.

$$i(\infty) = \frac{V}{R} = k_1 e^{-\frac{R}{L} \times \infty} + k_2 = k_2 \qquad (5.8)$$

Using Eqs. (5.7) and (5.8), we get $k_2 = \dfrac{V}{R}$, $k_1 = -\dfrac{V}{R}$. Thus, Eq. (5.6) can be written as

$$i(t) = \frac{V}{R}\left(1 - e^{-\frac{R}{L}t}\right) A \qquad (5.9)$$

The inverse of coefficient of $i(t)$ in Eq. (5.5) is known as *time constant* (τ) of the element. The unit of time constant is seconds. Thus

$$\tau = \frac{L}{R}$$

Equation (5.9) can be written in terms of time constant as

$$i(t) = \frac{V}{R}(1 - e^{-t/\tau}) A \qquad (5.10)$$

Equation (5.10) is known *transient response* of the circuit.

Alternate method:

Equation (5.5) can also be solved by arranging properly and integrating it. Equation (5.5) can be expressed as

$$\frac{di(t)}{dt} = \frac{V}{L} - \frac{R}{L} i(t)$$

Integrating both side as

$$\int di(t) = \int \left(\frac{V}{L} - \frac{R}{L} i(t)\right) dt$$

or

$$\int_0^i \frac{di(t)}{\left(\dfrac{V}{L} - \dfrac{R}{L}i(t)\right)} = \int_0^t dt \qquad (5.11)$$

Equation (5.11) can be expressed as

$$\int_0^i \frac{di(t)}{\left(\dfrac{V}{R} - i(t)\right)} = \frac{R}{L}\int_0^t dt$$

or

$$\ln\left(\frac{\dfrac{V}{R}}{\dfrac{V}{R} - i(t)}\right) = \frac{R}{L}t$$

Thus

$$i(t) = \frac{V}{R}(1 - e^{-t/\tau}) \qquad (5.12)$$

The voltage across resistance and inductance will be written as

$$v_R(t) = V(1 - e^{-t/\tau}) \qquad (5.13)$$

and

$$v_L(t) = V - V(1 - e^{-t/\tau}) = Ve^{-t/\tau} \qquad (5.14)$$

The plot of current and voltages are shown in Figure 5.3.

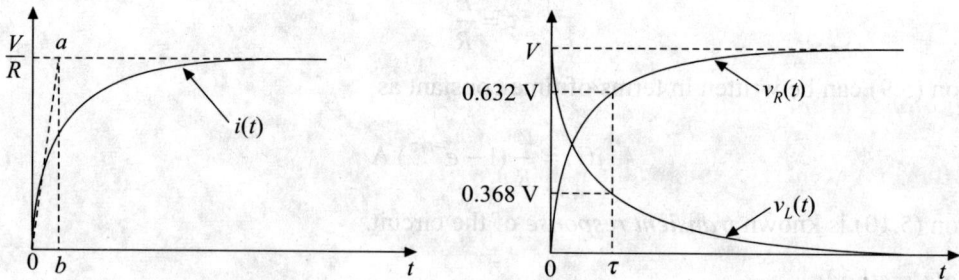

FIGURE 5.3   Transient current and voltages in series R-L circuit.

At $t = \tau$, the voltage across the inductor will be

$$v_L(\tau) = Ve^{-\tau/\tau} = \frac{V}{e} = 0.36788 \text{ V}$$

and

$$v_R(\tau) = V(1 - e^{-\tau/\tau}) = 0.63212 \text{ V}$$

Equation (5.14) also confirms that $v_L(t) = V$ at $t = 0$.

The curves showing the variations of current and voltage across resistor with time, are called an *exponential growth curves* and graphs are called *resistor voltage/time* and *current/time characteristics*, respectively. The curve showing the variation of $v_L(t)$ with time is called an *exponential decay curve* and the graph is called the *inductor voltage/time characteristic*.

The time constant τ of a first-order differential equation may be found graphically from the response curve (Figure 5.3). A tangent drawn to the exponential curve at time $t = 0$ intersects the final steady state value at point *a*. The perpendicular line drawn from the point *a* on time axis intersects at point *b*. The time *ob* is the time constant of the circuit. Mathematically, it can also be proved. The equation of a straight line tangent to the current curve at $t = 0$ is given by $y = mt$, where *m* is the slope of straight line and expressed as

$$m = \frac{di(t)}{dt}\bigg|_{t=0} = \frac{V}{R}\,\tau e^{-t/\tau}\bigg|_{t=0} = \frac{V}{L}$$

The value of vertical axis is steady state current value (V/R). Thus, we have

$$\frac{V}{R} = m \times t = \frac{V}{L} \times t \quad \Rightarrow \quad t = \frac{L}{R} = \tau$$

Inductor having current before switch S closed

Let current in the inductor be i_0 before switch was closed. Equation (5.7) will become

$$k_1 + k_2 = i_0 \tag{5.15}$$

From Eqs. (5.15) and (5.8), we get $k_2 = \dfrac{V}{R}$, $k_1 = \left(i_0 - \dfrac{V}{R}\right)$. Thus Eq. (5.6) can be written as

$$i(t) = \frac{V}{R}\left(1 - e^{-\frac{R}{L}t}\right) + \left(i_0 e^{-\frac{R}{L}t}\right)\,\text{A} \tag{5.16}$$

Energy stored in the inductor

At any time *t*, the energy stored in the inductor $w_L(t)$ is given by $\dfrac{1}{2}Li^2(t)$.

$$w_L(t) = \frac{1}{2}Li^2(t)$$

The resistor stored no energy. The power dissipated by the resistor is

$$p_R(t) = i^2(t)R$$

5.2.3 Transient in Series *RC* Circuit

Practically, an uncharged capacitor cannot be suddenly connected to a dc source as an uncharged ideal capacitor is equivalent to a short circuit. Therefore, an external resistance is connected in series with capacitor to limit the charging current. In the absence of external resistor, the surge of current is only limited by the internal resistance of voltage source and wiring resistance, and may be enough to damage the capacitor. Figure 5.4 shows an *RC* circuit connected with a dc voltage source through a switch. The charging current is known as *transient current*.

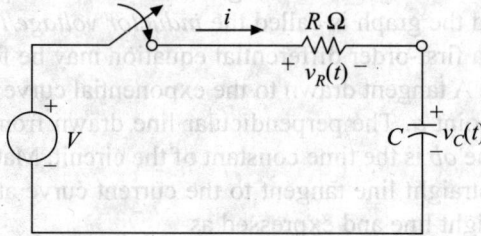

FIGURE 5.4   *RC* circuit.

Using KVL in the above circuit, we can write,

$$V = v_R(t) + v_C(t) \tag{5.17}$$

Using the relationships (voltage and current) for an ideal capacitor, we get

$$v_C(t) = \frac{1}{C} \int di(t) \quad \text{or} \quad i(t) = C \frac{dv_C(t)}{dt} \tag{5.18}$$

The voltage $v_R(t)$ can be written as

$$v_R(t) = Ri(t) = CR \frac{dv_C(t)}{dt} \tag{5.19}$$

Using Eqs. (5.18) and (5.19), Eq. (5.17) can be rewritten as

$$V = CR \frac{dv_C(t)}{dt} + v_C(t) \tag{5.20}$$

or

$$\frac{dv_C(t)}{dt} + \frac{1}{CR} v_C(t) = \frac{V}{CR} \tag{5.21}$$

The inverse of coefficient of $v_C(t)$ in Eq. (5.21) is known as *time constant*, of the *RC* circuit and represented as the Greek lower case letter τ (tau). The unit of time constant is seconds. Thus,

$$\tau = RC \tag{5.22}$$

Equation (5.21) is a first order, ordinary differential equation and can be solved by several methods. Using the calculus, the instantaneous voltage will be

$$v_C(t) = k_1 e^{-\frac{1}{RC}t} + k_2 \tag{5.23}$$

The value of constant k_1 and k_2 can be obtained using the initial ($t = 0$) and final ($t = \infty$) conditions. At $t = 0$, there is no voltage across the capacitor, thus

$$v_C(0) = 0 = k_1 e^{-\frac{1}{RC} \times 0} + k_2 = k_1 + k_2 \tag{5.24}$$

At final condition (also known as *steady-state condition*), the voltage across capacitor will be V. Thus,

$$v_C(\infty) = V = k_1 e^{-\frac{1}{RC} \times \infty} + k_2 = k_2 \tag{5.25}$$

From Eqs. (5.24) and (5.25), $k_1 = -V$
The Eq. (5.23) can be written as

$$v_C(t) = V(1 - e^{-t/\tau}) \tag{5.26}$$

Thus, voltage across the resistor (using Eq. (5.17)) will be

$$V = v_R(t) + V(1 - e^{-t/\tau})$$

or

$$v_R(t) = V e^{-t/\tau} \tag{5.27}$$

and current [using Eq. (5.18)] will be

$$i(t) = C \frac{dv_C(t)}{dt} = \frac{CV}{CR} e^{-t/\tau}$$

or

$$i(t) = \frac{V}{R} e^{-t/\tau} \tag{5.28}$$

The plot of current is shown in Figure 5.5 and voltages are shown in Figure 5.6.

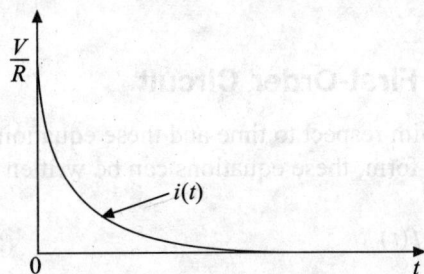

FIGURE 5.5   Transient current in *RC* circuit.   **FIGURE 5.6**   Transient voltage in *RC* circuit.

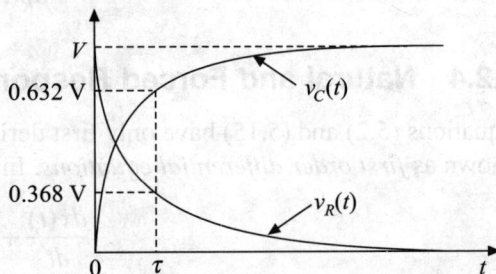

At $t = \tau$, the voltage across the capacitor will be

$$v_C(\tau) = V(1 - e^{-\tau/\tau}) = V - \frac{V}{e} = 0.63212 \text{ V}$$

and

$$i(\tau) = \frac{V}{R} e^{-\tau/\tau} = 0.36788 \frac{V}{R}$$

Thus, *after one time constant, the charging current has decayed to approximately 36.8% of its value at t = 0*. This statement is true regardless of the particular value of time constant. At $t = 5\tau$, charging current will be

$$i(5\tau) = \frac{V}{R}e^{-5\tau/\tau} \approx 0.0067\frac{V}{R}$$

This value is very small compared to the maximum value (V/R) at $t = 0$. Thus, it is assumed that capacitor is fully charged after five time constant.

There are some similarities in the equations for transients in RL circuit and those for transients in RC circuit. The transient voltage across inductor in an RL circuit has the same form as the transient current in an RC circuit $(ke^{-t/\tau})$. Furthermore, the transient current in an RL circuit is like the transient voltage across capacitor in an RC circuit $(k(1 - e^{-t/\tau}))$. The main difference between RL and RC circuits is the effect of resistance on the duration of transients. In an RL circuit, a large resistance shortens the transient (because time constant becomes small, $\tau = L/R$) whereas in an RC circuit, a large resistance prolongs the transient (because it makes the time constant large, $\tau = RC$).

Energy stored in the capacitor

At any time t, the energy stored in the capacitor $w_C(t)$ is given by

$$w_C(t) = \frac{1}{2}Cv^2(t)$$

The resistor stored no energy. The power dissipated by the resistor is

$$p_R(t) = \frac{v^2(t)}{R}$$

5.2.4   Natural and Forced Response of First-Order Circuit

Equations (5.2) and (5.15) have only first derivative with respect to time and these equations are known as *first order differential equations*. In general form, these equations can be written as

$$\frac{dx(t)}{dt} + ax(t) = f(t) \tag{5.29}$$

where $x(t)$ is a variable (voltage or current) and $f(t)$ is a *forcing function*. The constant a is the inverse of time constant (tau). When $f(t)$ is zero, the solution of Eq. (5.29) is known as *homogeneous solution* or *natural solution or complementary solution* or *natural response*. The solution of Eq. (5.29) in the absence of the source can be obtained as given below.

$$\frac{dx_N(t)}{dt} + ax_N(t) = 0$$

or

$$\frac{dx_N(t)}{dt} = -ax_N(t) \tag{5.30}$$

where subscript N denotes the natural solution. The solution of Eq. (5.30) can be obtained as

$$\int \frac{dx_N(t)}{x_N(t)} = -a\int dt$$

or

$$\ln(x_N(t)) = -at + k' = -at - \ln k \qquad (5.31)$$

Therefore,

$$x_N(t) = ke^{-at} \qquad (5.32)$$

The constant k can be found by using the initial condition.

The force response is obtained by using the forcing function which depends on the form of forcing function. The distinction between the natural and forced response is important because it clarifies the nature of transient response. The sum of these two responses forms the complete response of the circuit as

$$x(t) = x_N(t) + x_F(t) \qquad (5.33)$$

Let us consider a constant forcing function F. Equation (5.29) can be written as

$$\frac{dx_F(t)}{dt} + ax_F(t) = F \qquad (5.34)$$

During steady state condition with a constant forcing function (dc input), $\dfrac{dx_F(t)}{dt} = 0$, thus From Eq. (5.34), we get

$$x_F(t) = \frac{F}{a} \qquad (5.35)$$

The complete solution will be

$$x(t) = x_N(t) + x_F(t) = ke^{-at} + \frac{F}{a} \qquad (5.36)$$

The value of constant k can be obtained from the initial condition $x(t = 0) = x_0$

$$x_0 = k + F/a \quad \text{or} \quad k = x_0 - F/a \qquad (5.37)$$

Thus,

$$x(t) = \left(x_0 - \frac{F}{a}\right)e^{-at} + \frac{F}{a} = x_0 e^{-at} + \frac{F}{a}(1 - e^{-at}) \qquad (5.38)$$

The expression (5.36) is also expressed as *transient response* and *steady-state response* (at $t = \infty$).

$$x(t) = x_t(t) + x_{ss} \qquad (5.39)$$

From Eq. (5.38), we get

$$x_{ss} = x(t = \infty) = 0 + \frac{F}{a}(1 - 0) = \frac{F}{a} \qquad (5.40)$$

and

$$x_t(t) = \left(x_0 - \frac{F}{a}\right)e^{-at} \qquad (5.41)$$

It can be seen that the transient behaviour of a circuit depends on three factors.

1. Initial condition, x_0
2. Forcing function, F
3. Circuit parameters, a

Main steps to find the solution

1. *Write the differential solution for the circuit for $t \geq 0$*
2. *Find out the initial condition*
3. *Determine the solution of (natural and forced responses)*
4. *Apply the initial condition to get the complete solution by knowing the value of constant k.*

Table 5.1 shows the steady-state solution $x_F(t)$ of any order differential equation excited by some common forcing function.

TABLE 5.1   Forcing function and steady-state response

Type of Forcing Function $f(t)$ (Input)	Steady-State Solution $x_F(t)$ (Output)
$f(t) = A$ (DC source)	$x_F(t)$
$f(t) = At$ (Ramp function)	$x_F(t) = k_1 t + k_2$
$f(t) = At^2$ (Square function)	$x_F(t) = k_1 t^2 + k_2 t + k_3$
$f(t) = Ae^{at}$ (Exponential function)	$x_F(t) = ke^{at}$
$f(t) = A \sin \omega t$ (or $A \cos \omega t$) (Sinusoidal function)	$x_F(t) = k_1 \sin \omega t + k_2 \cos \omega t$
$f(t) = Ae^{at} \sin \omega t$ (Exponential Sine function)	$x_F(t) = e^{at}(k_1 \sin \omega t + k_2 \cos \omega t)$

5.2.5   Discharge Transients

Consider a circuit shown in Figure 5.7(a) where switch allows charging and discharging the capacitor at position 1 and position 2, respectively. When switch is at position 2, the voltage source is no longer in the circuit and current flows from the positive terminal of capacitor to negative side as shown on Figure 5.7(b).

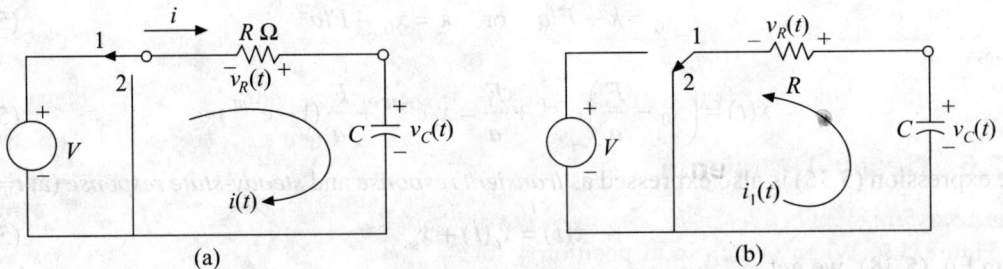

(a)   (b)

FIGURE 5.7   Capacitor charging and discharging.

The current, $i_1(t)$, which is in opposite direction from the original charging current, $i(t)$, is produced to transfer the charge. This process is called *the discharging of the capacitor*. The decaying voltage and current are known as *discharge transients*. The resistor during discharging will oppose the flow of current as shown polarity. Since, there is no voltage source, the sum of voltages across resistor and across capacitor will be zero (using KVL). The equation during the discharging can be written as

$$0 = v_R(t) + v_C(t) \tag{5.42}$$

or

$$v_R(t) = -v_C(t)$$

As current is opposite to the charging current, replacing the value of $v_R(t)$ as

$$v_R(t) = Ri_1(t) = RC \frac{v_C(t)}{dt}$$

Equation (5.42) becomes

$$RC \frac{v_C(t)}{dt} + v_C(t) = 0 \qquad (5.43)$$

Solution of Eq. (5.43) will be

$$v_C(t) = ke^{-t/\tau}$$

At $t = 0$, $v_C(t = 0) = k = V$
Thus

$$v_C(t) = Ve^{-t/t} \qquad (5.44)$$

and

$$v_R(t) = -Ve^{-t/\tau} \qquad (5.45)$$

$$i_1(t) = \frac{v_R(t)}{R} = -\frac{V}{R} e^{-t/\tau} \qquad (5.46)$$

The plots of voltages across resistor and capacitor during the discharge are shown in Figure 5.8.

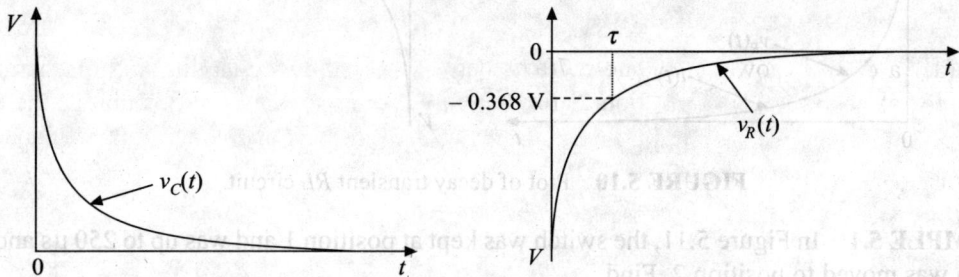

FIGURE 5.8 Discharge transient RC circuit.

5.2.6 Decay Transients

The decay transient in a series RL circuit is shown in Figure 5.9. When switch is in position 1 for long period, the current in the circuit is V/R. Due to change in switch position at 2, the voltage source is disconnected and current will try to decay. The resulting first-order differential equation will be

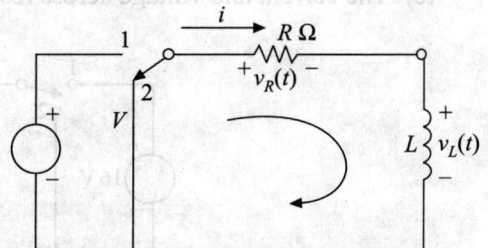

FIGURE 5.9 Decay transient in RL circuit.

$$0 = v_R(t) + v_L(t) = Ri(t) + L \frac{di(t)}{dt} \qquad (5.47)$$

The solution of Eq. (5.47) is,

$$i(t) = ke^{-t/\tau} \tag{5.48}$$

The value of constant k can be obtained from the initial condition. At $t = 0$, current was V/R, thus $k = V/R$.

$$i(t) = \frac{V}{R} e^{-t/\tau} \tag{5.49}$$

The voltage will be

$$v_R(t) = Ve^{-t/\tau} \tag{5.50}$$

and

$$v_L(t) = -Ve^{-t/\tau} \tag{5.51}$$

It can be seen that the polarity of voltage across inductor will be such that it will try to maintain the current flow in the direction of the original steady-state current. The polarity of voltage will be opposite of that which was induced when switch was in position 1. Please note that voltage across the resistor will be equal to the inductor at every instant of time. The plots of voltage and current are shown in Figure 5.10.

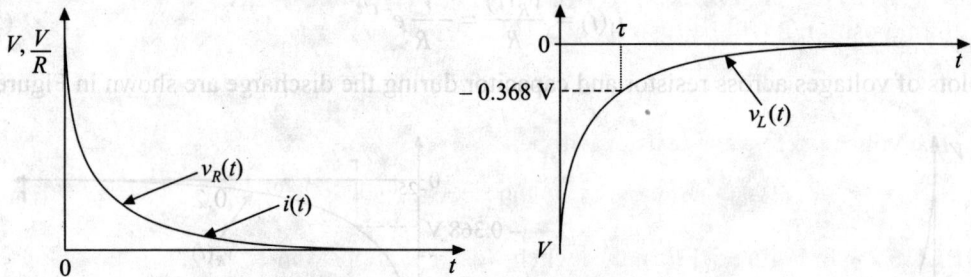

FIGURE 5.10   Plot of decay transient RL circuit.

EXAMPLE 5.1   In Figure 5.11, the switch was kept at position 1 and was up to 250 μs and then switch was moved to position 2. Find

(b) The current and voltage across resistor at $t = 100$ μs
(c) The current and voltage across resistor at $t = 300$ μs

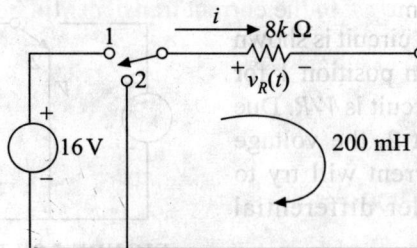

FIGURE 5.11   Example 5.1.

Solution: The time constant of circuit is

$$\tau = \frac{L}{R} = \frac{200 \text{ mH}}{8 \text{ k}\Omega} = 25 \text{ μs}$$

Switch is placed in position 1 at time $t = 0$, the current transient will be (using Eq. (5.10))

$$i(t) = \frac{V}{R}(1 - e^{-t/\tau}) = \frac{16}{8\,k\Omega}(1 - e^{-t/25\times10^{-6}}) = 2(1 - e^{-t/25\times10^{-6}})\,mA$$

The voltage across the resistor will be

$$v_R(t) = V(1 - e^{-t/\tau}) = 16(1 - e^{-t/25\times10^{-6}}) = 16(1 - e^{-t/25\times10^{-6}})\,V$$

(a) The current and voltage at $t = 100$ μs will be

$$i(t = 100\ \mu s) = 2(1 - e^{-100\times10^{-6}/25\times10^{-6}}) = 2(1 - e^{-4}) = 1.963369\,mA$$

and

$$v_R(t = 100\ \mu s) = 16(1 - e^{-100\times10^{-6}/25\times10^{-6}}) = 16(1 - e^{-4}) = 15.70695\,V$$

(b) The current at $t = 250$ μs when switch is moved to position 2,

$$i(t = 250\ \mu s) = 2(1 - e^{-250\times10^{-6}/25\times10^{-6}}) = 2(1 - e^{-10}) = 1.99909\,mA$$

The decay current at $t = 350$ μs from $t = 0$ (from position 1), i.e. $t = 100\ (=350 - 250)$ μs from position 2 will be (using Eq. 5.44)

$$i(t = 350\ \mu s) = 1.99909\ e^{-(350-250)\times10^{-6}/25\times10^{-6}} = 1.99909\ e^{-4} = 0.03663\,mA$$

The voltage across the resistor will be

$$v_R(t = 350\ \mu s) = 8 \times 1.99909\ e^{-(350-250)\times10^{-6}/25\times10^{-6}} = 0.293\,V$$

EXAMPLE 5.2   In Figure 5.12(a), the switch was kept at position 1 and was up to 100 μs and then switch was moved to position 2. Supply voltage is 5 volt dc. Find

(a) The current and voltage across capacitor at $t = 40$ μs
(b) The current and voltage across resistor at $t = 150$ μs

Solution: The time constant of circuit is

$$\tau = RC = 40 \times 10^3 \times 200 \times 10^{-12} = 8\ \mu s$$

Switch is placed in position 1 at time $t = 0$, the current transient will be [using Eq. (5.10)]

$$i(t) = \frac{V}{R}e^{-t/\tau} = \frac{5}{40\,k\Omega}e^{-t/8\times10^{-6}} = 125\ e^{-t/8\times10^{-6}}\,mA$$

The voltage across the capacitor will be

$$v_C(t) = V(1 - e^{-t/\tau}) = 5(1 - e^{-t/8\times10^{-6}})\,V$$

(a) The current and voltage at $t = 40$ μs will be

$$i(t = 40\ \mu s) = 125\ e^{-40\times10^{-6}/8\times10^{-6}} = 0.84224\,mA$$

and

$$v_C(t = 40\ \mu s) = 5(1 - e^{-40\times10^{-6}/8\times10^{-6}}) = 5(1 - e^{-5}) = 4.9663\,V$$

(b) The capacitor voltage at $t = 100$ μs when switch is moved to position 2, [Figure 5.12(b)]

$$v_C(t = 100 \text{ μs}) = 5(1 - e^{-100 \times 10^{-6}/8 \times 10^{-6}}) = 5(1 - e^{-12.5}) = 4.99998 \text{ V}$$

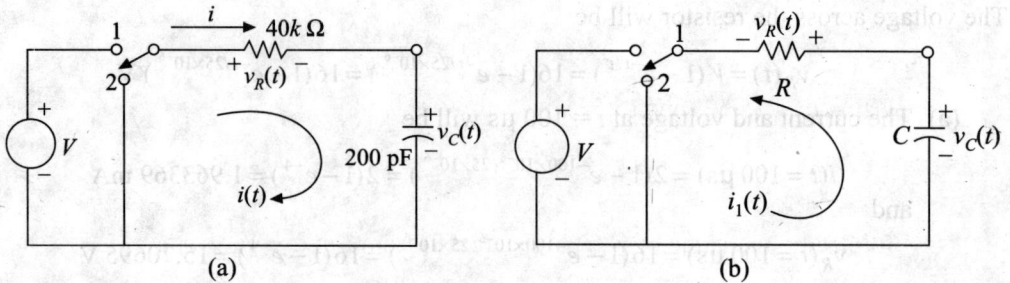

FIGURE 5.12   Example 5.2.

The discharge current at $t = 150$ μs from $t = 0$ (from position 1), i.e. $t = 50 (= 150 - 100)$ μs from position 2 will be

$$i(t = 150 \text{ μs}) = -\frac{4.99998}{40 \text{ } k\Omega} e^{-(150-100) \times 10^{-6}/8 \times 10^{-6}} = -0.125 \text{ } e^{-6.25} = -0.241 \text{ μA}$$

Please note that discharge current $i(t)$ will flow in opposite direction of charging current when switch position is moved to position 2 from position 1 as shown in Figure 5.12(b).
    The voltage across the resistor will be

$$v_R(t = 150 \text{ μs}) = 40 \times 0.241 = -0.9652 \text{ V}$$

This voltage is opposite to the shown polarity.

5.3 TRANSIENT RESPONSE OF SECOND-ORDER CIRCUITS

Second-order circuits should have two different energy storage elements (one inductor and one capacitor). The simple second-order circuits are shown in Figure 5.13.

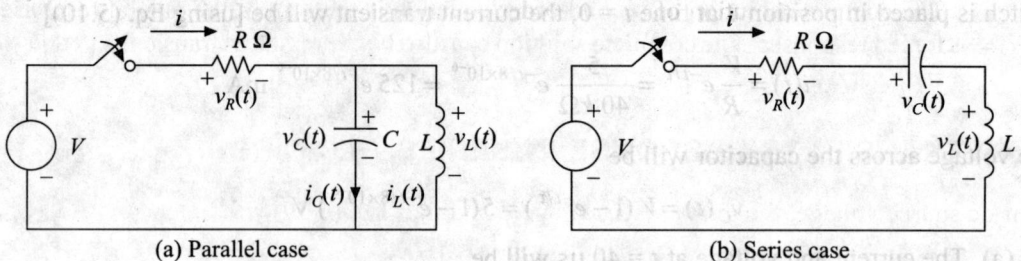

(a) Parallel case

(b) Series case

FIGURE 5.13   Second-order circuits.

(a) Parallel case: Using KVL and KCL in Figure 5.13(a), we get the following equations

$$i(t) = i_L(t) + i_C(t), \tag{5.52}$$

$$V = Ri(t) + v_C(t) \text{ and} \tag{5.53}$$

$$v_C(t) = v_L(t) = L\frac{di_L(t)}{dt} \tag{5.54}$$

Using Eqs. (5.52) and (5.53), we can write

$$i(t) = \frac{V - v_C(t)}{R} = i_L(t) + \frac{dv_C(t)}{dt} \tag{5.55}$$

Replacing the value of $v_C(t)$ in Eq. (5.55) from Eq. (5.54), we have

$$\frac{V}{R} - \frac{L}{R}\frac{di_L(t)}{dt} = i_L(t) + LC\frac{d^2i_L(t)}{dt^2}$$

or

$$\frac{d^2i_L(t)}{dt^2} + \frac{1}{RC}\frac{di_L(t)}{dt} + \frac{1}{LC}i_L(t) = \frac{V}{RLC} \tag{5.56}$$

(b) Series case: A series RLC circuit is shown in Figure 5.13(b). The KVL equation can be written as

$$V = Ri(t) + v_C(t) + v_L(t) \tag{5.57}$$

Representing the inductor voltage and capacitor voltage in terms of current, we can write Eq. (5.57) as

$$V = Ri(t) + \frac{1}{C}\int i(t)\,dt + L\frac{di(t)}{dt} \tag{5.58}$$

Differentiating Eq. (5.58) with respect to time, we get

$$L\frac{d^2i(t)}{dt^2} + R\frac{di(t)}{dt} + \frac{1}{C}i(t) = \frac{dV}{dt} \tag{5.59}$$

The complete solution of second-order equation is the sum of the natural response and forced responses. The complete solution can also be expressed as transient part and steady state part.

5.4 TRANSIENTS RESPONSE IN SERIES *RLC* CIRCUIT

For dc source voltage, the terms dV/dt will zero and thus Eq. (5.59) will become

$$L\frac{d^2i(t)}{dt^2} + R\frac{di(t)}{dt} + \frac{1}{C}i(t) = 0$$

or

$$\frac{d^2i(t)}{dt^2} + \frac{R}{L}\frac{di(t)}{dt} + \frac{1}{LC}i(t) = 0 \tag{5.60}$$

Let the solution of above equation is

$$i(t) = ke^{st} \tag{5.61}$$

Thus putting this equation in Eq. (5.60), we get

$$s^2 ke^{st} + bske^{st} + cke^{st} = 0$$

or

$$s^2 + bs + c = 0 \tag{5.62}$$

where $b = R/L$ and $c = 1/LC$. This polynomial of s is known as characteristic equation. Since it is a quadratic equation, s will have two values and the exponent s_1 and s_2 are found by solving the quadratic equation (5.62) as

$$s_1, s_2 = \frac{-b \pm \sqrt{b^2 - 4c}}{2} = -\frac{R}{2L} \pm \frac{1}{2}\sqrt{\left(\frac{R}{L}\right)^2 - \frac{4}{LC}} \tag{5.63}$$

where

$$s_1 = -\frac{R}{2L} + \frac{1}{2}\sqrt{\left(\frac{R}{L}\right)^2 - \frac{4}{LC}} \tag{5.64}$$

and

$$s_2 = -\frac{R}{2L} - \frac{1}{2}\sqrt{\left(\frac{R}{L}\right)^2 - \frac{4}{LC}} \tag{5.65}$$

The value of s_1 and s_2 depends on the $\left[\left(\frac{R}{L}\right)^2 - \frac{4}{LC}\right]$.

Case I: $\left(\frac{R}{L}\right)^2 > \frac{4}{LC}$

s_1 and s_2 are real and distinct roots: $s_1 = \alpha_1$ and $s_2 = \alpha_2$. This case is also known as *over-damped case* and the solution will be

$$i(t) = k_1 e^{\alpha_1 t} + k_2 e^{\alpha_2 t} \tag{5.66}$$

Case II: $\left(\frac{R}{L}\right)^2 = \frac{4}{LC}$

s_1 and s_2 are real and the same roots: $s_1 = s_2 = \alpha$. This case is known as *critically damped case* and the solution will be

$$i(t) = k_1 e^{\alpha t} + k_2 t e^{\alpha t} \tag{5.67}$$

Case III: $\left(\frac{R}{L}\right)^2 < \frac{4}{LC}$

s_1 and s_2 are the complex conjugate roots: $s_1 = s_2^* = \alpha \pm j\beta$. This case is known as *under-damped solution* and the solution will be

$$i(t) = k_1 e^{(\alpha + j\beta)t} + k_2 e^{(\alpha - j\beta)t} \tag{5.68}$$

The solution of Eqs. (5.66) to (5.68) requires two initial conditions as there are two unknown constants. These initial conditions are the values of $i(t)$ at $t = 0$ and of the derivative of $i(t)$, $di(t)/dt$, at $t = 0$. The value of $di(t = 0)/dt$ is obtained by the basic equation by knowing the value of $i(t = 0)$.

EXAMPLE 5.3 A series RLC circuit as shown in Figure 5.13(b) has $R = 5$ ohm, $L = 2$ H and $C = 0.5$ F. The supply voltage is 10 V dc. Find

  (a) The current when there is no charge on capacitor.
  (b) The current when the capacitor's initial voltage is 5 V.
  (c) Repeat the case (a) if resistance is changed to 4 ohm.
  (d) Repeat the case (a) if resistance is changed to 1 ohm

Solution: Using Eqs. (5.63) and (5.64), we get

$$s_{1,2} = -\frac{R}{2L} \pm \frac{1}{2}\sqrt{\left(\frac{R}{L}\right)^2 - \frac{4}{LC}} = -\frac{5}{2 \times 2} \pm \frac{1}{2}\sqrt{\left(\frac{5}{2}\right)^2 - \frac{4}{2 \times 0.5}} = -1.25 \pm 0.75$$

Thus, $s_1 = -0.5$ and $s_2 = -2.0$
The current in the circuit will be

$$i(t) = k_1 e^{-0.5t} + k_2 e^{-2t} \tag{i}$$

At time $t = 0$, there is no current in the circuit and thus,

$$i(t = 0) = k_1 e^{0.5 \times 0} + k_2 e^{-2.\times 0} = k_1 + k_2 = 0 \tag{ii}$$

The basic equation of voltage can be written as

$$V = Ri(t) + v_C(t) + L\frac{di(t)}{dt}$$

At time $t = 0$, we get

$$\frac{di(t = 0)}{dt} = \frac{1}{L}\left(V - Ri(t = 0) - v_C(t = 0)\right) \tag{iii}$$

  (a) At time $t = 0$, there is no voltage across the capacitor and thus

$$V = Ri(t = 0) + v_C(t = 0) + L\frac{di(t = 0)}{dt} = 0 + 0 + 2\frac{di(t = 0)}{dt} = 10$$

  or

$$\frac{di(t = 0)}{dt} = 5 \tag{iv}$$

Differentiating Eq. (i) and putting the value of Eq. (iv), we get

$$\frac{di(t = 0)}{dt} = -0.5\,k_1 e^{-0.5 \times 0} - 2k_2 e^{-2 \times 0} = -0.5k_1 - 2k_2 = 5 \tag{v}$$

Solving the Eqs. (iv) and (v), we have

$$k_1 = \frac{10}{3} \quad \text{and} \quad k_2 = -\frac{10}{3}$$

Thus, the current will be

$$i(t) = \frac{10}{3}(e^{-0.5t} - e^{-2t}) \text{ A}$$

(b) At time $t = 0$, there is 5 V voltage across the capacitor and thus from Eq. (iii)

$$\frac{di(t=0)}{dt} = \frac{1}{L}(V - Ri(t=0) - v_C(t=0)) = \frac{(10 - 0 - 5)}{2} = 2.5$$

Differentiating Eq. (i) and putting the value of $di(0)/dt$, we get

$$\frac{di(t=0)}{dt} = -0.5\,k_1 e^{-0.5\times0} - 2k_2 e^{-2\times0} = -0.5k_1 - 2k_2 = 2.5 \qquad \text{(vi)}$$

Solving Eqs. (ii) and (v), we have

$$k_1 = 5/3 \quad \text{and} \quad k_2 = -5/3$$

Thus the current will be

$$i(t) = \frac{5}{3}(e^{-0.5t} - e^{-2t}) \text{ A}$$

(c) The roots of characteristics equation will be

$$s_{1,2} = -\frac{R}{2L} \pm \frac{1}{2}\sqrt{\left(\frac{R}{L}\right)^2 - \frac{4}{LC}} = -\frac{4}{2\times2} \pm \frac{1}{2}\sqrt{\left(\frac{4}{2}\right)^2 - \frac{4}{2\times0.5}} = -1.0$$

Since s_1 and s_2 are the same, thus the solution will be

$$i(t) = k_1 e^{-1.0t} + k_2 t e^{-1.0t}$$

Using the initial conditions, at $t = 0$, $i(0) = 0$, therefore, $k_1 = 0$.
Using Eq. (iv), at $t = 0$,

$$\frac{di(t)}{dt} = -k_1 e^{-1.0t} + k_2(e^{-1.0t} - t e^{-1.0t})$$

or

$$\frac{di(t=0)}{dt} = -k_1 e^{-1.0\times0} + k_2(e^{-1.0\times0} - 0 \times e^{-1.0\times0}) = -k_1 + k_2 = 5$$

Thus, $k_2 = 5$ and the current will be

$$i(t) = 5t e^{-1.0t}$$

(d) The roots of characteristics equation will be

$$s_{1,2} = -\frac{R}{2L} \pm \frac{1}{2}\sqrt{\left(\frac{R}{L}\right)^2 - \frac{4}{LC}} = -\frac{1}{2\times2} \pm \frac{1}{2}\sqrt{\left(\frac{1}{2}\right)^2 - \frac{4}{2\times0.5}} = -0.25 \pm j1.9465$$

Since, s_1 and s_2 are the same, thus the solution will be

$$i(t) = k_1 e^{(-0.25+j1.9465)t} + k_2 e^{(-0.25-j1.9465)t} = e^{-0.25t}(k_1 e^{j1.9465t} + k_2 e^{-j1.9465t})$$

Using the initial conditions, at $t = 0$, $i(0) = 0$, therefore, $k_1 + k_2 = 0$.
Using Eq. (iv), at $t = 0$,

$$\frac{di(t=0)}{dt} = (-0.25 + j1.9465)k_1 e^{(-0.25+j1.9465)\times 0} + (-0.25 - j1.9465)k_2 e^{(-0.25-j1.9465)\times 0}$$

$$= (-0.25 + j1.9465)k_1 + (-0.25 + j1.9465)k_2 = 5$$

Thus,

$$k_2 = \frac{j5}{2 \times 1.9465} = j1.2844 \text{ and } k_1 = -j1.2844$$

Thus current,

$$i(t) = 1.2844 e^{-0.25t}(-je^{j1.9465t} + je^{-j1.9465t}) = 1.2844 e^{-0.25t}[2\sin(1.9465t)]$$

Cases (a), (c) and (d) are shown graphically in Figure 5.14.

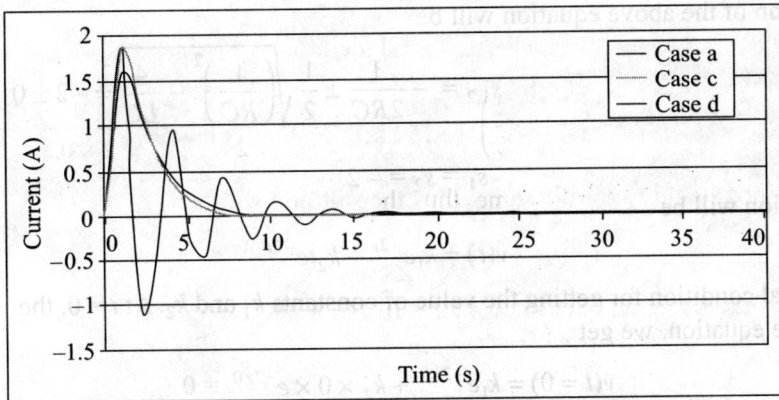

FIGURE 5.14   Plot of current response of Example 5.3.

EXAMPLE 5.4   Assuming the initial inductor current and capacitor voltage are both zero in a parallel *RLC* circuit, as shown in Figure 5.15, having $R = 1$ ohm, $L = 2$ H and $C = 0.5$ F. The supply current is 5 A. Find the voltage $v(t)$ for $t \geq 0$ in the circuit.

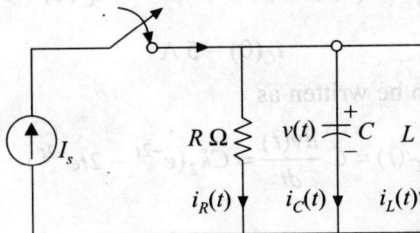

FIGURE 5.15   Example 5.4.

Solution: Using KCL, we have

$$i_R(t) + i_C(t) + i_L(t) = I_s$$

or

$$\frac{1}{R} v(t) + C \frac{dv(t)}{dt} + \frac{1}{L} \int v(t) = I_s$$

Differentiating the above equation with respect to time, we get

$$\frac{1}{R} \frac{dv(t)}{dt} + C \frac{d^2 v(t)}{dt^2} + \frac{1}{L} v(t) = \frac{dI_s}{dt} = 0$$

or

$$\frac{d^2 v(t)}{dt^2} + \frac{1}{RC} \frac{dv(t)}{dt} + \frac{1}{LC} v(t) = 0$$

This is a second-order differential equation having the characteristics polynomial,

$$s^2 + \frac{1}{RC} s + \frac{1}{LC} = 0$$

The solution of the above equation will be

$$s_{1,2} = -\frac{1}{2RC} \pm \frac{1}{2} \sqrt{\left(\frac{1}{RC}\right)^2 - \frac{4}{LC}} = -2 \pm 0$$

or

$$s_1 = s_2 = -2$$

Thus solution will be

$$v(t) = k_1 e^{-2t} + k_2 t e^{-2t}$$

Using initial condition for getting the value of constants k_1 and k_2. At $t = 0$, the $v(t = 0) = 0$. Thus, from above equation, we get

$$v(t = 0) = k_1 e^{-2 \times 0} + k_2 \times 0 \times e^{-2 \times 0} = 0$$

or

$$k_1 = 0.$$

Thus

$$v(t) = k_2 t e^{-2t}$$

The current in the capacitor at time $t = 0$ may be zero. Using the KVL at time $t = 0$, we have

$$\frac{v(0)}{R} + i_C(0) + i_L(0) = I_s = 0 + i_C(0) + 0$$

or

$$i_C(0) = 5 \text{ A}$$

The current in the capacitor can be written as

$$i_C(t) = C \frac{dv(t)}{dt} = C k_2 (e^{-2t} - 2t e^{-2t})$$

At $t = 0$, we have

$$i_C(0) = 5 = 0.5 \, k_2 (e^{-2 \times 0} - 2 \times 0 \times e^{-2 \times 0})$$

or

$$k_2 = 10$$

Thus,

$$v(t) = 10te^{-2t} \text{ volt}$$

5.5 LAPLACE TRANSFORM

For complex circuits, solving the differential equations is very cumbersome. The Laplace transform, named after the French mathematician and astronomer Pierre Simon de Laplace is very useful and defined as

$$F(s) = \int_0^\infty f(t)e^{-st}dt \tag{5.69}$$

The function $F(s)$ is the Laplace transform of function $f(t)$ and is a function of complex frequency, $s = \sigma + j\omega$. This definition of Laplace transform is known as *one-sided* or *unilateral Laplace transform*. Table 5.2 shows the Laplace transform of commonly used functions.

The computation of the inverse Laplace transform is, in general, complex, if one considers arbitrary function of s. Many applications, the transform pairs given in Table 5.2 are used for desired results.

TABLE 5.2 Laplace transform pair

$f(t)$ (Function)	$F(s)$ (Laplace Transform)
$u(t)$ (unit step)	$1/s$
$\delta(t)$ (unit impulse)	1
e^{-at}	$\dfrac{1}{(s+a)}$
$\sin \omega t$	$\dfrac{w}{(s^2+\omega^2)}$
$\cos \omega t$	$\dfrac{s}{(s^2+\omega^2)}$
$e^{-at}\sin \omega t$	$\dfrac{\omega}{(s+a)^2+\omega^2}$
$e^{-at}\cos \omega t$	$\dfrac{(s+a)}{(s+a)^2+\omega^2}$
t	$1/s^2$
$\dfrac{df(t)}{dt}$	$sF(s)$
$\int f(t)dt$	$F(s)/s$

EXAMPLE 5.5 Using the basic principle, find the Laplace transform of $\sin \omega t$.

Solution: Using Euler's formula, $\sin \omega t$ can be expanded into $(e^{j\omega t} - e^{-j\omega t})/2j$. Hence, Laplace transform can be obtained as

$$F(s) = \int_0^\infty \frac{1}{2j}(e^{j\omega t} - e^{-j\omega t})e^{-st}dt = \frac{1}{2j}\left(\int_0^\infty e^{-(s-j\omega t)}dt - \int_0^\infty e^{-(s+j\omega t)}dt\right)$$

or

$$F(s) = \frac{1}{2j}\left(\frac{1}{s-j\omega}e^{-(s-j\omega)t}\Big|_0^\infty - \frac{1}{s+j\omega}e^{-(s+j\omega)t}\Big|_0^\infty \right)$$

Thus,

$$F(s) = \frac{1}{2j}\left(\frac{1}{s-j\omega} - \frac{1}{s+j\omega} \right) = \frac{\omega}{s^2+\omega^2}$$

EXAMPLE 5.6   Find the inverse Laplace transform of $\dfrac{2}{s^2+3s+2}$.

Solution: The function can be written as

$$F(s) = \frac{2}{s^2+3s+2} = \frac{2}{(s+1)(s+2)} = \frac{A}{(s+1)} + \frac{B}{(s+2)}$$

The values of A and B are to determined* as

$$A = (s+1)F(s)\Big|_{s=-1} = \frac{2}{(s+2)}\Big|_{s=-1} = 2$$

$$B = (s+2)F(s)\Big|_{s=-2} = \frac{2}{(s+1)}\Big|_{s=-2} = -2$$

Thus, the Laplace function can be written as

$$F(s) = \frac{2}{(s+1)} - \frac{2}{(s+2)}$$

The Laplace inverse can be obtained looking at Table 5.2 as

$$f(t) = 2e^{-t} - 2e^{-2t}$$

5.6   TRANSIENT RESPONSE IN AC CIRCUITS

Figure 5.16 shows the series RL circuit excited by an ac voltage source. The corresponding voltage equation can be written as

$$L\frac{di(t)}{dt} + Ri(t) = E_m \sin(\omega t + \phi) = E_m \cos\phi \sin\omega t + E_m \sin\phi \cos\omega t \tag{5.70}$$

Taking Laplace transform of Eq. (5.70), we get

$$LsI(s) + RI(s) = \frac{E_m\omega \cos\phi}{s^2+\omega^2} + \frac{sE_m \sin\phi}{s^2+\omega^2} \tag{5.71}$$

---

* Please note that the values of A and B can also be obtained by equating the coefficients of s as

$$\frac{A}{(s+1)} + \frac{B}{(s+2)} = \frac{A(s+2)+B(s+1)}{(s+1)(s+2)} = \frac{2}{(s+1)(s+2)}$$

i.e. $A + B = 0$, $2A + B = 2$

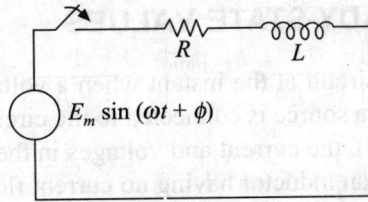

FIGURE 5.16 Series RL circuit excited by ac voltage source.

or

$$I(s) = \frac{E_m \omega \cos \phi}{(Ls + R)(s^2 + \omega^2)} + \frac{s E_m \sin \phi}{(Ls + R)(s^2 + \omega^2)} \qquad (5.72)$$

Simplifying Eq. (5.72), we have

$$I(s) = \frac{E_m \omega \cos \phi}{(Ls + R)(s^2 + \omega^2)} + \frac{s E_m \sin \phi}{(Ls + R)(s^2 + \omega^2)} = I_1(s) + I_2(s) \qquad (5.73)$$

Laplace inverse of $I_1(s)$ and $I_2(s)$ can be obtained separately,

$$I_1(s) = \frac{E_m \omega \cos \phi}{(Ls + R)(s^2 + \omega^2)} = \frac{E_m \omega \cos \phi}{L}\left(\frac{A}{s + R/L} + \frac{Bs + C}{(s^2 + \omega^2)}\right)$$

or

$$I_1(s) = \frac{E_m \omega \cos \phi}{(Ls + R)(s^2 + \omega^2)} = \frac{E_m \omega \cos \phi}{L}\left(\frac{(A + B)s^2 + (B(R/L) + C)s + (A\omega^2 + CR/L)}{(s + R/L)(s^2 + \omega^2)}\right)$$

Equating the nominator for various values of s components,
Thus, $A + B = 0$; $BR/L + C = 0$; and $A\omega^2 + CR/L = 1$

We have, $\qquad A = \dfrac{1}{\omega^2 + (R/L)^2}; \quad B = -A \quad \text{and} \quad C = AR/L$

$$I_1(s) = \frac{E_m \omega \cos \phi}{L(\omega^2 + (R/L)^2)}\left(\frac{1}{(s + R/L)} - \frac{s}{(s^2 + \omega^2)} + \frac{R/L}{(s^2 + \omega^2)}\right)$$

Thus,

$$i_1(t) = \frac{E_m \omega L \cos \phi}{(R^2 + (\omega L)^2)}\left(e^{-\frac{R}{L}t} - \cos \omega t + \frac{R}{\omega L}\sin \omega t\right) \qquad (5.74)$$

Similarly,

$$i_2(t) = \frac{E_m L \sin \phi}{(R^2 + (\omega L)^2)}\left(-\frac{R}{L}e^{-\frac{R}{L}t} + \frac{R}{L}\cos \omega t + \omega \sin \omega t\right) \qquad (5.75)$$

Thus,

$$i(t) = \frac{E_m}{\sqrt{R^2 + (\omega L)^2}}\left\{\sin(\omega t + \phi - \theta) - e^{-\frac{R}{L}t}\sin(\phi - \theta)\right\} \qquad (5.76)$$

where $\theta = \tan^{-1}\left(\dfrac{\omega L}{R}\right)$

5.7 INITIAL AND STEADY-STATE VALUES

The current, which flows in a circuit at the instant when a voltage source is switched into it, is called the *initial current*. When a source is connected to the circuit for a long period of time (i.e., after all transients have finished), the current and voltages in the circuit are said to have reached their steady-state values. When an inductor having no current flowing before it switched into the circuit, it behaves like an open circuit because at that instant the current cannot change instantaneously from its initial value of zero. After steady-state conditions have been reached, an inductor in a dc circuit behaves like a short circuit because the steady state voltage across it is zero.

An uncharged capacitor is equivalent to a short circuit at $t = 0$ because at that instant the voltage cannot change instantaneously from its initial value of zero. Note that initial current may not be zero. Under the steady-state conditions, all capacitors are fully charged when applied to the dc source and act as open circuits. We can obtain the steady-state voltages and currents in dc circuit using conventional analysis methods by replacing all capacitors with open circuits and all inductors with short circuit. Similarly, initial voltages and currents can be obtained using conventional analysis methods by replacing all capacitors with short circuits and all inductors with open circuit.

If an inductor is having current before ($t < 0$), the inductor will behave as a current source at $t = 0$ and the inductor can be replaced by the current source having the initial current value. However, at the steady-state condition, inductor in a dc circuit behaves like a short circuit whatever the initial current was flowing through the inductor. In case of capacitor having initial charge on it, capacitor is equivalent to a voltage source at $t = 0$. Whereas under steady state conditions, all capacitors in dc circuit are fully charged and equivalent to open circuits irrespective of initial charge on them.

EXAMPLE 5.7   Find the initial and steady-state currents in the inductor L and capacitor C as shown in Figure 5.17. There is no change on capacitor and current through inductor at time $t = 0$.

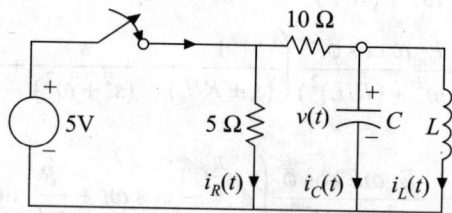

FIGURE 5.17 Example 5.7.

Solution: The circuit diagram for initial condition (capacitor short-circuited and inductor open-circuited) can be shown as

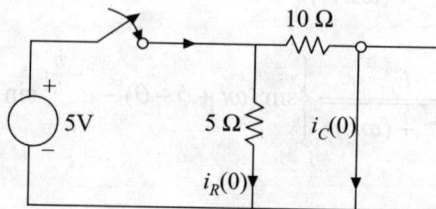

The current in 5 ohm resistor will be 5/5 = 1 A and in 10 ohm resistor will be 5/10 = 0.5 A. Thus initial current through inductor will be zero (seen from figure) and in capacitor, it will be 0.5 A. During the steady-state condition, capacitor will be open-circuited and inductor will short-circuited. Thus current in inductor will be 5/10 = 0.5 A.

EXAMPLE 5.8   Find the current $i(t)$ in the circuit of Figure 5.18.

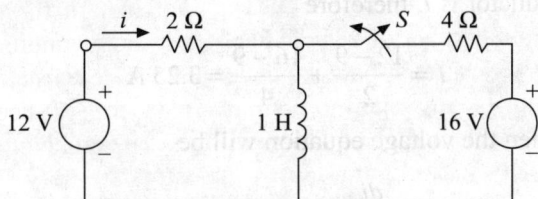

FIGURE 5.18   Example 5.8.

Solution: After steady state, the voltage across the inductor will be zero (i.e. inductor will behave like short circuit). Thus current through the inductor will be the sum of current (I_1) from 12 V source and current (I_2) from 16 V source as

$$I = I_1 + I_2 = \frac{12}{2} + \frac{16}{4} = 10 \text{ A}$$

At time $t = 0$, switch opens and 16 V source is disconnected. The voltage equation of remaining circuit (12 V source, 2 ohm resistor and 1 H inductor) will be

$$\frac{di}{dt} + 2i = 12 \tag{i}$$

The steady state current will be 12/2 = 6 A.
Thus solution of current Eq. (i) will be

$$i(t) = i_N(t) + i_{ss}(t) = ke^{-2t} + 6$$

The value of k can be obtained by using the initial condition as

$$i(t = 0) = ke^{-2 \times 0} + 6 = 10$$

Thus, $k = 4$ and current equation will be

$$i(t) = 4e^{-2t} + 6$$

EXAMPLE 5.9   Find the current $i(t)$ in the circuit of Figure 5.19 when the switch S is opened. Before opening of the switch, circuit was in the steady state condition.

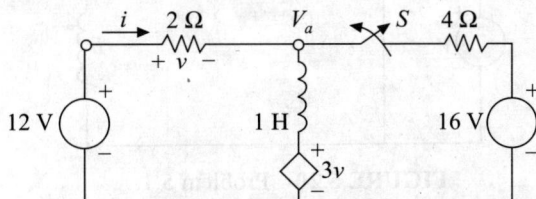

FIGURE 5.19   Example 5.9.

Solution: There are several ways to obtain the current through the inductor. It is true that during the steady state condition, the voltage across the inductor will be zero (short-circuited). The voltage at node-a will be

$$V_a = 3v = 6i = 6(12 - V_a)/2$$

Thus, $$V_a = 9 \text{ Volt.}$$

Let current through the inductor is I, therefore

$$I = \frac{12-9}{2} + \frac{16-9}{4} = 3.25 \text{ A}$$

When switch is opened, then the voltage equation will be

$$\frac{di}{dt} + 4v = 12$$

or,

$$\frac{di}{dt} + 8i = 12 \tag{i}$$

The final current will be 12/8 (= 1.5 A).

The solution of Eq. (i) will be

$$i(t) = i_N(t) + i_{ss}(t) = ke^{-8t} + 1.5$$

The value of k can be obtained by using the initial condition as

$$i(t = 0) = ke^{-2\times 0} + 1.5 = 3.25$$

Thus, $k = 1.75$ and current equation will be

$$i(t) = 1.75 \, e^{-8t} + 1.5$$

PROBLEMS

5.1 A coil of 1 H inductance and 5 ohm resistance was connected to a dc supply voltage as shown in Figure 5.20. When the current, i, was 5 A in the inductance, the switch was changed to position 2 from the position 1. Find the current magnitude and direction in the coil after one second when switch was moved to position 2.

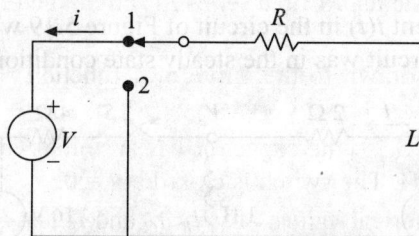

FIGURE 5.20   Problem 5.1.

[**Ans.** 0.0335 A in the same direction as shown in problem]

5.2 Determine the current expression in a series *RLC* circuit having inductance of 1 H, resistance of 4 ohm and capacitance of 1/3 F. The initial charge on capacitor is V_0 as shown in Figure 5.21.

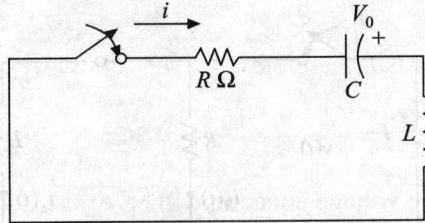

FIGURE 5.21 Problem 5.2.

$$\textbf{[Ans.} = 0.5 \ V_0 \ (e^{-t} - e^{-3t}) \, \text{A}\,]$$

5.3 Derive the expression for the natural response of the circuit shown in Figure 5.21 having 2 ohm resistance, 1 H inductance and 1/17 F capacitance.

$$\textbf{[Ans.} \ i = 0.25 V_0 e^{-t} \sin 4t \, \text{A}]$$

5.4 Switch *S* in the circuit shown in Figure 5.22 has been in position 1 for long time. At $t = 0$, the switch is changed to position 2. Assuming that capacitor C_2 is uncharged, find,

(a) the expression of current $i(t)$ when $C_1 = C_2$

(b) the current in resistance at $t = 0$ and $t = \infty$.

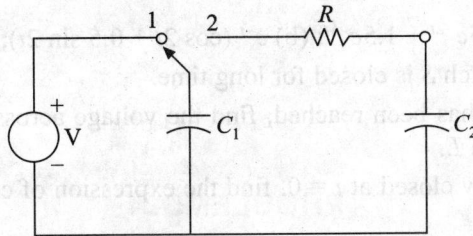

FIGURE 5.22 Problem 5.4.

$$\left[\textbf{Ans.} \ (a) \ i(t) = \frac{V}{R} e^{-t/R(C_1+C_2)} \text{A}, (b) \ V/R \text{ A, 0A}\right]$$

5.5 Show that the time constants, *RC* in a series *RC* circuit and *L/R* in series *RL* circuit have units of seconds.

5.6 Show that the tangent drawn to the curve of a function $f(t) = Ae^{-t/\tau}$ at $t = t_1$ always intersects the *t*-axis at $t = t + \tau$.

5.7 In the circuit of Figure 5.23, the capacitor has an initial voltage $V_0 = 12$ V, $C = 1 \, \mu\text{F}$, $R = 1 \, \text{k}\Omega$ and $L = 6.25$ H. The switch is closed at $t = 0$.

(a) Determine the numerical values of v, i_R, i_C and i_L just after switch is closed, i.e. at $t = 0^+$.

(b) Calculate $\dfrac{dv}{dt}$ and $\dfrac{d^2v}{dt^2}$ at $t = 0^+$.

(c) Find the expression of $v(t)$

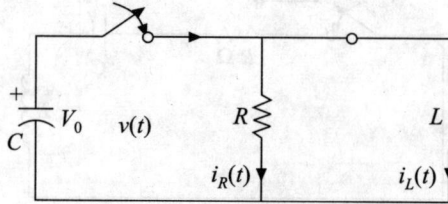

FIGURE 5.23   Problem 5.7.

[**Ans.** (a) V_0, V_0/R, V_0/R, 0 ; (b) -9000, 1080000; (c) $v(t) = (16e^{-800t} - 4e^{-200t})$]

5.8 Find the Laplace inverse of the following

(a) $\dfrac{(s+1)}{(s+2)(s+4)}$

(b) $\dfrac{(s+2)}{[(s+1)^2 + 4]}$

(c) $\dfrac{2s}{(s^2 + 2s + 5)}$

[**Ans.** (a) $-0.5e^{-2t} + 1.5e^{-4t}$; (b) $e^{-t}(\cos 2t + 0.5 \sin 2t)$; (c) $2e^{-2t}\cos t - 4e^{-2t}\sin t$]

5.9 In Figure 5.24, switch S is closed for long time.

(a) If steady state has been reached, find the voltage across L, current through R and current through L.

(b) If switch is now closed at $t = 0$, find the expression of current through and voltage across L.

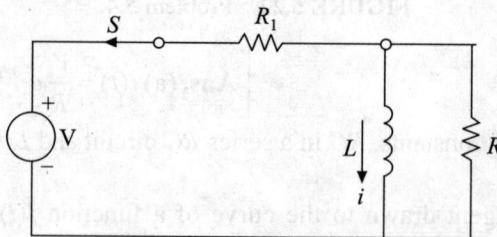

FIGURE 5.24   Problem 5.9.

$$\left[\textbf{Ans.}\ (a)\ 0, 0, V/R_1;\ (b)\ i(t) = \frac{V}{R_1} e^{-(R/L)t};\ (c)\ v(t) = -\frac{R}{R_1} V e^{-(R/L)t} \right]$$

5.10 Find the voltage expression v when switch in Figure 5.25 is opened after a long time.

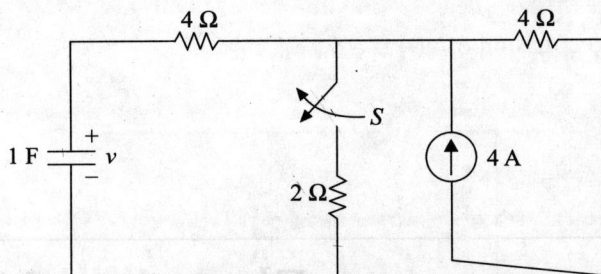

FIGURE 5.25   Problem 5.10.

[**Ans.** $12 - 8e^{-t/8}$ V]

6

Electrical Measuring Instruments and Measurements

6.1 INTRODUCTION

Instruments and the instrumentation systems are used for measuring, detecting, observing, controlling, computing, communicating and/or displaying physical qualities. It is very useful in system operation, control, technology development and future research. In any instrumentation system, there is a flow of information in the form of signals. Because of ease and precision of electrical signals, it is customary to convert physical variables into electrical form. In order to detect the electrical quantities such as current, voltage, power, resistance, frequency, energy, etc., it is necessary to transform an electrical quantity or condition into a visible indication. This is achieved with the help of instruments (or meters) that indicate the magnitude of quantities either by the position of pointer moving over a scale (called an *analog instrument*) or in the form of a decimal number (called a *digital instrument*). An *electric meter* is a device built to accurately detect and display an electrical quantity in a readable form. In the analysis and testing of circuits, there are meters designed to accurately measure the basic quantities of voltage, current, frequency, impedance, etc.

Most modern meters are "digital" in design, meaning that their readable display is in the form of numerical digits. Older designs of meters are mechanical in nature, using some kind of pointer device to show quantity of measurement. In either case, the principles applied in adapting a display unit to the measurement of (relatively) large quantities of voltage, current, or resistance are the same. The display mechanism of an analog meter is often referred to as a *movement*, borrowing from its mechanical nature to *move* a pointer along a scale so that a measured value may be read. The first electrical meter movements built were known as *galvanometers*, and were usually designed with maximum sensitivity. A very simple galvanometer may be made from a magnetized needle (such as the needle from a magnetic compass) suspended from a string, and positioned within a coil of wire. Current through the wire coil will produce a magnetic field which will deflect the needle from pointing towards the direction of earth's magnetic field.

6.2 TYPES OF ELECTRICAL MEASURING INSTRUMENTS

There are a large number of measuring instruments which can be classified into several ways as described below.

6.2.1 AC and DC Instruments

There are some instruments which can accurately measure only dc quantities or only ac quantities or both dc and ac quantities. When the electrical instrument is used for measuring only the dc quantities, these instruments are known as *dc instruments*. And when the instruments are used for only ac quantities, these instruments are known as *ac instruments*. There are several instruments, which are used for measuring both ac and dc quantities accurately, known as ac/dc instruments.

6.2.2 Absolute and Secondary Instruments

In non-electrical field of study, an absolute instrument measures a quantity (such as pressure and temperature) in absolute units by means of physical measurements. In electrical field, these instruments are not used in laboratories and calibrations of these instruments are not required. The quantities are measured in terms of the movements and instruments constants. Tangent galvanometer and Rayleigh current balance are the examples of absolute instruments.

On the other hand, the secondary instruments are very widely used in the laboratories. These instruments are equipped with scale on which deflection pointer moves. These instruments are calibrated with the standard instruments for its accuracy. Ammeter, voltmeter, wattmeter, etc. are the examples of secondary instruments.

6.2.3 Deflection and Null Type Instruments

Most of the secondary instruments are deflection types as they read the measurement values on the scale. Whenever these instruments are connected to measure the quantities, there will be a deflection on the calibrated scale to read the direct values. There are some instruments which are indicating the null (zero) for reading the values and therefore, some calculation is required to get the measured value.

6.2.4 Analog and Digital Instruments

When the instruments provide the readings due to movements of some part of the instruments, it is due the torque production by current and/or voltage signals. There are some analog instruments which provide the reading in digits such as energy meters. In digital instruments, the current and voltage signals are sampled and then these sampled signals are used to provide the digital values for the measured quantities. There is no movement part in the digital instruments.

6.2.5 Indicating, Recording and Integrating Instruments

Secondary instruments can be indicating, recording or integrating types. In indicating instrument, the readings are directly obtained by looking at the scale on which the pointer moves. The ammeters, voltmeters and wattmeters are the examples of indicating instruments. Indicating instruments give the instantaneous value of the quantities, i.e. due to change in the quantities, the indication changes instantaneously.

Recording instruments keep on recording the instantaneous values of quantities on the graph or on indicators. Demand meters are the recording type of instruments. Maximum demand meter records the maximum demand which occurs during a particular period. When the demand exceeds the previous value, the indicator moves to the new value till there is another value which exceeds this value.

Integrating instruments totalize the quantities over the time. It gives the product of time and electrical quantity. Ampere hour and watt-hour meter (energy meter) are the integrating instruments.

6.2.6   Direct Measuring and Comparison Instruments

As its name suggests, the direct measuring instruments give the measure quantity directly by displaying or recording the quantity. Direct measuring instruments convert the energy of the measurands for actuating the instruments. The actuation of the instruments measures the unknown quantity. Ammeter, voltmeter, wattmeters and energy meters are the examples of direct measuring instruments. These instruments are very commonly used as these are less expensive and very simple.

In comparison instruments, the unknown quantity is compared with the existing standards. Examples of comparison type instruments are the ac and dc bridges.

6.3   MAIN COMPONENTS OF INDICATING INSTRUMENTS

All analog indicating instruments possess three essential mechanisms with a few exceptions:

(a) *Deflecting or operating mechanism:*   It produces a mechanical force (also called *deflection torque*, T_d), which causes the pointer to deflect from its initial position, with help of current or voltage. The deflection torque is a function of current and/or voltage. A force is produced by the electromagnetic action between currents or between a current and the field of a permanent magnet. This force is used to deflect a pointer, on a calibrated scale, against a controlling torque. Normally electrical indicating instruments are classified, based on deflecting mechanism or measuring quantity.

(b) *Controlling mechanism:*   The controlling force or torque (T_c) acts in opposition to the deflecting force and ensures the deflection shown on the meter, is always the same for a given measured quantity. The controlling system produces a force equal and opposite to the deflection force at the final steady state position of the pointer. It is also useful to bring the pointer to zero when the measuring quantity is removed. In absence of the controlling force, the pointer will not come back to the initial position when the measuring quantity is zero. There are two types of controlling devices, namely spring control and gravity control*. In spring control, the controlling torque is directly proportional to the angular deflection of the pointer. Controlling torque in gravity control is proportional to the sine of the angular deflection.

---

* Gravity control is not used in the modern instruments.

$$T_c \propto \theta \text{ (with spring control)} \qquad (6.1)$$

$$T_c \propto \sin \theta \text{ (with gravity control)} \qquad (6.2)$$

At the final deflection,

$$T_d = T_c \qquad (6.3)$$

(c) *Damping mechanism:* The damping force produced by the device ensures that the pointer comes to rest in its final position quickly without undue oscillations when the deflecting force is applied. The combination of inertia of moving system and controlling torque gives moving system a natural frequency of oscillation. The pointer will oscillate about its mean position. Thus, it is desirable to provide a sufficient damping so that pointer should reach the steady-state position without oscillations. Eddy current damping and air damping are the two commonly used damping methods.

6.4 PRINCIPLE OF OPERATION OF INDICATING INSTRUMENTS

Secondary instruments work on different effect such as magnetic effect, thermal effect, electrostatic effect, electromagnetic effect, Hall effect, etc.

6.4.1 Magnetic Effect

When a current-carrying conductor is placed in a magnetic field either produced by permanent magnet, electromagnet or due to flow of current in another coil, it experiences a force which is used for the deflection of pointer in an instrument. Most of the common electrical instruments work on this principle. When a magnetic field is produced by a coil and a piece of soft iron (unmagnetized) is brought near the end of coil, it will be attraeted by the coil. This mechanism is used in attraction type of moving iron instrument. If two unmagantized pieces of soft iron placed near the coil, both the pieces of iron will be magnetized and will experience a force of repulsion between them. This effect is utilized in repulsion type moving iron instruments.

When a coil carrying a current is placed in a magnetic field produced by the permanent magnet, the coil will experience a force of attraction or repulsion depending on the current direction in the coil. This effect is utilized in permanent magnet moving coil instruments. When two coils carrying the currents are brought to each other, they will experience a force of repulsion or attraction. This effect is used in dynamometer type of instruments.

6.4.2 Thermal Effect

When a current, which is to be measured, is passed through a heating element, the change in temperature produces an *emf* by a thermocouple attached to this element. A thermocouple is made of two dissimilar conductors joined at the ends to form a closed loop. If the temperatures of the two ends are different, a current will flow in the closed path. The current is evaluated as *rms* value.

6.4.3 Electrostatic Effect

Two charged plates experience a force between them. If one plate is fixed, another one will experience a movement and this effect is known as *electrostatic effect*. High rating voltmeters normally use this effect.

6.4.4 Electromagnetic Effect

In a magnetic field, produced by an electromagnet having ac current, a non-magnetic disc or drum paced in it will have an *emf* induced in it. If a closed path is provided, a current will flow through the disc or drum which will produce another magnetic field. There will be a torque produced in the disc or drum and it will move or rotate. This is known as *induction effect* and ac energy meters utilize this principle.

6.4.5 Hall Effect

When a current flows in a strip of magnetic material in presence of a transverse magnetic field, an *emf* will be produced between two edges of strip conductor. The magnitude of *emf* depends on the current, flux density and property of conductor material. Normally the value of *emf* is very small and thus amplification is required. Poynting vector wattmeter is an instrument which uses Hall effect.

6.5 CONTROLLING MECHANISM

Frictional torque is one of the controlling torques but this torque is not sufficient to balance the deflecting torque at final value (steady state). Frictional torque, which is one of the factors deciding the performance of the indicating instruments, is dependent on the weight of the moving part. Hence, torque (deflection)/weight ratio of an instrument is a performance indicator. The higher value of ratio will give better performance.

To have better performance, there are two controlling mechanism used for indicating instruments mounted on pivoted spindle: gravity control and spring control.

6.5.1 Gravity Control

In gravity control mechanism, one small adjustable weight is attached to the moving system as shown in Figure 6.1. There is one balance weight attached. These weights produce the controlling torque which is proportional to the sine of deflection angle for a given control weight position. However, the controlling torque can also be varied by changing the control weight position. If the deflecting torque is proportional to the current flow, the scale of gravity control instruments will be non-uniform (compressed or crowded scale at lower end). The main advantages of gravity control are that it is cheap, independent of temperature variation and does not change with time. To operate successfully without error, instrument

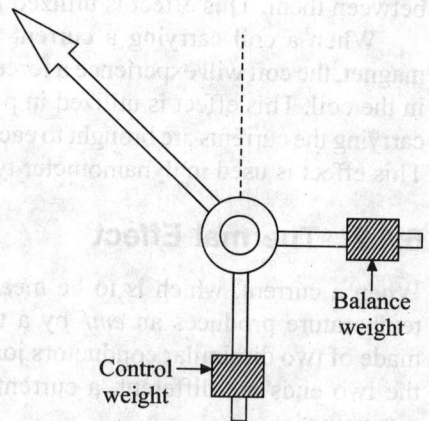

FIGURE 6.1 Gravity control mechanism.

having gravity control should be vertically level- positioned. The instruments having gravity control are obsolete now.

6.5.2   Spring Control

In this mechanism, a hair spring is attached to the moving system. This spring should have the following special electrical and mechanical properties as it carries current and stays in the magnetic field.

(a) It should be free from mechanical fatigue which can be avoided by proper annealing and ageing during manufacture.

(b) It should be non-magnetic such as silicon bronze, hard rolled copper, platinum silver, German silver, etc.

(c) When spring carried the current, the spring should have small resistance so that minimum loss (thus less heat) will be there. It should also have low resistance temperature coefficient. To avoid the effect of temperature variations upon the length, two springs coiled in opposite directions are used.

Advantages of spring control mechanism are the following:

(a) It requires less space,

(b) Scale is uniform, if the deflecting torque is proportional to the current flowing in the instruments.

6.6   DAMPING MECHANISM

The main requirement of damping mechanism is that the pointer to come to final value quickly without overshoot and oscillations. It is also necessary for the instrument that it should not take much time to come to the final value reading. Depending on the damping, the instruments are classified as underdamped, critically damped (or dead beat) and over-damped. Figure 6.2 shows the deflection of instrument in different types of damping. Critically damped instruments are preferred.

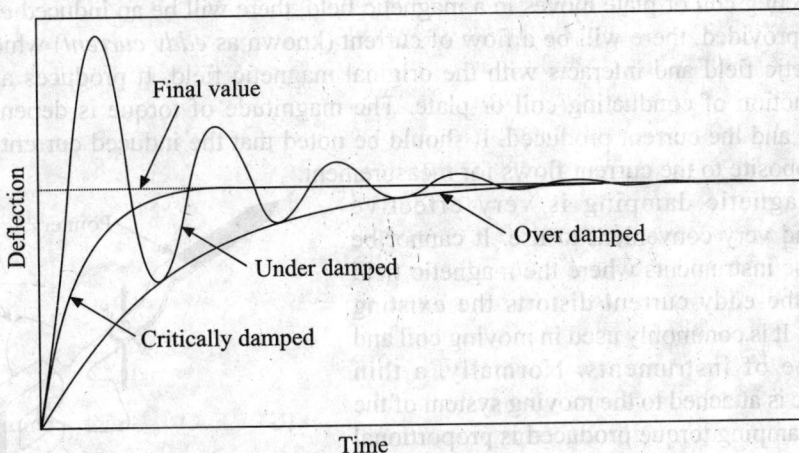

FIGURE 6.2   Damping.

The damping torque can be provided by the following methods.

6.6.1 Friction Damping

A light aluminium piston attached to the moving system swings in a closed fitting stationary chamber. The clearance between piston and chamber walls is uniform and is very small. The amount of damping action greatly depends on the clearance. When air is filled in the chamber, it is called *air friction damping*. During the oscillation of pointer, the air exerts damping force in the opposite direction of torque. Figure 6.3 shows an air friction damping mechanism. If any liquid (normally oil) having high viscosity* is filled, it is known as *liquid friction damping*. In liquid friction, there are various arrangements possible to provide damping.

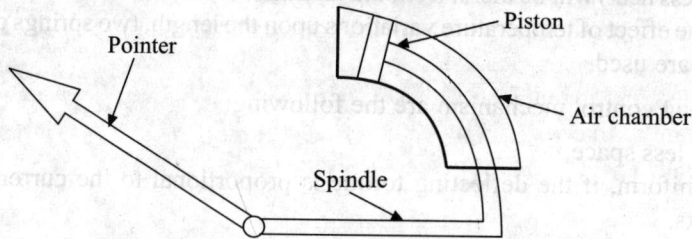

FIGURE 6.3   Air friction damping.

It should be noted that air friction damping is cheap and simple. Care should be taken about the piston and chamber. It is commonly used in moving iron instruments. Liquid friction damping provides good damping force and with proper arrangements, the friction force and load on bearing can be minimized. However, the major problem with liquid friction damping is the leakage of liquid and thus it is should be properly positioned.

6.6.2 Electromagnetic Damping

When a conducting coil or plate moves in a magnetic field, there will be an induced *emf* in it. If a closed path is provided, there will be a flow of current (known as *eddy current*) which produces another magnetic field and interacts with the original magnetic field. It produces a torque that opposes the motion of conducting coil or plate. The magnitude of torque is dependent on the magnetic field and the current produced. It should be noted that the induced current flow in the coil/plate is opposite to the current flows for measurement.

Electromagnetic damping is very effective instruments and very convenient to use. It cannot be applied in those instruments where the magnetic field produced by the eddy current distorts the existing magnetic field. It is commonly used in moving coil and induction type of instruments. Normally, a thin aluminium disc is attached to the moving system of the instruments. Damping torque produced is proportional to the velocity of pointer. Figure 6.4 shows the magnetic damping.

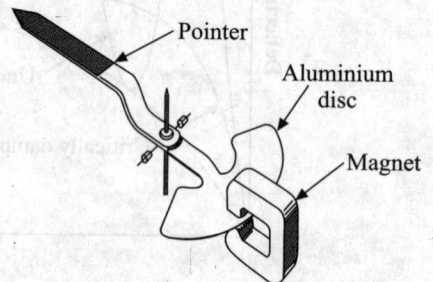

FIGURE 6.4   Electromagnetic damping.

---

* High viscosity liquid provides high damping force.

6.7 PERMANENT MAGNET MOVING COIL INSTRUMENTS

The permanent magnet moving coil (PMMC) instruments works on the d'Arsonval galvanometer principle. A small coil of fine wire is supported but can move in the magnetic field of a permanent magnet as shown in Figure 6.5. The coil is supported by either pivots in jewel bearings or by a thin ribbon held taut, under substantial tension, by springs. Current flows through the moving coil, produces a magnetic field which interacts with the magnetic field produced by the permanent magnet. As a result, a deflecting torque is exerted on the moving coil which turns until the deflecting torque is balanced by the restoring torque exerted by springs attached to the stationary part of the meter. The deflecting torque is proportional to the magnetic field of the coil, and this, in turn, is proportional to the current flowing through the coil. Consequently, the position at which equilibrium between the deflecting torque and the restoring torque is reached, is a measure of the current flowing through the coil. The current is indicated on a calibrated scale by a pointer fastened to the coil. In jewel-bearing instruments, helical springs provide both the restoring torque and a means of conducting current to the coil. In taut ribbon (called *taut band*) suspension instruments, there are no bearings as such. The band suspends the coil as well as provides restoring torque.

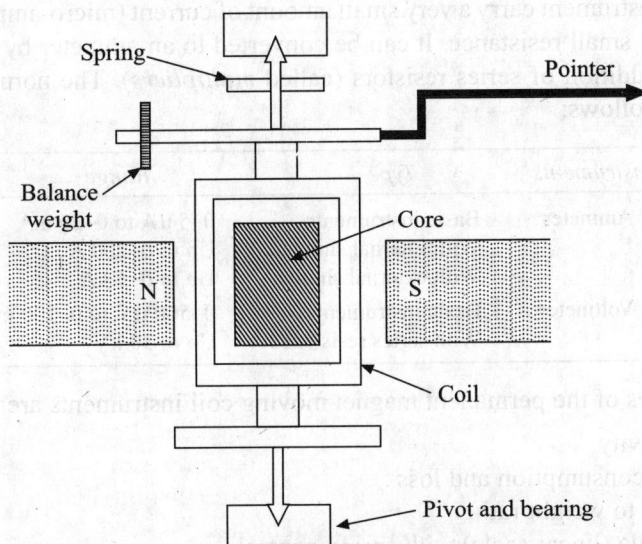

FIGURE 6.5   Permanent magnet moving coil arrangements.

Figure 6.6 shows a PMMC instrument. The deflection torque for instrument can be expressed as

$$T = NBldI \qquad (6.4)$$

where, l is the axial length and d the width of the coil, N the number of turns, I the current in the coil, and B the flux density in the air gap in which coil is positioned. If controlling force* ($T_c = K\theta$) is provided by the spring, for the final deflection, it holds the following relation

$$T (= NBldI) = K\theta \qquad (6.5)$$

---

* where θ is in radian and K is spring constant in Nm/radian.

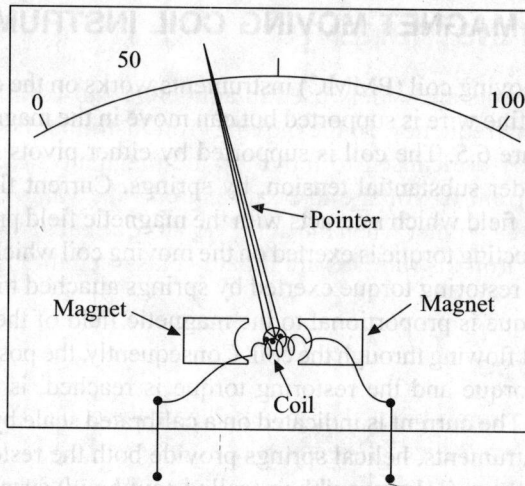

FIGURE 6.6   Permanent magnet moving coil instrument.

The basic PMMC instrument carry a very small amount of current (micro-ampere to milli-ampere) and these have very small resistance. It can be converted to an ammeter by the use of shunts or into voltmeter by addition of series resistors (called *multipliers*). The normal range of PMMC instruments are as follows:

Instruments	Type	Range
Ammeter	Basic instruments	0–5 μA to 0–20 mA
	With internal shunts	Up to 0–200 A
	With external shunts	Up to 0–5 kA
Voltmeter	Basic instruments	0–500 μV to 0–100 mV
	With series resistance	Up to 30 kV

The main advantages of the permanent magnet moving coil instruments are:

(a) high sensitivity
(b) low power consumption and loss
(c) high torque to weight ratio
(d) uniform scale (linear scale) with spring control
(e) well-shielded from any stray magnetic field.

The main disadvantages are:

(a) only suitable for dc voltages and currents
(b) weakening of magnet may induce errors
(c) more expensive than the moving-iron instruments.

The permanent-magnet moving-coil type of instrument is useless for alternating-current measurements because the deflecting torque with ac, reverses the direction too rapidly to permit movement of the needle. Alternating current should never be connected to such an instrument. It can cause no deflection, but can burn out the coil.

6.8   MOVING-IRON INSTRUMENTS

Moving iron (MI) instruments are widely used in ammeters and voltmeters for laboratory and switch-board use at power frequency. In the moving-iron instruments, one or several pieces of soft iron or magnetic alloy are caused to move by the magnetic field of a fixed coil or coil system excited by the current or voltage to be measured. The value of the current/voltage is indicated by the position of a pointer when the deflection torque caused by the current is balanced by spiral springs. Moving iron instruments are broadly classified into two types: attraction type and the repulsion type.

6.8.1   Attraction Type

When a small piece of soft iron is pivoted or mounted near the coil and current is passed through the coil, the iron vane is magnetized (become electro-magnet) and it is attracted and tends to move into the stronger magnetic field inside the coil because vane tries to occupy a position of minimum reluctance. Thus, the produced force or torque is always in such a way to increase the inductance of the coil*. Figire 6.7 shows an attraction type moving iron instrument.

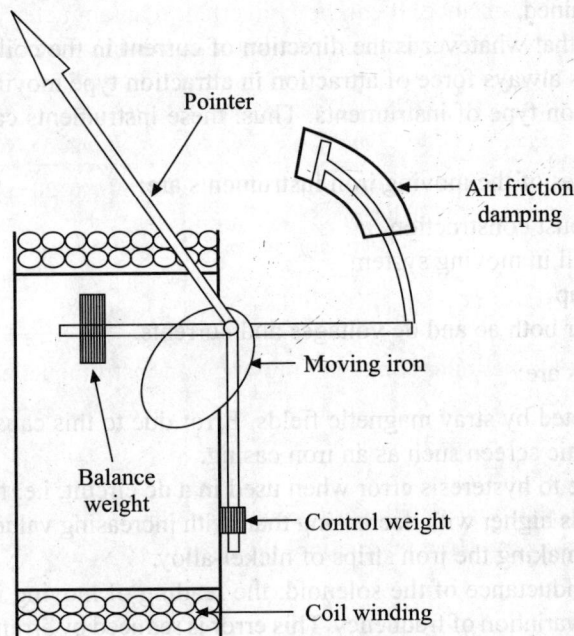

FIGURE 6.7   Attraction type moving iron instrument.

6.8.2   Repulsion Type

If two vanes or pieces are placed near the coil, and current passed through the coil, both iron vane are magnetized. Repulsion will take place between them because like poles are adjacent to one

---

* Inductance increases as reactance decreases.

another. The force or torque of repulsion can be used for measurement if one vane is fixed and other is allowed to move. Figure 6.8 shows the repulsion type moving iron instrument.

The deflection torque produced in the moving iron instruments can be given as

$$T = \frac{1}{2} I^2 \frac{dL}{d\theta} \qquad (6.6)$$

where, I is the current in the coil (A), L is the inductance (in Henry) of instrument and θ is the deflection angle in radian.

Normally attraction type instruments have lower inductance than the corresponding repulsion type instruments. Therefore, attraction

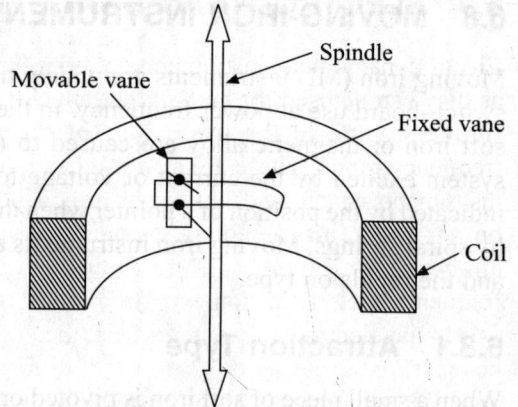

FIGURE 6.8 Repulsion type moving iron instrument.

type moving iron voltmeters are more accurate and there is greater flexibility using ammeter shunts. However, repulsion type of instrument is cheap in manufacturing and uniform (nearly) scale can be easily obtained.

It should be noted that whatever is the direction of current in the coil, the iron vanes are so magnetized that there is always force of attraction in attraction type moving iron instrument and repulsion in the repulsion type of instruments. Thus, these instruments can be used for both ac and dc.

The main advantages of the moving iron instruments are:

(a) Simple and robust construction
(b) No carrying coil in moving system
(c) Relatively cheap
(d) Can be used for both ac and dc voltages and currents.

The main disadvantages are:

(a) These are affected by stray magnetic fields. Error due to this cause is minimized by the use of a magnetic screen such as an iron casing.
(b) These are liable to hysteresis error when used in a dc circuit, i.e. for a given current, the instrument reads higher with decreasing than with increasing value of current. This error is reduced my making the iron strips of nickel-alloy.
(c) Owing to the inductance of the solenoid, the reading of moving iron voltmeter may be affected by the variation of frequency. This error is reduced by arranging for the resistance of the voltmeter to be large compared to the reactance of the solenoid.
(d) Moving iron voltmeters are liable to a temperature error owing to the solenoid being wound with copper wire. This error is minimized by connecting a resistor (in series with the solenoid) of a material having negligible temperature coefficient of resistance such as manganin.
(e) For current and voltage measurements, the scale is non-uniform as deflection torque is proportional to the square of current passing into the meter.

6.9 ELECTRODYNAMIC OR DYNAMOMETER INSTRUMENTS

Electrodynamometer instruments are very important as they can be used for both in the range of power frequency and in the lower part of the audio-frequency range. These are also used as a transfer instruments which are calibrated on dc and then used on ac. These instruments are used as voltmeter, ammeter, wattmeter, power factor meter and frequency meter with some modification.

In the electrodynamometer type, the current to be measured is passed through two coils, one fixed and one movable as shown in Figure 6.9. The movable coil is mounted on pivots and jewel bearings in the field of the fixed coil. The force exerted by the magnetic field produced by one coil on the current flowing in the other turns the moving coil until the deflecting torque is balanced by the restoring torque exerted by a spring.

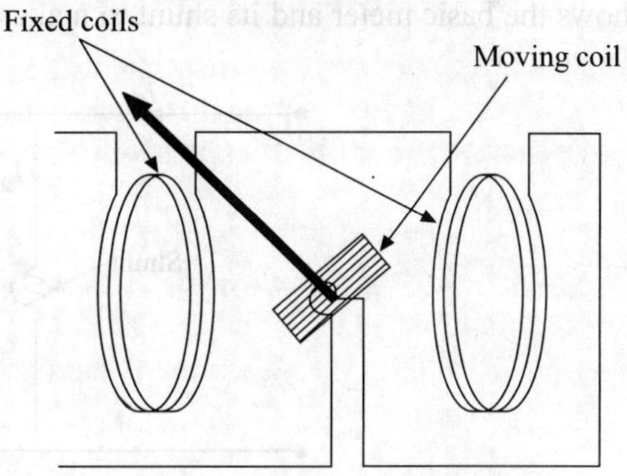

FIGURE 6.9 Electrodynamometer instrument.

The deflection torque produced is given by

$$T = i_1 i_2 \frac{dM}{d\theta} \tag{6.7}$$

where,

i_1 = Instantaneous value of current in the fixed coil (A)
i_2 = Instantaneous value of current in the moving coil (A)
M = Mutual inductance between moving and fixed coils (H)
θ = Deflection angle (radian)

The main advantages of the dynamometer type moving coil instruments are as follows:

(a) These can be used for both ac and dc voltages and currents.
(b) These are precision grade instruments.
(c) These are very useful for voltmeter for reading *rms* value irrespective of waveform.
(d) These are free from hysteresis and eddy current errors.

The main disadvantages are as follows:

(a) Scale is non-uniform for voltage and current measurements.
(b) These are more expensive than PMMC and MI instruments.
(c) There is more power loss and power consumption.
(d) Torque to weight ratio is low.
(e) Magnetic strength is small due to absence of iron and thus more number of ampere-turns is required.
(f) These instruments are sensitive to overload and mechanical impacts.

6.10 AMMETER DESIGN

A meter designed to measure electrical current is popularly known as "ammeter" because the unit of measurement is "ampere". As the moving coil in direct-current ammeters can carry only a small current because weight of moving coil, the greater part of the current to be measured is bypassed through a low resistance (called *shunt*) connected in parallel with the coil. Figure 6.10 shows the basic meter and its shunt to make a PMMC ammeter[*].

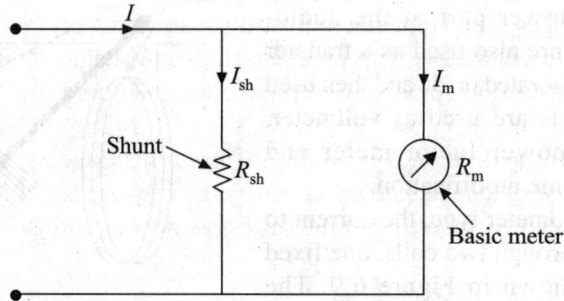

FIGURE 6.10   DC ammeter with shunt.

The value of shunt resistance (R_{sh}) can be calculated using the conventional circuit analysis. Let the meter resistance be R_m, I_m be full scale deflection current of meter and I be the current to be measured. Then,

$$R_m I_m = R_{sh} I_{sh} = R_{sh} (I - I_m) \qquad (6.8)$$

Thus,

$$R_{sh} = R_m \frac{I_m}{(I - I_m)} \qquad (6.9)$$

The ratio of current to be measured (I) to the current in meter (I_m) is known as *multiplying power* (*m*) of shunt and can be written as

$$m = \frac{I}{I_m} \qquad (6.10)$$

Using Eqs. (6.9) and (6.10), the value of shunt resistance can be written as

$$R_{sh} = \frac{R_m}{(m - 1)} \qquad (6.11)$$

The basic requirements of shunt are:

    (a) Temperature coefficient of shunt should be low.
    (b) Resistance should not vary with time.
    (c) The thermal electromotive force should be low.
    (d) The temperature rise in the shunt should be low.

Now-a-days, the ammeters are available in multi-range by using the shunt resistances in steps. The shunt resistance is varied to get the clear reading in different range. The universal shunt or Ayrton shunt is used for multi-range ammeters. It should be noted that ammeters are connected in the series with circuit element in which the current is to be measured. Due to low effective resistance (including the shunt), the voltage-drop across the ammeter is negligible. The suitable materials for shunt are manganin and constantan.

---

[*] For ac ammeters (moving iron or electrodynamometer instruments), the meter inductance is also to be considered. In Eq. (6.8), resistance is to be replaced with impedance.

6.11 VOLTMETER DESIGN

As it is stated earlier, most meter movements are sensitive devices. Some d'Arsonval movements have full-scale deflection current ratings as low as 50 µA, with an (internal) wire resistance of less than 1000 Ω. This makes for a voltmeter with a full-scale rating of only 50 millivolts (50 µA × 1000 Ω). In order to build voltmeters with practical (higher voltage) scales from such sensitive movements, it is required to find some way to reduce the measured quantity of voltage appear across the instrument.

Due to limited current and voltage capability of basic instruments, the series resistance is added in series with the instrument, known as *multipliers*, to reduce the current in meter and applied voltage across meter by diverting to the series element as shown in Figure 6.11.

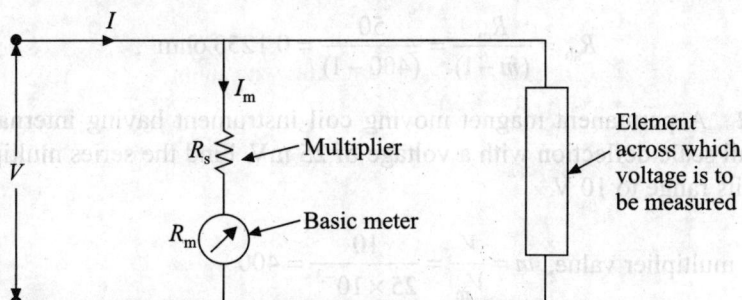

FIGURE 6.11 DC voltmeter with a multiplier.

Let

V_m = Voltage across the meter for current I_m

V = Full range voltage across the instrument

R_m = Internal resistance of the basic meter

R_s = Series resistance (multiplier resistance)

I_m = Full scale deflection current of meter

From Figure 6.11, we can write the following equations as

$$V = I_m (R_s + R_m) \qquad (6.12)$$

and

$$V_m = I_m R_m \qquad (6.13)$$

Using Eqs. (6.12) and (6.13), we can get

$$R_s = \frac{(V - I_m R_m)}{I_m} = \left(\frac{V}{I_m} - R_m \right) \qquad (6.14)$$

Multiplier factor (m) of voltmeter is defined as

$$m = \frac{V}{V_m} = \frac{I_m (R_m + R_s)}{I_m R_m} = 1 + \frac{R_s}{R_m} \qquad (6.15)$$

Thus, the value of multiplier in terms of multiply factor will be

$$R_s = (m - 1)R_m \tag{6.16}$$

The suitable materials for series resistance (multiplier) are manganin and constantan.

EXAMPLE 6.1 A permanent magnet moving coil instrument having internal resistance of 50 ohm gives full scale deflection with a current of 50 mA. Find the shunt resistance value to extend its range to 20 A.

Solution: Current multiplier value, $m = \dfrac{I}{I_m} = \dfrac{20}{50 \times 10^{-3}} = 400$

The value of shunt resistance [using Eq. (6.11)] will be

$$R_{sh} = \frac{R_m}{(m-1)} = \frac{50}{(400-1)} = 0.1253 \text{ ohm}$$

EXAMPLE 6.2 A permanent magnet moving coil instrument having internal resistance of 50 ohm gives full scale deflection with a voltage of 25 mV. Find the series multiplier resistance value to extend its range to 10 V.

Solution: The multiplier value, $m = \dfrac{V}{V_m} = \dfrac{10}{25 \times 10^{-3}} = 400$

The series resistance will be

$$R_s = (m-1)\,R_m = (400-1) \times 50 = 19950 \text{ ohm}$$

EXAMPLE 6.3 Find the value of shunt resistances to a multi-range dc ammeter having internal resistance of 25 ohm and full scale deflection current of 1 mA. The required ranges are 0–50 mA, 0–1 A and 0–5 A.

Solution: *For range of 0–50 mA:*

Multiply factor, $m = 50/1 = 50$

The value of shunt resistance, $R_{sh} = \dfrac{R_m}{(m-1)} = \dfrac{25}{(50-1)} = 0.5102$ ohm

For range of 0–1A:

Multiply factor, $m = \dfrac{1}{1 \times 10^{-3}} = 1000$

The value of shunt resistance, $R_{sh} = \dfrac{R_m}{(m-1)} = \dfrac{25}{(1000-1)} = 0.02503$ ohm

For range of 0–5 A:

Multiply factor, $m = \dfrac{5}{1 \times 10^{-3}} = 5000$

The value of shunt resistance, $R_{sh} = \dfrac{R_m}{(m-1)} = \dfrac{25}{(5000-1)} = 0.0050$ ohm

EXAMPLE 6.4   Find the value of shunt resistances to a multi-range dc voltmeter having internal resistance of 25 ohm and full scale deflection current of 1 mA. The required ranges are 0–500 mV, 0–1 V and 0–5 V.

Solution: The full scale deflection voltage = $25 \times 1 = 25$ mV

For range of 0–500 mV:

Multiply factor, $m = \dfrac{500}{25} = 20$

The value of series resistance, $R_s = (m - 1)\, R_m = (20 - 1) \times 25 = 475$ ohm

For range of 0–1 V:

Multiply factor, $m = \dfrac{1}{25 \times 10^{-3}} = 40$

The value of shunt resistance, $R_s = (m - 1)\, R_m = (40 - 1) \times 25 = 975$ ohm

For range of 0–10 V:

Multiply factor, $m = \dfrac{10}{25 \times 10^{-3}} = 400$

The value of shunt resistance, $R_s = (m - 1)\, R_m = (400 - 1) \times 25 = 9975$ ohm

6.12   WATTMETER OR POWER METER

Electrical power can be measured:

(a) In direct-current circuits, by measuring the current through and the voltage across the load with an ammeter and voltmeter. The power is the product of the two.

(b) In alternating-current circuits, by measuring the current through the load, the voltage across the load, and the power factor of the load, using an ammeter, voltmeter, and power-factor meter. The power can be computed from these quantities.

(c) In either direct or alternating-current circuits, by the use of a wattmeter.

Wattmeter, which is used to measure the power (real power in ac circuit), is normally of the electrodynamometer type instrument. In the single-phase wattmeter, which can also be used for dc power measurement, a moving coil is mounted on pivots and jewel bearings or a taut ribbon (band) so as to be free to move in the magnetic field produced by a fixed coil. The moving coil (also known as *potential* or *pressure* or *voltage coil*) is connected across the load to carry a current proportional to the line voltage, whereas fixed coil (also known as *current coil* or *field coil*) is connected in series with load to carry a current, proportional to line current. It should be noted that current coil should have minimum resistance and potential coil should have maximum resistance to reduce the loss in instrument and measurement error*. The torque exerted on the moving coil is balanced by spiral springs. A pointer fastened to the coil indicates the average real power as shown in Figure 6.12.

---

* The current coil has thick coil to reduce the resistance and pressure coil is made of thin wire to increase resistance. The current in potential coil is negligible.

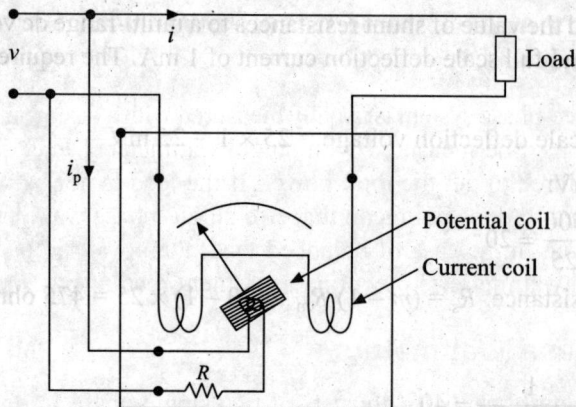

FIGURE 6.12   Single-phase wattmeter connection.

Let load current (through the current coil) be i and i_p be the current in potential circuit (v/R) where R is the total resistance in potential circuit and v is the instantaneous supply voltage. The torque developed will depend on the currents in both the current and voltage coils. Thus,

$$\text{Instantaneous torque} \propto v \times i \qquad (6.17)$$

$$\text{Average torque} \propto \frac{1}{(t_2 - t_1)} \int_{t_1}^{t_2} vi\, dt \qquad (6.18)$$

Assuming voltage and current vary sinusoidally in time with a phase angle difference of θ as

$$v = \sqrt{2}V \cos \omega t \quad \text{and} \quad i = \sqrt{2}I \cos(\omega t \pm \theta) \qquad (6.19)$$

The average power will be

$$W = VI \cos \theta \qquad (6.20)$$

In more general form, wattmeter reading can be expressed in the voltage and current phasors[*].

$$W = |\bar{V}|\,|\bar{I}| \cos \angle_{\bar{I}}^{\bar{V}} = |\bar{V}|\,|\bar{I}| \cos \left(\begin{array}{l} \text{Angle between current and voltage phasors} \\ \text{of current coil and potential coil, respectively} \end{array} \right) \qquad (6.21)$$

where V and I are the effective (or *rms*) values of voltage and current, respectively.

It should be noted that single-phase wattmeters have four terminals, whereas voltmeters and ammeters have only two terminals. A proper care should be taken while making connection of current and voltage coils. Interchange of these connections may damage the wattmeter due to excessive current in current coil connected across the load voltage. If pointer moves backward of zero point for any reason, the current coil connection should be reversed but not the potential coil connection. It is also important to see that voltage and current rating should not exceed, even if the pointer shows below the full scale deflection.

---

[*] It is always easy to find the power calculation for any connection of voltage and current coils by getting voltage phasor of potential coil connection and current phasor of current coil and then use Eq. (6.21).

6.12.1 Measurement of Three-Phase Power with Single-Phase Wattmeters

Electrical power in a three-phase system is usually measured with a three-phase wattmeter. However, three-phase power can also be measured with single wattmeter if load is balanced. Reading of meter is multiplied by three to get the total power. If the load is not balanced, two single-phase wattmeters are to be used. It does not mean that two single-phase wattmeters cannot be used for balanced load. Moreover, in the case of balanced load, the power factor can also be calculated using two single-phase wattmetrers' power measurement readings as discussed in Chapter 7.

6.12.2 Three-Phase Wattmeters

The three-phase wattmeter essentially consists of two single-phase wattmeters with the moving elements mounted on a common shaft. A single pointer shows the total power, the instrument itself making the necessary addition or subtraction of the two readings automatically and correctly, provided that the connections have been correctly made. The wiring diagram furnished with the instrument must, therefore, be strictly followed in making connections.

- Make sure that the potential coil of each single-phase wattmeter element is connected between the phase in which its current coil is connected and the common phase in which neither current coil is connected.
- If the load is balanced and the power factor is greater than 0.5, disconnect one potential coil, reconnect, and disconnect the other. If the wattmeter deflects in the same direction in both the cases when only one potential coil is connected, the connections are correct.
- If the power factor is unknown or the system unbalanced, interchange the single-phase wattmeter elements as described above. If the wattmeter reading has the same magnitude as before, the connections are correct.

6.13 ENERGY METER

Energy meter is similar to as wattmeter in several aspects such as it has potential (voltage) coil and current coil which are connected in the same way as wattmeters are connected. The major difference between the wattmeter and energy meter is that wattmeter is an indicating instrument, whereas energy meter is an integrating instrument which takes account of time duration of power. Energy meters are also known as *watt-hour meter* and most common energy meter is electromechanical induction type, invented by Elihu Thomson in 1888. The electromechanical induction meter operates by counting the revolutions of an aluminium disc which is made to rotate at a speed proportional to the power. The number of revolutions is, thus, proportional to the energy usage.

Figure 6.13 shows a single-phase energy meter. The metallic disc is acted upon by two coils. Voltage coil produces a magnetic flux in proportion to the voltage and current coil produces a magnetic flux in proportion to the current. The field of the voltage coil is delayed by 90 degree using a lag coil. This produces eddy currents in the disc and the effect is such that a force is exerted on the disc in proportion to the product of the instantaneous current and voltage. A permanent magnet, known as *breaking magnet*, exerts an opposing force proportional to the speed of rotation of the disc. The equilibrium between these two opposing forces results in the disc rotating at a speed proportional to the power being used. The disc drives a register mechanism which integrates the speed of the disc over time by counting revolutions in order to render a measurement of the total energy used over a period of time.

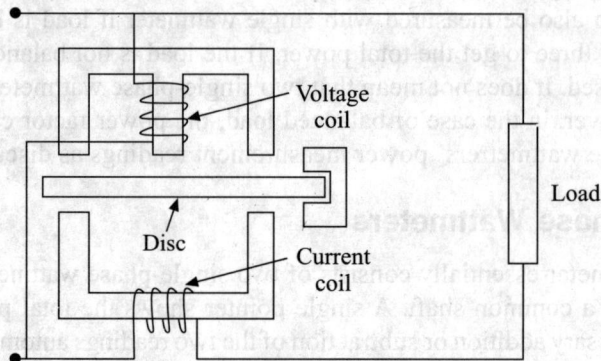

FIGURE 6.13   Single-phase energy meter.

The aluminium disc is supported by a spindle and having worm gear to drive the register. The register is a series of dials which record the amount of energy used. In an induction type meter, creep is a phenomenon adversely affecting the accuracy when the meter disc rotates continuously with potential applied and the load terminals open circuited (no-load).

6.14   PERFORMANCE OF INDICATING INSTRUMENTS

There are several factors which affect the performance of an indicating instrument. The most important among those are accuracy, precision, errors, loading effects and sensitivity. A proper precaution is also necessary for safe and reliable reading of instruments. The words accuracy and precision seem to be the same but these have different meaning in measurement system. Accuracy is the closeness to the true value of the quantity to be measured, whereas precision is a measure of the degree of agreement within a group of measurements. A precise reading may not be accurate.

6.14.1   Errors

Knowledge of different types of errors and how they vary is essential to the intelligent use of measuring device. The errors can be classified broadly into two categories:

Type-1:   These types of errors produce the same error at any point of scale such as

- Scale error
- Zero error
- Reading error
- Parallax error
- Friction error

Type-2:   These types of errors produce the errors proportional to the pointer deflection such as

- Effect of temperature in changing the resistance and control spring characteristic
- Effect of frequency in ac instruments

- Effect of waveform error in instruments
- Inconsistence resistance in instruments

6.14.2 Loading Effect

Normally the value of resistance in the current coil of ammeters and wattmeters is kept very small and it is very high in voltage or potential coil of voltmeters and watmetters. Since the ammeters and current coil of watmetters are connected in series with the load, there is some voltage drop in the circuit and the current flow is reduced. Similarly, due to connection of voltmeters and potential coil of watmmeters, the current through the load is reduced and this cause is known as *loading effect*. To understand this, let us take an example of a load having 30 ohm and connected through a source as shown in Figure 6.14(a). Using circuit analysis, the current through load will be 2 [= 100/(20 + 30)] A and voltage across the load will be 60 (= 2 × 30) V.

Impact of ammeter connection:

In Figure 6.14(b), an ammeter having effective resistance (including the shunt) of 5 ohm is connected to measure the load current. Now the current through the load will be 1.818 [= 100/(20 + 30 + 5)] A. It shows that ammeter will read only 1.81 A instead of 2 A.

Impact of voltmeter connection:

If a voltmeter load having resistance of 120 ohm is connected, as shown in Figure 6.14(c), to measure the voltage across, the current drawn from the source will be

$$2.727 \text{ A} = \left[\cfrac{100}{\left(20 + \cfrac{30 \times 120}{30 + 120} \right)} \right].$$

Thus, voltage across the load will be 54.55 V which is read by the instrument instead of 60 V. This is due to the effect of loading.

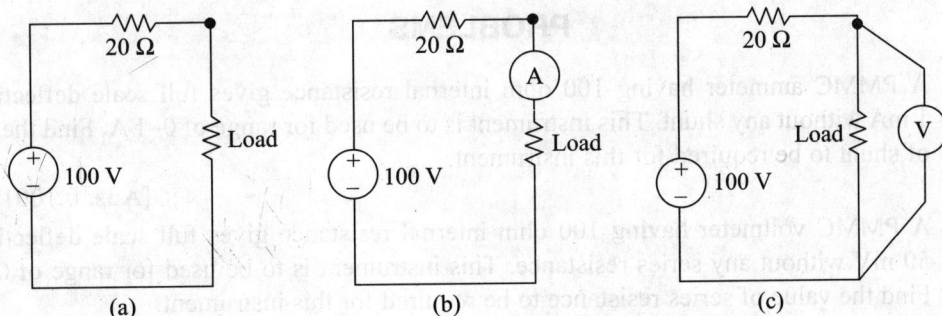

FIGURE 6.14   Loading effect.

6.14.3 Sensitivity and Efficiency

The sensitivity of an instrument is defined as a ratio of the magnitude of the output signal or response to the magnitude of input signal or the quantity being measured. The reciprocal of sensitivity is known as *deflection factor* or inverse sensitivity. Mathematically, it can be written as

$$\text{Sensititvity} = \frac{\text{Infinitesimal change in output}}{\text{Infinitesimal change in input}} \tag{6.22}$$

The efficiency of an instrument is defined as the ratio of the measured quantity as full scale to the power taken by the instrument at full scale. For voltmeter and ammeter, the efficiency will be $\frac{R_m}{V_{fs}}$ and $\frac{1}{I_{fs}R_m}$, respectively where R_m is the resistance of the instrument. V_{fs} and I_{fs} are the full scale voltage and current deflections, respectively.

6.14.4 Precautions

While using the instruments for measurement of any quantity following care, but not limited to, should be taken.

- In order to avoid damage of the instrument used, care must be taken to see that its range is greater than the line current in case of ammeter and wattmeter (current coil). In case of voltmeter and wattmeter (potential coil), voltage to be measured should be lower than instruments voltage rating.
- Special care must be exercised not to connect current-measuring instruments directly across a line or between any two points which differ in potential by more than a very small amount. Connection across a line results in practically a dead short circuit and immediate destruction of an instrument. The effect on the instrument is immediate and disastrous.
- Be sure that all of the connections at the terminals of the instruments are clean, tight, and well made to avoid contact resistance which would affect the accuracy of the readings.
- When using multi-range instruments on a circuit of unknown voltage/current, start with the highest range first and work down to a range which gives a deflection on the upper part of the scale.

PROBLEMS

6.1 A PMMC ammeter having 100 ohm internal resistance gives full scale deflection of 1 mA without any shunt. This instrument is to be used for range of 0–1 A. Find the value of shunt to be required for this instrument.

[**Ans.** 0.1001 ohm]

6.2 A PMMC voltmeter having 100 ohm internal resistance gives full scale deflection of 50 mV without any series resistance. This instrument is to be used for range of 0–1 V. Find the value of series resistance to be required for this instrument.

[**Ans.** 1900 ohm]

6.3 A basic ammeter with internal resistance of 500 ohm gives full scale deflection of 50 mA. It is connected to a circuit with a shunt and reads 4 A on 0–5 A scale. Another calibrated instrument is used for the same reading and it was found that reading is 4.2 A.

It is noticed that first instrument has faulty shunt. Find the faulty value of shunt. What should be the correct value of shunt for the first instrument?

[**Ans.** 5.051 ohm, 4.808 ohm]

6.4 A PMMC instrument gives full scale deflection of 5 mA when voltage across it is 100 mV. Find the value of

(a) Shunt resistance for full scale deflection corresponding to 10 A

(b) Series resistance for full scale deflection with 100 V.

[**Ans.** (a) 0.01 ohm (b) 19980 ohm]

6.5 An ammeter gives full scale deflection of 5 mA without shunt. When it is connected to a shunt of 0.2 ohm, its range becomes 0–1 A. Find the value of ammeter resistance.

[**Ans.** 39.8 ohm]

6.6 A voltmeter reads full scale deflection for 50 mV without any multiplier (series resistance). When a multiplier of 100 ohm added, its full scale deflection becomes 10 V. Find the value of new multiplier to give full scale deflection of 500 V.

[**Ans.** 5024.62 ohm]

7

Three-Phase AC Circuits

7.1 INTRODUCTION

It has been a well-established fact that generation and utilization of three-phase ac systems are more convenient, efficient and economical, compared to the single-phase ac system. Power in three-phase circuit is constant rather than pulsating as it is in single-phase circuit. The electrical loads are both single-phase and three-phase in nature. Low power domestic loads are mostly in single-phase, whereas the industrial loads are usually three-phase (due to higher power rating). Three-phase (3-ϕ) motors have better starting torque compared to the single-phase (1-ϕ) motors. Therefore, three-phase power generations are invariably produced by three-phase ac generators (also known as *alternators*) in the system.

To understand the three-phase concept, let us see single-phase power generation as shown in Figure 7.1. An elementary single-phase ac generator consists of a rotating magnet and a stationary coil *aa'*. The voltage induced in the coil will be proportional to the rate of change of flux, linking the coil. Flux will be the maximum when magnet is aligned with coil (*yy'*-axis) and minimum when it is 90° (aligned to *xx'*-axis).

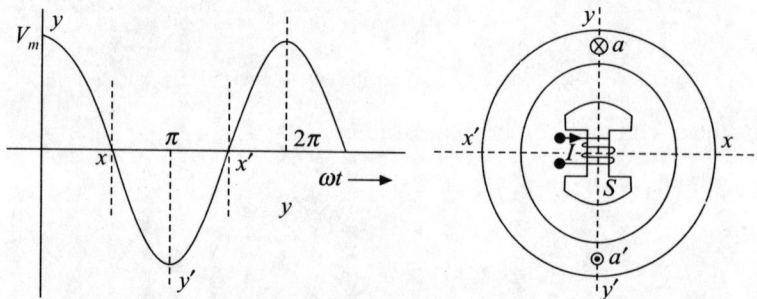

FIGURE 7.1 Generation of single-phase power.

The voltage generated in the coil, which is sinusiodally varying voltage, will be a function of time (t), speed of rotation (ω) and peak voltage magnitude ($V_m = \sqrt{2}\,V$), where V is the *rms* value. It can be expressed as

$$v = \sqrt{2}\,V \cos \omega t \qquad (7.1)$$

7.2 GENERATION OF THREE-PHASE VOLTAGE

Three-phase ac generator consists of three stationary windings (aa', bb', cc') as shown in Figure 7.2. These windings are displaced by $120°$ from each other in the space. The detailed theory behind the voltage generation will be explained in the later chapter of synchronous generator. The voltage in the coil aa' will be maximum when the N-S pole of magnet aligns with the aa' direction. Thus, peak voltage $V_{aa'}$ is at $\omega t = 0°$ and similarly, the peaks in coils bb' and cc' will be at $\omega t = 120°$ and $\omega t = 240°$, respectively.

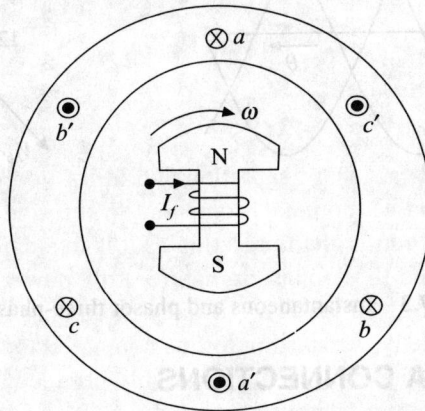

FIGURE 7.2 Three-phase ac generator.

Since, the magnet is rotating at an angular speed (clock-wise direction as shown in Figure 7.2) of ω rad/sec, the instantaneous voltage generated in the different coils (assumed identical) will be

$$v_{aa'} = \sqrt{2}\,V \cos \omega t \qquad (7.2)$$

$$v_{bb'} = \sqrt{2}\,V \cos (\omega t - 120) \qquad (7.3)$$

$$v_{cc'} = \sqrt{2}\,V \cos (\omega t - 240) \qquad (7.4)$$

In phasor forms, these voltages can be expressed as

$$\bar{V}_{aa'} = V \angle 0° \qquad (7.5)$$

$$\bar{V}_{bb'} = V \angle -120° \qquad (7.6)$$

$$\bar{V}_{cc'} = V \angle -240° = V \angle 120° \qquad (7.7)$$

The addition of these voltages equals zero. Mathematically, it can be written as

$$v_{aa'} + v_{bb'} + v_{cc'} = 0 \tag{7.8}$$

or

$$\bar{V}_{aa'} + \bar{V}_{bb'} + \bar{V}_{cc'} = 0 \tag{7.9}$$

The phasor and instantaneous voltages are shown in Figure 7.3. The phase rotation is $a \rightarrow b \rightarrow c \rightarrow$ (clockwise direction).

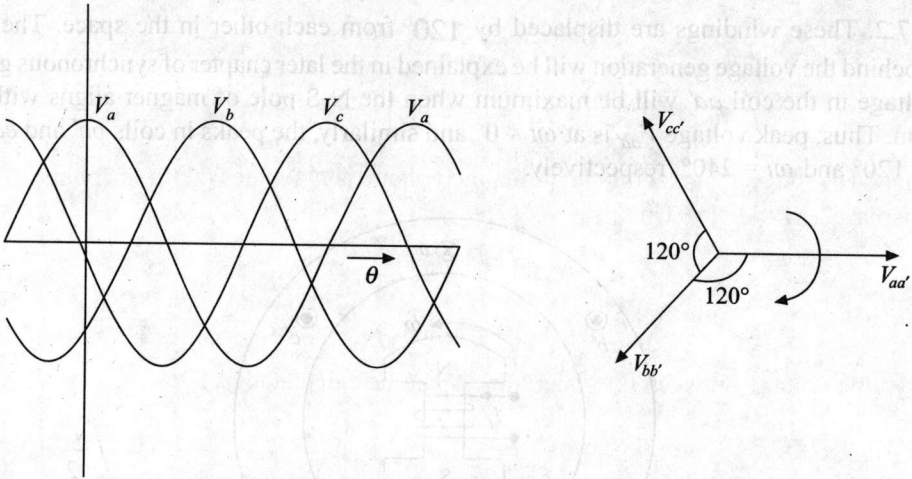

FIGURE 7.3   Instantaneous and phasor three-phase voltages.

7.3   STAR AND DELTA CONNECTIONS

Since the voltages are induced in the coils, to get the current (or power), they are connected either in star (Y) or delta (Δ) connection as explained below.

7.3.1   Star Connection

Figure 7.4 shows the star (or Wye) connection. One end of coils are connected together and normally known as *neutral point* (n). In the star connection, the line current (I_L), which is coming out from node a or b or c, is the same as the phase current (I_P) which is coming to nodes a or b or c from node-n. The phase current is the current which flows in the coils. However, line voltage (voltage between two lines) will be $\sqrt{3}$ times the phase voltage (coil voltage) which can be proved mathematically and graphically.

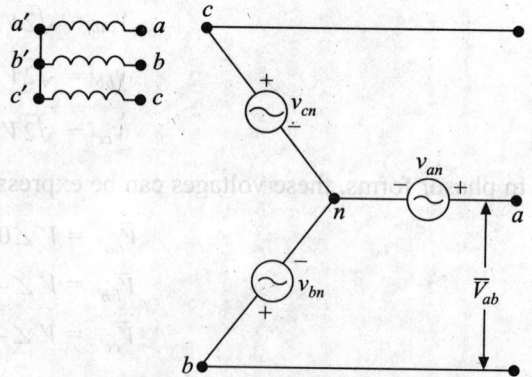

FIGURE 7.4   Three-phase star connection.

Let *rms* voltage generated in the coils (phase) be the same and equal to V_P. The phasor voltages[*] can be written as

$$\overline{V}_{an} = V_p \angle 0°$$

$$\overline{V}_{bn} = V_p \angle -120°$$

$$\overline{V}_{cn} = V_p \angle 120° = V_p \angle -240°$$

The line voltage between phase-*a* and phase-*b* will be calculated as follows:

$$\overline{V}_{ab} = \overline{V}_{an} - \overline{V}_{bn} = V_p \angle 0° - V_p \angle -120°$$

$$= V_p \left[1 - \left(\frac{-1}{2} \right) + j \frac{\sqrt{3}}{2} \right] = \sqrt{3} V_p \left[\frac{\sqrt{3}}{2} + j \frac{1}{2} \right] = \sqrt{3} V_p \angle 30° \qquad (7.10)$$

Equation (7.10) shows that magnitude of the line voltage is $\sqrt{3}$ times of phase voltage magnitude and is displaced by the 30°. Thus,

$$V_L = \sqrt{3} \times V_P \qquad (7.11)$$

and

$$\angle V_L = \angle V_P + 30° \qquad (7.12)$$

Similarly, line voltage between line-*b* and line-*c* can be calculated as

$$\overline{V}_{bc} = \overline{V}_{bn} - \overline{V}_{cn} = V_p \angle -120° - V_p \angle 120° = V_p \left[\left(\frac{-1}{2} \right) - j \frac{\sqrt{3}}{2} + \frac{1}{2} - j \frac{\sqrt{3}}{2} \right] \qquad (7.13)$$

$$= -j\sqrt{3} V_p = \sqrt{3} V_p \angle -90°$$

and line voltage between line-*c* and line-*a* can be obtained as

$$\overline{V}_{ca} = \overline{V}_{cn} - \overline{V}_{an} = \sqrt{3} V_p \angle -210° \qquad (7.14)$$

Figure 7.5 shows the phasor diagram of line and phase voltages. It can be seen from Figure 7.5 that the length of phasor $\overline{V}_{ab}$ is $\sqrt{3}$ time the length of phasor $\overline{V}_{an}$.

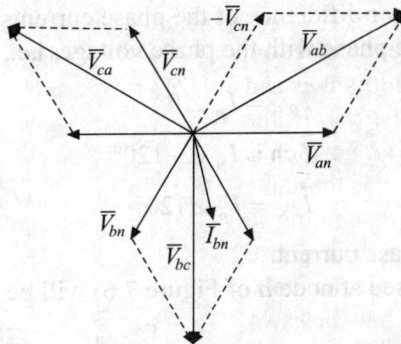

FIGURE 7.5   Phasor diagram with star connection.

[*] Phase voltage *a* is taken as reference.

In star connection, magnitudes of

    Line voltage (V_L) = $\sqrt{3}$ × Phase voltage (V_P)

    Line current (I_L) = Phase current (I_P)

7.3.2 Delta Connection

The coils connection diagram for delta configuration is shown in Figure 7.6. It can be seen from Figure 7.6 that line voltages are the same as the phase voltage.

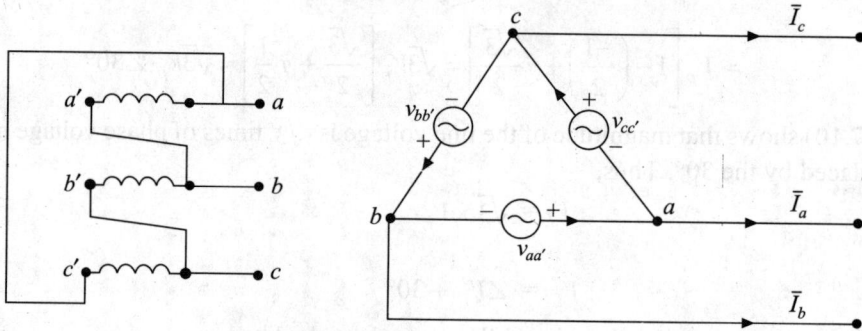

FIGURE 7.6 Three-phase delta connection.

If V_P is the magnitude of the phase voltage, the line voltage phasors taking phase-a as reference can be written as

$$\overline{V}_{ab} = \overline{V}_{aa'} = V_p \angle 0° \tag{7.15}$$

$$\overline{V}_{bc} = \overline{V}_{bb'} = V_p \angle -120° \tag{7.16}$$

$$\overline{V}_{ca} = \overline{V}_{cc'} = V_p \angle -240 = V_p \angle 120° \tag{7.17}$$

The line currents coming out from nodes a, b and c ($\overline{I}_a, \overline{I}_b$ and $\overline{I}_c$) are different than the phase currents ($\overline{I}_{a'a}, \overline{I}_{b'b}$ and $\overline{I}_{c'c}$)* which can be seen from Figure 7.6 as at node there are three currents. The line current will be the vector difference of the phase currents.

    Let phase currents be in the phase with the phase voltage, i.e., power factor is unity.

$$\overline{I}_{a'a} = I_p \angle 0°$$

$$\overline{I}_{b'b} = I_p \angle -120°$$

$$\overline{I}_{c'c} = I_p \angle 120°$$

where I_p is the magnitude of phase current.

    Thus the current in line-a (see at node a of Figure 7.6) will be

$$\overline{I}_a = \overline{I}_{a'a} - \overline{I}_{c'c} = I_p \angle 0 - I_p \angle 120° = I_p \left[1 + \frac{1}{2} - j\frac{\sqrt{3}}{2} \right] = \sqrt{3} I_p \angle -30° \tag{7.18}$$

---

*The phase current can also be written $\overline{I}_{ba}, \overline{I}_{ac}$ and $\overline{I}_{cb}$ in delta connection.

It is evident from Eq. (7.18), that the magnitude of line current is $\sqrt{3}$ times of the magnitude of phase current. It can also be seen from the phasor diagram of Figure 7.7.

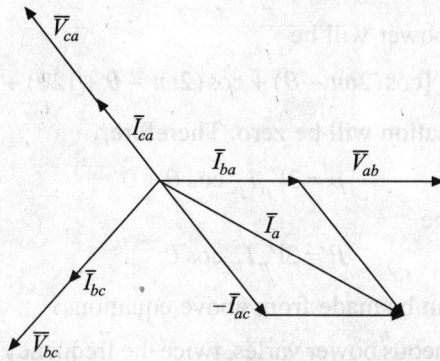

FIGURE 7.7   Phasor diagram of line and phase currents with delta connection.

Similarly, current in line-b and line-c can be obtained as

$$\bar{I}_b = \bar{I}_{cb} - \bar{I}_{ba} = \sqrt{3}\,I_p\ \angle -150°$$

$$\bar{I}_c = \bar{I}_{ac} - \bar{I}_{cb} = \sqrt{3}\,I_p\ \angle -270° = \sqrt{3}I_p\ \angle 90°$$

In delta connection, magnitudes of
   Line voltage (V_L) = Phase voltage (V_P)
   Line current (I_L) = $\sqrt{3}$ × Phase current (I_P)

7.4 POWER IN THREE-PHASE CIRCUIT

The instantaneous power in a single-phase circuit is written as

$$p(t) = vi = VI\,[\cos\theta + \cos(2\omega t - \theta)]$$

Similarly, the instantaneous power in three-phase ac circuit will be

$$p = p_a + p_b + p_c = v_a i_a + v_b i_b + v_c i_c \qquad (7.19)$$

If a balanced supply system feeding a balanced load bank connected in star, the phase voltages and currents can be written as

$$v_a = \sqrt{2}V_p \cos\omega t; \quad v_b = \sqrt{2}V_p \cos(\omega t - 120) \text{ and } v_c = \sqrt{2}V_p \cos(\omega t - 240)$$

$$i_a = \sqrt{2}I_p \cos(\omega t - \theta); \quad i_b = \sqrt{2}I_p \cos(\omega t - \theta - 120) \text{ and } i_c = \sqrt{2}I_p \cos(\omega t - \theta - 240)$$

The instantaneous power of phases a, b and c will be

$$p_a = v_a i_a = V_p I_p\,[\cos\theta + \cos(2\omega t - \theta)]$$

$$p_b = v_b i_b = V_p I_p \left[\cos\theta + \cos(2\omega t - \theta - 240)\right] = V_p I_p \left[\cos\theta + \cos(2\omega t - \theta + 120)\right]$$

$$p_c = v_c i_c = V_p I_p \left[\cos\theta + \cos(2\omega t - \theta + 240)\right] = V_p I_p \left[\cos\theta + \cos(2\omega t - \theta - 120)\right]$$

Thus, the total instantaneous power will be

$$p = 3V_p I_p \cos\theta + V_p I_p \left[\cos(2\omega t - \theta) + \cos(2\omega t - \theta + 120) + \cos(2\omega t - \theta - 120)\right]$$

The second term of above equation will be zero. Therefore,

$$p = 3V_p I_p \cos\theta + 0 \tag{7.20}$$

The total average power will be

$$P = 3V_p I_p \cos\theta \tag{7.21}$$

The following observations can be made from above equations:

1. Single-phase instantaneous power varies, twice the frequency of supply around its average power.
2. Total instantaneous power of three-phase system is constant and equal to the three-times average power per phase.
3. The total power of three-phase circuit is equal to the three times of the average power of single-phase.

Equation (7.21) is in terms of phase voltage and phase current. Using the phase and line quantities relations, the total power in a three-phase circuit can be expressed in terms of line voltage and line current as

Star Connection

In star connection, $V_L = \sqrt{3}\,V_p$ and $I_L = I_p$ and thus,

$$P = 3V_p I_p \cos\theta = 3\frac{V_L}{\sqrt{3}} I_L \cos\theta = \sqrt{3}V_L I_L \cos\theta \tag{7.22}$$

Delta Connection

In delta connection, $V_L = V_p$ and $I_L = \sqrt{3}\,I_p$ and thus,

$$P = 3V_p I_p \cos\theta = 3V_L \frac{I_L}{\sqrt{3}} \cos\theta = \sqrt{3}V_L I_L \cos\theta \tag{7.23}$$

From Eqs. (7.22) and (7.23), the following equation is true for any connection

$$P = 3V_p I_p \cos\theta = \sqrt{3}V_L I_L \cos\theta \tag{7.24}$$

EXAMPLE 7.1   Find the phase voltage of a three-phase, 440 V, 50 Hz supply system.

Solution: Normally generators are connected with star to get the neutral point for protection purpose. Moreover, if nothing is mentioned about the three-phase connection, star-connection is taken. Thus, for star-connection, phase voltage is

$$V_p = \frac{V_L}{\sqrt{3}} = \frac{440}{\sqrt{3}} = 231 \text{ V}$$

7.5 THREE-PHASE LOAD CIRCUIT

Single-phase load can be connected separately across each of the phases. But it is economical to connect loads in three-phase banks. Loads can also be connected to either star (Y) or delta (Δ). A three-phase load is said to be balanced if the load in each phase is the same not only in the magnitude but in phasor also. Thus, for balance load,

$$\bar{Z}_a = \bar{Z}_b = \bar{Z}_c = \bar{Z}$$

> *A three-phase supply system is said to be balanced, if the magnitudes of each phase voltage are equal and these voltage are displaced 120° electrically (in clockwise direction). Thus,*
>
> $$\left| \bar{V}_{an} \right| = \left| \bar{V}_{bn} \right| = \left| \bar{V}_{cn} \right| = V_p \text{ and}$$
>
> $$\bar{V}_{an} = V_p \angle 0; \quad \bar{V}_{bn} = V_p \angle -120; \quad \bar{V}_{cn} = V_p \angle -240 \text{ and } \bar{V}_{an} + \bar{V}_{bn} + \bar{V}_{cn} = 0$$
>
> *Please note that phasor addition of phase current may not be zero as currents depend on the loads and type of connection. If three-phase load is also balanced, the phasor addition of currents (both line as well as phase) will be zero.*

7.5.1 Star-Connected Three-Phase Load

Figure 7.8 shows the star connected balanced load having load impedance $\bar{Z}$.

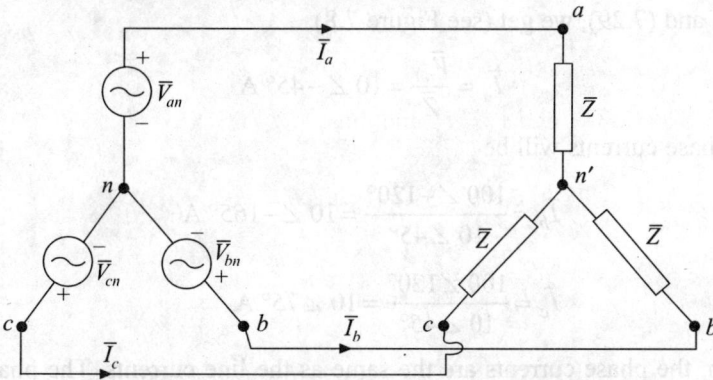

FIGURE 7.8 Star connected three-phase balanced load.

Consider a three-phase balanced supply system. The voltage equations in terms of line currents can be written as

$$\bar{V}_{an} = \bar{I}_a \bar{Z} + \bar{V}_{n'n} \tag{7.25}$$

$$\bar{V}_{bn} = \bar{I}_b \bar{Z} + \bar{V}_{n'n} \tag{7.26}$$

$$\bar{V}_{cn} = \bar{I}_c \bar{Z} + \bar{V}_{n'n} \tag{7.27}$$

where $\bar{V}_{n'n}$ is the voltage between neutral points of load and supply system.

Adding Eqs. (7.25), (7.26) and (7.27), we get

$$\bar{V}_{an} + \bar{V}_{bn} + \bar{V}_{cn} = \bar{Z}\left[\bar{I}_a + \bar{I}_b + \bar{I}_c\right] + 3\bar{V}_{n'n} \qquad (7.28)$$

Since the three-phase supply system is balanced, the phase summation of the phase voltages will be zero. The phase addition of line currents will also be zero as KCL at node-n. Thus, Eq. (7.28) can be written as

$$0 = 0 + 3\bar{V}_{n'n}$$

or

$$\bar{V}_{n'n} = 0 \qquad (7.29)$$

Therefore, it can be stated that if a three-phase balanced load is connected (star) to the balanced three-phase supply system, the potential difference between the neutrals of load and supply system will be zero. If a wire is connected between points n and n', there will be no current in that wire.

EXAMPLE 7.2   If a three-phase, $\dfrac{100}{\sqrt{3}}$ V, 50 Hz balanced supply system is feeding to a balanced star-connected load of $10\angle 45°$ ohm. Find the phase and line currents.

Solution: Since supply is balanced $(\bar{V}_{n'n} = 0)$ and line voltage magnitude is $\dfrac{100}{\sqrt{3}}$ V, thus, phase voltage, taking phase-a voltage as reference voltage, will be

$$\bar{V}_{an} = 100\angle 0°; \quad \bar{V}_{bn} = 100\angle -120° \text{ and } \bar{V}_{cn} = 100\angle -240° = 100\angle 120° \text{ V}$$

Using Eqs. (7.25) and (7.29), we get (see Figure 7.8)

$$\bar{I}_a = \frac{\bar{V}_{an}}{\bar{Z}} = 10\angle -45° \text{ A}$$

Similarly, other phase currents will be

$$\bar{I}_b = \frac{100\angle -120°}{10\angle 45°} = 10\angle -165° \text{ A}$$

$$\bar{I}_c = \frac{100\angle 120°}{10\angle 45°} = 10\angle 75° \text{ A}$$

In star connection, the phase currents are the same as the line currents. The phasor diagram of voltage and line currents is shown in Figure 7.9.

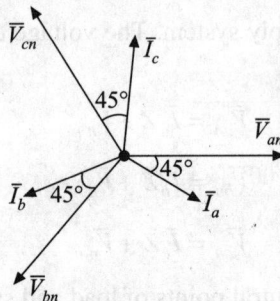

FIGURE 7.9   Phasor diagram of Example 7.2.

7.5.2 Delta-Connected Three-Phase Load

Figure 7.10 shows the delta-connected balanced load having load impedance $\bar{Z}$. From Figure 7.10, the following relations can be written

$$\bar{V}_{ab} = \bar{V}_{an} - \bar{V}_{bn} = \bar{I}_{ab}\bar{Z} \tag{7.30}$$

$$\bar{V}_{bc} = \bar{V}_{bn} - \bar{V}_{cn} = \bar{I}_{bc}\bar{Z} \tag{7.31}$$

$$\bar{V}_{ca} = \bar{V}_{cn} - \bar{V}_{cn} = \bar{I}_{ca}\bar{Z} \tag{7.32}$$

where $\bar{I}_{ab}, \bar{I}_{bc}$ and $\bar{I}_{ca}$ are the phase currents in the load. Knowing the voltages and the load impedance, the phase current in load can be obtained and thus, line current.

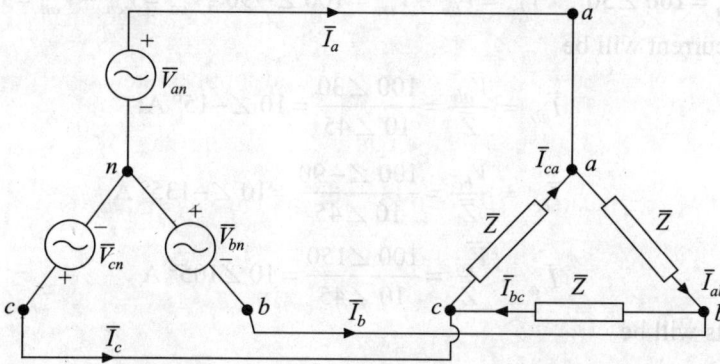

FIGURE 7.10 Delta-connected balanced load.

EXAMPLE 7.3 If a three-phase, 100 V, 50 Hz balanced supply system is feeding to a balanced delta-connected load of $10 \angle 45°$ ohm. Find the phase and line currents.

Solution: Since nothing is mentioned about the reference, the problem can be solved by taking either the line voltage or phase voltage as reference.

Taking line voltage as reference

The line voltages, taking $\bar{V}_{ab}$ as reference, can be expressed as

$$\bar{V}_{ab} = 100 \angle 0°; \quad \bar{V}_{bc} = 100 \angle -120°; \quad \bar{V}_{ca} = 100 \angle -240° = 100 \angle 120° \text{ V}$$

Thus load phase currents will be

$$\bar{I}_{ab} = \frac{\bar{V}_{ab}}{\bar{Z}} = \frac{100 \angle 0°}{10 \angle 45°} = 10 \angle -45° \text{ A};$$

$$\bar{I}_{bc} = \frac{\bar{V}_{bc}}{\bar{Z}} = \frac{100 \angle -120°}{10 \angle 45°} = 10 \angle -165° \text{ A and}$$

$$\bar{I}_{ca} = \frac{\bar{V}_{ca}}{\bar{Z}} = \frac{100 \angle 120°}{10 \angle 45°} = 10 \angle 75° \text{ A}$$

Line currents (see Figure 7.10) will be calculated as

$$\overline{I}_a = \overline{I}_{ab} - \overline{I}_{ca} = 10 \angle -45° - 10 \angle 75° = 17.321 \angle -75° \text{ A}$$

$$\overline{I}_b = \overline{I}_{bc} - \overline{I}_{ab} = 10 \angle -165° - 10 \angle -45° = 17.321 \angle -195° \text{ A}$$

$$\overline{I}_c = \overline{I}_{ca} - \overline{I}_{bc} = 10 \angle 75° - 10 \angle -165° = 17.321 \angle 45° \text{ A}$$

Taking phase voltage as Reference

$$\overline{V}_{an} = \frac{100}{\sqrt{3}} \angle 0°; \quad \overline{V}_{bn} = \frac{100}{\sqrt{3}} \angle -120°; \quad \overline{V}_{cn} = \frac{100}{\sqrt{3}} \angle 120° \text{ V}$$

Thus, line voltage which will be the phase voltage of load will be

$$\overline{V}_{ab} = \overline{V}_{an} - \overline{V}_{bn} = 100 \angle 30°; \quad \overline{V}_{bc} = \overline{V}_{bn} - \overline{V}_{cn} = 100 \angle -90°; \overline{V}_{ca} = \overline{V}_{cn} - \overline{V}_{an} = 100 \angle 150° \text{ V}$$

Now load phase current will be

$$\overline{I}_{ab} = \frac{\overline{V}_{ab}}{\overline{Z}} = \frac{100 \angle 30}{10 \angle 45} = 10 \angle -15° \text{ A};$$

$$\overline{I}_{bc} = \frac{\overline{V}_{bc}}{\overline{Z}} = \frac{100 \angle -90}{10 \angle 45} = 10 \angle -135° \text{ A}$$

and

$$\overline{I}_{ca} = \frac{\overline{V}_{ca}}{\overline{Z}} = \frac{100 \angle 150}{10 \angle 45} = 10 \angle 105° \text{ A}$$

Thus, line currents will be

$$\overline{I}_a = \overline{I}_{ab} - \overline{I}_{bc} = 10[1 \angle -15° - 1 \angle -135°] = 10\sqrt{3} \angle -45° \text{ A}$$

$$\overline{I}_b = \overline{I}_{bc} - \overline{I}_{ab} = 10 \angle -135° - 10 \angle -15° = 17.321 \angle -165° \text{ A}$$

$$\overline{I}_c = \overline{I}_{ca} - \overline{I}_{bc} = 10 \angle 105° - 10 \angle -135° = 17.321 \angle 75° \text{ A}$$

The phasor diagram taking phase-*a* voltage as reference is shown in Figure 7.11.

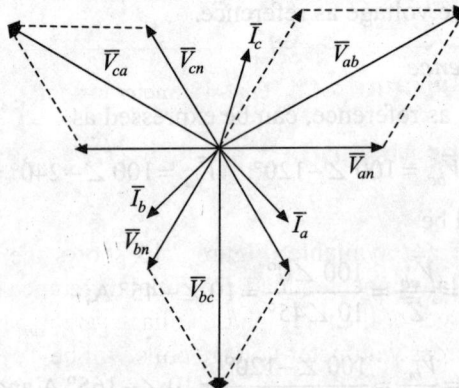

FIGURE 7.11 Phasor diagram.

Thus due to change in the reference, there will be no change in the magnitude of either voltages or currents. There will be only change in the phase angles.

7.6 MEASUREMENT OF REAL POWER

Real power in a single-phase ac circuit can be measured, using a single wattmeter (alternatively one ammeter and one voltmeter method). A wattmeter as described in Chapter 6 has a low resistance current coil (CC) connected in series to carry the load current and a high resistance voltage coil or pressure coil (PC) connected parallel to the load to measure the voltage across it. Very small current flows through pressure coil. Voltage coil is connected in parallel to the load. An external resistance of high value is connected in series with PC for having different tapings in series. Wattmeter reads real power (P) which is equal to the

$$P = VI \cos \angle_{\overline{I}}^{\overline{V}} = VI \cos \theta \qquad (7.33)$$

where V is the *rms* voltage across the pressure coil and I is the *rms* current flowing in the current coil and θ is the phase angle[*] between voltage phasor ($\overline{V}$) and current phasor ($\overline{I}$).

7.6.1 One-Wattmeter Method

Three-phase power[**] can be measured with one-wattmeter only if the three-phase load and three-phase supply are balanced, and neutral point is available which is only possible in the case of star connection as shown in Figure 7.12.

Total power consumed by the load = 3 × wattmeter reading

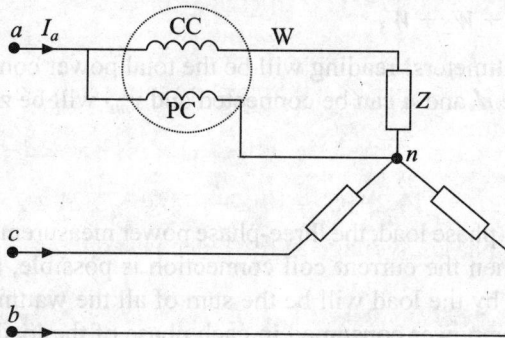

FIGURE 7.12 One-wattmeter method.

7.6.2 Three-Wattmeter Method

(a) *Star-connected load*

Let star point of the load be not available. Figure 7.13 shows the power measurement using three-wattmeter method for star-connected load. Potential difference between the nodes n' and n is $v_{nn'}$. If the instantaneous values of the three-phase voltages are v_{an}, v_{bn} and v_{cn} at load terminals, the instantaneous voltages across wattmeter voltage coils will be

$$v_{an'} = v_{an} + v_{nn'};\ v_{bn'} = v_{bn} + v_{nn'} \text{ and } v_{cn'} = v_{cn} + v_{nn'} \qquad (7.34)$$

---

[*] Notation $\angle_{I}^{\overline{V}}$ shows the angle between phasor $\overline{V}$ and current phasor $\overline{I}$ taken in the same reference.

[**] Three-phase power means three-phase ac power.

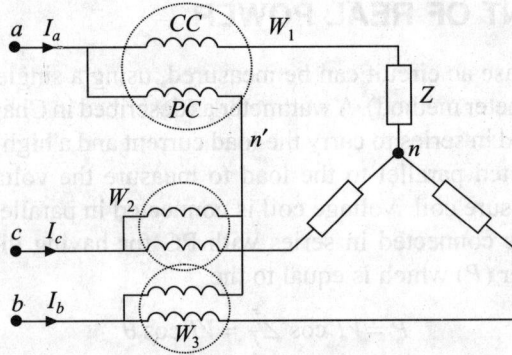

FIGURE 7.13 Three-wattmeter method.

Total instantaneous power measured by the three wattmeters will be

$$p = v_{an'}i_a + v_{bn'}i_b + v_{cn'}i_c \tag{7.35}$$

Putting the value of voltages in Eq. (7.35) from Eq. (7.34), we have

$$\begin{aligned}
p &= v_{an}i_a + v_{bn}i_b + v_{cn}i_c + (i_a + i_b + i_c)v_{nn'} \\
&= v_{an}i_a + v_{bn}i_b + v_{cn}i_c + 0 \\
&= W_1 + W_2 + W_3
\end{aligned} \tag{7.36}$$

Thus, the summation of wattmeters' reading will be the total power consumed by the load. If the star point is available, node n' and n can be connected and $v_{nn'}$ will be zero. The total power will be the same as Eq. (7.36).

(b) *Delta-connected load*

In the delta-connected three-phase load, the three-phase power measurement using three-wattmeter method is possible only when the current coil connection is possible, as shown in Figure 7.14. The total power consumed by the load will be the sum of all the wattmeter readings. Moreover, each wattmeter will read the power consumed in each phase of the load.

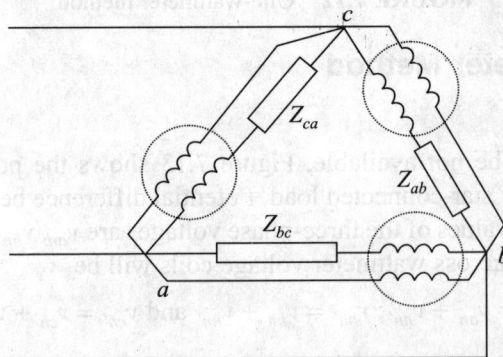

FIGURE 7.14 Three-wattmeter method of delta-connected load.

7.6.3   Two-Wattmeter Method

The three-phase ac power measurement for the balanced or unbalanced loads, connected in either star or delta configuration, is possible with only two wattmeters as shown in Figure 7.14. The total instantaneous power will be

$$p_T = v_{an}i_{an} + v_{bn}i_{bn} + v_{cn}i_{cn} \tag{7.37}$$

If there is no neutral wire, the instantaneous current at neutral point in star connection will be

$$i_{an} + i_{bn} + i_{cn} = 0 \tag{7.38}$$

However, this relation is also true in the case of delta connection. Using Eqs. (7.37) and (7.38), we get

$$p_T = v_{an}i_{an} + v_{bn}(-i_{an} - i_{cn}) + v_{cn}i_{cn} = (v_{an} - v_{bn})i_{an} + (v_{cn} - v_{bn})i_{cn}$$

or

$$p_T = v_{ab}i_{an} + v_{cb}i_{cn} = W_1 + W_2 \tag{7.39}$$

Thus, the power measured by both the wattmeters is the total three-phase power.

Using the phasor diagram, it can also be proved that two wattmeters read the total three-phase power of load. For simplicity, consider the balanced load (phasor diagram is shown in Figure 7.15), the power measured by first wattmeter will be

$$P_1 = \left|\bar{V}_{ab}\right|\left|\bar{I}_{an}\right|\cos \angle^{\bar{V}_{ab}}_{\bar{I}_{an}} = \left|\bar{V}_{ab}\right|\left|\bar{I}_{an}\right|\cos(30 + \theta) = V_L I_L \cos(30 + \theta) \tag{7.40}$$

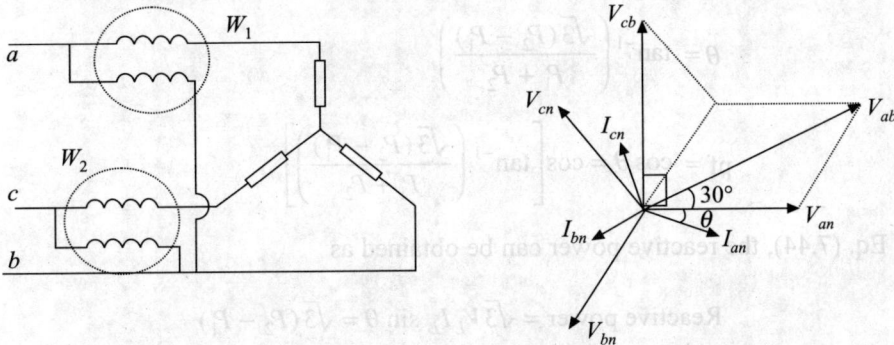

FIGURE 7.15   Two-wattmeter method of power measurement.

The power measured by second wattmeter will be

$$P_2 = \left|\bar{V}_{cb}\right|\left|\bar{I}_{cn}\right|\cos \angle^{\bar{V}_{cb}}_{\bar{I}_{cn}} = \left|\bar{V}_{cb}\right|\left|\bar{I}_{cn}\right|\cos(30 - \theta) = V_L I_L \cos(30 - \theta) \tag{7.41}$$

The total power will be

$$P_T = P_1 + P_2 = V_L I_L \left[\cos(30 + \theta) + \cos(30 - \theta)\right] = 2V_L I_L \cos 30 \cos \theta = \sqrt{3}V_L I_L \cos \theta \tag{7.42}$$

= Total three-phase power

For the balanced load, the readings of both the wattmeter readings depend on the power factor of the load. Table 7.1 shows the readings of wattmeters for different power factors.

TABLE 7.1   Wattmeter readings with different power factors of load

	Power factor angle				
	0°	30°	60°	90°	−60°
$\cos \theta$	1	0.866 (lag)	0.5 (lag)	0	0.5 (lead)
W_1	positive	positive	zero	negative	positive
W_2	positive $(P_1 = P_2)$	positive $(P_1 < P_2)$	positive	positive $(P_1 = -P_2)$	zero

7.6.4   Power Factor Determination for Three-Phase Balanced Load

With two wattmeters readings the power factor of load can be determined as follows:
The sum of the wattmeter reading will be [refer to Eq. (7.42)],

$$P_1 + P_2 = \sqrt{3} V_L I_L \cos \theta \tag{7.43}$$

From Eqs. (7.40) and (7.41), we can get

$$P_1 - P_2 = -V_L I_L (2 \sin 30 \sin \theta) = -V_L I_L \sin \theta \tag{7.44}$$

From Eqs. (7.43) and (7.44), we have

$$\tan \theta = \frac{\sqrt{3}(P_2 - P_1)}{P_1 + P_2} \tag{7.45}$$

The power factor angle (θ) and power factor will be

$$\theta = \tan^{-1}\left(\frac{\sqrt{3}(P_2 - P_1)}{P_1 + P_2} \right) \tag{7.46a}$$

$$\text{pf} = \cos \theta = \cos\left[\tan^{-1}\left(\frac{\sqrt{3}(P_2 - P_1)}{P_1 + P_2} \right) \right] \tag{7.46b}$$

Using the Eq. (7.44), the reactive power can be obtained as

$$\text{Reactive power} = \sqrt{3} V_L I_L \sin \theta = \sqrt{3}(P_2 - P_1) \tag{7.47}$$

7.7   SINGLE-PHASE ANALYSIS OF 3φ CIRCUIT

Supply system is normally balanced and always assumed to be balanced unless it is stated. If load is balanced, the analysis can be done, based on the single-phase basis. Load can be connected in either delta or star forms.

7.7.1   Star-Connected Balanced Load

Figure 7.16 shows a star connected balanced load having impedance $\bar{Z}$ in all three phases. The single-phase representation of this is shown in Figure 7.17.

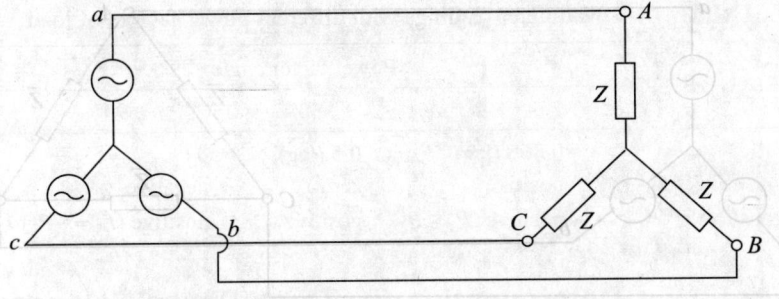

FIGURE 7.16   Balanced star-connected load.

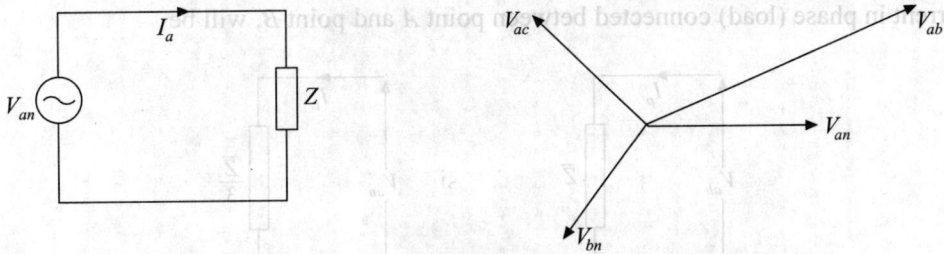

FIGURE 7.17   Single-phase representation of Figure 7.16.

The magnitude of phase current will be

$$I_p = \left| \frac{\overline{V}_{an}}{\overline{Z}} \right| \tag{7.48}$$

Taking phase-a voltage as reference and $\overline{Z} = Z \angle \theta$, the phase currents would be

$$\overline{I}_a = I_p \angle -\theta; \quad \overline{I}_b = I_p \angle (-120 - \theta) \quad \text{and} \quad \overline{I}_c = I_p \angle (120 - \theta) \text{ A} \tag{7.49}$$

The line voltage will be

$$\overline{V}_{ab} = \sqrt{3} V_{an} \angle 30; \quad \overline{V}_{bc} = \sqrt{3} V_{an} \angle (-120 + 30) \quad \text{and} \quad \overline{V}_{ca} = \sqrt{3} V_{an} \angle (120 + 30) \tag{7.50}$$

The three-phase complex power will be

$$\overline{S} = 3 \overline{V}_{an} \overline{I}_a \text{ VA} \tag{7.51}$$

7.7.2   Delta-Connected Balanced Load

Figure 7.18 shows the three-phase delta connected load having impedance $\overline{Z}$ in each phase (balanced). The voltage across each phase of the load is the line to line voltage of supply. Thus, the single phase equivalent (taking phase between node A and node B) has been shown in Figure 7.19(a).

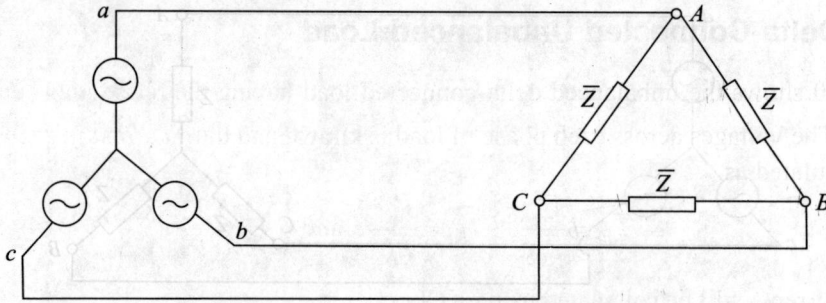

FIGURE 7.18 Balanced delta-connected load.

The current in phase (load) connected between point A and point B, will be

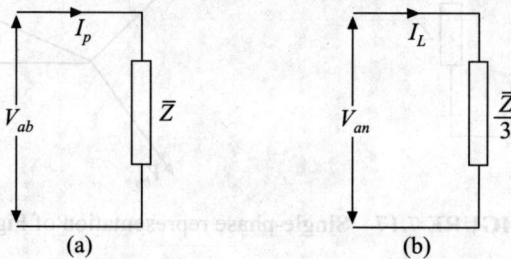

| (a) | (b) |

FIGURE 7.19 Single-phase representation of Figure 7.18.

$$I_p = \left|\frac{\bar{V}_{ab}}{\bar{Z}}\right| \tag{7.52}$$

Taking line voltage (V_{ab}) as reference and $\bar{Z} = Z \angle\theta$ ohm, the load phase currents would be

$$\bar{I}_{AB} = I_p \angle-\theta; \quad \bar{I}_{BC} = I_p \angle(-120-\theta) \quad \text{and} \quad \bar{I}_{CA} = I_p \angle(120-\theta) \text{ A} \tag{7.53}$$

The line currents will be

$$\bar{I}_a = \sqrt{3}I_p \angle(-30-\theta); \quad \bar{I}_b = \sqrt{3}I_p \angle(-150-\theta) \quad \text{and} \quad \bar{I}_c = \sqrt{3}I_p \angle(90-\theta) \text{ A} \tag{7.54}$$

Another way to solve the delta-connected load by transforming the delta into star connection and take the phase voltage as shown in Figure 7.19(b). It should be noted that the current is in now line current and the phase current can be obtained.

7.8 UNBALANCED LOADS

If load is not balanced, the phase current and line currents will be different in each phase and line, respectively. The load may be connected in either delta or star configuration. The analysis is different for different connections.

7.8.1 Delta-Connected Unbalanced Load

Figure 7.20 shows the unbalanced delta-connected load having the phase impedances $\bar{Z}_{ab}, \bar{Z}_{bc}$ and $\bar{Z}_{ca}$. The voltages across each phase of load is known and thus, current in each phase can be easily calculated as

$$\bar{I}_{ab} = \frac{\bar{V}_{ab}}{\bar{Z}_{ab}}; \quad \bar{I}_{bc} = \frac{\bar{V}_{bc}}{\bar{Z}_{bc}} \quad \text{and} \quad \bar{I}_{ca} = \frac{\bar{V}_{ca}}{\bar{Z}_{ca}} \quad (7.55)$$

The line currents will be calculated using KCL as

$$\bar{I}_a = \bar{I}_{ab} - \bar{I}_{ca}; \quad \bar{I}_b = \bar{I}_{bc} - \bar{I}_{ab} \quad \text{and} \quad \bar{I}_c = \bar{I}_{ca} - \bar{I}_{bc} \quad (7.56)$$

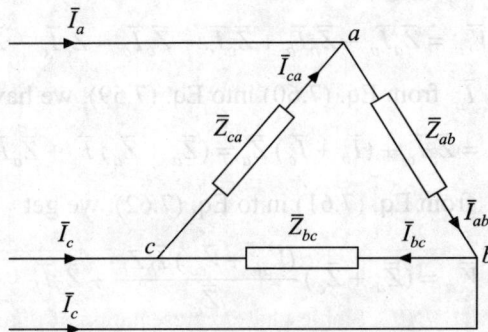

FIGURE 7.20 Unbalanced delta-connected load.

7.8.2 Star-Connected Unbalanced Load

If the unbalanced load, which has phase impedances $\bar{Z}_a, \bar{Z}_b$ and $\bar{Z}_c$ as shown in Figure 7.21, is connected in star, the voltages across each phase of load and current in each phase can be determined using the following two approaches.

(a) Using Kirchhoff's law
(b) By converting Y into equivalent Δ configuration

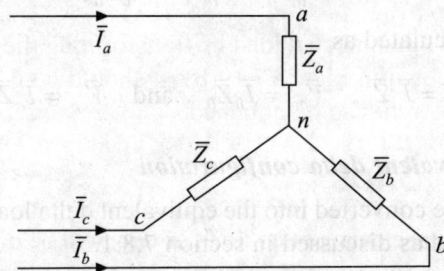

FIGURE 7.21 Unbalanced star-connected load.

(a) Using Kirchhoff's law

The voltage Eq. (Figure 7.21) can be written as

$$\bar{V}_{ab} = \bar{Z}_a \bar{I}_a - \bar{Z}_b \bar{I}_b \qquad (7.57)$$

$$\bar{V}_{bc} = \bar{Z}_b \bar{I}_b - \bar{Z}_c \bar{I}_c \qquad (7.58)$$

$$\bar{V}_{ca} = \bar{Z}_c \bar{I}_c - \bar{Z}_a \bar{I}_a \qquad (7.59)$$

And the current equation at node-n will be

$$\bar{I}_a + \bar{I}_b + \bar{I}_c = 0 \qquad (7.60)$$

Adding Eqs. (7.57) and (7.59), we get

$$\bar{V}_{ab} + \bar{V}_{ca} = \bar{Z}_a \bar{I}_a - \bar{Z}_b \bar{I}_b + \bar{Z}_c \bar{I}_c - \bar{Z}_a \bar{I}_a = \bar{Z}_c \bar{I}_c - \bar{Z}_b \bar{I}_b \qquad (7.61)$$

Putting the value of current $\bar{I}_a$ from Eq. (7.60) into Eq. (7.59), we have

$$\bar{V}_{ca} = \bar{Z}_c \bar{I}_c + (\bar{I}_b + \bar{I}_c)\,\bar{Z}_a = (\bar{Z}_a + \bar{Z}_c)\,\bar{I}_c + \bar{Z}_a \bar{I}_b \qquad (7.62)$$

Substituting the value of $\bar{I}_c$ from Eq. (7.61) in to Eq. (7.62), we get

$$\bar{V}_{ca} = (\bar{Z}_a + \bar{Z}_c)\frac{(\bar{V}_{ab} + \bar{V}_{ca})\,\bar{I}_b \bar{Z}_b}{\bar{Z}_c} + \bar{Z}_a \bar{I}_b \qquad (7.63)$$

Solving above equation, we get

$$\bar{I}_b = \frac{\bar{Z}_a \bar{V}_{bc} - \bar{V}_{ab} \bar{Z}_c}{(\bar{Z}_a \bar{Z}_b + \bar{Z}_b \bar{Z}_c + \bar{Z}_c \bar{Z}_a)} \qquad (7.64)$$

Similarly, we can get the other line/phase currents,

$$\bar{I}_a = \frac{\bar{Z}_c \bar{V}_{ab} - \bar{V}_{ca} \bar{Z}_b}{(\bar{Z}_a \bar{Z}_b + \bar{Z}_b \bar{Z}_c + \bar{Z}_c \bar{Z}_a)} \qquad (7.65)$$

$$\bar{I}_c = \frac{\bar{Z}_b \bar{V}_{ca} - \bar{Z}_a \bar{V}_{bc}}{(\bar{Z}_a \bar{Z}_b + \bar{Z}_b \bar{Z}_c + \bar{Z}_c \bar{Z}_a)} \qquad (7.66)$$

The phase voltage will be calculated as

$$\bar{V}_{an} = \bar{I}_a \bar{Z}_a; \quad \bar{V}_{bn} = \bar{I}_b \bar{Z}_b \quad \text{and} \quad \bar{V}_{cn} = \bar{I}_c \bar{Z}_c \qquad (7.67)$$

By converting star into equivalent delta configuration

The star-connected load can be converted into the equivalent delta load and then line currents and phase currents can be obtained as discussed in section 7.8.1.

EXAMPLE 7.4   If a three-phase, 100 V, 50 Hz balanced supply system is feeding to a balanced delta-connected load of $10 \angle 45°$ ohm. Find the phase and line currents. Take phase-a voltage as reference. Draw the phasor diagram.

Solution: Taking phase a voltage as reference

$$\bar{V}_{an} = \frac{100}{\sqrt{3}} \angle 0°; \bar{V}_{bn} = \frac{100}{\sqrt{3}} \angle -120°; \quad \bar{V}_{cn} = \frac{100}{\sqrt{3}} \angle 120°$$

Thus line voltage which will be the phase voltage of load will be

$$\bar{V}_{ab} = \bar{V}_{an} - \bar{V}_{bn} = 100 \angle 30°; \quad \bar{V}_{bc} = \bar{V}_{bn} - \bar{V}_{cn} = 100 \angle -90°; \quad \bar{V}_{ca} = \bar{V}_{cn} - \bar{V}_{an} = 100 \angle 150°$$

Now load phase current will be

$$\bar{I}_{ab} = \frac{\bar{V}_{ab}}{\bar{Z}} = \frac{100 \angle 30°}{10 \angle 45°} = 10 \angle -15° \text{ A};$$

$$\bar{I}_{bc} = \frac{\bar{V}_{bc}}{\bar{Z}} = \frac{100 \angle -90°}{10 \angle 45°} = 10 \angle -135° \text{ A}$$

and

$$\bar{I}_{ca} = \frac{\bar{V}_{ca}}{\bar{Z}} = \frac{100 \angle 150°}{10 \angle 45°} = 10 \angle 105° \text{ A}$$

Thus, line currents will be

$$\bar{I}_a = \bar{I}_{ab} - \bar{I}_{bc} = 10[1 \angle -15° - 1 \angle -135°] = 10\sqrt{3} \angle -45° \text{ A}$$

$$\bar{I}_b = \bar{I}_{bc} - \bar{I}_{ab} = 10 \angle -135° - 10 \angle -15° = 17.321 \angle -165° \text{ A}$$

$$\bar{I}_c = \bar{I}_{ca} - \bar{I}_{bc} = 10 \angle 105° - 10 \angle -135° = 17.321 \angle 75° \text{ A}$$

The phasor diagram taking phase-*a* voltage as reference is shown in Figure 7.22.

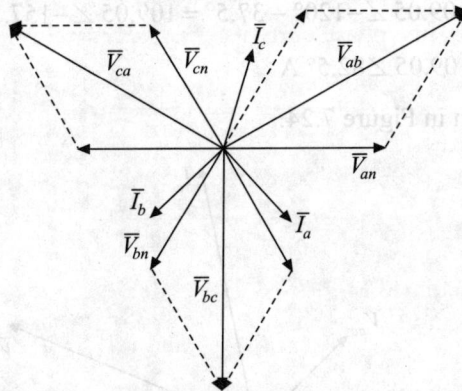

FIGURE 7.22   Phasor diagram of Example 7.4.

EXAMPLE 7.5   Two balanced three-phase loads are connected in parallel to the three-phase supply system of 220 V (phase to neutral) as shown in Figure 7.23. The phase impedance of delta-connected load is $12 \angle 30°$ Ω whereas the phase impedance of star-connected load is $4 \angle 45°$ Ω. Find the line currents and total power drawn by the loads. The phase sequence in *abc*.

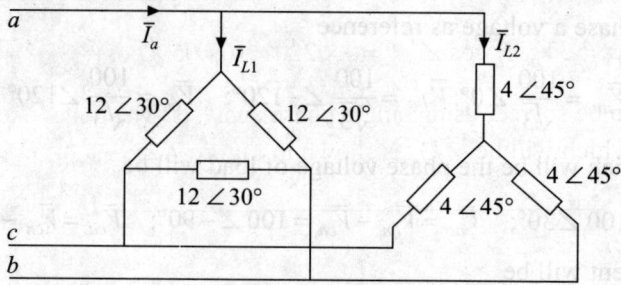

FIGURE 7.23 Example 7.5.

Solution: Since load is balanced, the only current for one line needs to be calculated. Taking phase-*a* voltage as reference,

$$\overline{V}_a = 230 \angle 0 \text{ V}$$

$$\overline{I}_{L2} = \frac{220 \angle 0}{4 \angle 45°} = 55 \angle -45° \text{ A}$$

$$\overline{I}_{P1} = \frac{220 \times \sqrt{3} \angle 30°}{12 \angle 30°} = 31.75 \angle 0° \text{ A}$$

$$\overline{I}_{L1} = \overline{I}_{ab} - \overline{I}_{ca} = 31.75 \times \sqrt{3} \angle -30° = 55 \angle -30° \text{ A}$$

$$\overline{I}_a = \overline{I}_{L1} + \overline{I}_{L2}$$

Thus,

$$\overline{I}_a = 55 \angle -45° + 55 \angle -30° = 109.05 \angle -37.5 \text{ A}$$

$$\overline{I}_b = 109.05 \angle -120° - 37.5° = 109.05 \angle -157.5 \text{ A}$$

$$\overline{I}_c = 109.05 \angle 82.5° \text{ A}$$

The phasor diagram is shown in Figure 7.24.

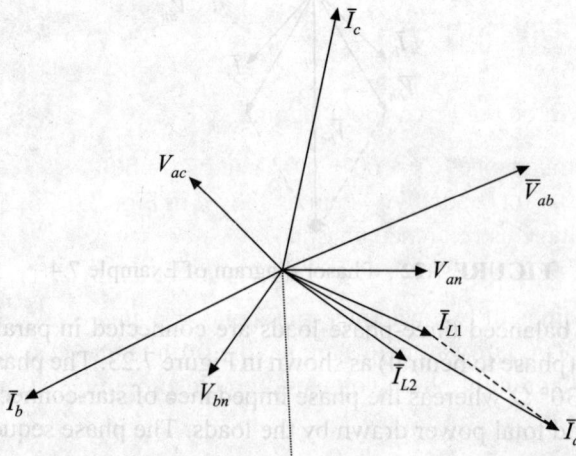

FIGURE 7.24 Phasor diagram of Example 7.5.

$$\text{Total power} = 3\,V_{an}I_a \cos \angle \frac{V_{an}}{I_a} = 3 \times 220 \times 109.05 \times \cos(-37.5) = 57.1\,\text{kW}$$

This problem can also be solved, using single phase analysis. The delta load can be converted to equivalent star load which will be $\overline{Z}/3$ as shown in Figure 7.25.

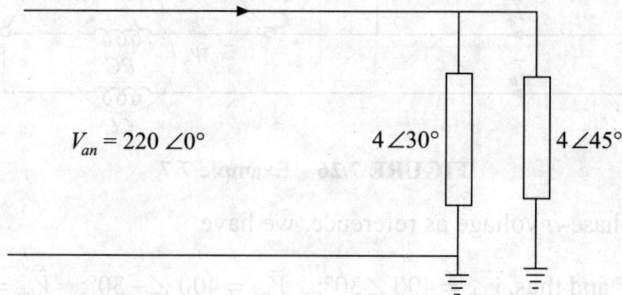

$V_{an} = 220 \angle 0°$ $4\angle 30°$ $4\angle 45°$

FIGURE 7.25   Single circuit equivalent of Example 7.5.

Thus, the equivalent load impedance will be

$$Z_{eq} = \frac{4\angle 30 \times 4\angle 45}{4\angle 30 + 4\angle 45} = 2.02 \angle 37.5° \,\Omega = R_{eq} + jX_{eq}$$

$$I_L = \frac{220 \angle 0}{2.017 \angle 37.5} = 109.05 \angle -37.5° \,\text{A}$$

EXAMPLE 7.6   In three-phase system (balanced load and supply) power measurement using two wattmeters method, one wattmeter reads zero, whereas second reads 100 kW. Find the total power and factor of load.

Solution: Total power $(W) = W_1 + W_2 = 100 + 0 = 100\,\text{kW}$
Thus the power factor angle (θ) will be

$$\theta = \tan^{-1}\sqrt{3}\left(\frac{W_1 - W_2}{W_1 + W_2}\right) = \tan^{-1}\sqrt{3} = 60°$$

Thus, the power factor will be 0.5 (= cos 60°) lagging.

EXAMPLE 7.7   Two impedances $Z_1 = 60 + j80\,\Omega$ and $Z_2 = 50\,\Omega$ are connected with a three-phase 400 V (line-to-line), 50 Hz supply system as shown in Figure 7.26. The phase sequence is *abc*. Assume phase-*a* voltage as reference phasor.

   (a) Find out the line currents.
   (b) Find out the readings of two wattmeters (W_1 and W_2) connected as shown. Assume that the voltage coils (*PC*) of wattmeters are having infinite impedance.
   (c) Draw a phasor diagram of phase voltages and line currents.

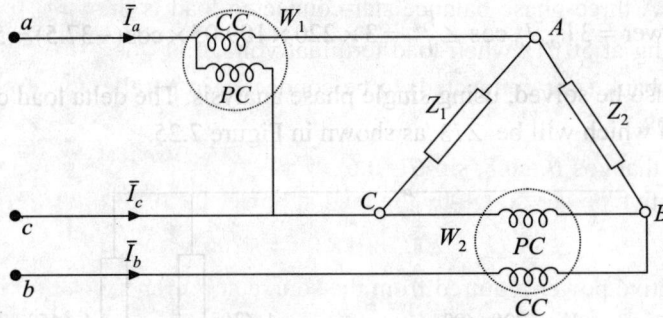

FIGURE 7.26 Example 7.7.

Solution: Taking phase-*a* voltage as reference, we have

$$\overline{V}_a = \frac{400}{\sqrt{3}} \angle 0° \text{ and thus, } \overline{V}_{ab} = 400 \angle 30°; \quad \overline{V}_{ac} = 400 \angle -30°; \quad \overline{V}_{bc} = 400 \angle -90°$$

(a) The current in line-*a* will be

$$\overline{I}_a = \frac{400 \angle 30°}{50} + \frac{400 \angle -30°}{60 + j80} = 7.406 \angle 0.22° \text{ A}$$

Since phase *bc* is open (infinite impedance), the current in this phase will be zero. Thus,

$$\overline{I}_c = -\frac{400 \angle -30°}{60 + j80} = -4 \angle -83.13° = 4 \angle 96.87° \text{ A}$$

$$\overline{I}_b = -\frac{400 \angle 30°}{50} = -8 \angle 30° = 8 \angle -150° \text{ A}$$

(b) Wattmeters readings will be

$$W_1 = |\overline{V}_{ac}||\overline{I}_a| \cos \angle^{\overline{V}_{ac}}_{\overline{I}_a} = 400 \times 7.406 \times \cos(30° + 0.22°) = 2571.18 \text{ W}$$

$$W_2 = |\overline{V}_{bc}||\overline{I}_b| \cos \angle^{\overline{V}_{bc}}_{\overline{I}_b} = 400 \times 8 \times \cos(90° - 150°) = 1600 \text{ W}$$

(c) Phasor diagram is shown in Figure 7.27.

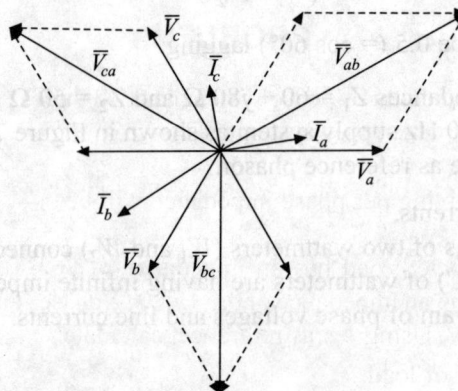

FIGURE 7.27 Phasor diagram of Example 7.7.

EXAMPLE 7.8 A three-phase balance star-connected load is drawing 10 A per phase at 0.8 power factor lagging at 50 Hz when load terminal voltage is $200\sqrt{3}$ V (line-to-line). Find the value of capacitor bank (star-connected) to be connected in parallel to improve the load power factor to 0.9 lagging.

Solution: Given that cos θ = 0.8, sin θ = 0.6

The per phase reactive power drawn by the load at power factor of 0.8 lagging will be

$$Q_0 = 200 \times 10 \times 0.6 = 1200 \text{ VAr}$$

The per phase reactive power required from the source at power factor of 0.9 lagging

$$Q_1 = 200 \times 10 \times \sin(\cos^{-1} 0.9) = 871.8 \text{ VAr}$$

Thus, the capacitor should provide the remaining reactive power (per phase) of 328.2 (= 1200 – 871.8) VAr. The single-phase equivalent is shown in Figure 7.28.

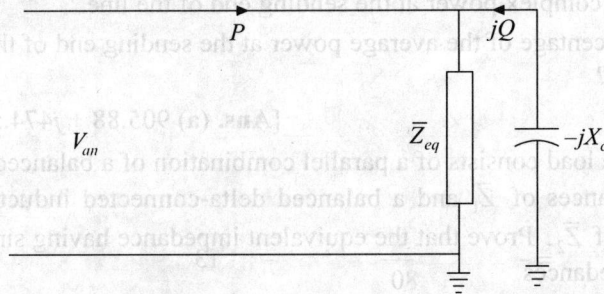

FIGURE 7.28 Single-phase equivalent circuit.

The value of capacitor will be calculated as

$$328.2 = \frac{V^2}{X_c} = V^2 \omega C$$

$$C = \frac{328.2}{2\pi \times 50 \times 200^2} = 1305.95 \text{ } \mu\text{F per phase}$$

PROBLEMS

7.1 Show that if a balanced three-phase system has two three-phase balanced load connected in parallel, having per phase impedances $\bar{Z}_1$ and $\bar{Z}_2$, respectively, the equivalent three-phase load having per phase impedance will be $\dfrac{\bar{Z}_1 \bar{Z}_1}{\bar{Z}_1 + \bar{Z}_2}$.

7.2 A balance Δ-connected load has an impedance of 864-j252 Ω/phase. The load is fed through a line having an impedance of 0.5 + j4.0 Ω/phase. The line voltage at the terminals of the load is 69 kV. Using V_{ab} as reference, calculate

(a) Phase currents of load

(b) Line currents

(c) Sending end line voltage

(d) Power consumed by load, power loss in line.

[**Ans.** (a) 76.67 $\angle 16.26°$, 76.67 $\angle -103.74°$, 76.67 $\angle 136°$ A

(b) 132.8 $\angle -13.73°$, 132.8 $\angle -133.74°$, 132.8 $\angle 106.26°$ A

(c) 68.86 $\angle 0.76°$, 68.86 $\angle -119.24°$, 68.86 $\angle 120.76°$ kV

(d) 15.235 MW, 8.82 kW/phase]

7.3 Three balanced three-phase loads are connected in parallel. Load-1 is Y-connected with an impedance of $300 + j150$ Ω/phase; load-2 is Δ-connected with an impedance of $3600-j2700$ Ω/phase and load-3 is 450 kVA at 0.8 p.f. lagging. The load is fed from a distribution line having an impedance of $1 + j8$ Ω/phase. The magnitude of the line voltage with neutral voltage at load end of the line is 7.5 kV. Calculate

(a) The total complex power at the sending end of the line.

(b) What percentage of the average power at the sending end of the line is delivered to the loads?

[**Ans.** (a) $905.88 + j474.56$ kVA; (b) 99.35%]

7.4 A three-phase load consists of a parallel combination of a balanced star-connected with phase impedances of $\overline{Z}_1$ and a balanced delta-connected inductive load with phase impedances of $\overline{Z}_2$. Prove that the equivalent impedance having single three-phase load of phase impedances

(a) $\dfrac{\overline{Z}_1\overline{Z}_1}{3\overline{Z}_1 + \overline{Z}_2}$ connected in star (b) $\dfrac{3\overline{Z}_1\overline{Z}_1}{3\overline{Z}_1 + \overline{Z}_2}$ connected in delta.

7.5 A balanced delta connected load has per phase impedances of $8 + j6$ ohm and the line voltage is 400 V at the load terminals. Find the total power delivered to the load.

[**Ans.** 38.4 kW]

7.6 A balanced delta-connected load has a line voltage of 400 V at the load terminals and absorbs a total power of 48 kW. If the power factor of the load is 0.8 leading, find the phase impedance.

[**Ans.** $6.4 - j\,4.8$ ohm]

7.7 A balanced delta-connected load has real part of 14 ohm. If the load consumes 1 kW at a power factor of 0.5 leading, find the load impedance and reactive power delivered to the load.

[**Ans.** $14 - j14\sqrt{3}$ ohm, $\sqrt{3}$ kVA]

7.8 Calculate the reading of each wattmeter in the circuit shown in Figure 7.29. The value of $\overline{Z} = 40 \angle -30°\ \Omega$. Verify the sum of wattmeter readings that equals the total average power delivered to the Δ-connected load.

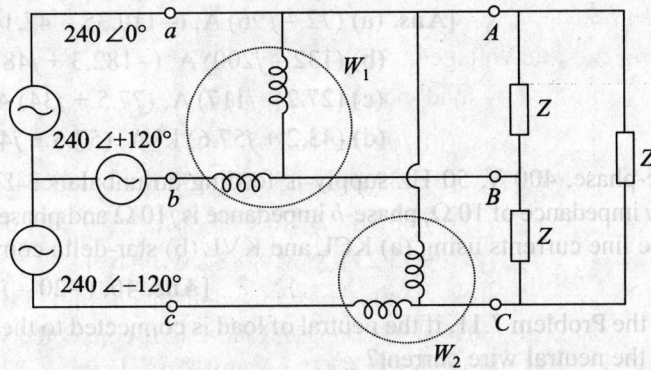

FIGURE 7.29   Problem 7.8.

[Ans. 7482 W, 3741.3 W]

7.9 In balance three-phase circuit, the current and voltage coils of the wattmeter are connected as shown in Figure 7.30. Show that $\sqrt{3}$ times of wattmeter reading equals the total reactive power consumed by the load.

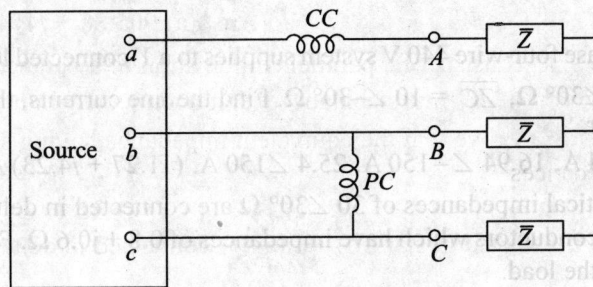

FIGURE 7.30   Problem 7.9.

7.10 For the circuit shown in Figure 7.31, calculate the following, if $\bar{Z}_1 = 3 + j4$ ohm, $\bar{Z}_2 = 4 + j3$ ohm $\bar{Z}_3 = 5$ ohm,

(a) Load phase currents
(b) Line currents
(c) Source phase currents
(d) Complex power in each phase of the load

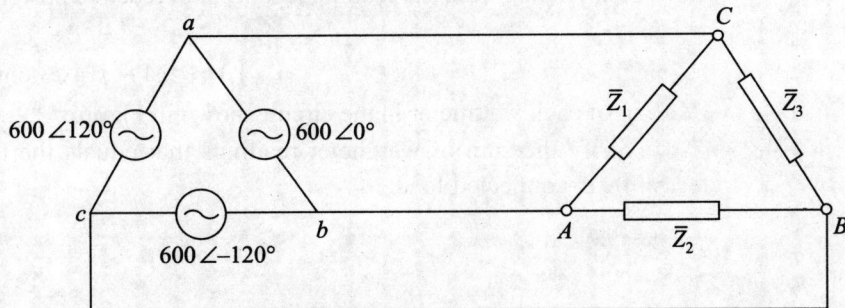

FIGURE 7.31   Problem 7.10.

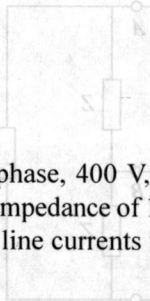

[**Ans.** (a) $(72 - j96)$ A, $(-110.35 - 47.14)$ A, $(-60 + j103.9)$ A

(b) $(132 - j200)$ A, $(-182.3 + j48.8)$ A, $(50.35 + j15)$ A

(c) $(27.2 - j117)$ A, $(77.5 + j34)$ A, $(-104.8 + j83)$ A

(d) $(43.2 + j57.6)$ kVA, $(57.6 + j43.2)$ kVA, 72 kVA]

7.11 A three-phase, 400 V, 50 Hz supply is feeding an unbalance Y-connected load having phase-a impedance of 10 Ω, phase-b impedance is $j10$ Ω and phase-c impedance is $-j10$ Ω. Find the line currents using (a) KCL and KVL (b) star-delta conversion.

[**Ans.** 40, $(-20 - j5.36)$, $(-20 + j5.36)$ A]

7.12 Repeat the Problem 7.11, if the neutral of load is connected to the neutral of supply. What will be the neutral wire current?

[**Ans.** 23.1 A, $(-20 + j11.55)$ A, $(-20 - j11.55)$ A, -16.9 A]

7.13 A three-phase balance supply is fed to a balance load having impedance Z per phase, connected in delta. A wattmeter with current coil connected in phase A and voltage coil connected between phases B and C reads W watt. Find the total reactive power consumed by the load in terms of wattmeter reading.

[**Ans.** $\sqrt{3}$ W]

7.14 A three-phase four-wire 440 V system supplies to a Y-connected load with $\overline{ZA} = 10 \angle 0°$ Ω, $\overline{ZB} = 15 \angle 30°$ Ω, $\overline{ZC} = 10 \angle -30°$ Ω. Find the line currents, the neutral current and the total power.

[**Ans.** 25.4 A, $16.94 \angle -150$ A, $25.4 \angle 150$ A, $(11.27 + j4.23)$ A, $(15.77 - j1.073)$ kVA]

7.15 Three identical impedances of $30 \angle 30°$ Ω are connected in delta to a three-phase 220 V system by conductors which have impedances of $0.8 + j0.6$ Ω. Find the magnitude of line voltage at the load.

[**Ans.** $V_{ab} = 213.32 \angle 0.59°$ V, taking V_{ab} as reference]

8

Resonance

8.1   INTRODUCTION

Resonance is not a property of electrical circuits only. It can occur in any physical system, provided that the system has opposite types of energy storage devices. Whenever the energy stored in these devices is exchanged between them so that added energy is not taken from a source, a resonant condition exists. In electrical system, capacitor and inductor are two such devices which store energy in different forms. The inductive and capacitive reactances are opposite to each other in two senses: (a) the magnitude of inductive reactance (X_L) increases with frequency, while the magnitude of capacitive reactance (X_C) decreases with frequency, (b) the angle of X_L is $+\,90°$ whereas angle of X_C is $-\,90°$. At some frequency, the inductive reactance is equal to the capacitive reactance. When circuit is operated at a frequency where the reactive (imaginary) component of the total impedance is zero, the circuit is said to be resonant. Note that the impedance circuit at resonance is purely resistive, (i.e. the voltage and current are in phase). Resonance is a very useful property for many electrical applications. In a radio receiver, a resonant circuit is used to select a single signal from many radio waves which come to antenna. The extreme degree of frequency selectivity achievable is one of the key properties of electrical circuits.

8.2   SERIES *RLC* RESONANCE

One circuit capable of exhibiting resonance is the series *RLC* circuit as shown in Figure 8.1. A variable frequency voltage source is applied to the circuit. The total impedance of this circuit at a frequency $\omega\,(=2\pi f)$ will be

$$\bar{Z} = R + j\omega L - j\,\frac{1}{\omega C} \tag{8.1}$$

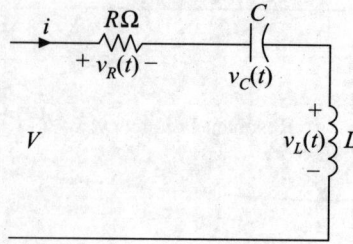

FIGURE 8.1   Series *RLC* circuit.

The frequency, ω_r, at which resonance occurs, (i.e. the imaginary part of impedance is zero) can be obtained as

$$\omega_r L - \frac{1}{w_r L} = 0 \tag{8.2}$$

or

$$\omega_r = \frac{1}{\sqrt{LC}} \text{ rad/s} \tag{8.3}$$

Converting angular frequency to frequency in Hz, we get

$$f_r = \frac{1}{2\pi\sqrt{LC}} \text{ Hz} \tag{8.4}$$

The frequency specified by Eq. (8.3) or (8.4) are called resonant frequency of series *RLC* circuit. It can be seen that resonance frequency does not depend on the circuit resistance. At resonance, in a series *RLC* circuit, the following observations can be made:

(a) The impedance of the circuit becomes resistive and impedance becomes minimum $(\bar{Z} = R)$.

(b) The current in the circuit becomes maximum.

(c) Voltage across the capacitor becomes equal and opposite of the voltage across inductor.

(d) The magnitude of capacitive reactance becomes equal to the magnitude of inductive reactance.

Current in the circuit, connected with the supply voltage V, at resonance can be written as

$$\bar{I} = \frac{V \angle 0}{\bar{Z}} = \frac{V \angle 0}{R} \tag{8.5}$$

The series resonant circuit is often described as an *acceptor circuit* since it has its minimum impedance and thus maximum current at the resonant frequency. Figure 8.2 shows the graphs of current I and impedance with respect to the frequency.

The voltage across inductor is

$$\bar{V}_L = jIX_L = \frac{VX_L}{R} \angle 90° \text{ and} \tag{8.6}$$

voltage across capacitor is

$$\bar{V}_C = -jIX_C = \frac{VX_C}{R} \angle -90° \tag{8.7}$$

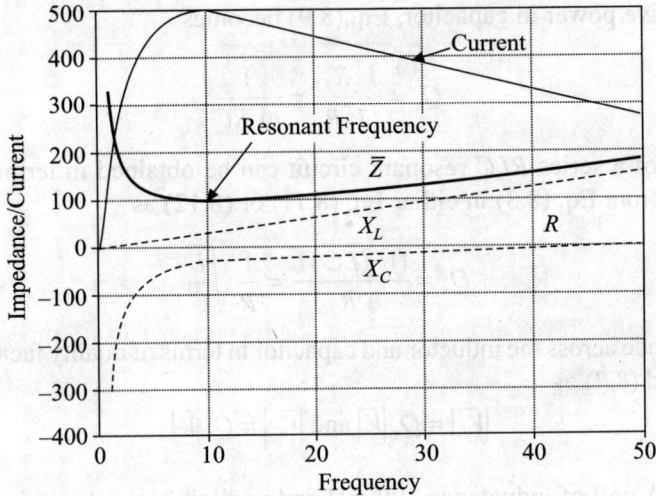

FIGURE 8.2 Impedances and current in series *RLC* circuit.

8.2.1 Quality Factor

If R is small compared with X_L and X_C, it is possible for V_L and V_C to have voltage many times greater than the supply voltage. The voltage magnification at resonance or quality factor is defined as

$$\text{Quality factor } (Q_s) = \frac{\text{Voltage across } L \text{ (or } C)}{\text{Supply voltage } V} = \frac{V_L}{V} = \frac{X_L}{R} = \frac{\omega_r L}{R} = \frac{1}{\omega_r RC} \qquad (8.8)$$

In other way, the quality factor of a series *RLC* circuit is defined as the ratio of the reactive power in either the inductor or the capacitor to the average power at resonance and can be written as

$$Q_s = \frac{\text{reactive power at resonance}}{\text{average power at resonance}} \qquad (8.9)$$

From Eq. (8.9), it is clear that quality factor is large when the average power is small. It indicates that quality of a circuit is high when energy dissipated in the circuit is small. However, the quality factor is used as an indicator of other circuit characteristics than energy dissipation. The reactive power in inductor and capacitor can be written as

$$\text{Reactive power} = \begin{cases} I^2 X_L & \text{VAr (inductor)} \\ I^2 X_C & \text{VAr (capacitor)} \end{cases} \qquad (8.10)$$

The average power (real) is $I^2 R$. Thus, Eq. (8.9) using reactive power in the inductor can be written as

$$Q_s = \frac{I^2 X_L}{I^2 R} = \frac{\omega_r L}{R} \qquad (8.11)$$

Similarly, for reactive power in capacitor, Eq. (8.9) becomes

$$Q_s = \frac{I^2 X_C}{I^2 R} = \frac{1}{\omega_r RC} \qquad (8.12)$$

The quality factor of a series RLC resonant circuit can be obtained in terms of R, L, and C by substituting for ω_r from Eq. (8.3) in either Eq. (8.11) or (8.12) as

$$Q_s = \frac{(1/\sqrt{LC})L}{R} = \frac{1}{R}\sqrt{\frac{L}{C}} \qquad (8.13)$$

The voltage magnitude across the inductor and capacitor in terms of quality factor can be expressed, using Eqs. (8.6) and (8.7) as

$$|\bar{V}_L| = Q_s|\bar{V}| \text{ and } |\bar{V}_C| = Q_s|\bar{V}| \qquad (8.14)$$

EXAMPLE 8.1   A coil of inductance 100 mH and negligible resistance is connected in series with a resistor of 12.5 ohm and a capacitor of 2.5 µF. The supply voltage is 25 V with variable frequency. Find

(a) The resonant frequency
(b) The current at resonant
(c) Voltage across coil and capacitor at resonant
(d) Quality factor of the circuit.

Solution:

(a) resonant frequency $f_r = \dfrac{1}{2\pi\sqrt{LC}} = \dfrac{1}{2\pi\sqrt{100\times10^{-3}\times2.5\times10^{-6}}}$

or $f_r = \dfrac{1}{2\pi\times5\times10^{-4}} = 318.31\,\text{Hz}$

(b) The current at resonance (V/R) = 25/12.5 = 2 A

(c) The voltage across inductor $(V_L) = IX_L = 2\times2\pi\times318.31\times100\times10^{-3} = 400$ V

The voltage across capacitor $(V_C) = IX_C = 2/(2\pi\times318.31\times2.5\times10^{-6}) = 400$ V

(d) Quality factor = 400/25 = 16.

8.2.2   Bandwidth and Selectivity

It is established that the current in a series RLC circuit is maximum at the resonance and falls off at frequencies above and below resonance frequency. Bandwidth of any resonant circuit is defined as the range of frequencies between half-power points (also known as *cut-off frequencies*). These are the points around the resonant frequency where the current is $1/\sqrt{2}$ (= 0.707) of maximum current, (i.e gain of 3dB above or below the resonant frequency). For this to occur, the impedance must be equal to $\sqrt{2}$ times the impedance at resonance. Thus, the total impedance will be $\sqrt{2}\,R$. In order for Z to be equal to $\sqrt{2}\,R$, the net reactance must be equal to R. If f_1 is the frequency below f_r for which $I = \sqrt{2}\,I_{max}$ and frequency above the resonance frequency at which current again falls to $\sqrt{2}\,I_{max}$ is f_2, we get

$$f = f_1: \quad \overline{Z}_T = \sqrt{2}\,R = R - j(X_C - X_L) \tag{8.15}$$

$$f = f_2: \quad \overline{Z}_T = \sqrt{2}\,R = R + j(X_L - X_C) \tag{8.16}$$

Mathematically, bandwidth (BW) is defined as

$$\mathrm{BW} = f_2 - f_1 = \frac{f_r}{Q_s} = \frac{R}{2\pi L} \tag{8.17}$$

Figure 8.3 shows the current and phase shift variation with frequency in a series *RLC* circuit. At the frequencies f_1 and f_2, the net reactance equals the resistance of the circuit. The frequencies f_1 and f_2 define the bandwidth of the resonant circuit. The resonant frequency is often called the centre frequency of the circuit, although it is not necessarily midway between the cut-off frequencies. The resonant frequency f_r is at the geometric centre of response and is defined as

$$f_r = \sqrt{f_1 f_2} \tag{8.18}$$

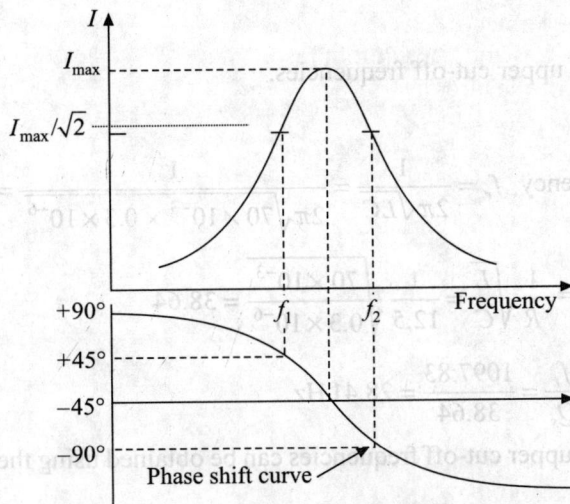

FIGURE 8.3   Current and phase shift variation with frequency.

Bandwidth is simply one means of defining the selectivity of a resonant circuit. Selectivity is a measure of the ability of a circuit to reject frequencies above and below resonant frequency. Selectivity is proportional to quality factor Q_s, while bandwidth is inversely proportional to the quality factor Q_s, as shown in Figure 8.4. Sometimes extra resistance is added to reduce the quality factor of circuit, and thus the more bandwidth.

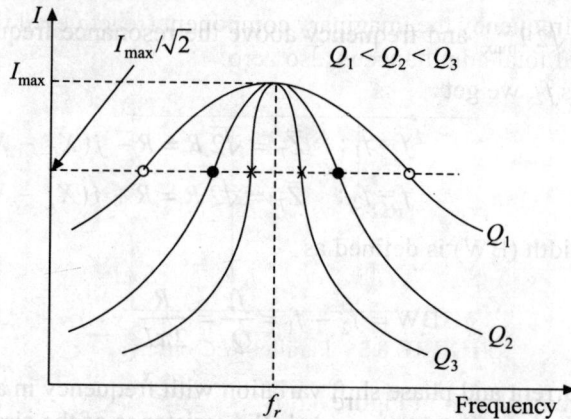

FIGURE 8.4 Effect of quality factor on bandwidth.

EXAMPLE 8.2 A coil of inductance 70 mH and negligible resistance is connected in series with a resistor of 12.5 ohm and a capacitor of 0.30 µF. Find

(a) The resonant frequency
(b) Quality factor
(c) Bandwidth
(d) The lower and upper cut-off frequencies.

Solution:

(a) resonant frequency $f_r = \dfrac{1}{2\pi\sqrt{LC}} = \dfrac{1}{2\pi\sqrt{70\times10^{-3}\times0.3\times10^{-6}}} = 1097.83$ Hz

(b) Quality factor $= \dfrac{1}{R}\sqrt{\dfrac{L}{C}} = \dfrac{1}{12.5}\sqrt{\dfrac{70\times10^{-3}}{0.3\times10^{-6}}} = 38.64$

(c) Bandwidth $= \dfrac{f_r}{Q_s} = \dfrac{1097.83}{38.64} = 28.41$ Hz

(d) The lower and upper cut-off frequencies can be obtained using the impedance equation $R = (X_L - X_C)$
Using Eq. (8.18), we get

$$f_1 f_2 = f_r^2 = 1205234.16 \text{ and from bandwidth BW} = f_2 - f_1 = 28.41 \text{ Hz}$$

Thus, we get, $f_1 = 1083.72$ Hz and $f_2 = 1112.13$ Hz.

8.3 PARALLEL *RLC* RESONANCE

A parallel *RLC* circuit as shown in Figure 8.5 is often called *tank circuit*. It is assumed that inductor coil has zero resistance (ideal inductor). As per definition of resonance in the previous section that at resonance frequency the imaginary component (reactive) of the total impedance is zero, the imaginary part of total admittance is also zero.

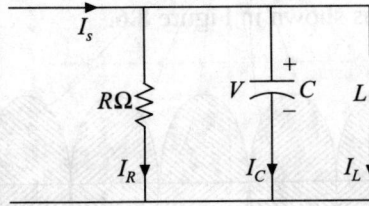

FIGURE 8.5   Parallel *RLC* circuit.

The admittance of the circuit shown in Figure 8.5 is

$$\bar{Y}_T = \frac{1}{R} + j\omega C - \frac{j}{\omega L} \tag{8.19}$$

At resonant frequency $\omega_r (= 2\pi f_r)$, we get

$$\omega_r C - \frac{1}{\omega_r L} = 0 \tag{8.20}$$

or

$$\omega_r = \frac{1}{\sqrt{LC}} \text{ rad/s} \tag{8.21}$$

and

$$f_r = \frac{1}{2\pi \sqrt{LC}} \tag{8.22}$$

From Eqs. (8.21) and (8.22), it is clear that resonance frequency is calculated in the same way as in a series *RLC* circuit resonance because at resonance, the inductive reactance (X_L) equals the capacitive reactance (X_C). However, as frequency changes, the impedance variation of the two circuits are quite different. It can be seen from the following expressions, that at the resonance, the total impedance reaches a maximum and the total current reaches a minimum value, whereas these variations are opposite in a series *RLC* circuit.

The parallel equivalent impedance of the inductor and capacitor will be

$$X_L \| X_C = \frac{(jX_L) \times (-jX_C)}{jX_L - jX_C} = \frac{-jX_L X_C}{(X_L - X_C)} \tag{8.23}$$

Since at resonance, $X_L = X_C$, the denominator of Eq. (8.23) becomes zero, thus, the *LC* combination has theoretically infinite impedance. In practical circuit, it is not true as inductor will have some internal resistance. At resonance, in a parallel *RLC* circuit, the following observation can be made:

(a) The impedance of the circuit becomes resistive and impedance becomes maximum
(b) The current in the circuit reaches a minimum value.
(c) Current through the capacitor becomes equal and opposite of the current through inductor.
(d) The capacitive reactance becomes equal to the inductive reactance.

(e) Voltage across each element of *RLC* circuit is equal.

(f) *LC* circuit does not draw any current from the source. Instead, the branch currents circulate between the inductance and the capacitance. This current which is known as *tank current*, is a circulating current as shown in Figure 8.6.

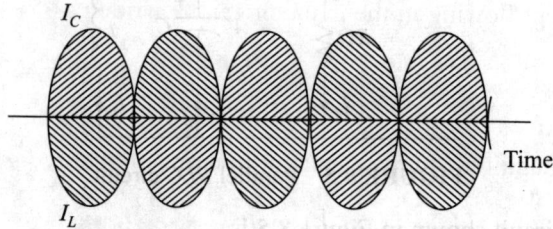

FIGURE 8.6   Circulating tank current.

At frequencies less than resonant frequency, the current in the inductive branch is greater than current in the capacitive branch and *LC* circuit appears as a pure inductance. However, at frequencies more than resonant frequency, the current in the inductive branch is lesser than current in the capacitive branch and *LC* circuit appears as a pure capacitance.

8.3.1 Quality Factor

Using the general definition of quality factor as defined in Eq. (8.9), the quality factor of a parallel *RLC* circuit will be the ratio of reactive power in either *L* or *C* at resonant frequency to the average power in *R*. Since the voltage across each element of parallel *RLC* circuit is the same, the reactive power and real power across the element can be expressed as

$$\text{Reactive power} = \begin{cases} V^2/X_L & \text{VAr (inductor)} \\ V^2/X_C & \text{VAr (capacitor)} \end{cases} \tag{8.24}$$

The average power (real) is V^2/R. Thus, Eq. (8.24) using reactive power in the inductor can be written as

$$Q_p = \frac{V^2/X_L}{V^2/R} = \frac{R}{\omega_r L} \tag{8.25}$$

Similarly, for reactive power in capacitor, Eq. (8.24) becomes

$$Q_p = \frac{V^2/X_C}{V^2/R} = \omega_r RC \tag{8.26}$$

The quality factor of a parallel resonant circuit can be obtained in terms of *R, L*, and *C* by substituting for ω_r from Eq. (8.21) in either Eq. (8.25) or (8.26) as

$$Q_p = \frac{R}{(1/\sqrt{LC})L} = R\sqrt{\frac{C}{L}} \tag{8.27}$$

It should be noted that Q_p is computed differently compared to Q_s. The total current drawn from the source is

$$I_s = I_R = \frac{V}{R} \tag{8.28}$$

The magnitude of current flowing in the inductor can be written as

$$|\overline{I}_L| = \left|\frac{V}{j\omega L}\right| = \left|\frac{\overline{I}_s R}{j\omega L}\right| = Q_p |\overline{I}_s| \tag{8.29}$$

Similarly, the current magnitude in the capacitor in terms of quality factor can be expressed as

$$|\overline{I}_C| = |j\omega CV| = |j\omega C\overline{I}_s R| = Q_p |\overline{I}_s| \tag{8.30}$$

From Eqs. (8.29) and (8.30), it can be seen that the magnitude of currents in the reactive components at resonance is Q_p times the magnitude of the total current at resonance. It is similar to the relation between Q_s and voltage magnitudes in a series resonant circuit. Note that in a parallel circuit, the Q-factor is a measure of current magnification, whereas in a series circuit, it is a measure of voltage magnification. At supply frequency, the Q-factor of a parallel circuit is usually low (typically less than 10) but in radio-frequency circuits, the Q-factor can be very high.

8.3.2 Selectivity and Bandwidth

Bandwidth, as always, is measured between half-power points (cut-off frequencies). These frequencies are calculated when the voltage-drop across the parallel circuit is $\sqrt{2}$ of the voltage as resonance. The solution for f_1 (lower frequency) and f_2 (upper frequency) is similar to the calculation for series resonant circuits.

$$BW = f_2 - f_1 = \frac{f_r}{Q_p} \tag{8.31}$$

The cut-off frequencies are related to the resonant (centre) frequency as

$$f_r = \sqrt{f_1 f_2} \tag{8.32}$$

If Q_p is large, the cut-off frequencies are at approximately one-half the bandwidth below and above the resonant (centre) frequency and thus,

$$f_1 \approx f_r - \frac{BW}{2} \tag{8.33}$$

and

$$f_2 \approx f_r + \frac{BW}{2} \tag{8.34}$$

Thus,

$$f_r = \frac{f_1 + f_2}{2} \tag{8.35}$$

EXAMPLE 8.3 In an ideal parallel resonant circuit, the resistance, inductance and capacitance are of 500 ohm, 0.1 mH and 0.1 μF, respectively. The circuit is connected through a 10 mA source. Find

(a) Resonant frequency and quality factor
(b) Magnitude of maximum output voltage
(c) Cut-off frequencies and bandwidth.

Solution:

(a) Resonance frequency f_r will be

$$f_r = \frac{1}{2\pi\sqrt{LC}} = \frac{1}{2\pi\sqrt{0.1\times10^{-3}\times0.1\times10^{-6}}} = 50.33 \text{ kHz}$$

$$Q_p = R\sqrt{\frac{C}{L}} = 500\times\sqrt{\frac{0.1\times10^{-6}}{0.1\times10^{-3}}} = 15.81$$

(b) The output voltage is maximum at resonance and will be

$$I\times R = 10\times10^{-3}\times500 = 5 \text{ V}$$

(c) From Eqs. (8.32) and (8.35), we get

$$f_1 = 48.76 \text{ kHz} \quad \text{and} \quad f_2 = 51.94 \text{ kHz}$$

Thus, the bandwidth will be $\dfrac{f_r}{Q_p} = \dfrac{50.33}{15.81} = 3.18$ kHz.

8.4 PRACTICAL PARALLEL RESONANT CIRCUITS

In an ideal parallel resonance circuit, an ideal inductor was assumed but it has some winding resistance which makes the ideal equation deviates significantly from reality. Figure 8.7 shows an inductor with its winding resistance in a parallel resonance circuit.

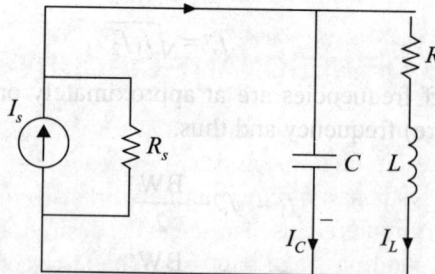

FIGURE 8.7 Series-parallel (practical) resonant circuit.

The total admittance of the circuit shown in Fig. 8.7 is

$$\bar{Y}_T = \frac{1}{R_s} + j\omega C + \frac{1}{R + j\omega L} = \frac{1}{R_s} + j\omega C + \frac{R - j\omega L}{R^2 + (\omega L)^2} \tag{8.36}$$

At resonant frequency $\omega_r(= 2\pi f_r)$ (imaginary part will be zero), we get

$$\omega_r C - \frac{\omega_r L}{R^2 + (\omega_r L)^2} = 0 \quad \Rightarrow \quad R^2 C + \omega_r^2 L^2 C = L \tag{8.37}$$

or

$$\omega_r = \frac{1}{\sqrt{LC}}\sqrt{1 - \frac{R^2 C}{L}} = \frac{1}{L}\sqrt{\frac{L}{C} - R^2} \tag{8.38}$$

and

$$f_r = \frac{1}{2\pi L}\sqrt{\frac{L}{C} - R^2} \tag{8.39}$$

For small value of R, Eq. (8.39) becomes Eq. (8.22) which is for ideal parallel resonance. Note that the effect of R can be so great as to prevent resonance. Whenever $R^2 \geq L/C$ the circuit cannot resonate. This is the reason why a high L to C ratio is desirable in a practical parallel resonant circuit.

The quality factor (Q_{pr}) is defined as the ratio of the reactive power in the equivalent inductance to the average power in the equivalent resistance at resonance and can be written as

$$Q_{pr} = \frac{V^2 \omega_r L/[R^2 + (\omega_r L)^2]}{V^2 R/[R^2 + (\omega_r L)^2]} = \frac{\omega_r L}{R} \tag{8.40}$$

Equivalently, the ratio of the reactive power in the capacitor to the average power in the equivalent resistance at resonance is

$$Q_{pr} = \frac{V^2 \omega_r C}{V^2 R/[R^2 + (\omega_r L)^2]} = \frac{\omega_r C}{R}[R^2 + (\omega_r L)^2] \tag{8.41}$$

From Eqs. (8.37), (8.40) and (8.41), we can get

$$f_r = \frac{1}{2\pi L}\sqrt{\frac{Q_{pr}^2}{Q_{pr}^2 + 1}} \text{ Hz} \tag{8.42}$$

8.5 APPLICATIONS OF RESONANT CIRCUITS

In high-frequency applications, resonant circuits have a variety of applications when frequency selectivity is needed. These circuits are used as frequency selective filters and loads in radio-frequency amplifier circuits as a part of coupling networks. To tune a radio receiver or to change channels in TV set, resonant circuits are used to select the desired station and reject others. Resonant circuits are widely used as band-pass and band-stop (band-reject) filters.

EXAMPLE 8.4 A coil of inductance 100 mH and resistance 800 ohm is connected with a variable capacitor across a 10 V, 5 kHz. Find

(a) Capacitance value when supply current is minimum
(b) Quality factor
(c) Minimum supply current.

Solution:

(a) The supply current is a minimum when the parallel circuit is at resonance and resonant frequency,

$$f_r = \frac{1}{2\pi L}\sqrt{\frac{L}{C} - R^2}$$

Thus,

$$C = \frac{1}{L\left\{(2\pi f_r)^2 + \left(\dfrac{R}{L}\right)^2\right\}} = \frac{1}{0.1\left\{4\pi^2 \times 5000^2 + \dfrac{800^2}{0.1^2}\right\}} = 9.515 \text{ nF}$$

(b) Quality factor $= \dfrac{2\pi f_r L}{R} = \dfrac{2\pi \times 5000 \times 0.1}{800} = 3.93$

(c) Supply current at resonance $= \dfrac{VR}{R^2 + (2\pi f_r L)^2} = \dfrac{10 \times 800}{800^2 + (2\pi \times 5000 \times 0.1)^2} = 0.761 \text{ mA.}$

PROBLEMS

8.1 A series circuit having resistance, and capacitance of 100 ohm and 0.159 pF, respectively, resonate at 10 MHz. Find the value of inductance. Calculate the quality factor of circuit.

[**Ans.** 1.59 mH, 1000]

8.2 A series resonant circuit of TV tuner operates from 4.1 to 5.8 MHz. The coil used in the tuner circuit has an inductance of 25 μH. Find the range of capacitive value of this circuit.

[**Ans.** 30.15–60.33 pF]

8.3 A coil of resistance 10 ohm and inductance 100 mH is connected in series with a capacitance of 2 μF across a 10 V variable frequency supply. Find (a) the resonant frequency, (b) the current at resonance, (c) voltage across capacitance at resonance, and (d) the quality factor of the circuit.

[**Ans.** 355.9 Hz, 1 A, 223.59 V, 22.36]

8.4 A 10 mH pure inductor and a capacitor of 0.15 μF are connected in parallel across 1 mA supply of variable frequency having 10 kΩ internal resistance. Find (a) resistant frequency, (b) quality factor and (c) the current circulating in the capacitor and inductor.

[**Ans.** 4.11 kHz, 38.74, 38.73 mA]

8.5 A series resonant *RLC* circuit is having resistance of 75 ohm, ideal inductor of 15 mH and capacitor of 20 nF. Find (a) resonant frequency, (b) quality factor, (c) the lower and upper cut-off frequencies, and (d) bandwidth .

[**Ans.** 9.189 kHz, 11.55, 8.8 kHz, 9.595 kHz, 795 Hz]

8.6 In Problem 8.5, find the voltage across resistance, inductor and capacitor at resonance. Take the supply voltage as 10 V. Draw the phasor diagram.

[**Ans.** 10 V, 115.5 V, 115.5 V]

8.7 In Problem 8.4, find the upper and lower cut-off frequencies, and bandwidth.

[**Ans.** 4163 Hz, 4057 Hz, 106.1 Hz]

9

Magnetic Circuit

9.1 INTRODUCTION

One of the main components of electric machines, apparatus and measuring instruments is permanent magnet or electromagnet which is a source of *magnetic field*. The intensity and direction of magnetic field at each of its points is defined by the flux density vector $\bar{B}$. The magnetic quantities such as magnetic flux (Φ), flux density (B), field intensity (H), permeability (μ) can be defined based on Ampere's law and his observation on forces between two current-carrying conductors.

Consider a single long conductor carrying current I as shown in Figure 9.1. Due to the current in the conductor, a magnetic field is setup around the conductor and force can be experienced if a unit charge or another current-carrying conductor is placed around it. The direction of the flux lines is given by *right-hand rule*, which states that if the conductor is held by right hand, with thumb pointing in the current direction, the flux direction will be in the direction of fingers. A line of flux is a closed path around the current such that magnetic force is tangential to it at all points around the line. It should be noted that the density of flux lines or flux density will often change from point to point. For example, flux density decreases

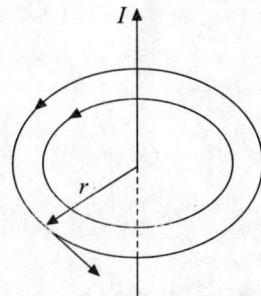

FIGURE 9.1 Line of fluxes.

as the point moves radially outward from the conductor, whereas flux density is constant at any point on a circle concentric with the conductor. Mathematically, magnetic flux is defined as

$$\Phi = \oint_S \bar{B} \cdot d\bar{A} \tag{9.1}$$

where Φ is the total number of lines (Wb), $\bar{B}$ is the flux density vector (in Wb/m² or Tesla). The vector $d\bar{A}$ has a magnitude dA and a direction normal to area dA.

If at all the points of a surface A, the magnetic field has the same induction B and is directed normally to this surface, the magnetic flux in that area is

$$\Phi = B \times A \tag{9.2}$$

9.2 AMPERE'S LAW

The line integral of the magnetic field intensity H around a closed path is equal to the total current enclosed by the path. Mathematically, it is defined as

$$\oint_l \bar{H} \cdot d\bar{l} = \Sigma i = i_1 - i_2 + i_3$$

or

$$\oint H \cdot dl \cdot \cos \theta = \Sigma i$$

where θ is the angle between vectors $\bar{H}$ and $d\bar{l}$ as shown in Figure 9.2. The quantity $\oint \bar{H} \cdot d\bar{l}$ is called the *magnetomotive force* (*mmf*) which is analogous to the electromotive force in the electric circuits.

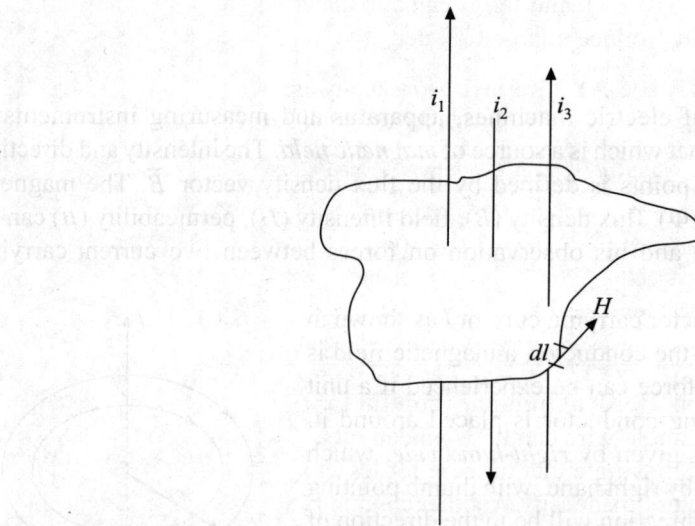

FIGURE 9.2   Magnetic flux intensity.

If a magnetic field is set up by a current i in coil of N turns, then for any contour embracing all the turns of coil

$$\sum i = Ni \tag{9.3}$$

For a long wire carrying a current i, the magnetic field intensity at distance r will be obtained as

$$H \cdot 2\pi r = i$$

Because of symmetry, the intensity of all the points of circle of radius r must be same. The vector $\bar{H}$ and $d\bar{l}$ have identical directions. The length of path of integration is $2\pi r$. Thus, the magnetic field intensity will be

$$H = i/2\pi r \tag{9.4}$$

Please note that field intensity is independent of medium and its unit is Ampere-turn (AT)/unit length. Magnetic field intensity H produces a magnetic flux density B everywhere it exists. It is related as

$$B = \mu H \qquad (9.5)$$

where μ is the characteristic of medium, known as *permeability* of medium.

> *Permeability is a measure of the ability of a material to conduct magnetic line of forces.*

The permeability of any material is expressed in terms of a product $\mu_r \cdot \mu_0$ where the dimensionless quantity μ_r is known as *relative permeability* and defined as

$$\mu_r = \frac{\mu}{\mu_0} \qquad (9.6)$$

where, μ_0 is the permeability of free space (vacuum) and is equal to $4\pi \times 10^{-7}$ Wb/m.

The μ_r (relative permeability) for electrical conductor material (copper and aluminium) is unity. However, for various types of steel, it has an order of magnitude 300 to 80000 as shown in Table 9.1. Based on the relative permeability, a material can be classified as diamagnetic materials ($\mu_r \leq 1$), paramagnetic materials ($\mu_r \geq 1$), and ferromagnetic materials ($\mu_r \gg 1$). The large value of μ_r implies a small current can produce a large flux density.

TABLE 9.1 Relative permeability of ferromagnetic materials

Material	Relative Permeability
Nickel	50
Cast iron	1000
Cast steel	2000
Transformer steel	5000
Permalloy	30,000 to 8,0000

To create a strong magnetic field using one conductor configuration, many turns of a single wire can be wounded on a cylindrical form such a solenoid having either air core for lesser flux density or iron core for greater flux density. Another common form is the toroid in which the wire is wound on a doughnut-shaped core as shown in Figure 9.3.

When current i flows through a coil of N turns as shown in Figure 9.3, the magnetic flux is mostly confined to in the core material (flux outside the core is negligible). The total current encompassed by any circle within the body of the toroid is equal to Ni which is called *magneto-motive force* (*mmf*). Thus using relation (9.2), we can have

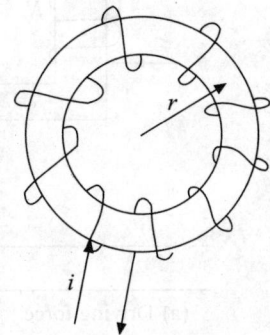

FIGURE 9.3 Toroidal coil.

$$H \cdot 2\pi r = Ni = mmf = F \qquad (9.7)$$

or

$$H = \frac{Ni}{2\pi r} \qquad (9.8)$$

The concepts developed from the toroidal ring are applicable with sufficient accuracy to widely differing geometries.

9.3 MAGNETIC RESISTANCE OR RELUCTANCE

The reluctance of any material is the ratio of the magnetizing force (*mmf*) to the magnetic flux and is defined as,

$$\Re = \frac{Ni}{\Phi} = \frac{F}{\Phi} \tag{9.9}$$

Using Eq. (9.5), we get

$$\Re = \frac{Ni}{B \times A} = \frac{Ni}{\mu H \times A} = \frac{l}{\mu \times A} \tag{9.10}$$

where A is the area of magnetic core, l is the length of the flux path in the material having magnetic permeability μ. The unit of reluctance is AT/Wb.

9.4 MAGNETIC CIRCUIT ANALYSIS

Consider a magnetic core of length l having relative permeability μ_r as shown in Figure 9.4. It has a small air gap of length l_a. The core is having N turns in which current i flows. Magnetic circuit has some analogy with electrical circuit. If the reluctance of the magnetic core is $\Re_i$, reluctance of air gap is $\Re_a$, and the *mmf* (F) is Ni, the equivalent electrical circuit can be shown as Figure 9.4. Now the analysis of equivalent electrical circuit is very easy. The analogies between magnetic and electrical quantities are shown in Table 9.2.

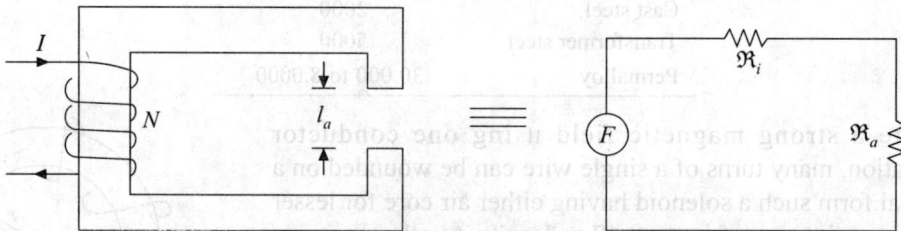

FIGURE 9.4 Magnetic circuit analysis.

TABLE 9.2 Magnetic and electrical analogy

	Electrical Case	Magnetic Case
(a) Driving force	Battery voltage (EMF) = V	Applied ampere turn (MMF) = F
(b) Response	Current = EMF/Resistance = V/R	Flux Φ = MMF/$\Re$ = $F/\Re$
(c) Impedance	• Resistance $R = \rho l/A$ • Conductance $(1/\rho)$, σ	• Reluctance $\Re = l/\mu A$ • Permeability (μ)
(d) Field	Electric field $E = V/l$ V/m, $\oint \overline{E} \cdot d\overline{l} = V$	Magnetic field intensity $H = F/l$ $\oint \overline{H} \cdot d\overline{l} = F$
(e) Density	Current density, $J = I/A$	Flux density, $B = \Phi/A$

It must be noted that nothing flows in the magnetic circuit. Another difference between electrical and magnetic circuit is that in the ferromagnetic materials permeability varies widely with flux density, whereas in most of the conductors, conductivity is independent of current density within normal operating range. However, analogy between electrical and magnetic circuit is very useful for analyzing the magnetic circuit.

The flux in the core of Figure 9.4 will be calculated as

$$\Phi = \frac{F}{\mathfrak{R}_i + \mathfrak{R}_a} = \frac{Ni}{\mathfrak{R}_i + \mathfrak{R}_a} \tag{9.11}$$

Thus, the reluctance of each section of the core should be calculated. If at any point in the flux direction, there is any change in area or permeability, the reluctance will be different. In the above circuit, it is assumed that area of all the sides of the core is the same and all the sides are having the same permeability, (i.e. core is made of the same material). It should be noted that area of core is the area perpendicular to the flux flow.

The behaviour of ferromagnetic material is shown in Figure 9.5 which is known as *hysteresis loop*. It is clear from the figure that the relation between B and H is not linear which shows that the permeability of material changes with different excitation. Retentivity

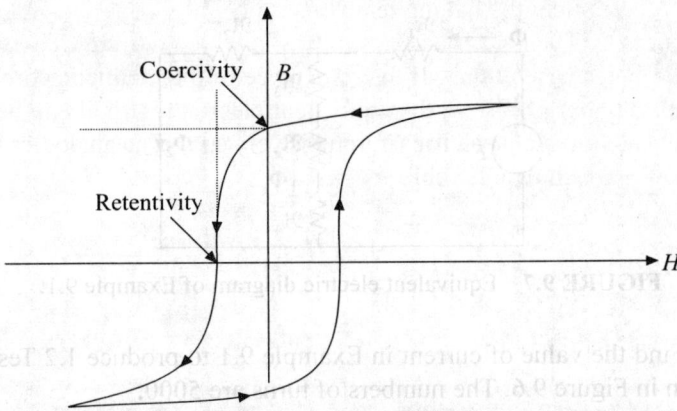

FIGURE 9.5 Hysteresis loop.

EXAMPLE 9.1 Draw the equivalent electrical diagram of the magnetic circuit shown in Figure 9.6 with the reluctances. The area of the core is (2×2) cm^2. The length of air gap is 1 cm and lengths of other limbs are shown in the figure. The relative permeability of the core is 4000.

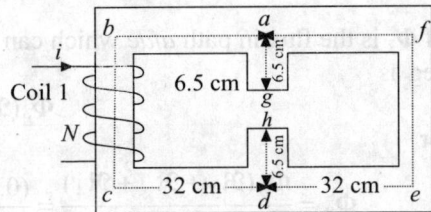

FIGURE 9.6 Magnetic circuit of Example 9.1.

Solution: The reluctance of path *abcd* will be

$$\mathfrak{R}_1 = \frac{l_1}{\mu_0 \mu_r A} = \frac{32 \times 10^{-2}}{\mu_0 \mu_r 4 \times 10^{-4}}$$

$$= \frac{32 \times 10^{-2}}{4\pi \times 10^{-7} \times 4000 \times 4 \times 10^{-4}}$$

$$= 1.59 \times 10^5 \text{ AT/Wb}$$

The reluctance of path *afed* will be the same as path *abcd* has the same length, area, and permeability. Thus, $\mathfrak{R}_2 = \mathfrak{R}_1$.

The reluctance of path *ag* ($\mathfrak{R}_3$) will be equal to the path *dh* ($\mathfrak{R}_4$) and can be calculated as

$$\mathfrak{R}_3 = \frac{6.5 \times 10^{-2}}{\mu_0 \mu_r 4 \times 10^{-4}} = \mathfrak{R}_4 = 0.32 \times 10^5 \text{ AT/Wb}$$

The reluctance of the air gap will be calculated as

$$\mathfrak{R}_g = \frac{1 \times 10^{-2}}{\mu_0 4 \times 10^{-4}} = 198.94 \times 10^5 \text{ AT/Wb}$$

The equivalent electrical diagram is shown in Figure 9.7.

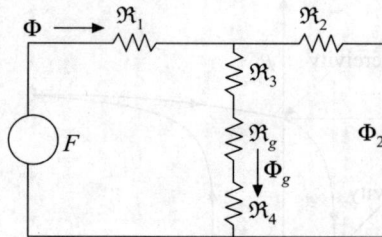

FIGURE 9.7   Equivalent electric diagram of Example 9.1.

EXAMPLE 9.2   Find the value of current in Example 9.1 to produce 1.2 Tesla flux density in the air gap as shown in Figure 9.6. The numbers of turns are 5000.

Solution: The flux in the air gap (Φ_g) will be

$$\Phi_g = B \times A = 1.2 \times 4 \times 10^{-2} = 4.8 \times 10^{-2} \text{ Wb}$$

If Φ_2 is the flux in path *afde* which can be obtained by equating the *mmf* of inner leg to the outer leg as

$$\Phi_g(\mathfrak{R}_3 + \mathfrak{R}_g + \mathfrak{R}_4) = \Phi_2\mathfrak{R}_2$$

or

$$\Phi_2 = \frac{\Phi_q(\mathfrak{R}_3 + \mathfrak{R}_g + \mathfrak{R}_4)}{\mathfrak{R}_2} = \frac{(0.32 + 198.94 + .32) \times 10^5 \times 4.8 \times 10^{-2}}{1.59 \times 10^5} = 6.025 \text{ Wb}$$

The total flux produced by the winding will be the some of air gap flux and the flux in the outer leg and is equal to

$$\Phi = \Phi_g + \Phi_2$$

Alternate method using field intensity can be used to compute the flux Φ_2 *as mmf drops in the circuit,* $F_3 + F_g + F_4 = F_2$

or

$$H_3 l_3 + H_g l_g + H_4 l_4 = H_2 l_2$$

where $H_3 = H_4 = \dfrac{B_g}{\mu}$

Thus, $H_2 l_2 = \dfrac{B_g}{\mu} l_3 + \dfrac{B_g}{\mu_0} l_g + \dfrac{B_g}{\mu} l_4$

or

$$H_2 = \frac{B_g}{l_2 \mu_0}\left(\frac{l_3}{\mu_r} + l_g + \frac{l_4}{\mu_r}\right) = \frac{1.2}{32 \times 4\pi \times 10^{-7}}\left(\frac{6.5}{4000} + 1.0 + \frac{6.5}{4000}\right)$$

and

$$\Phi_2 = \mu H_2 A = \frac{4\pi \times 10^{-7} \times 1.2 \times (4000 \times 4 \times 10^{-2})}{32 \times 4\pi \times 10^{-7}}\left(\frac{6.5}{4000} + 1.0 + \frac{6.5}{4000}\right) = 6.025 \text{ Wb}$$

Using outer loop of the equivalent circuit, the *mmf* will be

$$F = Ni = \Phi \mathfrak{R}_1 + \Phi_2 \mathfrak{R}_2$$

Thus,

$$\Phi = \Phi_g + \Phi_2 = 6.073 \text{ Wb}$$

$$i = \frac{(\Phi R_1 + \Phi_2 R_2)}{N} = \frac{[6.073 \times 1.59 \times 10^5 + 6.025 \times 1.59 \times 10^5]}{5000}$$

or

$$i = 384.72 \text{ A}$$

EXAMPLE 9.3   Find the value of current required to establish a flux density of 1.2 Wb/m² in the air gap as shown in Figure 9.8.

Solution: The flux produced by the leg *bc* will be from *c* to *b* direction (using right-hand rule), whereas the flux produced by the leg *ef* will be from *f* to *e* direction. The equivalent circuit is displayed in Figure 9.8.

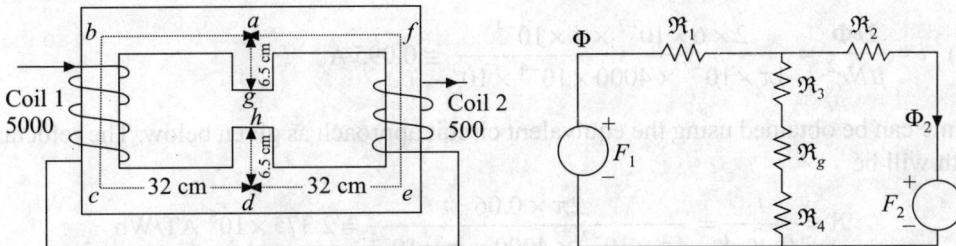

FIGURE 9.8   Equivalent circuit of Example 9.3.

The reluctances will be the same as calculated in Example 9.1. The *mmf* generated by coil 1 (F_1) will be 5000*i* and the *mmf* produced by coil 2 (F_2) will be 500*i*.

The flux in the air gap (Φ_g) will be

$$\Phi_g = B \times A = 1.2 \times 4 \times 10^{-2} = 4.8 \times 10^{-2} \text{ Wb}$$

Since the direction of flux in the air gap is not mentioned, let us consider it from g to h. If Φ_2 is the flux in path *afde* which can be obtained by equating the *mmf* of inner leg to the outer leg. Using the circuit equations, we can have the following equations:

$$\Phi_g(\mathfrak{R}_3 + \mathfrak{R}_g + \mathfrak{R}_4) = 5000i - (\Phi_g + \Phi_2)\mathfrak{R}_1 \tag{i}$$

$$\Phi_g(\mathfrak{R}_3 + \mathfrak{R}_g + \mathfrak{R}_4) = 500i + \Phi_2\mathfrak{R}_2 \tag{ii}$$

Using the values of reluctances calculated in Example 9.1 and air gap flux, Eqs. (i) and (ii) can be solved for current i as

$$2\Phi_g(\mathfrak{R}_3 + \mathfrak{R}_g + \mathfrak{R}_4) + \Phi_g\mathfrak{R}_1 = 5500i$$

or

$$i = \frac{2 \times 4.8 \times 10^{-2} \times (0.32 + 198.94 + 0.32) \times 10^5 + 4.8 \times 10^{-2} \times 1.59 \times 10^5}{5500}$$

Thus,

$$i = 349.45 \text{ A}$$

EXAMPLE 9.4   A toroidal core of radius 6 cm is having 1000 turns on it. The radius of core cross-section is 1 cm. Find the current required to establish a total magnetic flux of 0.4 mWb, if

(a) the core is nonmagnetic.
(b) The core is made of iron having the relative permeability of 4000.

Solution: For nonmagnetic core, $\mu = \mu_0$. Using Eqs. (9.5) and (9.8), we have

$$B = \mu H = \mu \frac{Ni}{2\pi r} = \frac{\Phi}{A(=\pi r_c^2)}$$

or

$$i = \frac{2r\Phi}{\mu N r_c^2}$$

(a) $i = \dfrac{2r\Phi}{\mu N r_c^2} = \dfrac{2 \times 6 \times 10^{-2} \times 4 \times 10^{-4}}{4\pi \times 10^{-7} \times 10^{-4} \times 10^3} \cong 380 \text{ A}$

(b) $i = \dfrac{2r\Phi}{\mu N r_c^2} = \dfrac{2 \times 6 \times 10^{-2} \times 4 \times 10^{-4}}{4\pi \times 10^{-7} \times 4000 \times 10^{-4} \times 10^3} \cong 0.095 \text{ A}$

The same can be obtained using the equivalent circuit approach as given below. The reluctance of the path will be

$$\mathfrak{R} = \frac{l}{\mu_0 \mu_r A} = \frac{2\pi \times 0.06}{4\pi \times 10^{-7} \times 4000 \times \pi \times 10^{-4}} = 2.375 \times 10^5 \text{ AT/Wb}$$

Using relation, $Ni = \Phi\mathfrak{R}$, we have

$$i = \frac{\Phi\mathfrak{R}}{N} = \frac{0.4 \times 10^{-3} \times 2.375 \times 10^5}{1000} = 0.095 \text{ A.}$$

9.5 INDUCTANCE

The total flux of the vector B through N turns of a coil can be easily approximated by multiplying the flux, Φ, in the coil (in core or air) by the number of turns. This total flux linking the coil is usually called the *flux linkage* and denoted by the Greek letter lambda, λ. Thus,

$$\lambda = N\Phi \tag{9.12}$$

Using Eq. (9.9), the value of Φ in Eq. (9.12) can be replaced by $\dfrac{Ni}{\mathfrak{R}}$. The above equation becomes,

$$\lambda = \frac{N^2}{\mathfrak{R}} i \tag{9.13}$$

The coefficient of i in Eq. (9.13) is known as the *self-inductance* or *inductance* (in short) and represented by letter L. Thus, we have

$$\lambda = Li \tag{9.14}$$

where

$$L = \frac{N^2}{\mathfrak{R}} \tag{9.15}$$

According to the Faraday's law of induction, any change in magnetic flux linked with a circuit produces an electromotive force (*emf*) in that circuit. Mathematically, it can be expressed as

$$e = -\frac{d\lambda}{dt} = -\frac{d(N\Phi)}{dt} \tag{9.16}$$

Negative sign in Eq. (9.16) is due to Lenz's law which states that voltage induced in a coil opposes to which it has been produced. It should be noted that it does not matter whether the flux is produced by a current in another conductor, by a magnet or even by the current in the same electric current.

Change in flux linkage ($N\Phi$) of a coil may occur in two ways:

1. Coil remains stationary and flux changes with time. (Due to this action, there is a statically induced *emf* as in the case of transformers).
2. Flux density distribution remains constant and stationary in space, but the coil moves relative to it, so as to change the flux linkage of the coil. (Due to this action, there is a dynamically induced *emf* as in the case of rotating machines).

Inductance, L, is independent of excitation and depends on geometry, μ and N. However, the value of μ is not constant for all the value of excitation. It is constant only in the linear region of B-H curve where equation $B = \mu H$ satisfies. In general case, if both configuration and current vary, the counter *emf*, $v\,(=-e)$, will be

$$v = \frac{d\lambda}{dt} = \frac{d(Li)}{dt} = L\frac{di}{dt} + i\frac{dL}{dt} \tag{9.17}$$

$$= \text{statically induced } emf + \text{dynamically induced } emf$$

If the coil, thus core, remains stationary and flux changes, Eq. (9.17) becomes

$$v = L \frac{di}{dt} \quad (9.18)$$

Equation (9.17) states that the voltage induced in a coil, if an alternating current (time varying) flows through coil, is proportional to the rate of change of current. The proportionality constant is called *self-inductance* of coil. If there are two coils having turn N_1 and N_2, the inductances of the coils will be defined as

$$L_1 = \frac{N_1^2}{\mathfrak{R}_1} \text{ and } L_2 = \frac{N_2^2}{\mathfrak{R}_2}.$$

9.6 MUTUAL INDUCTANCE

If the flux linking the coil is due to its own current, the coil is a simple or uncoupled coil. Sometimes the turns of a coil are linked by the flux produced by the currents of one or more coils, and these are coupled coils. If two or more coils are present on the same magnetic core or very near to each other, there will have flux in one coil set up by current in another. Let us suppose that a current i_1 in coil 1 having N_1 turns causes a magnetic flux Φ_{21} in coil 2 having N_2 turns as shown in Figure 9.9. The flux linkage with coil 2

$$\lambda_{21} = N_2 \Phi_{21} \quad (9.19)$$

is proportional to the current i_1 in the coil 1 and can be expressed as

$$\lambda_{21} = M_{21} i_1 \quad (9.20)$$

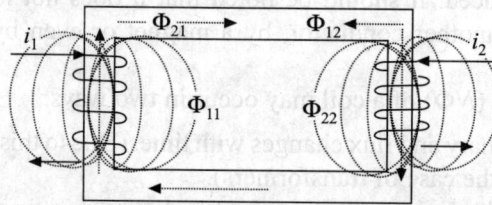

FIGURE 9.9 Mutually coupled circuit.

If the kth fraction (k is always less than unity) of flux produced in coil 1 links coil 2, then $\Phi_{21} = k\Phi_1$. The factor $(1 - k)$ is the leakage flux, i.e.

$$\Phi_1 = \Phi_{11} + \Phi_{21} \text{ and } (1 - k) \Phi_1 = \Phi_{11}$$

From Eqs. (9.19) and (9.20), we get,

$$M_{21} \times i_1 = N_2 \times k \times \Phi_1$$

or

$$M_{21} = \frac{N_2 \times k \times \Phi_1}{i_1} \quad (9.21)$$

Thus,

$$M_{21} = \frac{\text{Flux linking second coil due to current in the first coil}}{\text{Current in first coil}}$$

Similarly, if current in coil 2 causes a magnetic flux Φ_{12} in coil 1, we get

$$\lambda_{12} = M_{12}i_2 = N_1\Phi_{12} \qquad (9.22)$$

If the kth fraction of flux produced in coil 2 links coil 1, then $\Phi_{12} = k\Phi_2$. The Eq. (9.22) can be written as

$$M_{12} = \frac{N_1 \times k \times \Phi_2}{i_2} \qquad (9.23)$$

The self-inductances of coil 1 (L_1) and of coil 2 (L_2) will be given by

$$L_1 = \frac{\Phi_1 N_1}{i_1} \quad \text{and} \quad L_2 = \frac{\Phi_2 N_2}{i_2} \qquad (9.24)$$

where $\Phi_2 = \Phi_{22} + \Phi_{12}$.

Multiplying Eqs. (9.21) and (9.23) and substituting the value of Eq. (9.24), we have

$$M_{12} \cdot M_{21} = \frac{k^2 \Phi_1 \Phi_2 N_1 N_2}{i_1 i_2} = k^2 L_1 L_2 \qquad (9.25)$$

Normally, M_{12} is equal to M_{21}. Let $M_{12} = M_{21} = M$, then Eq. (9.25) can be written as

$$M = k\sqrt{L_1 L_2} \qquad (9.26)$$

where k is also known as coefficient of coupling ($0 \leq k \leq 1$). When $k = 1$, circuits are said to be tightly coupled ($M = \sqrt{L_1 L_2}$) and when $k = 0$, circuits are having no coupling ($M = 0$).

The flux linkage in coil 1 due to currents in coil 1 and coil 2 will be

$$\lambda_1 = N_1(\Phi_1 \pm \Phi_{12})$$

The sign $\pm$ is used because flux produced due to coil 2 may be additive or opposite to the flux produced due to coil 1 current. The voltage induced in the coil 1 will be written as

$$v_1 = L_1 \frac{di_1}{dt} \pm M_{12} \frac{di_2}{dt} \qquad (9.27)$$

Similarly, the voltage induced in the coil 2 can be written as

$$v_2 = L_2 \frac{di_2}{dt} \pm M_{21} \frac{di_1}{dt} \qquad (9.28)$$

Let us consider three coils wounded on the same core as shown in Figure 9.10. The current i_1 is flowing in coil 1 having N_1 turns. The voltage induced in coil 2 having N_2 turns and in coil 3 having N_3 turn will be proportional to $\frac{di_1}{dt}$. The induced voltages in the coil 2 and coil 3 will be given by

$$v_{21} = M_{21} \frac{di_1}{dt} = \left(\frac{N_2 \Phi_{21}}{i_1} \right) \frac{di_1}{dt} \tag{9.29}$$

and

$$v_{31} = M_{31} \frac{di_1}{dt} = \frac{\Phi_{31} N_3}{i_1} \frac{di_1}{dt} \tag{9.30}$$

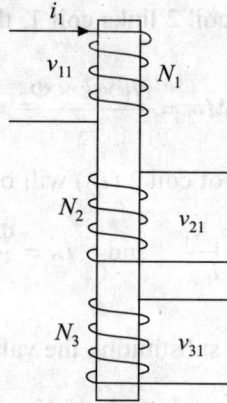

FIGURE 9.10   Induced voltages in the mutually-coupled circuits.

The voltage induced in the coil 1 will be

$$v_{11} = L_1 \frac{di_1}{dt} = \left(\frac{N_1 \Phi_1}{i_1} \right) \frac{di_1}{dt} \tag{9.31}$$

EXAMPLE 9.5   When current in coil 1 is $5 \sin 2t$ and coil 2 is opened circuited, the voltage induced in the coil 1 and coil 2 are $2 \cos 2t$ and $5 \cos 2t$. With the current of $4 \sin 2t$ in coil 2, the induced voltage in coil 2 is $4 \cos 2t$ with coil 1 open circuited. Find the value of self-inducatnces and mutual inductance of the coil.

Solution: When the coil 2 is open circuited ($i_2 = 0$), the voltages in coil 1 can be written, using Eq. (9.27) as

$$2 \cos 2t = L_1 \frac{d(5 \sin 2t)}{dt} = 10 \, L_1 \cos 2t$$

Thus, $L_1 = 0.2$ H.

Using Eq. (9.28) using the excitation current in coil 1, we have

$$5 \cos 2t = M_{21} \frac{d(5 \sin 2t)}{dt} = 10 \, M_{21} \cos 2t$$

Thus, $M_{21} = 0.5$ H $= M$

Using second relation with excitation in coil 2 and coil 1 is open circuited ($i_1 = 0$), we get

$$4 \cos 2t = L_2 \frac{d(4 \sin 2t)}{dt} = 8 L_2 \cos 2t$$

Thus, $L_2 = 0.5$ H.

9.6.1 Sign Convention of Mutually Induced Voltages

The sign of mutually induced voltage depends on the geometry of the coil formation and is positive or negative depending on whether fluxes are additive or opposing each other. In electrical circuit, the physical formation of coils is not identifiable, hence a dot convention is followed.

If both the currents are leaving or entering the dot, the fluxes produced add/help each other.

> *When a reference current enters the marked (dotted) end of a coil, the voltage of mutual induction in another coil is $+M\,di/dt$ in terms of a reference polarity of plus (+) on the marked (dotted) end of the second coil.*
> *Each reversal of any reference (current) multiples, the terms $M\,di/dt$ by minus one (−1).*

Let us consider Figure 9.11 to see the sign convention of mutually induced volatges.

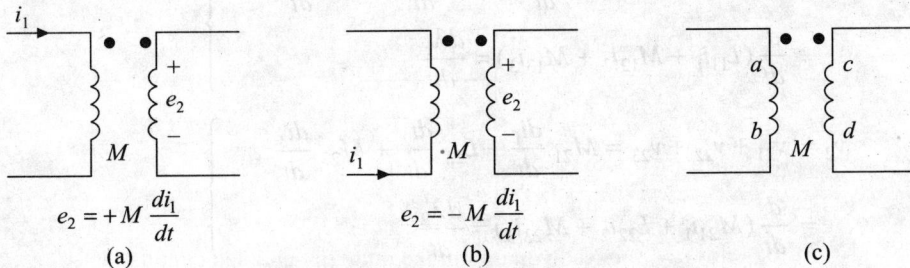

$$e_2 = + M\frac{di_1}{dt} \qquad e_2 = -M\frac{di_1}{dt}$$

(a)   (b)   (c)

FIGURE 9.11   Mutual inductance effect.

As per dot rule, the current in coil 1 (i_1) in Figure 9.11(a) is entering through dot mark, thus the induced voltage in coil 2 (e_2) will be $+\,M di_1/dt$, whereas the current in coil 1 (i_1) in Figure 9.11(b) is leaving through dot mark and therefore the induced voltage in coil 2 (e_2) will be $-M di_1/dt$. Thus, the induced voltage in Figure 9.11(c) could be written as

$$e_{cd} = + M\frac{di_{ab}}{dt} = -e_{dc} \quad \text{or} \quad e_{cd} = - M\frac{di_{ba}}{dt} = -e_{dc} \tag{9.32}$$

9.6.2 Induced Voltage Phasor

Let a sinusoidal varying current $i_1 = I_{m1}\cos(\omega t + \theta)$ flows in the coil 1 as shown in Figure 9.10, where I_{m1} is the magnitude of the current and $\omega\,(= 2\pi f)$ is the angular frequency. In phasor form, it can be represented as

$$\bar{I} = I\,\angle\theta$$

The induced voltages in the phasor form using Eqs. (9.29), (9.30) and (9.31), can be written as

$$\bar{V}_{11} = j\omega L_1\bar{I}_1,\, \bar{V}_{21} = j\omega M_{21}\bar{I}_1 \text{ and } \bar{V}_{31} = j\omega M_{31}\bar{I}_1 \tag{9.33}$$

For a bilateral magnetic circuit,

$$M_{12} = M_{21},\, M_{23} = M_{32} \text{ and } M_{13} = M_{31}$$

When current, i_2, flows in coil 2, the induced voltages in the coils will be

$$v_{12} = M_{12}\frac{di_2}{dt}, \quad v_{22} = L_{22}\frac{di_2}{dt}, \text{ and } v_{32} = M_{32}\frac{di_2}{dt} \tag{9.34}$$

where L_{22} is the self-inductance of coil 2 (L_2).

Similarly, if current, i_3, flows in coil 3, the induced voltages in the coils will be

$$v_{13} = M_{13}\frac{di_3}{dt}, \quad v_{23} = M_{23}\frac{di_3}{dt}, \text{ and } v_{33} = L_{33}\frac{di_3}{dt} \tag{9.35}$$

When these currents flow in all the coils such that the fluxes produced by these currents are additive, the voltage induced in coil 1 (v_1), coil 2 (v_2) and coil 3 (v_3) can be expressed as

$$\left.\begin{aligned}
v_1 = v_{11} + v_{12} + v_{13} &= L_{11}\frac{di_1}{dt} + M_{12}\frac{di_2}{dt} + M_{13}\frac{di_3}{dt} \\
&= \frac{d}{dt}(L_{11}i_1 + M_{12}i_2 + M_{13}i_3) = \frac{d\lambda_1}{dt}
\end{aligned}\right\} \tag{9.36}$$

$$\left.\begin{aligned}
v_2 = v_{21} + v_{22} + v_{23} &= M_{21}\frac{di_1}{dt} + L_{22}\cdot\frac{di_2}{dt} + M_{23}\frac{di_3}{dt} \\
&= \frac{d}{dt}(M_{21}i_1 + L_{22}i_2 + M_{23}i_3) = \frac{d\lambda_2}{dt}
\end{aligned}\right\} \tag{9.37}$$

$$\left.\begin{aligned}
v_3 = v_{31} + v_{32} + v_{33} &= M_{31}\frac{di_1}{dt} + M_{32}\frac{di_2}{dt} + L_{33}\frac{di_3}{dt} \\
&= \frac{d}{dt}(M_{31}i_1 + M_{32}i_2 + L_{33}i_3) = \frac{d\lambda_3}{dt}
\end{aligned}\right\} \tag{9.38}$$

Equations (9.36) to (9.38) can be expressed in the phasor form as

$$\left.\begin{aligned}
\overline{V}_1 &= j\omega L_{11}\overline{I}_1 + j\omega M_{12}\overline{I}_2 + j\omega M_{13}\overline{I}_3 \\
\overline{V}_2 &= j\omega M_{21}\overline{I}_1 + j\omega L_{22}\overline{I}_2 + j\omega M_{23}\overline{I}_3 \\
\overline{V}_3 &= j\omega M_{31}\overline{I}_1 + j\omega M_{32}\overline{I}_2 + j\omega L_{33}\overline{I}_3
\end{aligned}\right\} \tag{9.39}$$

9.6.3 Two-Port Representation

The magnetic circuit shown in Figure 9.9 can be represented as electrical equivalent circuit as shown in Figure 9.12.

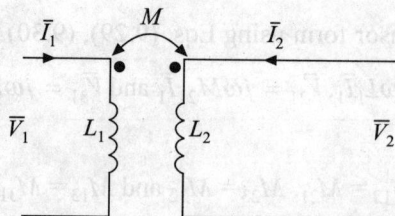

FIGURE 9.12   Electrical equivalent circuit of magnetic circuit.

Circuit shown in Figure 9.12 can be treated as two-port network consisting of two pairs of terminals or two-port entries to which sources or loads are connected. One terminal is called *input* and other is called *output* terminal. The current entering the network is taken as positive. In general form, the two-port network can be represented as shown in Figure 9.13. There are four variables, out of which two can be independent Z parameters (also called open-circuit parameters).

$$\bar{V}_1 = \bar{Z}_{11}\bar{I}_1 + \bar{Z}_{12}\bar{I}_2 \tag{9.40}$$

$$\bar{V}_2 = \bar{Z}_{21}\bar{I}_1 + \bar{Z}_{22}\bar{I}_2 \tag{9.41}$$

where $\bar{Z}_{11} = \dfrac{\bar{V}_1}{\bar{I}_1}\Bigg]_{I_2=0}$ (also known as *input* or *self-impedance*)

$\bar{Z}_{22} = \dfrac{\bar{V}_2}{\bar{I}_2}\Bigg]_{I_1=0}$ (also known as *output impedance*)

$\bar{Z}_{12} = \dfrac{\bar{V}_1}{\bar{I}_2}\Bigg]_{I_1=0}$ (also known as *forward impedance*)

$\bar{Z}_{21} = \dfrac{\bar{V}_2}{\bar{I}_1}\Bigg]_{I_2=0}$ (also known as *backward impedance*)

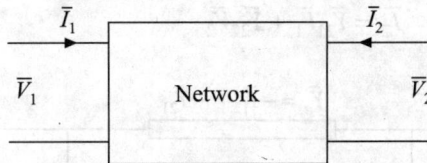

FIGURE 9.13   Two-port network.

Using Eqs. (9.40) and (9.41), any two-port network (as shown in Figure 9.13) can be represented as star or T configuration as shown in Figure 9.14.

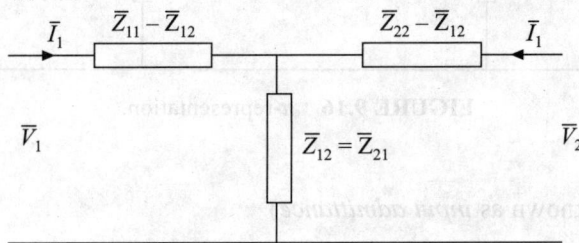

FIGURE 9.14   Star (T)-representation.

From Figure 9.12, the open-circuit parameters can be derived as

$$\bar{Z}_{11} = \frac{\bar{V}_1}{\bar{I}_1}\bigg]_{I_2=0} = j\omega L_1, \bar{Z}_{22} = \frac{\bar{V}_2}{\bar{I}_2}\bigg]_{I_1=0} = j\omega L_2 \tag{9.42}$$

$$\bar{Z}_{12} = \frac{\bar{V}_1}{\bar{I}_2}\bigg]_{I_1=0} = j\omega M_{12} = j\omega M, \bar{Z}_{21} = \frac{\bar{V}_2}{\bar{I}_1}\bigg]_{I_2=0} = j\omega M_{21} = j\omega M \tag{9.43}$$

Thus, Figure 9.13 can be represented as Figure 9.15 using star-(T) representation.

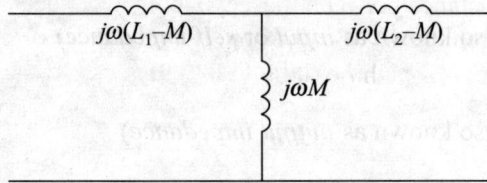

FIGURE 9.15   Equivalent star circuit of mutually-coupled circuit.

Two-port network as shown in Figure 9.13 can also be analyzed with the help of π-equivalent representation as shown in Figure 9.16. The parameters are normally represented in the terms of admittances (also knows as *short circuit parameters*).

$$\bar{I}_1 = \bar{Y}_{11}\bar{V}_1 + \bar{Y}_{12}\bar{V}_2 \tag{9.44}$$

$$\bar{I}_2 = \bar{Y}_{21}\bar{V}_1 + \bar{Y}_{22}\bar{V}_2 \tag{9.45}$$

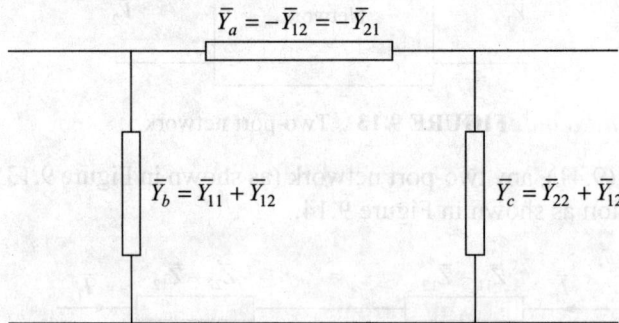

FIGURE 9.16 π-representation.

where

$$\bar{Y}_{11} = \frac{\bar{I}_1}{\bar{V}_1}\bigg|_{V_2=0} \quad \text{(also known as \textit{input admittance})}$$

$$\bar{Y}_{22} = \frac{\bar{I}_2}{\bar{V}_2}\bigg|_{V_1=0} \quad \text{(also known as \textit{output admittance})}$$

$$\overline{Y}_{12} = \frac{\overline{I}_1}{\overline{V}_2}\bigg|_{V_1=0} \qquad \text{(also known as } \textit{transfer admittance}\text{)}$$

$$\overline{Y}_{21} = \frac{\overline{I}_2}{\overline{V}_1}\bigg|_{V_2=0} \qquad \text{(also known as } \textit{transfer admittance}\text{)}.$$

9.6.4 Computation of Inductances

Let us consider a circuit shown in Figure 9.17 for computation of inducatnce of coil 1 (L_1), coil 2 (L_2), and mutual inductance between these two coils wounded on a core having reluctance of path $bcde$ is equal to the reluctance of path $bafe$. The reluctance of the middle leg (be) is half of the outer paths. Let reluctance of path be (middle leg) $\mathfrak{R}$.

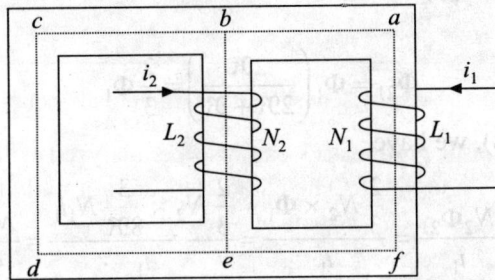

FIGIURE 9.17 Computation of inductances.

The self-inductance of coils is defined as

$$L_1 = \frac{N_1\Phi_1}{i_1}, \; L_2 = \frac{N_2\Phi_2}{i_2} \text{ and } M_{21} = \frac{\text{Flux in coil 2 due to current in coil 1}}{i_1} \qquad (9.46)$$

Calculation of self-inductance of coil 1

The equivalent reluctance ($\mathfrak{R}_{eq1}$) seen by coil 1 will be

$$\mathfrak{R}_{eq1} = [\mathfrak{R}_{bafe} + (\mathfrak{R}_{be} \| \mathfrak{R}_{bcde})] = 2\mathfrak{R} + \frac{\mathfrak{R} \times 2\mathfrak{R}}{\mathfrak{R} + 2\mathfrak{R}} = \frac{8\mathfrak{R}}{3} \qquad (9.47)$$

Thus flux produced by the coil 1 with current i_1 will be

$$\Phi_1 = \frac{N_1 i_1}{\mathfrak{R}_{eq1}} = \frac{3N_1 i_1}{8\mathfrak{R}} \qquad (9.48)$$

or

$$L_1 = \frac{N_1\Phi_1}{i_1} = \frac{N_1}{i}\frac{3N_1 i}{8\mathfrak{R}} = \frac{N_1^2}{\mathfrak{R}_{eq1}} = \frac{3N_1^2}{8\mathfrak{R}} \qquad (9.49)$$

Calculation of self-inductance of coil 2

The equivalent reluctance ($\Re_{eq2}$) seen by coil 2 will be

$$\Re_{eq2} = (\Re_{bafe} \parallel \Re_{bcde}) + \Re_{be} = \frac{2\Re \times 2\Re}{2\Re + 2\Re} + \Re = 2\Re \qquad (9.50)$$

Thus,

$$L_2 = \frac{N_2^2}{\Re_{eq2}} = \frac{N_2^2}{2\Re} \qquad (9.51)$$

Calculation of mutual-inductance between coil 1 and coil 2

The flux produced (Φ_1) by the current in coil 1 will be divided between path *bcde* and path *be* (middle leg). The distribution of fluxed produced will be according to the reluctances seen by the paths (electrical current analogy). Thus, the flux flows in the coil 2 (middle leg) due to flux produced by coil 1 will be

$$\Phi_{21} = \Phi_1\left(\frac{2\Re}{2\Re + \Re}\right) = \frac{2}{3}\Phi_1 \qquad (9.52)$$

Using Eqs. (9.46) and (9.48), we have

$$M_{21} = \frac{N_2\Phi_{21}}{i_1} = \frac{\frac{2}{3}N_2 \times \Phi_1}{i_1} = \frac{\frac{2}{3}N_2 \times \frac{3}{8\Re}N_1 i_1}{i_1} = \frac{N_1 N_2}{4\Re} \qquad (9.53)$$

Similarly, the mutual inductance due to the current in coil 2 is calculated as follows. The flux produced in the middle leg (Φ_2) is shared equally in the outer legs. Thus the flux linkage in coil 1 will be $\Phi_2/2$ and

$$\Phi_2 = \frac{N_2 i_2}{\Re_{eq2}} = \frac{N_2 i_2}{2\Re} \qquad (9.54)$$

Therefore,

$$M_{12} = \frac{N_1\Phi_{12}}{i_2} = \frac{N_1 \times \Phi_2/2}{i_2} = \frac{N_1 \times (N_2 i_2)/2\Re}{2i_2} = \frac{N_1 N_2}{4\Re} \qquad (9.55)$$

From Eqs. (9.53) and (9.55), it can be seen that

$$M_{21} = M_{12} \qquad (9.56)$$

EXAMPLE 9.6   Find the Thevenin equivalent between points *a* and *b* as shown in Figure 9.18. The mutual inductance between the windings is *M*.

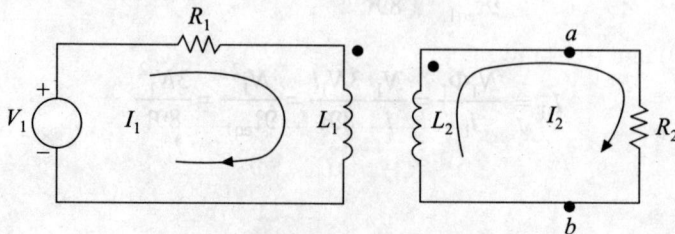

FIGURE 9.18   Example 9.6.

Solution: Since the assumed direction of currents is in such a way that induced voltage due to the mutual inductance will be negative (one current entering the dot point and other is leaving the dot point). The voltage equations can be written as

$$\bar{V}_s = R_1\bar{I}_1 + j\omega L_1\bar{I}_1 - j\omega M\bar{I}_2 \tag{i}$$

$$0 = R_2\bar{I}_2 + j\omega L_2\bar{I}_2 - j\omega M\,\bar{I}_1 \tag{ii}$$

For a given value of $\bar{V}_s$, $\bar{I}_1$ and $\bar{I}_2$ can be obtained from Eqs. (i) and (ii).

Determination of Thevenin voltage (V_{th}) between points a and b

The open-circuit voltage between ponits a and b is obtained by Eq. (i) putting $\bar{I}_2 = 0$ as

$$\bar{V}_{ab} = j\omega M\bar{I}_1 = \frac{j\omega M\bar{V}_s}{R_1 + j\omega L_1} = \bar{V}_{th} \tag{9.57}$$

(Since there is no current in the second coil and current in coil 1 is entering through the dotted terminal, the voltage induced in the coil 2 will be positive at dotted terminal as per sign convcention).

Calculation of short-circuit current (I_{sc}) between points a and b

To caluctate the I_{sc}, the terminal a and b must be short circuited. Please note that there will be no current in R_2. Now the voltage equation can be written as

$$\bar{V}_s = R_1\bar{I}_1 + j\omega L_1\bar{I}_1 - j\omega M\bar{I}_2 \tag{9.58}$$

$$0 = j\omega L_2\bar{I}_2 - j\omega M\bar{I}_1 \tag{9.59}$$

It should be noted that currents in Eqs. (9.58) and (9.59) are different than the currents in Eqs. (i) and (ii) as curcuit condition is changed. The current I_2 is I_{sc}. From Eq. (9.59), we have

$$\bar{I}_2 = \frac{M}{L_2}\bar{I}_1 = \bar{I}_{sc} \tag{9.60}$$

Putting the value of current of Eq. (9.60) in Eq. (9.58), we get

$$\bar{V}_s = R_1\bar{I}_1 + j\omega L_1\bar{I}_1 - j\omega\,\frac{M^2}{L_2}\bar{I}_1$$

or

$$\bar{I}_1 = \frac{\bar{V}_s}{R_1 + j\omega L_1 - j\omega\,\dfrac{M^2}{L_2}} \tag{9.61}$$

Thus, from Eq. (9.60) and (9.61), we get

$$I_{sc} = \frac{M}{L_2}\,\frac{\bar{V}_s}{R_1 + j\omega L_1 - j\omega\,\dfrac{M^2}{L_2}} \tag{9.62}$$

Calculation of Thevenin impedance

From Eqs. (9.57) and (9.62), the Thevenin impedance can be calculated as follows

$$\bar{Z}_{th} = \frac{\bar{V}_{ab}}{\bar{I}_{sc}} = (j\omega L_2) \times \frac{\left(R_1 + j\omega L_1 - j\omega \dfrac{M^2}{L_2}\right)}{(R_1 + j\omega L_1)} \tag{9.63}$$

PROBLEMS

9.1 The magnetic circuit of Figure 9.19, a ferromagnetic material having the relative permeability of 1200. The dimensions are shown in Figure 9.19. Neglecting the leakage and fringing, determine the value of current i to establish a flux of 0.8 Wb in the air gap.

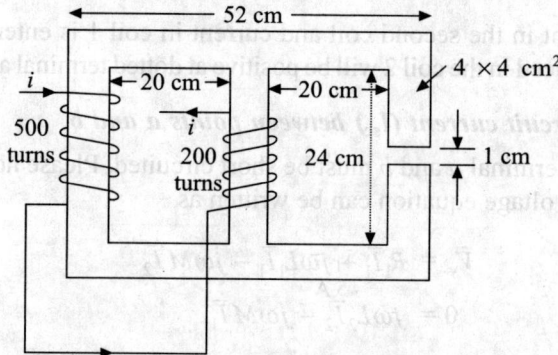

FIGURE 9.19 Problem 9.1.

[**Ans.** 15.3 kA]

9.2 Two coils are wound on a toroidal core as shown in Figure 9.20 core is made of silicon steel and has square cross-section. The coil current is 1 ampere. *B-H* data is given in below:

H(AT/m)	100	150	200	300	400	450	700
B(T)	0.70	0.90	1.00	1.12	1.20	1.25	1.30

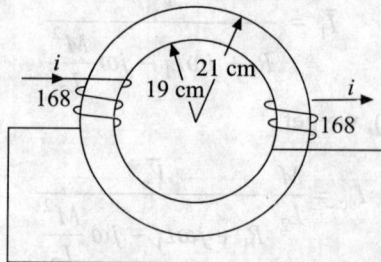

FIGURE 9.20 Problem 9.2.

(a) Determine the flux density at the mean radius of the core.

(b) Assuming constant flux density (same as at the mean radius) over the cross-section of the core, determine the flux in core.

(c) Determine the relative permeability μ_r of the core.

[**Ans.** (a) 1.081 Tesla, (b) 43.24 mWb, (c) 3217.27]

9.3 In Figure 9.21, each air gap length is $g = 0.1$cm and cross-sectional area $A_g = 4$ cm^2. $\mu = \infty$ in iron. Find the fluxes through the coils and their directions.

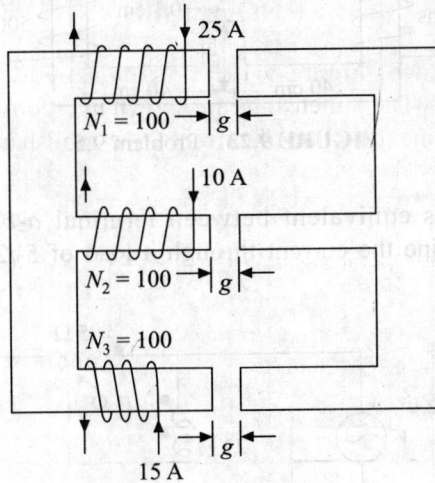

FIGURE 9.21   Problem 9.3.

[**Ans.** 10^{-3} Wb, 0 Wb, -10^{-3} Wb]

9.4 In Figure 9.22, the closed magnetic path lengths from a to b are $l_1 = l_3 = 30$ cm, $l_2 = 10$ cm. Cross-sectional areas are $A_1 = A_3 = 2$ cm^2, $A_2 = 4$ cm^2, the relative permeabilities are $\mu_{r1} = \mu_{r3} = 2250$, $\mu_{r2} = 1350$. For $N = 25$ and $I = 0.5$ A,

(a) Find B_1, B_2, B_3.

(b) Find the self-inductance of the coil.

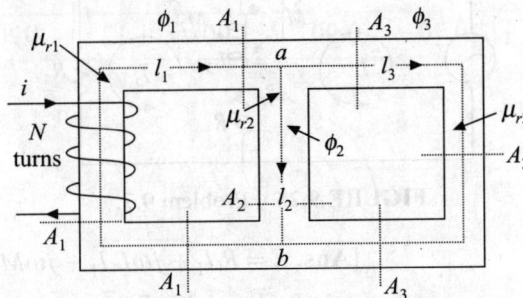

FIGURE 9.22   Problem 9.4.

[**Ans.** (a) 0.0966 T, 0.0377 T, 0.021 T, (b) 0.966 mH]

9.5 A magnetic circuit consists of an iron core of relative permeability 2300 and has 1 mm air gap as shown in Figure 9.23. With only coil 1 of 1000 turns energized by 3 A, calculate the flux in the air gap. The flux in the gap is to be increased by 40 percent by a second coil 2 of 500 turns. Determine the necessary current and its direction. Cross-sectional area of iron core is 20 cm² throughout.

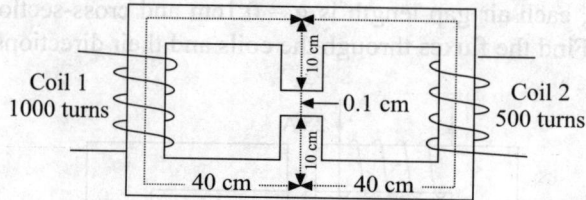

FIGURE 9.23   Problem 9.5.

[**Ans.** 3.33619 mWb, 2.4 A]

9.6 Find the Thevenin's equivalent between terminal *a-b* of the network shown in Figure 9.24. Determine the current through a load of 5-*j*20 ohms connected between terminal *a-b*.

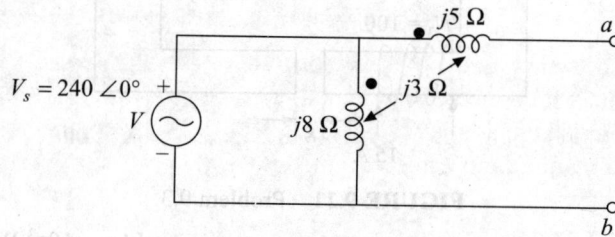

FIGURE 9.24   Problem 9.6.

[**Ans.** 8.88 $\angle$ 72.77° A]

9.7 Write down mesh equations for the circuit shown in Figure 9.25.

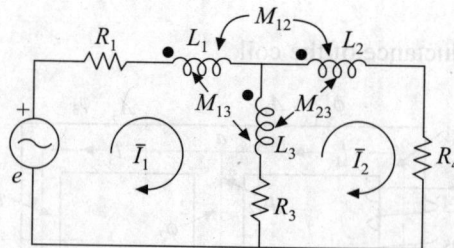

FIGURE 9.25   Problem 9.7.

$$[\textbf{Ans. } \bar{E} = R_1\bar{I}_1 + j\omega L_1\bar{I}_1 + j\omega M_{31}(\bar{I}_1 - \bar{I}_2) + j\omega M_{12}\bar{I}_2$$
$$+ j\omega L_3(\bar{I}_1 - \bar{I}_2) + R_3(\bar{I}_1 - \bar{I}_2) + j\omega M_{23}\bar{I}_2 + j\omega M_{31}\bar{I}_1,$$
$$j\omega L_2\bar{I}_2 + R_4\bar{I}_2 + R_3(\bar{I}_2 - \bar{I}_1) + j\omega L_3(\bar{I}_2 - \bar{I}_1) + j\omega M_{12}\bar{I}_1$$
$$+ j\omega M_{23}(\bar{I}_1 - \bar{I}_2) - j\omega M_{31}\bar{I}_1 - j\omega M_{23}\bar{I}_2 = 0]$$

9.8 In the circuit diagram of Figure 9.26, the source voltage, V_s, is $200\sqrt{2}\cos(400t)$ volt. The coupling coefficient (k) of the coupled inductors is 0.707.

(a) Obtain the Thevenin equivalent at the terminal a, b.

(b) A variable impedance Z_0 is connected across terminal ab. Find the value of Z_0 to obtain the maximum power transfer across this. What is maximum power under this condition?

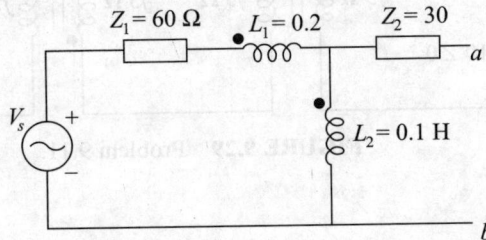

FIGURE 9.26 Problem 9.8.

[**Ans.** $(a)\ \bar{V}_{th} = 76.63\ \angle 16.7°$ V, $\bar{Z}_{th} = 40.22\ \angle 15.336°\ \Omega$;

$(b)\ \bar{Z}_0 = 40.22\ \angle -15.336°\ \Omega,\ 75.696\ W$]

9.9 An ac circuit contains a coil of self-inductance L_1. Near this but not connected to it, another coil of inductance L_2 is kept as shown in Figure 9.27. If mutual inductance between two coils is M. Show that the effective resistance and inductance as experienced by the source ($\bar{V}$) connected across the first coil will be

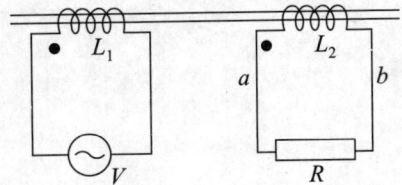

FIGURE 9.27 Problem 9.9.

$$R_{eq} = \frac{\omega^2 M^2 R}{R^2 + \omega^2 L_2^2} \quad \text{and} \quad L_{eq} = L_1 - \frac{\omega^2 M^2 L_2}{R^2 + \omega^2 L_2^2}$$

Will the above value of R_{eq} and L_{eq} change, if the dot in coil 2 corresponds to terminal b?

9.10 In Figure 9.28, find the current i_1 for $t > 0$ if the circuit is in steady state at $t = 0$. Take $M = 1/\sqrt{2}$ H, $L_1 = 0.5$ H, $L_2 = 1$ H and $V = 12$ V.

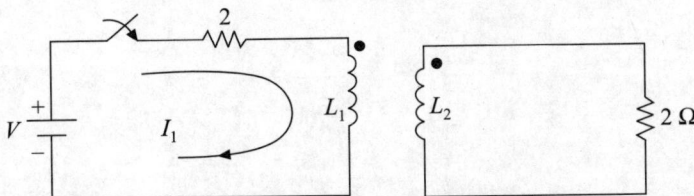

FIGURE 9.28 Problem 9.10.

[**Ans.** $6 - 2e^{-4t/3}$ A]

9.11 Find the current I_3 in Figure 9.29 using mesh analysis. Mutual impedances between coils are $j4$ ohm.

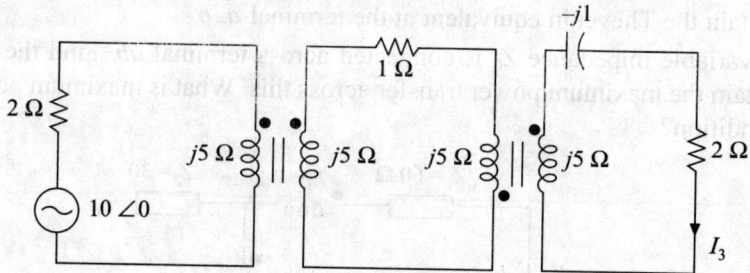

FIGURE 9.29   Problem 9.11.

[**Ans.** 1.21 $\angle -179.1°$ A]

10

Transformers

10.1 INTRODUCTION

Transformer, which is an important device used in power, communication and electronic circuits, is a two-port circuit containing coupled coils around a common core. It is also known with different names such as power transformer, distribution transformer, generating transformer, impedance matching transformer, isolation transformer, filament transformer, instrument transformer (current and potential transformer), etc. depending on the applications. Transformer is a static electromagnetic device (static in the sense that it has no moving part and does not try to rotate) operated on the basis of mutual induction. It is constructionally similar to the magnetic coupled circuit. In transformer, an alternating current of one voltage is transformed into different alternating current of the same frequency but of a different voltage.

In electrical power network, the usage of transformers, which vary in size, rating and location, is very high. Due to technical and economical reasons, the generation voltage is limited to 35 kV. Since the generating stations may be far from the load centres where used voltages are less than 33 kV, power must be transmitted to the load centres through high voltage power lines. It is well established that power transfer is more economical at high voltage, which is constrained by several operational and technical constraints. Therefore, generated voltage of alternators must be increased by step-up transformers [also called *generating transformers* (GTs)]. Since utilization of power is at low voltage (again due to technical, economical and safety reasons), the transmitted power voltage must be lowered using step-down transformers (also known as *distribution transformers*). Due to different power requirements at different load centres, the voltages of transmission lines are also different. Thus, various transformers are used at different locations and these are normally called *power transformers*, which handle several MWs of power. These can be step-up or step-down transformers. Electric power network is also interconnected as it offers several advantages. This is accomplished through transformers of different voltage levels. Figure 10.1 shows the various transformers used in an interconnected power system.

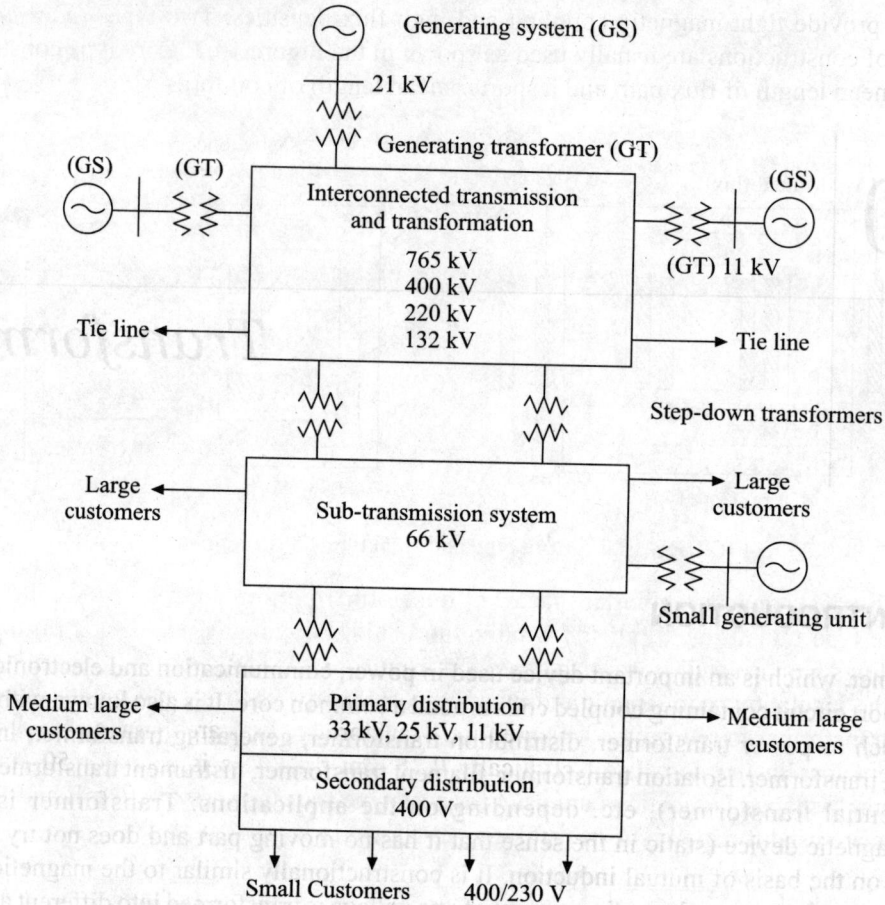

FIGURE 10.1   Various transformers in interconnected power system.

10.2 CONSTRUCTIONAL FEATURES

A transformer has several windings (at least two) electrically insulated from each other but magnetically coupled. The winding connected to the energy source (supply) is called the *primary winding* and the other windings, which deliver energy to the external circuit, are called *secondaries*. If there are two secondaries, the high voltage winding is called *secondary winding* and low voltage winding is called *tertiary winding*. If the voltage across the primary winding is less than the secondary winding, the transformer is known as *step-up transformer* and in the reverse case it is known as *step-down transformer*. Corresponding to the relative voltage magnitude of the rated voltage of windings, it is common to differentiate between the *high-voltage* or *high tension winding* and *low-voltage* or *low-tension winding*.

In a transformer, the magnetic coupling may be through air (called *air-core transformer*) or iron core (called *iron-cored transformer*). For power from one voltage level to another voltage level, iron-cored transformer is used having magnetic core made of high permeability ferro-magnetic material to provide tight magnetic coupling and high flux densities. Two types (*core type* and *shell type*) of constructions are usually used as shown in the Figure 10.2. Core type construction has larger mean length of flux path and a shorter mean length of coil turns.

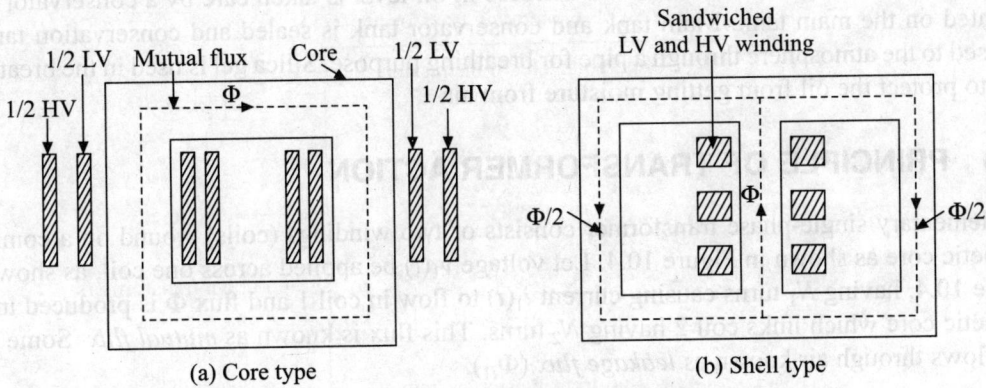

FIGURE 10.2   Core type and shell type constructions.

Since the flux leakage is an important factor of transformer performance, to reduce the leakage flux in core type transformer, half LV winding and half HV winding are wound on one limb. Since windings are insulated from each other and from core, the LV winding is placed near core to reduce the insulation requirement and thus the cost. Sandwiched type of winding is used in shell type transformer to reduce the leakage flux. Transformer cores are laminated to reduce the eddy current loss in the core itself. Typically 0.35 mm thickness is used for 50 Hz power transformers. L-shaped laminations are used for core type transformers whereas E and I-shaped laminations are used for shell type transformers as shown in Figure 10.3.

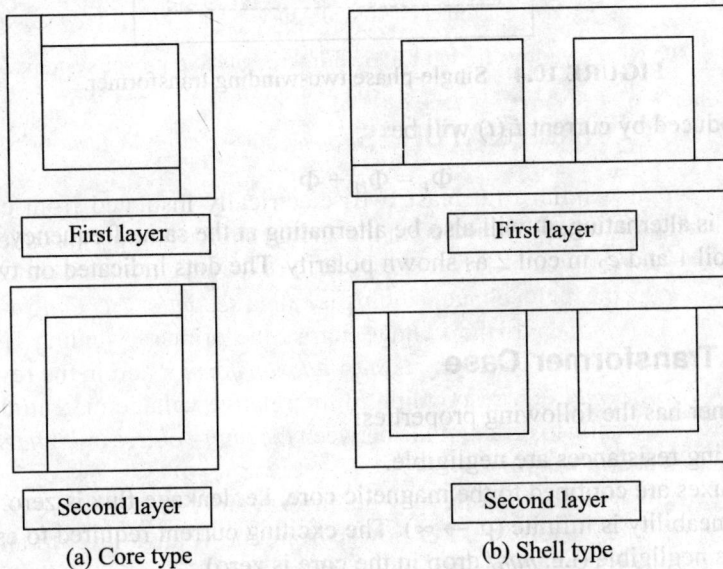

FIGURE 10.3   Shape of laminations.

Transformer core along with windings are placed in a tank, filled with transformer oil to prevent the moisture and deterioration of winding insulation. Some tubes are provided on the outside surface of tank to provide circulation of oil which removes the heat (produced due to losses) of the core and winding. The decrease and increase in oil level is taken care by a conservator tank mounted on the main tank. Main tank and conservator tank is sealed and conservation tank is exposed to the atmosphere through a pipe for breathing purpose. Silica gel is used in the breathing pipe to protect the oil from getting moisture from air.

10.3   PRINCIPLE OF TRANSFORMER ACTION

An elementary single-phase transformer consists of two windings (coils) wound on a common magnetic core as shown in Figure 10.4. Let voltage $v_1(t)$ be applied across one coil, as shown in Figure 10.4, having N_1 turns causing current $i_1(t)$ to flow in coil 1 and flux Φ is produced in the magnetic core which links coil 2 having N_2 turns. This flux is known as *mutual flux*. Some flux also flows through air known as *leakage flux* (Φ_{l1}).

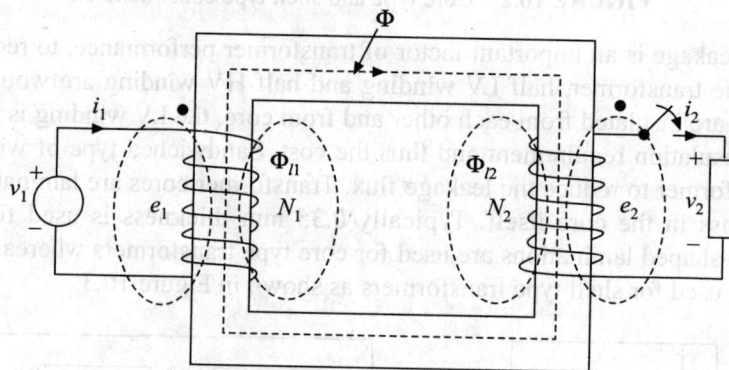

FIGURE 10.4   Single-phase two-winding transformer.

The total flux produced by current $i_1(t)$ will be

$$\Phi_1 = \Phi_{l1} + \Phi \tag{10.1}$$

If applied voltage is alternating, Φ will also be alternating at the same frequency. Thus a voltage e_1 is induced in coil 1 and e_2 in coil 2 as shown polarity. The dots indicated on two coils are the polarity marks.

10.3.1   Ideal Transformer Case

An ideal transformer has the following properties:

(a) The winding resistances are negligible.
(b) All the fluxes are confined to the magnetic core, i.e. leakage flux is zero.
(c) Core permeability is infinite ($\mu \to \infty$). The exciting current required to establish flux in the core is negligible (i.e. *mmf* drop in the core is zero).

Under these conditions, when switch of secondary side is opened (load is not connected), the induced voltage e_1 in primary coil will be equal and opposite to the applied voltage and can be written as

$$v_1 = -e_1 = N_1 \frac{d\Phi}{dt} = \text{Supply voltage} \qquad (10.2)$$

The core flux also induces voltage e_2 in secondary winding. Thus, it can be written as

$$v_2 = -e_2 = N_2 \frac{d\Phi}{dt} = \text{Terminal voltage} \qquad (10.3)$$

From Eqs. (10.2) and (10.3), we get

$$\frac{e_1}{e_2} = \frac{v_1}{v_2} = \frac{V_1}{V_2} = \frac{N_1}{N_2} = a \qquad (10.4)$$

where, v_1 and v_2 are instantaneous voltages whose ratio is identical with the ratio of *rms* voltage V_1 to V_2. a is called *turns ratio* or *transformation ratio*.

The reluctance of the magnetic circuit having cross-sectional area A and magnetic flux path length (mean) l will be

$$\mathfrak{R} = \frac{l}{\mu A} = 0 \ (\text{because } \mu = \infty) \qquad (10.5)$$

Thus, the inductance of primary coil will be

$$L = \frac{N^2}{\mathfrak{R}} = \infty \qquad (10.6)$$

Since secondary current is zero, the total *mmf* will be $N_1 i_1$. Thus,

$$N_1 i_1 = \Phi \mathfrak{R} \quad \Rightarrow \quad i_1 = 0 \qquad (10.7)$$

Equation (10.7) states that when secondary current is zero (no load on secondary side), the primary current is ideally zero for an ideal transformer.

When switch is closed, i.e. load is connected to the secondary winding. The secondary winding will produce an *mmf* equal to $N_2 i_2$ in the core. The flux produced by the secondary current will oppose the flux produced by the primary winding current (Lenz's law). Due to this reduction, e_1 will be reduced and therefore more current will flow in primary winding of transformer which will increase the flux in the core. It should be noted that core flux remains constant even under different loading conditions of given excitation voltage and frequency.

The total *mmf* in the magnetic circuit producing flux Φ will be,

$$F = N_1 i_1 - N_2 i_2 \qquad (10.8)$$

Since *mmf* required is zero ($F = \Phi \mathfrak{R} = 0$ as $\mathfrak{R}$ is zero), Eq. (10.8) will become

$$N_1 i_1 - N_2 i_2 = 0 \qquad (10.9)$$

or

$$\frac{i_1}{i_2} = \frac{N_2}{N_1} = \frac{1}{a} \qquad (10.10)$$

From Eqs. (10.4) and (10.10), we get

$$\frac{i_1}{i_2} = \frac{v_2}{v_1} = \frac{1}{a} \tag{10.11}$$

pr

$$v_1 i_1 = v_2 i_2 \tag{10.12}$$

> *In ideal transformer,*
> *Primary side total mmf = Secondary side total mmf*
> *Primary side volt-ampere = Secondary side volt-ampere*

Equation (10.12) states that the instantaneous power input to the transformer is equal to instantaneous power output from the transformer. This is true because power losses are assumed to be zero. Thus transformer is not an energy-storing device.

If supply voltage is sinusoidal, we write Eq. (10.11) in *rms* quantities as

$$\frac{I_1}{I_2} = \frac{N_2}{N_1} = \frac{V_2}{V_1} \tag{10.13}$$

or

$$V_1 I_1 = V_2 I_2 \tag{10.14}$$

Thus,

$$\text{Input volt-ampere} = \text{Output volt-ampere}$$

10.3.2 Induced *emf*–Flux Relationship

Let us consider applied voltage v_1 is

$$v_1 = -V_m \sin \omega t = V_m \cos (\omega t + \pi/2)$$

The flux Φ will be [using Eq. (10.2)]

$$\Phi = \frac{V_m}{N_1 \omega} \cos \omega t = \Phi_m \cos \omega t \tag{10.15}$$

The induced *emf* in primary winding,

$$e_1 = N_1 \frac{d\Phi}{dt} = -N_1 \Phi_m \omega \sin \omega t$$
$$= N_1 \Phi_m \omega \cos (\omega t + \pi/2) = E_{m1} \cos (\omega t + \pi/2) \tag{10.16}$$

Thus,

$$E_{rms} = \frac{E_{m1}}{\sqrt{2}} = \frac{N_1 \cdot \Phi_m \cdot 2\pi f}{\sqrt{2}} = E_1$$

$$\therefore \qquad E_1 = 4.44 \, N_1 f \Phi_m = 4.44 \, N_1 f B_m A \tag{10.17}$$

where A is the cross-sectional area of core.

The induced voltage in secondary winding will be

$$e_2 = N_2 \frac{d\Phi}{dt} = -N_2 \Phi_m \omega \sin \omega t = N_2 \Phi_m 2\pi f \cos \left(\omega t + \frac{\pi}{2} \right) \tag{10.18}$$

or

$$\therefore \qquad E_2 = E_{rms} = \frac{N_2 \Phi_m \cdot 2\pi f}{\sqrt{2}} = 4.44 N_2 f \Phi_m \qquad (10.19)$$

Thus, the secondary voltage is of the same frequency [Eq. (10.18)] as of primary applied voltage.

> In an ideal transformer, voltages are transformed in the direct ratio of turns, currents in the inverse ratio, power, frequency and volt-amperes are unchanged.

It should be noted that the flux used in Eqs. (10.17) and (10.19) is the peak flux rather *rms* flux value.

The phasor diagram as shown in Figure 10.5 is taken Φ as reference. The flux produced in the core is not a perfect sinusoidal as shown in Figure 10.6 which can be analyzed using B-H (Φ-i) curve.

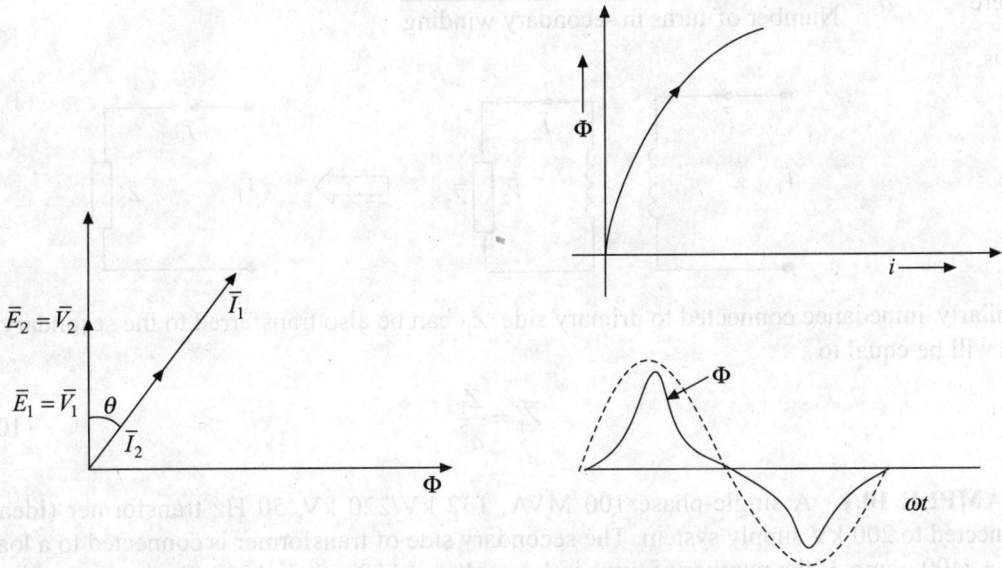

FIGURE 10.5 Phasor diagram of ideal transformer. **FIGURE 10.6** Time-varying flux in the core.

10.3.3 Impedance Transformation

Due to the presence of load in secondary side, the input impedance ($\overline{Z}_1$) will be

$$\overline{Z}_1 = \frac{\overline{V}_1}{\overline{I}_1} \qquad (10.20)$$

Replacing $\overline{V}_1$ and $\overline{I}_1$ in above equation using the following relations

$$\frac{\overline{V}_1}{\overline{V}_2} = \frac{N_1}{N_2} = \frac{\overline{I}_2}{\overline{I}_1}$$

We get

$$\bar{V}_1 = \left(\frac{N_1}{N_2}\right)\bar{V}_2, \quad \bar{I}_1 = \left(\frac{N_2}{N_1}\right)\bar{I}_2 \tag{10.21}$$

Using Eq. (10.21), Eq. (10.20) can be written as

$$\bar{Z}_1 = \frac{\left(\dfrac{N_1}{N_2}\right)\bar{V}_2}{\left(\dfrac{N_2}{N_1}\right)\bar{I}_2} = \left(\frac{N_1}{N_2}\right)^2 \cdot \frac{\bar{V}_2}{\bar{I}_2} = a^2 \bar{Z}_2 \tag{10.22}$$

$$= \bar{Z}_2' = \text{seocondary impedance referred (transferred) to as the primary side}$$

where $a = \dfrac{\text{Number of turns in primary winding}}{\text{Number of turns in secondary winding}}$

Thus,

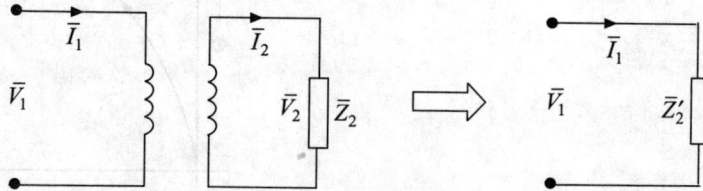

Similarly, impedance connected to primary side $\bar{Z}_1$ can be also transferred to the secondary side and will be equal to

$$\bar{Z}_1' = \frac{\bar{Z}_1}{a^2} \tag{10.23}$$

EXAMPLE 10.1 A single-phase 100 MVA, 132 kV/220 kV, 50 Hz transformer (ideal) is connected to 200 kV supply system. The secondary side of transformer is connected to a load of $300 + j400$ ohms. If the number of turns in low voltage (LV) side is 1000, find

(a) Turns ratio
(b) Secondary side voltage
(c) Number of turns on high voltage side
(d) The maximum value of core flux
(e) Primary (source) and secondary (load) currents
(f) Power supplied by source
(g) The value of impedance seen by source
(h) Phasor diagram taking source voltage as reference.

Solution: Transformer 220 kV (HV) side will be connected to the supply as shown in Figure 10.7 because connecting 200 kV on 132 kV side (LV) will damage the transformer insulation.

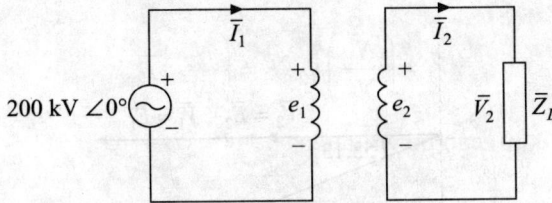

FIGURE 10.7   Example 10.1.

(a) Turns ratio = (Primary side turns)/ (Secondary side turns)

= (Primary voltage)/ (Secondary voltage)

$$= \frac{220}{132} = 1.667 = \frac{N_1}{N_2}$$

(b) Secondary side (load or low voltage side) will be calculated as

$$V_2 = \frac{N_2}{N_1} V_1 = \frac{1}{1.667} \times 200 = 120 \text{ kV}$$

(c) Number of turns in HV side $(N_1) = N_2 \times 1.667 = 1667$

(d) Using equation, $E_1 = 4.44 \, \Phi_m f N_1$

$$\Phi_m = \frac{E_1(=V_1)}{4.44 f N_1} = \frac{200 \times 10^3}{4.44 \times 50 \times 1667} = 0.54 \text{ Wb}$$

(e) Secondary or LV side current (I_2) will be

$$\bar{I}_2 = \frac{\bar{V}_2}{\bar{Z}_2} = \frac{120 \times 10^3 \angle 0°}{(300 + j400)} = 240 \angle -53.13° \text{ A}$$

High voltage side current,

$$\bar{I}_1 = \frac{N_2}{N_1} \bar{I}_2 = \frac{V_2}{V_1} \bar{I}_2 = 144 \angle -53.13° \text{ A}$$

(f) Power supplied = Power consumed (due to ideal transformer)

$$\bar{S}_2 = \bar{V}_2 \bar{I}_2^* = 1728 + j23.04 \text{ MVA}$$

(g) Impedance seen from source end

$$\bar{Z}_1 = \frac{\bar{V}_1}{\bar{I}_1} = 833.33 + j1111.11 \, \Omega$$

(h) Phasor diagram taking voltage as reference is shown in Figure 10.8.

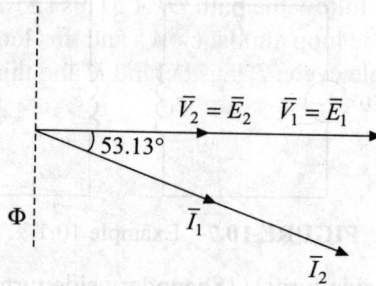

FIGURE 10.8

10.4   TRANSFORMER RATING

In theory, an ideal transformer might be capable of handling any voltage or any current, no matter how large it is. There are practical limits to both voltage (insulation problem) and current (heating problem). Since both voltage and current have upper limits, a transformer capability is given by the product of voltage and current in volt-amperes (S) which is equal for both the windings. In practical transformer, the relative phase angle of voltage and current has almost no effect on the voltage and current capabilities of the windings, and hence the magnitude of complex power, S (kVA or MVA), is used as rating of transformer along with voltages of primary and secondary sides. It is also mentioned in the nameplate rating whether transformer is single-phase or three-phase along with the operating frequency at which transformer performs satisfactorily. For example,

> *Single (or Three)-phase*, 100 MVA, 220 kV/132 kV, 50 Hz

10.5   LOSSES IN TRANSFORMER

In a practical transformer, it is not possible to have a core of infinite permeability and windings of zero resistance. Due to finite value of permeability, energy transferred during one half of current is not returned back during other half cycle of current. This is due to magnetic property of the material.

Consider a core which is initially un-magnetized. A coil of N turns is wounded on it. The magnetic intensity is increased slowly by increasing coil current i. The variation of flux density (B) will follow OA for H_1 as shown in Figure 10.9. The relationship of current i and flux intensity H for a toroidal will be

$$H = \frac{Ni}{l} \tag{10.24}$$

If magnetic field intensity is decreased slowly by decreasing exciting current i, the flux density will follow path $ABCD$. At the zero current ($H = 0$), the flux density is not zero. The magnetic material has retained flux density OB which is known as *residual flux density*. The flux density only becomes zero when flux intensity becomes negative (by reversing the current i). This magnetic flux intensity OC is known as the *coercively* or *coercive force* of magnetic material. If H varies from $-H_1$ to H_1, B-H curve will follow the path DEA'. This shows that loop does not close. After a few cycles of magnetization, the loop almost closes and this loop is known as *hysteresis loop*. It should be noted that during whole cycle, B lags behind H and this lagging phenomenon in core is called *hysterisis*.

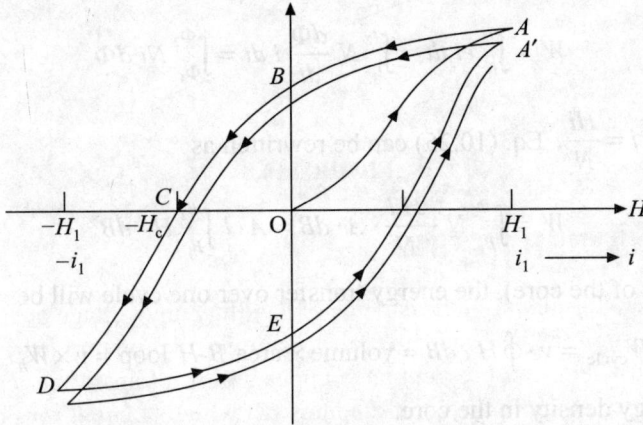

FIGURE 10.9   Typical hysteresis loop.

For different range of H, the different sizes of hysteresis loops are obtained. The locus of hysteresis loop is called *magnetization curve* as shown in Figure 10.10. The shape of hysteresis loop is different for different magnetic material. The area of loop is the hysteresis loss.

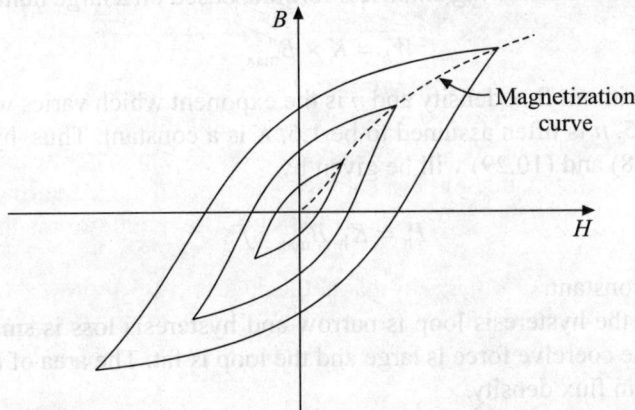

FIGURE 10.10   Hysteresis loop with magnetization curve.

10.5.1 Hysteresis Loss

Energy flows from the source to the coil-core assembly during one-half cycle is not returned completely during next half cycle due to loss in core. The loss of power in the core due to hysteresis effect is called *hysteresis loss*.

The induced *emf* in the core winding having N turns due to flux Φ in the core will be

$$e = N \frac{d\Phi}{dt}$$

The energy transfer during interval t_1 to t_2 due to current flow i will be

$$W = \int_{t_1}^{t_2} ei \, dt = \int_{t_1}^{t_2} N \frac{d\Phi}{dt} i \, dt = \int_{\Phi_1}^{\Phi_2} Ni \, d\Phi \qquad (10.25)$$

Since $\Phi = B \cdot A$ and $i = \dfrac{Hl}{N}$, Eq. (10.25) can be rewritten as

$$W = \int_{B_1}^{B_2} N \cdot \frac{Hl}{N} \cdot A \cdot dB = A \cdot l \int_{B_1}^{B_2} H \cdot dB \qquad (10.26)$$

Let $A \cdot l = v$ (volume of the core), the energy transfer over one cycle will be

$$W_{\text{cycle}} = v \cdot \oint H \cdot dB = \text{volume} \times \text{area } B\text{-}H \text{ loop} = v \times W_h \qquad (10.27)$$

where W_h is the energy density in the core.

Thus, the power loss due to hysteresis effect will be

$$P_h = v \cdot W_h \cdot f \qquad \left(\because P = \frac{W}{t} \right) \qquad (10.28)$$

Since B-H curve is a non-linear and multi-valued, the simple mathematical formula is not possible. Charles Steinmetz gave the following empirical formula based on a large number of experiments.

$$W_h = K \times B_{\text{max}}^n \qquad (10.29)$$

where B_{max} is the maximum flux density and n is the exponent which varies with core material in the range of 1.5 to 2.5. n is often assumed to be 1.6. K is a constant. Thus, hysteresis loss in the core using Eqs. (10.28) and (10.29) will be given by

$$P_h = K_h \, B_{\text{max}}^n \cdot f \qquad (10.30)$$

where K_h is another constant.

For silicon steel, the hysteresis loop is narrow and hysteresis loss is small. However, for a permanent magnet, the coercive force is large and the loop is fat. The area of loop increases non-linearly with maximum flux density.

10.5.2 Eddy Current Loss

Due to rapid change of Φ (time varying) in the core, the voltage will be induced in the core and eddy current flows around the path as shown in the Figure 10.11.

The direction of i_e will be such that it will oppose the flux Φ. Since core will have some resistance R_e, there will be a power loss $i_e^2 R_e$, known as *eddy current loss* (P_e). Eddy current i_e depends on the induced voltage e_e which is proportional to the $f \cdot B_{\text{max}}$.

Thus,

$$P_e = i_e^2 R_e = \frac{e_e^2}{R_e} = K_e \, B_{max}^2 \cdot f^2 \qquad (10.31)$$

where K_e is a constant which depends on the type of material and its lamination thickness.

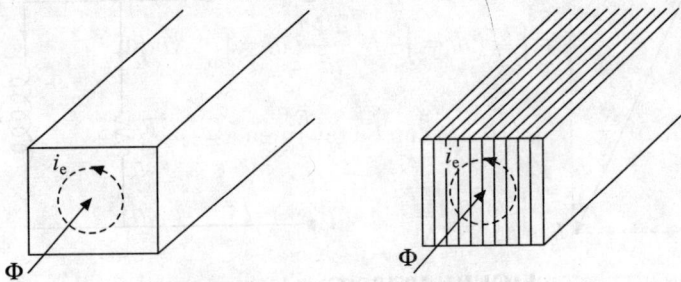

FIGURE 10.11   Eddy current flow.

Eddy current loss can be reduced in two ways as

1. By using core of high resistivity (a few % of silicon with iron increases the resistivity significantly).
2. By using laminated core: Instead of using one solid core, thin laminated cores (insulated core) are used as shown in Figure 10.11. By doing this, the resistance of eddy current path is increased and thus, the eddy current loss is reduced significantly. The thickness of laminations varies from 0.5 to 5 mm in electrical apparatus, whereas it varies from 0.01 mm to 0.5 mm in the electronic circuits devices using higher frequencies.

10.5.3   Core Loss

It should be noted that hysteresis loss is due to slow variation of flux whereas eddy current loss is due to fast variation of flux. At slow variation of Φ, the P_e is very small and at large variation of Φ, hysteresis loss is very small. At any rate of change of flux, the total loss including hysteresis loss (P_h) and eddy current (P_e) are known as *core loss* (P_c). Thus,

$$P_c = P_h + P_e \qquad (10.32)$$

When there is no loss in the core, the exciting current I_m (also known as *magnetizing current*) will be in phase with flux Φ which lags by 90° by induced voltage E. The equivalent circuit and phasor diagram is shown in Figure 10.12. L_m is magnetization inductance.

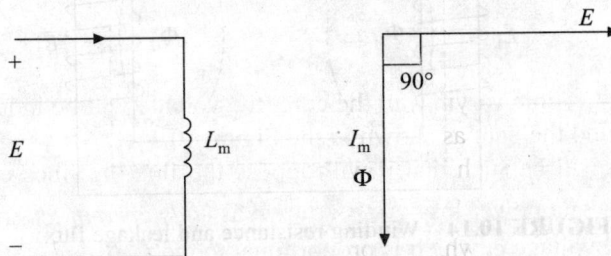

FIGURE 10.12   Magnetization current.

Due to core loss, a current I_c flows in phase with E along with magnetizing current I_m which lags E as shown in figure given in Figure 10.13. It should be noted that in practical transformer, the exciting current I_0 is different than the magnetizing current I_m.

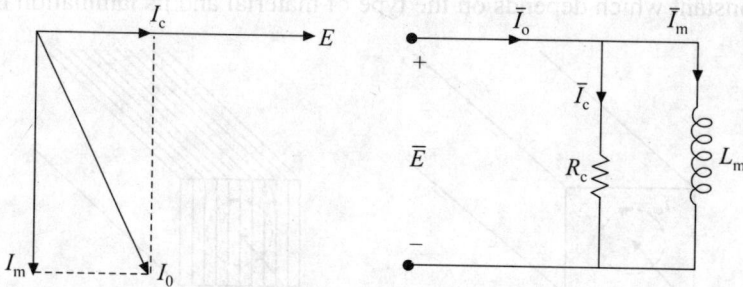

FIGURE 10.13   Core loss current.

10.5.4   Copper Loss

Due to finite resistances of windings, there are losses (real power) in both the windings (i^2R) which is known as *copper loss*.

10.6   PRACTICAL TRANSFORMER

The assumptions made for ideal transformer are not valid for practical transformers as in practical transformers,

- winding resistances are not zero.
- there are leakage fluxes. (The flux produced by the source is not only confined to the core. Leakage flux is magnetic flux in the space between windings).
- core has finite permeability.

A single-phase practical transformer having primary winding resistance r_1 ohm and secondary winding resistance r_2 ohm is shown in Figure 10.14.

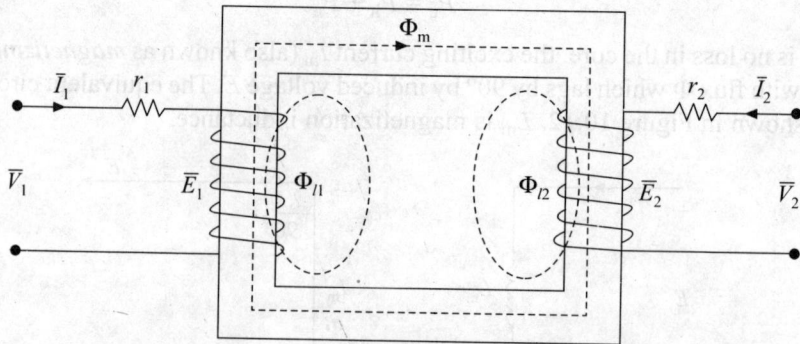

FIGURE 10.14   Winding resistance and leakage flux.

Due to leakage flux, the leakage inductance of primary winding (L_{l1}) will be defined as

$$L_{l1} = \frac{N_1 \Phi_{l1}}{I_1} \qquad (10.33)$$

Similarly leakage inductance of secondary winding

$$L_{l2} = \frac{N_2 \Phi_{l2}}{I_2} \qquad (10.34)$$

The equivalent circuit with leakage flux is shown in Figure 10.15.

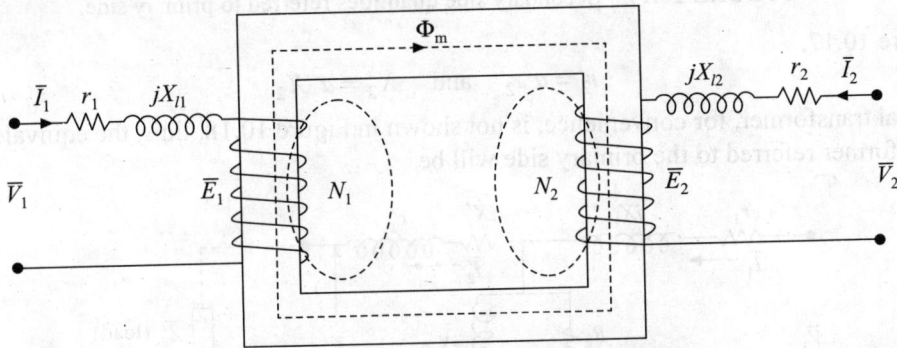

FIGURE 10.15   Equivalent circuit with leakage flux.

where X_{l1} (= X_1) and X_{l2} (= X_2) are leakage reactances of primary and secondary winding, respectively. Applying 3rd condition (the *mmf* drop in the core is not zero), some magnetizing current is required to flow the flux. Considering core loss and magnetization, the equivalent circuit can be shown (Figure 10.16) as

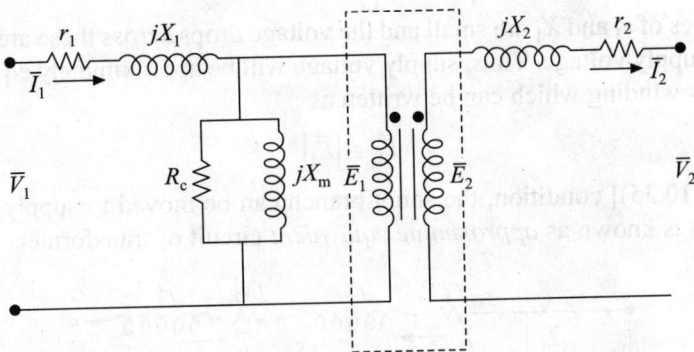

FIGURE 10.16   Equivalent circuit of a practical transformer.

Secondary side impedance and voltage can be transferred to the primary side as shown in the equivalent circuit of transformer (Figure 10.17).

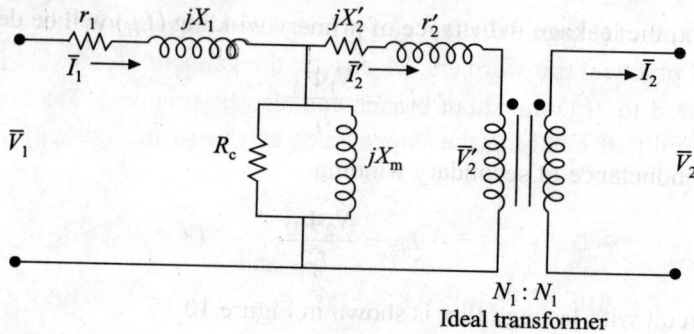

FIGURE 10.17   Secondary side quantities referred to primary side.

In Figure 10.17,

$$r_2' = a^2 r_2 \quad \text{and} \quad X_2' = a^2 X_2$$

The ideal transformer, for convenience, is not shown in Figure 10.18, thus, the equivalent circuit of transformer referred to the primary side will be

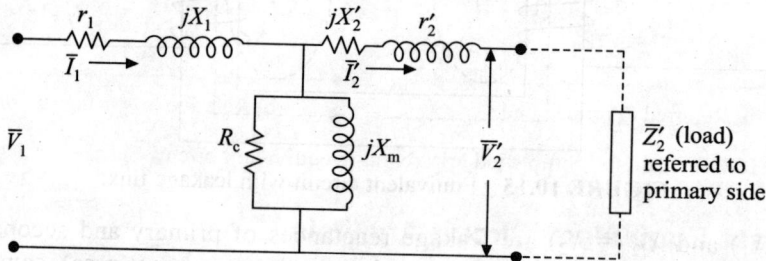

FIGURE 10.18   Equivalent circuit (referred to primary side).

10.6.1   Approximate Equivalent Circuit

Normally the values of r_1 and X_1 are small and the voltage drops across these are significantly less compared to the supply voltage. Thus, supply voltage will be approximately equal to the induced voltage in primary winding which can be written as

$$|\overline{V}_1| \cong |\overline{E}_1| \tag{10.35}$$

With above [Eq. (10.35)] condition, the shunt branch can be moved to supply side as shown in Figure 10.19. This is known as *approximate equivalent* circuit of transformer.

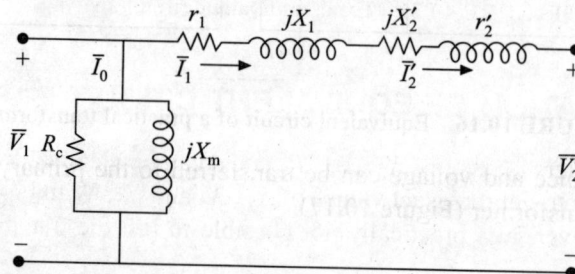

FIGURE 10.19   Approximate equivalent circuit of transformer.

This approximate circuit (Figure 10.19) is very widely used for performance analysis (efficiency and regulation) of practical transformers. Moreover, the value of $|\bar{I}_0|$ (shunt branch current) is very small compared to $|\bar{I}_1|$, the shunt branch can also be removed. The approximate circuit can be further simplified clubbing the impedances as shown in Figure 10.20. The variables used are

$$R_{eq1} = r_1 + a^2 r_2; \quad X_{eq1} = X_1 + a^2 X_2$$

$$R_{eq2} = r_2 + r_1/a^2; \quad X_{eq2} = X_2 + X_2/a$$

and

$$\bar{V}_2' = a\bar{V}_2, \quad \bar{V}_1' = \frac{\bar{V}_1}{a},$$

$$\bar{I}_1' = a\bar{I}_1 = \bar{I}_2, \bar{I}_2' = \bar{I}_2/a$$

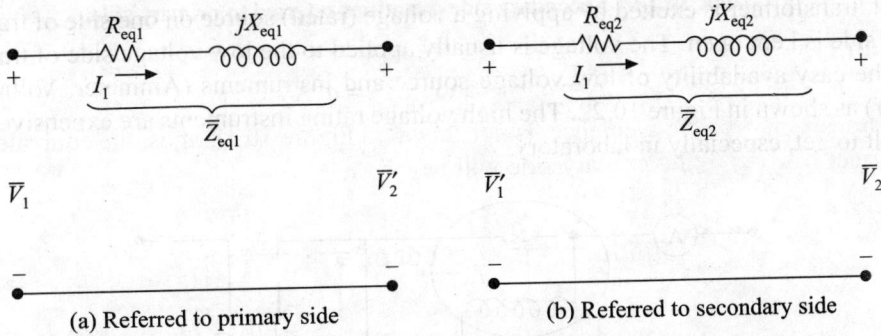

(a) Referred to primary side   (b) Referred to secondary side

FIGURE 10.20   Simplified approximate equivalent circuit of transformer.

10.6.2 Exact Equivalent Circuit Analysis

Thevenin equivalent between terminals a and b can be obtained and then the secondary current and other quantities can be calculated as shown in Figure 10.21.

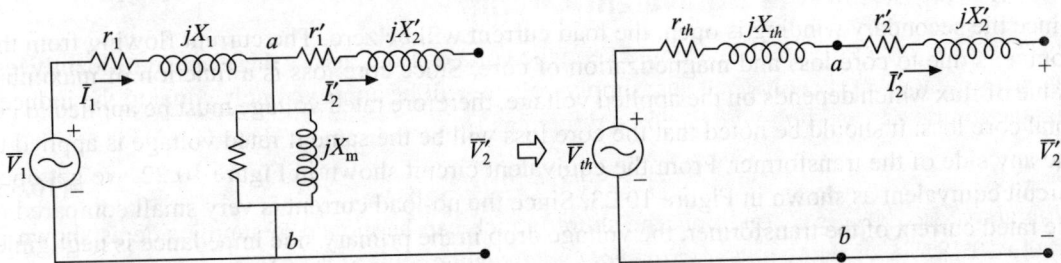

FIGURE 10.21   Exact equivalent circuit analysis.

10.7 DETERMINATION OF EQUIVALENT CIRCUIT PARAMETERS

Parameters of transformers r_1, X_1, r_2, X_2, X_m, R_c and a can be obtained from the design data but it is sometimes difficult to get the exact values of X_2, X_1 and X_m. With help of tests, these values can be obtained. However, it is practically not possible to test the transformer in real practical system to determine the equivalent parameters which are useful for determination of efficiency and regulation of the transformers. Direct load testing has the following three problems.

1. There is huge energy loss during testing.
2. Arrangement of large load is difficult.
3. Errors in the losses determined by direct load testing are higher.

Without direct load testing, the transformer losses and parameters are determined by two tests as explained below.

10.7.1 No Load Test or Open Circuit (OC) Test

In this test, transformer is excited by applying a voltage (rated) source on one side of transformer and other side is kept open. The voltage is usually applied to the low voltage side of transformer because the easy availability of low voltage source and instruments (Ammeter, Voltmeter and Wattmeter) as shown in Figure 10.22. The high voltage rating instruments are expensive and may be difficult to get, especially in laboratory.

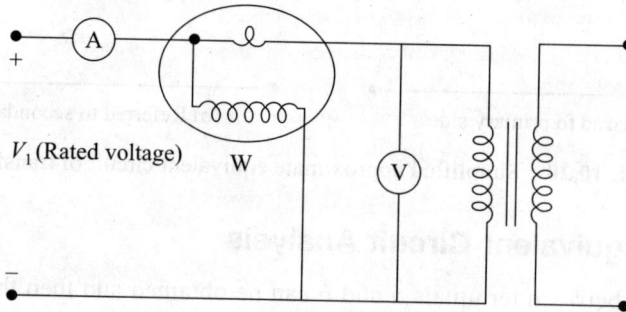

FIGURE 10.22   Open-circuit test.

Since the secondary winding is open, the load current will be zero. The current flowing from the source is due to core loss and magnetization of core. Since core loss is a function of maximum value of flux which depends on the applied voltage, therefore rated voltage must be applied to get total core loss. It should be noted that the core loss will be the same if rated voltage is applied to the any side of the transformer. From the equivalent circuit shown in Figure 10.22, we get open circuit equivalent as shown in Figure 10.23. Since the no-load current is very small compared to the rated current of the transformer, the voltage drop in the primary side impedance is negligible.

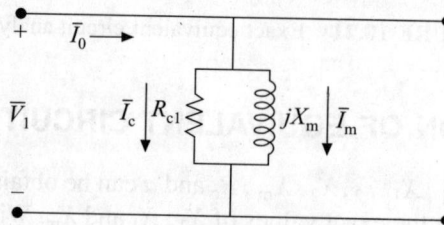

FIGURE 10.23   Equivalent circuit of open-circuit test.

Let the voltmeter reading[*] be V_1, ammeter reading I_0 and wattmeter readings P_c watt. Wattmeter reads the total core loss (P_c) in the transformer at applied voltage because the copper loss is negligible. Thus,

$$R_{c1} = \frac{V_1^2}{P_c}, \quad I_c = \frac{V_1}{R_c} \tag{10.36}$$

Hence,

$$I_m = \sqrt{I_0^2 - I_c^2} \quad \text{and} \quad X_{m1} = \frac{V_1}{I_m} \tag{10.37}$$

Therefrom no-load test we get,

(i) Core loss and
(ii) Shunt branch parameters (R_{c1} and X_{m1}) of transformer referred to the primary side of the transformer[**].

10.7.2 Short Circuit (SC) Test

As its name suggests, one winding of the transformer is short circuited and applying voltage so that rated current flows in other winding as shown in Figure 10.24. The excitation voltage is applied to high voltage side because the required rated current is low in high voltage side and thus the required excitation (V_{sc}) is also less (normally 5–10 % of rated voltage). Normally, voltage source is connected to the transformer through an autotransformer so that voltage can be varied slowly till ammeter reads the rated current of primary winding.

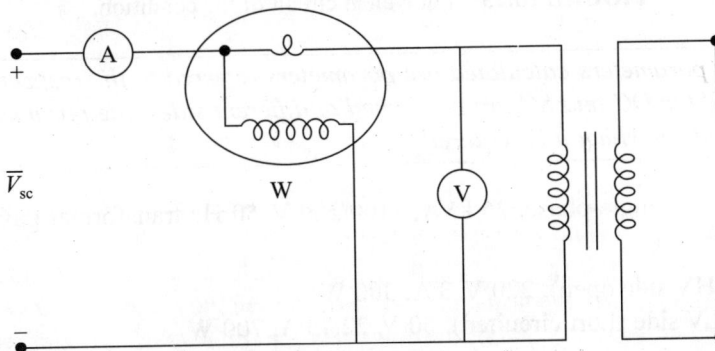

FIGURE 10.24 Short-circuit test.

Let the voltmeter reading is V_{sc}, ammeter reading is I_{sc} and wattmeter readings is P_{sc} watt. Wattmeter reads the total copper loss (P_{sc}) in the transformer because the core loss is negligible at small applied voltage. Since reduced voltage is applied for even full-load current, the following assumption can be taken.

(a) Shunt branch current is very small and therefore, shunt branch can be neglected.
(b) Due to reduced voltage, wattmeter reads the copper loss of both the winding. The core loss is very small.

---

[*] Please note that voltage and ammeter read the *rms* values of voltage and current, respectively.
[**] Primary side indicates the side where instruments are connected. It is normally low voltage side.

With these above assumptions, the equivalent circuit of SC test is shown in Figure 10.25. The equivalent impedance (referred to the primary side of the transformer) will be calculated as*

$$Z_{eq1} = \frac{V_{sc}}{I_{sc}} = \sqrt{R_{eq1}^2 + X_{eq1}^2} \qquad (10.38)$$

where $R_{eq1} = r_1 + r_2'$ and $X_{eq1} = x_1 + x_2'$
and

$$R_{eq1} = \frac{P_{sc}}{I_{sc}^2} \text{ and } X_{eq1} = \sqrt{Z_{eq1}^2 - R_{eq1}^2} \qquad (10.39)$$

By knowing any side parameters (r and X), the other side impedance can be calculated.

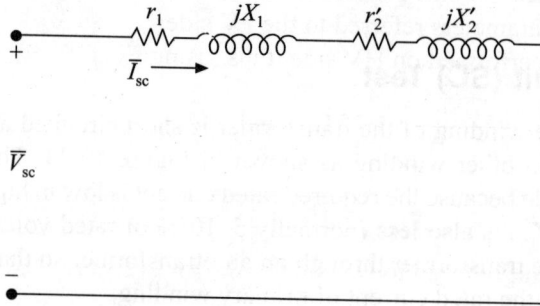

FIGURE 10.25   Equivalent circuit of SC condition.

Please note that parameters calculated are parameters referred to the respective side of the test performed. If the OC and SC are performed at different sides, the parameters should be changed to the side at which it is required.

EXAMPLE 10.2   A single-phase, 25 kVA, 1100/220 V, 50 Hz transformer gave the following test results

OC Test (HV side open): 220 V, 3 A, 300 W
SC Test (LV side short circuited): 50 V, 22.73 A, 700 W

Find,

(a) Rated current for HV and LV side
(b) Derive approximate equivalent circuit referred to the HV side and LV side.

Solution:

(a) Rated HV side current $= \dfrac{25 \times 10^3}{1100} = 22.73$ A

Rated LV side current $= \dfrac{25 \times 10^3}{220} = 113.64$ A

$* Z_{eq1} = |\bar{Z}_{eq1}|$

(b) $R_{c1} = \dfrac{(220)^2}{300} = 161.33 \ \Omega$

$I_{c1} = \dfrac{220}{R_{c1}} = 1.36 \ A$

Using Eq. (10.37), we get

$$I_{m1} = \sqrt{3.0^2 - 1.36^2} = 1.067 \ A$$

$$X_{m1} = \dfrac{V_1}{I_{m1}} = \dfrac{220}{1.067} = 187.32 \ \Omega$$

Suffix 1 denotes the LV side parameters and 2 denoted HV side. Please note that OC test will give shunt parameters referred to the LV side.

Since SC test is performed on HV side, thus, using Eq. (10.39), we get

$$R_{eq2} = \dfrac{700}{(22.73)^2} = 1.35 \ \Omega \quad \text{and} \quad Z_{eq2} = \dfrac{50}{22.73} = 2.2 \ \Omega$$

Thus,

$$X_{eq2} \sqrt{2.2^2 - 1.35^2} = 1.7 \ \Omega$$

(i) Equivalent circuit referred to HV side is shown in Figure 10.26.

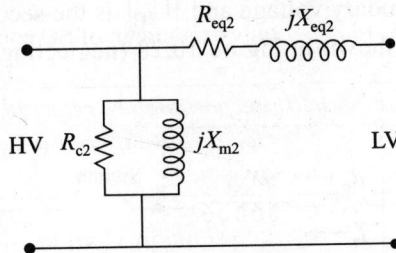

FIGURE 10.26

where

$$R_{c2} = \left(\dfrac{11000}{220}\right)^2 R_{c1} = 2500 \times 161.33 = 403325 \ \Omega$$

$$X_{m2} = 2500 \times 187.32 = 468300 \ \Omega$$

(ii) LV side equivalent circuit is shown in Figure 10.27.

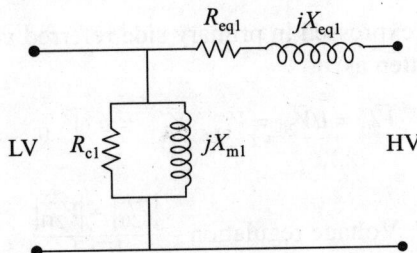

FIGURE 10.27

where

$$R_{eq1} = \frac{R_{eq2}}{a^2} = 0.00054 \ \Omega$$

$$X_{eq1} = \frac{X_{eq2}}{a^2} = 0.00088 \ \Omega$$

All other quantities are already computed.

10.8 VOLTAGE REGULATION

Most of the loads operate satisfactorily if they are connected to its rated voltage supply. The performance of load deteriorates if the terminal voltage varies from its rated voltage and frequency. Due to the internal drop of transformer, the load terminal voltage changes when current is drawn by the load.

The voltage regulation is defined as the ratio of change in the secondary voltage from the no-load to full-load condition keeping the primary voltage constant. Mathematically, it is expressed as

$$\text{Voltage regulation} = \frac{\left|\overline{V}_{20}\right| - \left|\overline{V}_{2,fl}\right|}{\left|\overline{V}_{2,fl}\right|} \tag{10.40}$$

where $\left|\overline{V}_{20}\right|$ is the no-load secondary voltage and $\left|\overline{V}_{2fl}\right|$ is the secondary voltage at full load.

Consider a transformer as shown in Figure 10.28 (neglecting shunt branch) connected to a load with switch.

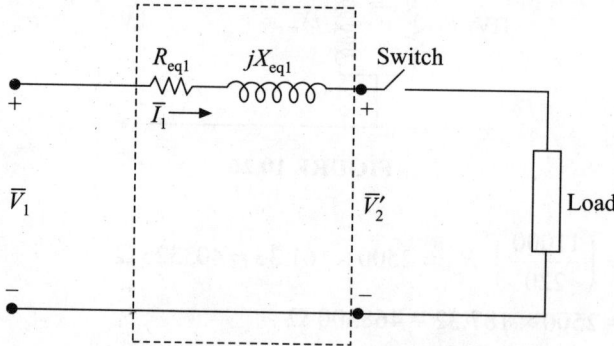

FIGURE 10.28 Transformer equivalent circuit connected with load.

The voltage regulation can be expressed in primary side referred voltages. If turns ratio is a, the secondary voltages can be written as

$$\overline{V}_{20}' = a\overline{V}_{20} = \overline{V}_1, \quad \overline{V}_{2,fl}' = a \cdot \overline{V}_{2,fl}$$

Thus,

$$\text{Voltage regulation} = \frac{\left|\overline{V}_{20}'\right| - \left|\overline{V}_{2fl}'\right|}{\left|\overline{V}_{2fl}'\right|} \tag{10.41}$$

It should be noted that secondary terminal voltage may go up or down from the rated voltage depending on the power factor of load.

Consider current $\overline{I}_1$ flow in the primary winding when switch S is closed (All the quantities are referred to the primary side). Taking V_2' as reference, the phasor diagram is shown in Figure 10.29.

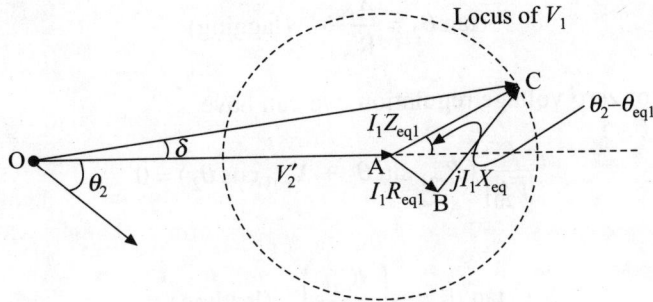

FIGURE 10.29 Phasor diagram.

From phasor diagram, we can write

$$\overline{V}_1 = \overline{I}_1(R_{eq1} + jX_{eq1}) + \overline{V}_2' \qquad (10.42)$$

when

$$\overline{I}_2' = 0 = \overline{I}_1 \text{ then } \overline{V}_1 = \overline{V}_2'.$$

In determining the voltage regulation, it is customary to assume that V_1 is varied to keep the load voltage to its rated voltage $\left(\left|\overline{V}_{2fl}'\right|\right)$ when rated current flows through it. From the locus of V_1 in Figure 10.29, it can be seen that $\left|\overline{V}_1\right|$ can be less than $\left|\overline{V}_2'\right|$ for the some value of load power factor. The value of $\left|\overline{V}_1\right|$ is maximum when $\theta_2 - \theta_{eq1} = 0$.

Thus, voltage regulation is maximum when power factor angle is

$$\theta_2 = \theta_{eq1} = \tan^{-1}\left(\frac{X_{eq1}}{R_{eq1}}\right) \qquad (10.43)$$

From phasor diagram, it can be written as, $\left(\text{please note that } \left|\overline{V}_{2fl}'\right| = \left|\overline{V}_2'\right|\right)$

$$\left|\overline{V}_1\right| \cong \left|\overline{V}_{2,fl}'\right| + (I_1 R_{eq1} \cos\theta_2 + I_1 X_{eq1} \sin\theta_2) \qquad (10.44)$$

or

$$\left|\overline{V}_1\right| - \left|\overline{V}_{2fl}'\right| = (I_1 R_{eq1} \cos\theta_2 + I_1 X_{eq1} \sin\theta_2) \qquad (10.45)$$

From Eq. (10.45), we can write

$$\text{Regulation} = \frac{\left|V_{20}'\right| - \left|V_{2fl}'\right|}{\left|V_{2fl}'\right|} = \left(\frac{I_1 R_{eq1} \cos\theta_2}{\left|\overline{V}_{2fl}'\right|} + \frac{I_1 X_{eq1} \sin\theta_2}{\left|\overline{V}_{2fl}'\right|}\right) \qquad (10.46)$$

For maximum value of regulation for a power factor angle, $\dfrac{d(\text{Regulation})}{d\theta_2} = 0$

or

$$-R_{eq1} \sin \theta_2 + X_{eq1} \cos \theta_2 = 0$$

Therefore,

$$\tan \theta_2 = \frac{X_{eq1}}{R_{eq1}} \quad \text{(lagging)} \tag{10.47}$$

Using Eq. (10.46), for zero voltage regulation, we can have

$$\frac{I_1}{|\overline{V}'_{2fl}|}(R_{eq1} \sin \theta_2 + X_{eq1} \cos \theta_2) = 0$$

or

$$\tan \theta_2 = -\left(\frac{R_{eq1}}{X_{eq1}}\right) \quad \text{(leading)} \tag{10.48}$$

For this condition, the sign of θ_2 is opposite as that of shown in phasor diagram. When regulation is multiplied by 100, the regulation is termed as percentage regulation.

EXAMPLE 10.3 Find the voltage regulation of a transformer of Example 10.2 for the following load conditions (LV side)

(a) 80% of full load at 0.8 pf lagging
(b) 100% of full load at 0.6 pf leading
(c) Draw the phasor diagram for conditions (a) and (b)

Solution: Consider an approximate equivalent circuit of transformer referred to primary (HV) side (Figure 10.30).

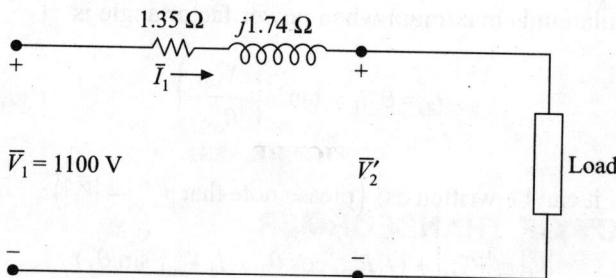

FIGURE 10.30

At no load, $\qquad\qquad\qquad\qquad V'_2 = 1100 \text{ V}$

Rated (full load) primary current $I_{fl} = \dfrac{25 \times 10^3}{1100} = 22.73 \text{ A} = I'_2$

(a) 80% of full load Case:

$$I_1 = 0.8 \times 22.73 = 18.18 \text{ A}$$

p.f. angle $= \cos^{-1}(0.8) = 36.87°$ (lagging)

Calculating V_1' with 80% loading at 0.8 p.f. lagging (Taking V_2' as reference)

$$\bar{V}_1 = \bar{I}_1(R_{eq1} + jX_{eq1}) + \bar{V}_2' = 18.18 \angle -36.87° \times (1.35 + j1.74) + 1100 \angle 0°$$

$$\bar{V}_1 = 1138.67 + j10.58 = 1138.67 \angle 0.53° \text{ V}$$

Thus,

$$\% \text{ voltage regulation} = \frac{1138.67 - 1100}{1100} \times 100 = 3.52\%$$

(b) Regulation at full load and 0.6 p.f. leading

$$\bar{I}_1 = 22.73 \angle 53.13° \text{ A}$$

$$\bar{V}_1 = 1100 \angle 0° + 22.73 \angle 53.13° \times (1.35 + j1.74) = 1086.77 + j48.28 \text{ V}$$

$$= 1087.84 \angle 2.54° \text{ V}$$

$$\% \text{ voltage regulation} = \frac{1087.84 - 1100}{1100} \times 100 = -1.11\%$$

It should be noted that voltage regulation for leading p.f. is negative but it is always not true.

(c) Phasor diagram (Figure 10.31)

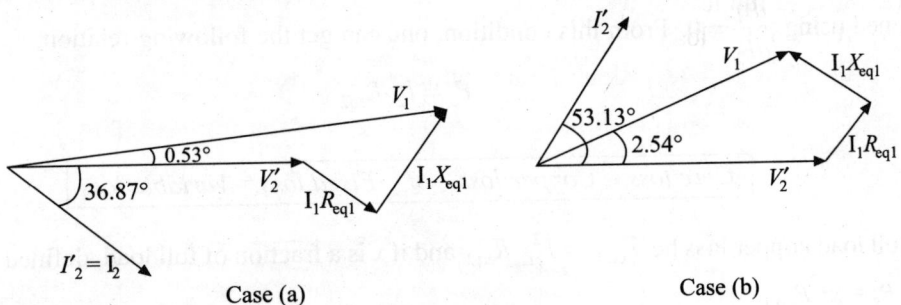

Case (a)   Case (b)

FIGURE 10.31

10.9 EFFICIENCY OF TRANSFORMER

The efficiency (η) of a transformer is defined as the ratio of the output power to the input power. Mathematically, it can be written as,

$$\eta = \frac{\text{Output power}}{\text{Input power}} = \frac{\text{Output power}}{\text{Output power} + \text{Losses}} \qquad (10.49)$$

or

$$\eta = \frac{\text{Input power} - \text{Losses}}{\text{Input power}} = 1 - \frac{\text{Losses}}{\text{Output power} + \text{Losses}} \qquad (10.50)$$

There are basically two types of losses, i.e. core loss (P_c) and copper loss (P_{cu}). The total copper loss can be obtained, if winding resistances and currents are known. The copper loss can be calculated as

$$P_{cu} = I_1^2 r_1 + I_2^2 r_2 = I_1^2 R_{eq1} = I_2^2 R_{eq2} \qquad (10.51)$$

Core loss is a function of the peak flux density (B_{max}) in core which depends on the applied voltage to the transformer. Normally transformers are connected to the constant voltage and thus the core loss is fixed. There are some other losses, which are very less and are normally neglected, such as:

(a) Dielectric loss due to leakage current in the insulating materials.
(b) Stray loss resulting from the voltage flux inducing eddy currents in tank wall and conductors.

From the OC and SC tests, we can determine core loss and equivalent resistance/reactances of winding. Knowing the load current, total losses can be calculated and thus, the efficiency of transformer.

$$\text{Power output} = V_2 I_2 \cos \theta_2$$

Therefore,
$$\eta = \frac{V_2 I_2 \cos \theta_2}{V_2 I_2 \cos \theta_2 + P_c + I_2^2 R_{eq2}} \qquad (10.52)$$

From above Eq. (10.52), it can be seen that the η of transformer depends on the load current (I_2), load power factor because voltage is normally fixed. The maximum η for load current I_2 can be obtained using $\dfrac{d\eta}{dI_2} = 0$. From this condition, one can get the following relation,

$$P_c = I_2^2 R_{eq2}$$

i.e.

$$\boxed{Core\ loss = Copper\ loss \quad or \quad Fixed\ loss = Variable\ loss}$$

Let full load copper loss be $P_{cu,fl} = I_{2fl}^2 R_{eq2}$ and if x is a fraction of full load, defined as $x = \dfrac{I_2}{I_{2fl}}$, then $P_c = x^2 P_{cu,fl}$

or
$$x = \sqrt{\frac{P_c}{P_{cu,fl}}} \qquad (10.53)$$

At any fractional loading x, the efficiency is defined as

$$\eta = \frac{x \times kVA \times p.f.}{x \times kVA \times p.f. + \text{Losses (in kVA)}} = \frac{x V_2 I_{2,fl} \cos \theta_2}{x V_2 I_{2,fl} \cos \theta_2 + P_c + x^2 (P_{cu,fl})} \qquad (10.54)$$

where kVA is the rated value (full load).

The maximum efficiency for a power factor at the constant value of voltage V_2 and load current I_2 will be

$$\frac{d\eta}{d\theta_2} = 0 \qquad (10.55)$$

From Eqs. (10.52) and (10.55), we get,

$$\theta_2 = 0$$

or

$$\text{p.f.} = \cos \theta_2 = 1 \qquad (10.56)$$

Thus, the maximum efficiency in a transformer occurs when core loss is equal to copper loss and power factor of load is unity.

Figure 10.32 shows the variation of efficiency of transformer with loading and power factor.

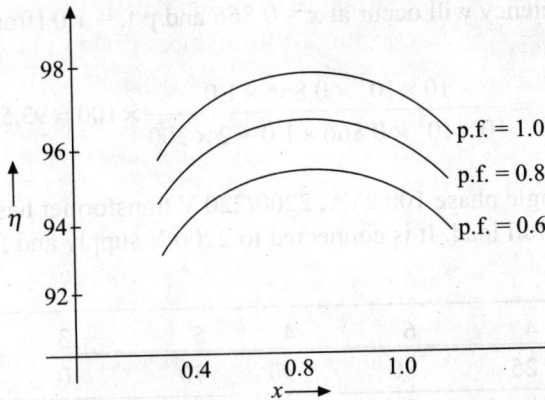

FIGURE 10.32   Variation of efficiency with load and power.

10.10   ALL DAY (ENERGY) EFFICIENCY

Efficiency defined in section (10.9) deals with power. The load on distribution transformers varies widely throughout the day and thus, the efficiency. It is, therefore, necessary for the distribution transformers to have high all day efficiency, (i.e. efficiency at average output power). All day (energy) efficiency is defined as

$$\eta_e = \frac{\text{Average power output}}{\text{Average power output} + \text{Average power loss}} \qquad (10.57)$$

or $\eta_e = \dfrac{\text{Energy output over 24 hours}}{\text{Energy input over 24 hours}} = \dfrac{\text{Energy output over 24 hrs}}{\text{Energy output over 24 hrs} + \text{Energy losses in 24 hrs}}$

$$(10.58)$$

EXAMPLE 10.4   A single-phase 10 kVA, 11000/220 V transformer has core loss $P_c = 300$ W at rated voltage and a copper loss $P_{cu} = 400$ W at full load. Find the efficiency of transformer feeding to a load 8 kVA at 0.8 p.f. lagging. What will be the maximum η of the transformer?

Solution: Loading of transformer $= \dfrac{8}{10} = 0.8$ of full load

Thus, copper loss at 80% loading will be $(0.8)^2 \times 400 = 256$ W

$$\eta = \frac{kVA \times \text{power factor}}{kVA \times \text{power factor} + P_c + P_{cu}} = \frac{8 \times 10^3 \times 0.6}{8 \times 10^3 \times 0.6 + 300 + 256} = 0.8962$$

or $\eta = 89.62\%$

For the maximum efficiency, $P_{cu} = P_c = 300$ W
P_{cu} at full load = 400 W

Loading of transformer $(x) = \sqrt{\left(\dfrac{300}{400}\right)} = 0.866$

Thus the maximum efficiency will occur at $x = 0.866$ and p.f. = 1.0 [from Eq. (10.56)]
Thus,

$$\eta_{max} = \frac{10 \times 10^3 \times 0.866 \times 1.0}{10 \times 10^3 \times 0.866 \times 1.0 + 2 \times 300} \times 100 = 93.52\%$$

EXAMPLE 10.5   A single phase 100 kVA, 2200/220 V transformer has core loss of 400 W and copper loss of 600 W at full load. It is connected to 2200 V supply and feeding load at 220 V as per following schedule.

Hours	4	6	4	5	3	2
Load(kVA)	25	80	90	70	0	switched off
p.f.	0.6 lag	0.8 lag	1	0 .8 lead	0	0

Find the all-day energy efficiency (η_e).

Solution: Energy output for 24 hrs (= Σ kVA × p.f. × time) will be

$= 25 \times 0.6 \times 4 + 80 \times 6 \times 0.8 + 90 \times 1.0 \times 4 + 70 \times 0.8 \times 5 + 0 + 0$
$= 1084$ kWh

Core loss for 24 hrs $= (24 - 2) \times 400 = 8.8$ kWh

Copper loss for 24 hrs $\left(\Sigma \left(\dfrac{kVA}{100} \right)^2 \times \text{time} \times 600 \right)$

$= (0.25)^2 \times 600 \times 4 + (0.8)^2 \times 600 \times 6 + (0.9)^2 \times 600 \times 4 + (0.70)^2 \times 600 \times 5$
$= 5.868$ kWh

Therefore,

$$\eta_e = \frac{1084}{1084 + 8.8 + 5.868} \times 100 = 98.66\%.$$

10.11   AUTOTRANSFORMER

Two windings in the conventional transformers as discussed earlier are electrically isolated whereas in autotransformer primary and secondary windings are electrically connected. However, the basic operation principle is the same as two-winding transformer. An ideal autotransformer is shown in Figure 10.33.

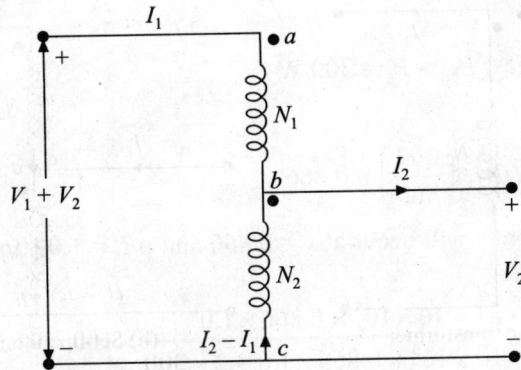

FIGURE 10.33   Autotransformer.

Since, the same flux links all the turns in the transformer core and polarity of induced voltages in both the windings are additive, we can write,

$$\frac{V_2 + V_1}{V_2} = \frac{N_1 + N_2}{N_2} = 1 + a = a' \tag{10.59}$$

where a' is the turns ratio of autotransformer.

Ampere-turn (*mmf*) provided by winding *ab* is

$$F_{ab} = N_1 I_1$$

Ampere turn provided by *bc* is

$$F_{bc} = N_2 (I_2 - I_1)$$

For no *mmf* drop (ideal autotransformer case),

$$F_{ab} = F_{bc} = N_1 I_1 = N_2 (I_2 - I_1) \tag{10.60}$$

or

$$(N_1 + N_2) I_1 = N_2 I_2$$

Thus,

$$\frac{I_1}{I_2} = \frac{N_2}{N_1 + N_2} = \frac{1}{a'} \tag{10.61}$$

Equation (10.61) is similar to two-winding transformer.

The main advantages of autotransformer connection are as follows:

(a) Lower leakage reactance

(b) Lower exciting current

(c) Increased kVA rating

(d) Variable voltage with sliding contact.

However, the major disadvantage is the direct connection between primary and secondary windings.

If two-winding transformer V_1/V_2 as shown in Figure 10.34(a) is connected as an autotransformer, the following configurations are normally used as shown in Figure 10.34.

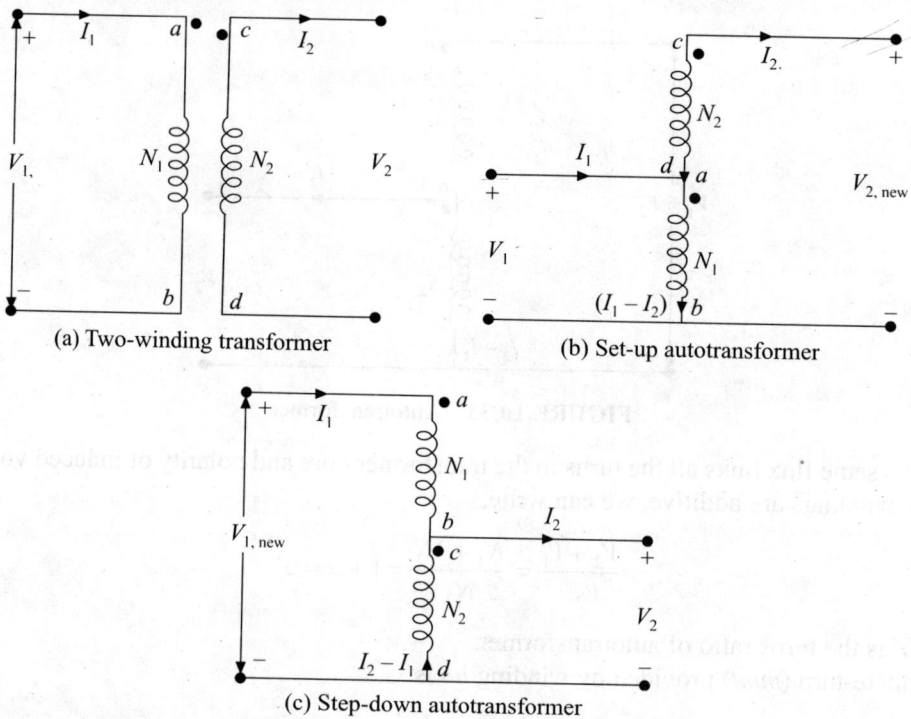

(a) Two-winding transformer

(b) Set-up autotransformer

(c) Step-down autotransformer

FIGURE 10.34   Autotransformer connections of two-winding transformer.

Using above connection, a 2-winding, 2500/250 V transformer can be changed either to 2500/2750 V or 2750/250 V auto transformer. Note that current in windings are such that flux produced by them must oppose each other, i.e. when one current leaves the dot then other must enter the dot or vice versa.

10.11.1   Volt-Ampere Rating of Autotransformer [Figure 10.34(b)]

Let volt-ampere of two-winding transformer $(VA)_{tw}$ is S and the rated current in the windings are the same in two-winding transformer and autotransformer connections. In two-winding transformer, the current in winding ab (from a to b) and in winding dc (from d to c) will be

$$I_{ab} = \frac{S}{V_1} \quad \text{and} \quad I_{dc} = \frac{S}{V_2} \tag{10.62}$$

Autotransformer output voltage $V_{2,\,new}$ will be additive of V_1 and V_2, and the current will be

$$I_1 = \frac{S}{V_1} + \frac{S}{V_2} \quad \text{and} \quad I_2 = \frac{S}{V_2} \tag{10.63}$$

Thus, volt-ampere of auto transformer (VA)$_{auto}$ of Figure 10.34(b) will be calculated as

$$(VA)_{auto} = V_1 I_1 = V_1 \left(\frac{S}{V_1} + \frac{S}{V_2} \right) = S \left(1 + \frac{V_1}{V_2} \right) \tag{10.64}$$

The same can also be computed using secondary side quantities (primary side VA will be equal to the secondary side VA).

$$(VA)_{auto} = (V_1 + V_2) I_2 = (V_1 + V_2) \frac{S}{V_2} = S \left(1 + \frac{V_1}{V_2} \right) \tag{10.65}$$

From Eqs. (10.64) and (10.65), it is clear that primary (VA)$_{auto}$ is equal to the secondary side (VA)$_{auto}$ which shows the transformer action. From these equations, it is also clear that (VA)$_{auto}$ > (VA)$_{tw}$.

10.11.2 Volt-Ampere Rating of Autotransformer [Figure 10.34(c)]

Using the autotransformer primary side quantities, the volt-ampere rating will be

$$(VA)_{auto} = (V_1 + V_2) \frac{S}{V_1} = S \left(1 + \frac{V_2}{V_1} \right) \tag{10.66}$$

In this connection, $I_2 - I_1 = \dfrac{S}{V_2}$ (= rating of secondary winding current).

Thus,
$$I_2 = \frac{S}{V_2} + \frac{S}{V_1} \tag{10.67}$$

Using the autotransformer primary side quantities, the volt-ampere rating using Eq. (10.67) will be

$$(VA)_{auto} = V_2 I_2 = V_2 \left(\frac{S}{V_2} + \frac{S}{V_1} \right) = S \left(1 + \frac{V_2}{V_1} \right)$$

In this case also, (VA)$_{auto}$ > (VA)$_{tw}$

The total VA of an autotransformer has two components: Magnetically transferred and conductively transferred. Please note that magnetically transferred VA is through the winding common to primary and secondary.

For the connection shown in Figure 10.34(c), the magnetically transferred VA = $V_2(I_2 - I_1)$ and conductively transferred VA = $V_2 I_2 - V_2(I_2 - I_1) = V_2 I_1$.

For the connection shown in Figure 10.34(b), magnetically transferred VA= $V_1(I_1 - I_2)$ and conductively transferred = $V_1 I_1 - V_1(I_1 - I_2) = V_1 I_2$.

EXAMPLE 10.6 A two winding, 20 kVA, 2000/200 V transformer is connected to form autotransformer to obtain voltage rating of

   (a) 2200/200 V   (c) 2000/1800 V
   (b) 2000/2200 V   (d) 1800/200 V

Find the kVA rating, VA transferred magnetically and conductively in all the cases.

Solution: Rated current in both the windings of the two-winding transformer

$$I_H = \frac{20 \times 10^3}{2000} = 10 \text{ A}$$

$$I_L = \frac{20 \times 10^3}{200} = 100 \text{ A}$$

(a) Connection diagram of autotransformer to get 2200/200 V is shown in Figure 10.35

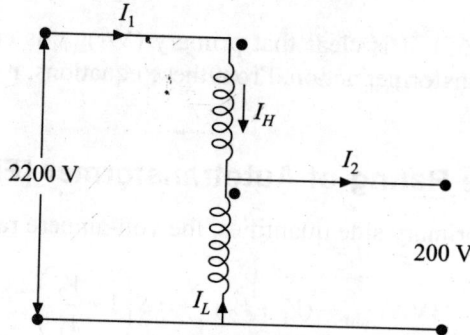

FIGURE 10.35

VA rating of autotransformer = $V_1 I_1$ = 2200 × 10 = 22 kVA
Alternatively, using secondary side quantities of autotransformer,

$$(VA)_{Auto} = (I_H + I_L) \times 200 = (10 + 100) \times 200 = 22 \text{ kVA}$$

Magnetically transferred VA,

$$(VA)_{mag} = I_L \times 200 = 20 \text{ kVA}$$

Conductively transferred

$$(VA)_{cond} = 10 \times 200 = 2 \text{ kVA}$$

(b) Step up configuration as shown in Figure 10.36.

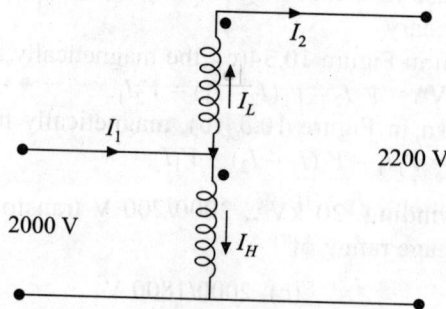

FIGURE 10.36

$$I_1 = 10 + 100 = 110 \text{ A}$$

$$(VA)_{auto} = V_1 I_1 = 2000 \times 110 = 220 \text{ kVA}$$

$$(VA)_{mag} = V_1 I_H = 2000 \times 10 = 20 \text{ kVA}$$

$$(VA)_{cond} = 220 - 20 = 200 \text{ kVA}$$

(c) To obtain voltage 2000/1800 V, the following connection can be made (Figure 10.37).

$$I_2 = 100 \text{ A}$$

$$I_2 - I_1 = 10 \text{ A}$$

$$\therefore \qquad I_1 = 90 \text{ A}$$

$$(VA)_{auto} = 90 \times 2000 = 180 \text{ kVA}$$

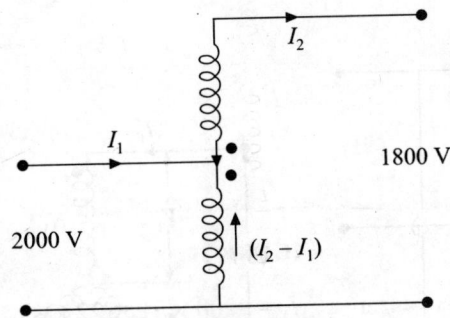

FIGURE 10.37

(d) To obtain voltage 1800/200 V, the following connection can be made (Figure 10.38).

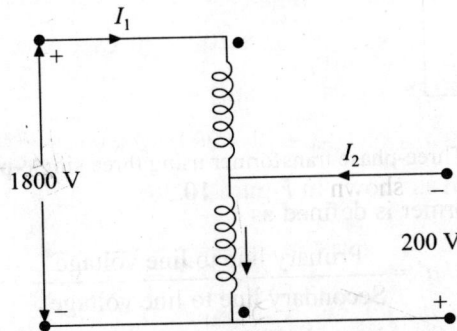

FIGURE 10.38

$$I_1 = 10 \text{ A}$$

$$I_1 + I_2 = 100 \text{ A}$$

$$I_2 = 90 \text{ A}$$

$$(VA)_{auto} = 18 \text{ kVA}$$

10.12 THREE-PHASE TRANSFORMERS

Three-phase transformers are widely used in electric power system to step-up the voltages (generating transformer and interconnecting transformers) or to step-down the voltages (distribution transformers and interconnecting transformers). 3-Φ transformers can have four types of connections viz. Y-Y, Δ-Δ, Y-Δ or Δ-Y which can be realized by connecting three single-phase transformers as shown in Figure 10.39 or constructing a 3-Φ transformer on a common magnetic structure.

Please note that in star side, the identical polarity winding are connected together to form the neutral point. Figure 10.39 shows the Y-Δ connection from three 1-Φ transformers having turn ratio $a (= N_1/N_2)$ and V_1/V_2 voltage rating. The 3-Φ transformer rating would be $\sqrt{3}\,V_1/V_2$.. Please also note that voltage rating of 3-Φ transformer voltage is represented in terms of line to line voltages of both sides.

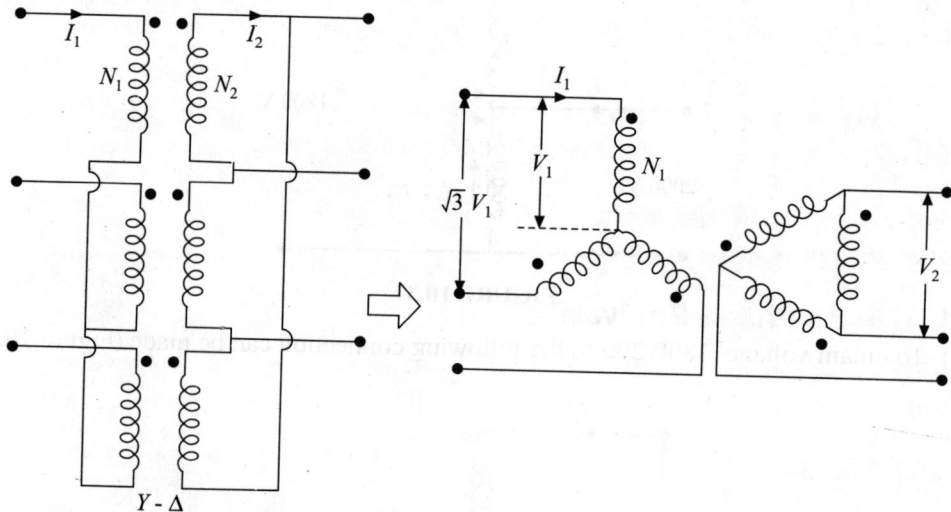

FIGURE 10.39   Three-phase transformer using three single-phase transformers.

The turn ratio of 3-Φ transformer is defined as

$$a' = \frac{\text{Primary line to line voltage}}{\text{Secondary line to line voltage}} \tag{10.68}$$

Thus, the turn ratio of connection shown in Figure 10.39 will be

$$a' = \frac{\sqrt{3}\,V_1}{V_2} = \sqrt{3}\,a$$

Other possible connections of three single-phase transformers to form a three-phase transformer are shown in Figure 10.40.

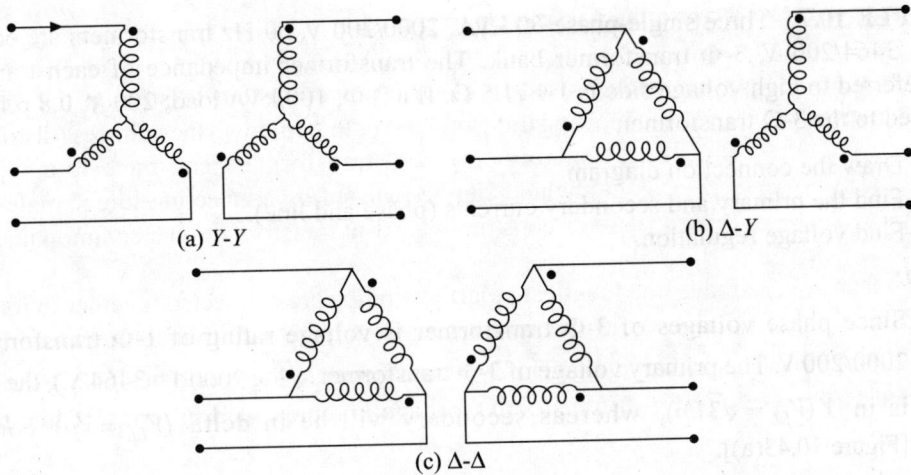

FIGURE 10.40 Different three-phase transformers using three single-phase transformers.

It should be remembered that there is a phase shift of 30° in Δ-Y connection and −30° in Y-Δ connection. Y-Δ connection is used normally to step down the voltage where neutral of star point is grounded for protection and other purposes. Δ-Y connection is used for stepping up the voltage. If any side of transformer is having delta winding, the triple harmonic current circulates in the winding and does not flow into the line. Δ-Δ connection has an advantage to operate at 58% loading of original bank in the event of one transformer is under repair or outage.

10.12.1 Single-Phase Equivalent

If supply and load is balanced and three transformers are identical, a single-phase equivalent can be obtained. Δ-connected load and transformer winding can be transformed into Y-connection using Δ-Y transformation. A Y-Δ arrangement is shown in Figure 10.41 and single-phase equivalent is shown in Figure 10.42.

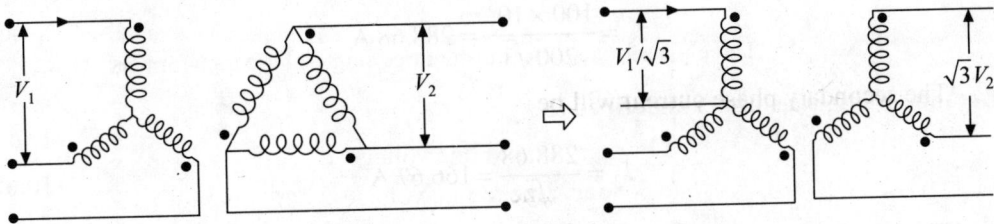

FIGURE 10.41 Delta-star transformation.

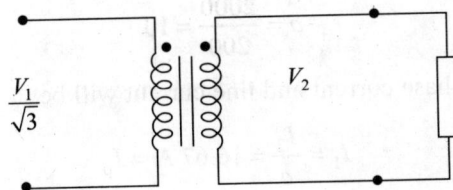

FIGURE 10.42 Single phase equivalent of Y Δ.

EXAMPLE 10.7   Three single-phase 50 kVA, 2000/200 V, 50 Hz transformers are connected to form 3464/200 V, 3-Φ transformer bank. The transformer impedance of each transformer (1-Φ) referred to high voltage side is $1 + j1.5$ Ω. If a 3-Φ, 100 kVA load, 200 V, 0.8 p.f. (lag) is connected to the 3-Ω transformer,

   (a) Draw the connection diagram
   (b) Find the primary and secondary currents (phase and line)
   (c) Find voltage regulation.

Solution:

   (a) Since phase voltages of 3-Φ transformer is voltage rating of 1-Φ transformer, i.e. 2000/200 V. The primary voltage of 3-Φ transformer $\sqrt{3} \times 2000$ (= 3464 V), the primary is in $Y (V_{LL} = \sqrt{3} Vp)$, whereas secondary will be in delta $(V_{LL} = Vp)$ connected [Figure 10.43(a)].

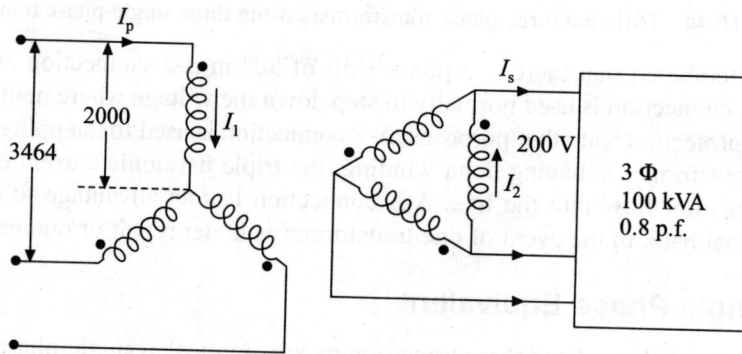

FIGURE 10.43(a)

   (b) Secondary side line current will be

$$I_s = \frac{100 \times 10^3}{200\sqrt{3}} = 288.68 \text{ A}$$

The secondary phase current will be

$$I_2 = \frac{288.68}{\sqrt{3}} = 166.67 \text{ A}$$

The transformer ratio (single-phase transformer [Figure 10.43(b)]) will be

$$a = \frac{2000}{200} = 10$$

Thus, the primary phase current and line current will be

$$I_1 = \frac{I_2}{a} = 16.67 \text{ A} = I_p$$

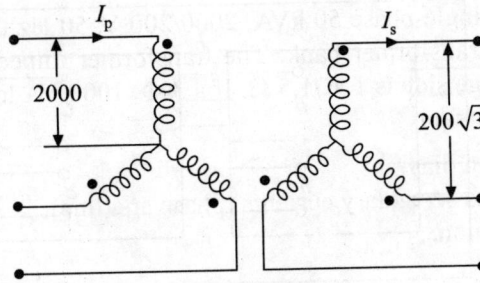

FIGURE 10.43(b)

Figure 10.43(b) shows Δ to γ transformation of secondary side.

(c) Per-phase equivalent [Figure 10.43(c)]

$$\overline{V}_1 = 2000 \angle 0 + \overline{I}_1(1 + j1.5)$$

$$= 20\ 28.35 \angle 0.28 \text{ volt}$$

$$\text{Regulation} = \frac{2028.35 - 2000}{2000} \times 100 = 1.42\%$$

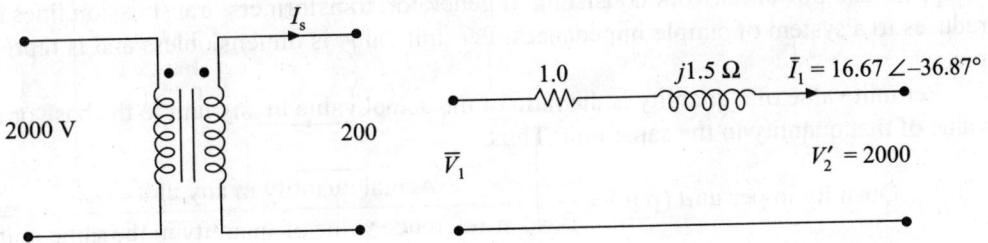

FIGURE 10.43(c)

10.12.2 Three-Phase Transformer on a Single Magnetic Core

A three-phase transformer constructed on a single magnetic core having three legs is shown in Figure 10.44. The total flux in the common legs will be zero as these legs are displaced by 120° and thus, common leg can be removed. However, this structure is not convenient to build. Leg of winding b is pushed as shown in Figure 10.45 which is convenient to use.

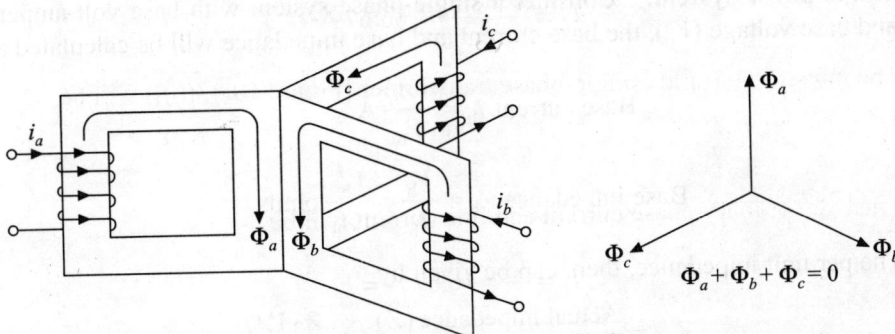

FIGURE 10.44 Three-phase transformer magnetic core and flux distribution.

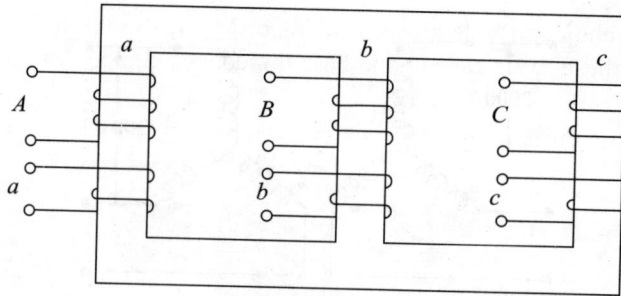

FIGURE 10.45   Three-phase transformer magnetic core without common leg.

10.13   PER UNIT (PU) REPRESENTATION

Due to cost and technical reasons, the different operating voltages (with the presence of transformers, different ratings of generators, etc.) exist in the power system and therefore calculation becomes difficult. Moreover, machine and system data are generally available in per unit value on its own base. Therefore, it is essential to use the per unit system of various physical quantities such as power, voltage, current and impedance. In the per unit system, the different voltage levels disappear and power network consisting of generator, transformers, transmission lines and loads reduces to a system of simple impedances. Per unit value is dimensionless and is represented as p.u.

Per unit value of a quantity is the ratio of the actual value in any unit to the base or reference value of that quantity in the same unit. Thus,

$$\text{Quantity in per unit (p.u.)} = \frac{\text{Actual quantity in any unit}}{\text{Base or reference value of quantity in the same unit}}$$

A well-chosen per unit system can reduce the computational effort, simplify evaluation and facilitate the understanding of system characteristics. The selections of base quantities are also very important. Some of the base quantities are chosen independently and arbitrarily while others follow automatically depending upon the fundamental relationships between system variables. Out of the four power system quantities, viz. power (VA), voltage (V), current (A) and impedance (Ω), only two are independent. The universal practice is to use machine rating, power and voltage, as base values and base values of current and impedance are calculated.

(a) **Single-phase system:**   Consider a single-phase system with base volt-amperes (VA_b) and base voltage (V_b), the base current and base impedance will be calculated as,

$$\text{Base current } I_b = \frac{VA_b}{V_b} \text{ A} \tag{10.69}$$

$$\text{Base impedance } Z_b = \frac{V_b}{I_b} = \frac{V_b^2}{VA_b} \text{ ohms} \tag{10.70}$$

The per unit impedance, then, can be given by

$$Z_{pu} = \frac{\text{Actual impedance } (Z)}{\text{Base impedance } (Z_b)} = \frac{Z \cdot VA_b}{V_b^2} \tag{10.71}$$

The practical choice of base power is kVA_b or MVA_b, however, base voltage is selected in kV_b. With these, base current and base impedance can be derived as

$$\text{Base current } I_b = \frac{kVA_b}{kV_b} = \frac{1000 \times MVA_b}{kV_b} \text{ A} \tag{10.72}$$

$$\text{Base impedance } Z_b = \frac{1000 \times kV_b}{I_b} = \frac{kV_b^2}{MVA_b} = \frac{1000 \times kV_b^2}{kVA_b} \text{ ohms} \tag{10.73}$$

Similarly, per unit impedance can be derived as,

$$Z_{pu} = \frac{Z \times MVA_b}{kV_b^2} = \frac{Z \times kVA_b}{kV_b^2 \times 1000} \tag{10.74}$$

(b) Three-phase system: In three-phase systems, the base values usually chosen are three-phase kVA_b or MVA_b and line-to-line voltage (kV_b). The base current can be found as,

$$\text{Base current } I_b = \frac{kVA_b}{\sqrt{3}\,kV_b} = \frac{1000 \times MVA_b}{\sqrt{3}\,kV_b} \text{ A} \tag{10.75}$$

$$\text{Base impedance } Z_b = \frac{1000 \times kV_b}{\sqrt{3}\,I_b} = \frac{kV_b^2}{MVA_b} = \frac{1000 \times kV_b^2}{kVA_b} \text{ ohm} \tag{10.76}$$

$$\text{Per unit impedance } Z_{pu} = \frac{Z \times MVA_b}{kV_b^2} = \frac{Z \times kVA_b}{kV_b^2 \times 1000} \tag{10.77}$$

It is necessary to change per unit impedance from one set of base values to the new set of base values. When base MVA is changed from $MVA_{b,old}$ to $MVA_{b,new}$ and base voltage from $kV_{b,old}$ to $kV_{b,new}$, per unit impedance can be calculated from the following equation as

$$\left(\begin{array}{l}\text{Per unit impedance referred}\\ \text{to new base } Z_{pu,new}\end{array}\right) = \left(\begin{array}{l}\text{Per unit impedance referred}\\ \text{to old base } Z_{pu,old}\end{array}\right)\left(\frac{MVA_{b,new}}{MVA_{b,old}}\right)\left(\frac{kV_{,old}}{kV_{b,new}}\right)^2 \tag{10.78}$$

10.13.1 Per Unit Representation of Transformers

A three-phase transformer can be represented by a single-phase transformer in getting the per phase solution of the system. The delta-winding transformer is replaced by an equivalent star so that the transformation ratio of equivalent single-phase transformer is always line-to-line voltage ratio of three-phase transformer. Figure 10.46 represents the single-phase equivalent transformer with primary winding impedance Z_p and secondary winding impedance Z_s. The turn ratio is N.

If primary winding voltage base is V_{1B} and the secondary side base voltage is V_{2B}, these can be written in the following form,

$$\frac{V_{1B}}{V_{2B}} = N = \frac{I_{2B}}{I_{1B}} \tag{10.79}$$

where I_{1B} and I_{2B} are the current bases of primary and secondary sides, respectively.

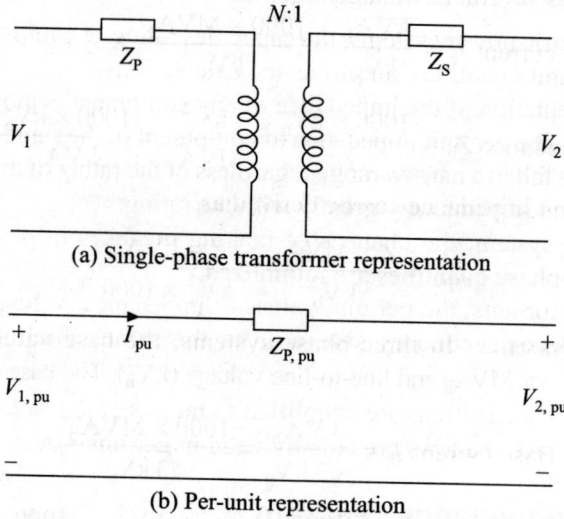

(a) Single-phase transformer representation

(b) Per-unit representation

FIGURE 10.46   Single-phase equivalent transformer representation.

Base impedance can be calculated as

$$Z_{1B} = \frac{V_{1B}}{I_{1B}}, \quad Z_{2B} = \frac{V_{2B}}{I_{2B}}$$

and their ratio can be defined as

$$\frac{Z_{1B}}{Z_{2B}} = \frac{V_{1B}}{I_{1B}} \frac{I_{2B}}{V_{2B}} = \frac{V_{1B}}{V_{2B}} \frac{I_{2B}}{I_{1B}} = N \times N = N^2 \qquad (10.80)$$

Impedance referred to primary side (Z_P') and secondary side (Z_S') will be

$$Z_P' = Z_P + N^2 Z_S \quad Z_S' = Z_S + Z_P/N^2$$

and per unit impedance on primary side $(Z_{P,pu}')$ will be

$$Z_{P,pu}' = \frac{Z_P + N^2 Z_S}{Z_{1B}} = \frac{Z_P}{Z_{1B}} + N^2 \frac{Z_S}{Z_{1B}} = Z_{P,pu} + N^2 Z_{S,pu} \frac{Z_{2B}}{Z_{1B}} = Z_{P,pu} + Z_{S,pu} \qquad (10.81)$$

On secondary side,

$$Z_{S,pu}' = \frac{Z_S + Z_P/N^2}{Z_{2B}} = \frac{Z_S}{Z_{2B}} + \frac{1}{N^2} \frac{Z_P}{Z_{2B}} = Z_{S,pu} + \frac{1}{N^2} Z_{P,pu} \frac{Z_{1B}}{Z_{2B}} = Z_{P,pu} + Z_{S,pu} \qquad (10.82)$$

From Eqs. (10.81) and (10.82), we have

$$Z_P' = Z_S' \qquad (10.83)$$

This shows that per unit impedance referred to the primary side will be equal to per unit impedance referred to the secondary side.

10.13.2   Advantages of Per Unit System

Per unit representation has several advantages such as

(i) In the large electric power systems, the capacities rating of equipments are different and the use of per unit quantities simplifies the calculations.

(ii) Per unit representation of the impedance of any equipment is more meaningful than its absolute value. The per unit impedance of equipment of the same general type based on their own rating fall in a narrow range regardless of the rating of the equipment. Whereas their impedances in ohms vary greatly with the rating.

(iii) Using per unit system, the chances of making mistakes in phase and line voltages, single or three-phase quantities are minimized.

(iv) In case of transformers, the per unit values of impedance, voltage and current referred to primary side or secondary side will be the same which further simplifies the calculations.

(v) Power and voltage equations are simplified as factors of $\sqrt{3}$ and 3 are eliminated in the per unit system. All the laws are equally valid in per unit systems.

10.13.3   Per Unit Impedance Diagram

Single-line diagram of power system can be represented by the per unit impedance diagram as shown in Figure 10.47.

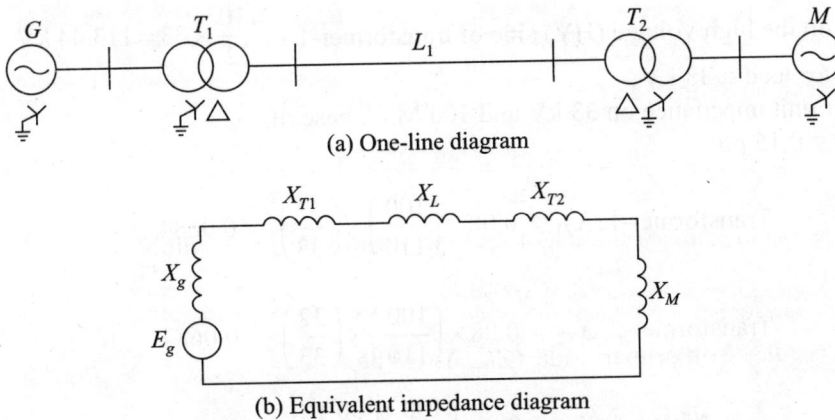

(a) One-line diagram

(b) Equivalent impedance diagram

FIGURE 10.47   Per-unit impedance diagram.

The following procedure is normally used for calculating the per unit impedances, which are used in drawing the impedance diagram.

(a) Choose an appropriate MVA or kVA base for the system. (Generally, 100 MVA is used in power system.)

(b) Choose the voltage base and calculate the appropriate voltage for other sections according to the transformation ratio.

(c) Calculate the per unit impedance of each equipment and connect them as per topology of the single-line diagram.

EXAMPLE 10.8   A 100 MVA, 33 kV, three-phase generator has a subtransient reactance of 15 %. The generator is connected to the motors through a transmission line and transformers as shown in Figure 10.48. The motors have rated inputs of 30 MVA and 50 MVA at 30 kV and 31 kV, respectively each with 20% subtransient reactance. The three-phase transformers are rated at 110 MVA, 32 kV Δ /110 kV Y with leakage reactance 8%. The line has reactance of 50 ohms. The three-phase load absorbs 20 MW, 0.8 power factor lagging at 30 kV. Selecting the generator rating as the base quantities, determine the base quantities in other part of the system and evaluate the corresponding p.u. values.

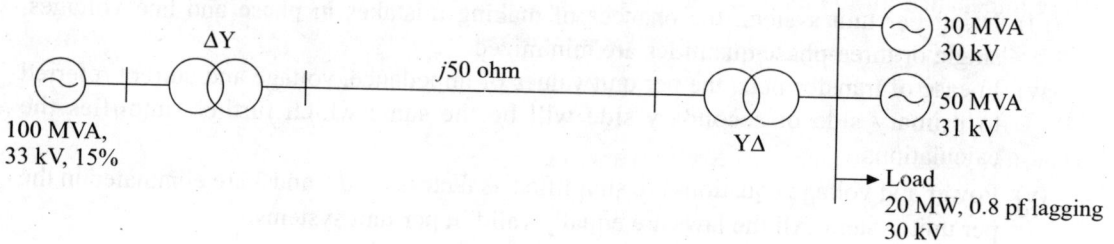

FIGURE 10.48   Example 10.8.

Solution: The voltage base for different section of the system taking generator voltage as base will be

Base voltage on the low voltage (LV) side of transformer-1 = 33 kV

Base voltage on the high voltage (HV) side of transformer-1 = $\dfrac{110}{32} \times 33 = 113.44$ kV

Base voltage for load = 33 kV

Therefore, per unit impedance on 33 kV and 100 MVA base of
Generator $Xg = 0.15$ pu;

$$\text{Transformer-1}, X_{T1} = 0.08 \times \left(\frac{100}{110}\right) \times \left(\frac{32}{33}\right)^2 = 0.0684$$

$$\text{Transformer-2}, X_{T2} = 0.08 \times \left(\frac{100}{110}\right) \times \left(\frac{32}{33}\right)^2 = 0.0684$$

$$\text{Motor-1}, X_{m1} = 0.2 \times \left(\frac{100}{30}\right)\left(\frac{30}{33}\right)^2 = 0.55$$

$$\text{Motor-2}, X_{m2} = 0.2 \times \left(\frac{100}{50}\right)\left(\frac{31}{33}\right)^2 = 0.353$$

Base impedance of line will be

$$Z_b = \frac{kV_b^2}{MVA} = \frac{(113.44)^2}{100} = 128 \cdot 68 \ \Omega$$

Per unit impedance of line, $Z_{pu} = \dfrac{50}{Z_b} = 0.388$

The three-phase load apparent power at 0.8 power factor lagging is given by

$$S_L = \frac{20}{0.8} \angle \cos^{-1} 0.8 = 25 \angle 36.87° \text{ MVA}$$

Hence load impedance in ohm will be

$$Z_L = \frac{(V_L)^2}{S_L^*} = \frac{30^2}{25 \angle -36.87} = 28.8 + j21.6 \, \Omega$$

Base impedance for the load will be

$$Z_{bL} = \frac{kV_b^2}{\text{MVA}} = \frac{(33)^2}{100} = 10.89 \, \Omega$$

Therefore, load impedance in per unit would be $\dfrac{28.8 + j021.6}{10.89} = 2.645 + j1.983$ pu

Impedance diagram is shown in Figure 10.49.

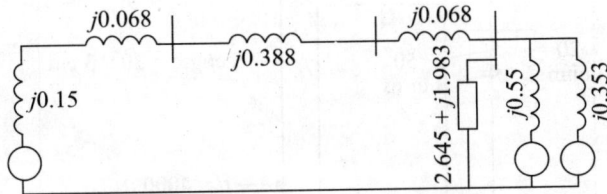

FIGURE 10.49   Impedance diagram.

EXAMPLE 10.9   A transformer having a square core cross-sectional area of 4 cm^2 is to be designed for 230 V/110 V and a further centre-tapped, 6-0-6 V winding is to be provided. If the flux density is not to exceed 1 Wb/m^2 (Tesla), find the suitable number of turns for each winding for 50 Hz frequency. Consider the ideal transformer case.

Solution: The low voltage winding is always designed first because the voltage ratio can be rarely obtained exact since low voltage winding has the fewest turns and the actual turn ratio must always be integer.

The maximum allowed flux $\Phi_m = B_{max} \times A = 1.0 \times 4 \times 10^{-4} = 4 \times 10^{-4}$ Wb

Using *emf* Eq. (10.17), $6 = 4.44 \times 4 \times 10^{-4} \times 50 \times N_3$

Thus, $N_3 = 67.57.$

Since the maximum flux density should not exceed to the given value, the number of turns be more than or equal to N_3. The nearest integer number is 68.

With this number of turns, the flux density will be slightly less than 1.0 Tesla. With more than 68 turns, the flux density will be further reduced and it is not good as material is not utilized properly.

The number of turns in secondary winding will be

$$N_2 = \frac{110}{6} \times N_3 = \frac{110}{6} \times 68 = 1246.7, \text{ say 1247 turns}$$

The number of turns in primary winding will be

$$N_1 = \frac{230}{6} \times N_3 = \frac{230}{6} \times 68 = 2606.7, \text{ say } 2607 \text{ turns}$$

The tertiary winding requires 2×68 turns.

EXAMPLE 10.10   Two coils are wound on an iron core as shown in Figure 10.50. A 220V, 50 Hz, source is connected to one of the coils and 20 ohm resistor to the other coil.

(a) Put polarity marks on the resistor, such that the voltages on both coils are in phase.
(b) Draw an equivalent circuit without using an ideal transformer that permits calculation of in-phase and out-phase components of the input current. Ignore iron and copper losses and leakage inductances.
(c) In this situation, are the coils attracted or repelled by the current in the wires? Explain.

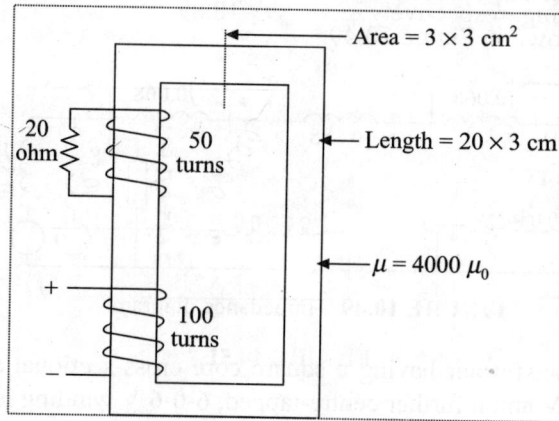

FIGURE 10.50   Example 10.10.

Solution:

(a) Dot markings either at the top or bottom terminal of both the coils

(b) Reluctance $\Re = \dfrac{l}{\mu A} = \dfrac{20 \times 3 \times 10^{-2}}{4000 \mu_0 \times 3 \times 3 \times 10^{-4}}$

Magnetizing inductance of primary winding

$$= \frac{100 \times 100}{\Re} = \frac{10000}{20 \times 3 \times 10^{-2}} \times 4\pi \times 10^{-7} \times 4000 \times 3 \times 3 \times 10^{-4}$$

$$= 0.0754 \ H$$

$$x_{max} = 2\pi \times 50 \times 0.0754 = 23.69 \ \Omega$$

Thus, load resistance referred to primary side $= 20 \times \left(\dfrac{100}{50}\right)^2 = 80 \ \Omega$

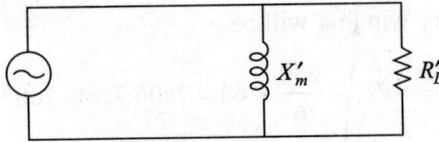

(c) Repelled as the currents are the following in opposite directions.

EXAMPLE 10.11 Two identical 5 kVA, 230/460 V transformers are connected in open delta to supply a balanced load at 460 V and p.f. of 0.8 lagging (Figure 10.51). Determine

(a) The maximum secondary line current without overloading the transformer.
(b) The primary line current.
(c) The real power delivered by each transformer.
(d) If a similar transformer is now added to complete the delta, find the percentage increase in real power that can be supplied if the load voltage and p.f. remain unchanged at 460 V and 0.8 p.f. lagging, respectively.

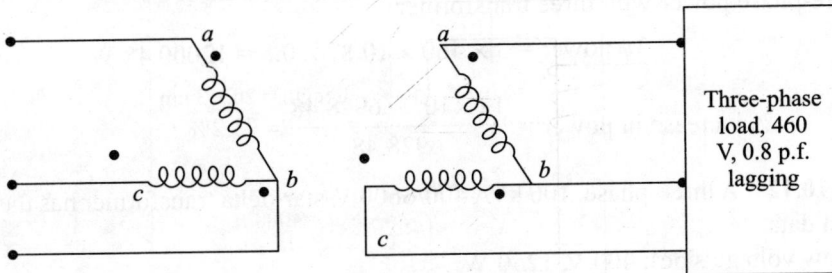

FIGURE 10.51

Solution:

(a) Max. secondary phase current of transformer

$$= \frac{5 \times 10^3}{460} = 10.87 \text{ A}$$

(b) Primary current

$$= 10.87 \times \left(\frac{460}{230}\right) = 21.74 \text{ A}$$

(c) Power drawn from phase-ab (see the angles from Figure 10.52)

$$P_{ab} = \left|\bar{V}_{ab}\right|\left|\bar{I}_a\right| \cos \angle_{\bar{I}_a}^{\bar{V}_{ab}}$$

$$= 460 \times 10.87 \times \cos(30 + 36.87) = 1964.1 \text{ W}$$

From phase *bc* will be

$$P_{bc} = \left|\bar{V}_{bc}\right|\left|\bar{I}_c\right| \cos \angle_{\bar{I}_c}^{\bar{V}_{bc}}$$

$$= 460 \times 10.87 \times \cos(30 - 36.87) = 4964.3 \text{ W}$$

Total power in open delta $= P_{ab} + P_{bc} = 6928.48$ W

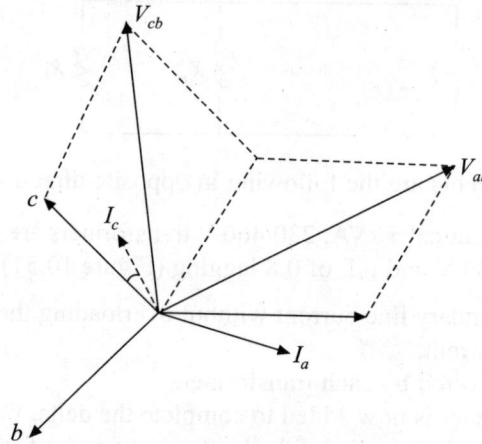

FIGURE 10.52

(d) Three-phase power with three transformer

$$3\phi \text{ power} = 3 \times 460 \times 10.87 \times 0.8 = 12000.48 \text{ W}$$

$$\% \text{ increase in power} = \frac{12 \times 10^3 - 6928.48}{6928.48} = 73.2\%$$

EXAMPLE 10.12   A three-phase, 100 kVA 400/6600 V, star-delta transformer has the following light load test data:

OC (at low voltage side): 400 V, 1250 W

SC (at high voltage side): 314 V, 1600 W, full load current.

Calculate the efficiency at full load, 0.8 p.f. and half load, unity *pf*. Find the maximum efficiency. What is percentage leakage impedance based on 100 % rating.

Solution: From the test data, it is clear that core loss $P_c = 1250$ and full load copper loss $P_{cu,fl} = 1600$. Using Eq. (10.54), the efficiency at full load ($x = 1$) and 0.8 p.f. lagging will be

$$\eta = \frac{1 \times 100 \times 10^3 \times 0.8}{1 \times 100 \times 10^3 \times 0.8 + 1250 + 1^2 \times 1600} = 0.9656 = 96.56\%$$

Efficiency at half load ($n = 0.5$), unity power factor will be

$$\eta = \frac{0.5 \times 100 \times 10^3 \times 1.0}{0.5 \times 100 \times 10^3 \times 1.0 + 1250 + 0.5^2 \times 1600} = 0.968 = 96.8\%$$

Using relation (10.53) for maximum efficiency, $x = \sqrt{\dfrac{1250}{1600}} = 0.884$ and unity power factor,

$$\eta = \frac{0.884 \times 100 \times 10^3 \times 1.0}{0.884 \times 100 \times 10^3 \times 1.0 + 1250 + 0.884^2 \times 1600} = 0.9725 = 97.25\%$$

Leakage impedance can be calculated as follows:

$$\text{Rated secondary current} = \frac{100 \times 10^3}{6600} = 5.05 \text{ A}$$

$$\text{Short circuit power factor} = \frac{1600/3}{314 \times 5.05} = 0.336$$

(SC power is divided by 3 because SC power is for three phases).

$$Z_{sc} \text{ (referred to secondary i.e. HV side)} = \frac{314}{5.05} (0.336 + j0.9417) = 20.89 + j58.55 \ \Omega$$

$$\text{Base impedance of HV side} = \frac{6600}{5.05} = 1306.9 \ \Omega$$

$$\text{Thus percentage impedance} = \frac{1}{1306.9} (20.89 + j58.55) \times 100 = 1.5984 + j4.4801\%.$$

PROBLEMS

10.1 A transformer having 500 and 100 turns in primary and secondary windings respectively, is wounded on a magnetic core having reluctance $\Re = 2.5 \times 10^4$ AT/Wb. Assume no leakage flux, no iron losses, and ignore wire resistance.

(a) If the primary winding is connected to 220 V, 50 Hz supply system while keeping the secondary open circuited, find the primary winding current.

(b) If the secondary winding is connected to a resistance of 50 ohm, determine the current in the resistor.

(c) For the circuit in part (b) what would be the magnitude of the current in the primary?

[**Ans.** (a) $-j0.07$ A; (b) $0.88 \angle 0°$A; (c) 0.1894 A]

10.2 A 50 Hz, single phase iron core transformer with $\mu_i = 4000 \ \mu_0$, area $= 0.05$ m^2 and nominal length of iron 2.5 m has 500 turns on the high voltage side and 200 turns on low voltage side. At the rated voltage, the flux density (rms) in the iron is 1.0 Tesla.

(a) Find the rated voltage at both the HV and LV sides.

(b) Find the magnetizing current required on the HV side.

(c) The rated current in the transformer is 50 A on the HV side. The losses at the rated current are 3000 watt. Find an equivalent circuit for transformer, neglecting iron losses and leakage inductance.

[**Ans.** (a) Rated voltage at HV side = 7848.88 V,
Rated voltage at LV side = 3139.55 V;
(b) $I_m = 0.9941$ A; (c) $R_{eq} = 1.2$ ohm]

10.3 An open-circuit (OC) test and short-circuit (SC) test are performed on a single-phase transformer in standard manner, with the results in table. Quantities not measured are noted "NM".

Quantity	OC Side 1	OC Side 2	SC Side 1
Voltage V	7500	220	210
Current, A	NM	3.00	2
Power, W	NM	250	200

(a) What is voltage rating of the transformer?
(b) What is transformer apparent power rating?
(c) Give an equivalent circuit for the transformer referred to the HV side.
(d) Find the magnetizing current, I_ϕ, if excited on the high voltage side.

[**Ans.** (a) 7500/220 V; (b) Trans. Apparent power rating = 15 kVA;
(c) R_c referred to HV side = 2.25×10^5 ohm,
X_m referred to HV side = 9.2089×10^4 ohm,
R_{eq} = 50 ohm, X_{eq} = 92.33 ohm; (d) I_ϕ = 0.081 A]

10.4 The equivalent circuit for a 25 kVA, 2200/220 V, 50 Hz single-phase transformer is shown in Figure 10.53. Assume that the transformer is operating at rated secondary voltage and apparent power with a 0.8 power factor (lagging).

(a) What is the turns ratio?
(b) Estimate the iron losses.
(c) Estimate the copper losses.
(d) What would be the exciting current, I_e?
(e) Estimate the input power factor.

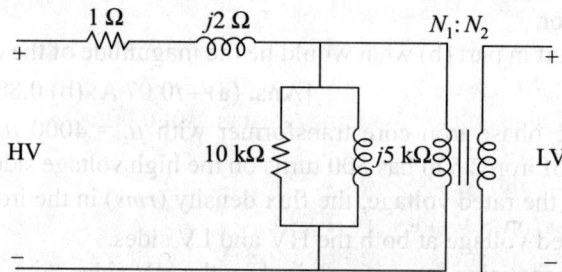

FIGURE 10.53 Problem 10.4.

[**Ans.** (a) Turns ratio = 10; (b) Iron losses = 484 watt;
(c) Copper losses = 139.24 watt; (d) I_e = 0.48 $\angle$ −63.43° A;
(e) Input power factor = 0.786 (lagging)]

10.5 The following test data were taken on a single-phase,15 kVA, 2200/440 V 50 Hz transformer:

Short circuit test: 620 W, 40A, 25V
Open circuit test: 320 W, 1A, 440V

Find the equivalent circuit. Also calculate the voltage regulation of this transformer when it supplies full load at 0.8 p.f. lagging. Neglect the magnetizing current.

[**Ans.** R_{eq} = 0.388 ohm, X_{eq} = 0.490 ohm, Voltage regulation = 4.7%]

10.6 A single-phase, 25 kVA 1100/220V, 50 Hz transformer has a high voltage winding resistance of 0.5 ohm and low winding resistance of 0.005 ohm. Laboratory tests showed the following results:

Open circuit test: 220 V, 5.7 A, 190 W

Short circuit test: 41.5 V, 22.73 A, $P = x$ watt

(a) Find the value of x.

(b) Find the equivalent circuit.

(c) Compute the value of primary voltage needed to give rated secondary voltage when the transformer is connected step-up delivering 20 kVA at a power factor of 0.8 lagging.

[**Ans.** (a) $x = 322.91$ watt; (b) $R_{c2} = 254.74$ ohm, $X_{m2} = 39.05$ ohm, $R_{eq2} = 0.025$ ohm, $X_{eq2} = 0.0686$ ohm; (c) $V_1 = 255.75 \angle 5.08°$ V]

10.7 An ideal three-winding transformer is rated 2300 V (50 Hz) on the high voltage side with a total turns of 300. Of the secondary windings each designed to handle 200 kVA, one is rated at 575 V and the other at 230 V. Determine primary current when rated current in 230 V winding is at unity power factor and that in 575 V winding is at 0.6 p.f. lagging.

[**Ans.** Total primary current $I_1 = 155.54 \angle -26.57°$ A]

10.8 The core loss for a given specimen of magnetic material is found to be 2000 W at 50 Hz. Keeping the flux density constant, the frequency of the supply is raised to 75 Hz resulting in core loss of 3200 W. Compute separately hysteresis and eddy current losses at both frequencies. What will be these losses at 60 Hz.

[**Ans.** At 50Hz : Hysteresis loss = 1733 watt, Eddy current loss = 267 watt;

At 75 Hz: Hysteresis loss = 2599.5 watt, Eddy current loss = 600.75 watt;

At 60 Hz: Hysteresis loss = 2080 watt, Eddy current loss = 384 watt]

10.9 A single-phase, 20 kVA 1000/200 V, 50 Hz transformer has a high voltage winding resistance of 0.5 ohm and low winding resistance of 0.0005 ohm. Laboratory tests showed the following results:

Open circuit test: 200 V, 2 A, 200W

Short circuit test (on HV side): 16.0 V, 8 A

(a) Find the equivalent circuit referred to HV side of transformer.

(b) Determine the efficiency at the full load, 0.8 power factor lagging.

(c) Voltage regulation at 0.8 power factor lagging if load is connected to LV side.

Above single phase transformer is connected to form a 200/1200 V autotransformer.

(d) Draw the connection diagram with the direction of current at full load.

(e) Find the efficiency of this autotransformer at full load 0.8 power factor lagging.

[**Ans.** (a) $R'_{c2} = 5000 \ \Omega$, $X'_{m2} = 2886.75 \ \Omega$; $R_{eq1} = 0.5125 \ \Omega$, $X_{eq1} = 1.93 \ \Omega$; (b) $\eta = 97.53\%$; (c) 3.16% (e) $\eta = 97.93\%$]

10.10 Core loss at rated voltage and copper loss at 80% of full load of a 10 kVA, 2200/220 V, 50 Hz, single phase transformer is 100 W and 250 W respectively. Calculate the efficiencies of the transformer if the output power of transformer is

(a) 2.5 kVA at 0.8 pf lagging

(b) 5.0 kVA at 0.9 pf lagging

(c) 10 kVA at unity power factor

(d) 5 kVA at 0.8 pf leading

(e) What will be the loading at which the efficiency of the transformer is maximum? Find the maximum efficiency at 0.8 pf lagging.

[**Ans.** (a) 94.14%, (b) 95.79%, (c) 95.325, (d) 95.29%, (e) 95.29]

10.11 Single phase 20 kVA, 2200/220 V transformer has the following parameters: Z_{eq} (HV) = $(4.0 + j5.0)$ ohm. The transformer is connected to a load whose power factor may vary. Determine the worst-case voltage regulation at full load.

[**Ans.** Voltage regulation = 2.64%]

10.12 A single phase, 5 kVA, 2300/230 V transformer has the following losses:
Core loss at full voltage = 100 W, copper loss at half load = 60 W.

Determine the efficiency of the transformer when it delivers full load at 0.8 power factor (lagging).

[**Ans.** Efficiency = 92.98%]

10.13 10 kVA, 2300/230 V transformer has the following load during the 24 hours of a day.

Duration (Hrs)	4	8	5	3	4
Load (kVA)	10	8	6	1	Switched off
pf	0.9 lagging	0.8 lagging	0.6 lagging	0.5 lagging	—

If the core loss at rated voltage is 200 W and copper loss at half load is 100 W, find the all-day (energy) efficiency of the transformer.

[**Ans.** All day efficiency = 92.72%]

10.14 A single phase 50 kVA, 2200/220 transformer is connected to form an autotransformer. Find the possible different ratings (kVA and voltage) of the autotransformers. Draw the schematic connection diagram with the direction and values of corrects in all windings.

[**Ans.** (a) $(VA)_{auto}$ = 55 kVA; (b) $(VA)_{auto}$ = 45.01 kVA; (c) $(VA)_{auto}$ = 550 kVA; (d) $(VA)_{auto}$ = 51.16 kVA]

10.15 Three single phase, 50 kVA, 11000/231 V, 50 Hz transformers are connected to form a three-phase, 1100/400 V transformer bank. The equivalent impedance of each transformer referred to the high voltage side is $300 + j400$ ohm. The transformer delivers 20 kW load at 0.8 p.f. (lagging) at rated voltage. The magnetizing and core loss components can be neglected.

(a) Draw a schematic diagram showing the transformer connection.

(b) Determine the transformer winding current.

(c) Determine the primary voltage.

(d) Determine the voltage regulation.

[**Ans.** (a) Secondary winding current = 36.08 A, Primary winding current = 0.76 A (c) Primary voltage line to line = 12.095 kV; (d) Voltage regulation = 9.96%]

10.16 The impedance of a three-phase, 240 MVA, 400/220 kV is 10 percent on transformer's own rating as base. Find the impedance of transformer in pu at base of 100 MVA and 420 kV (420 kV side of the transformer). What is the base current on the low-voltage side on new base?

[**Ans.** $(Z_{pu})_{new}$ = 0.0378 pu, (I_{base}) = 250 A]

10.17 A single phase 110 kVA, 11 kV/ 231 V transformer has impedance of $0.05 + j\,0.1$ pu. Find the full load voltage regulation at 0.8 pf lagging.

[**Ans.** Voltage regulation = 10%]

10.18 A single-phase, 10 kVA 2000/200 V, 50 Hz transformer has the following test results:

Open circuit test: 200 V, 1 A, 100 W

Short circuit test: 8 V, 10 A, 20 W

(a) Find the equivalent circuit referred to LV side of transformer.

(b) Determine the efficiency at the full load, unity power factor lagging.

(c) Voltage regulation at unity power factor lagging if load is connected to LV side.

Above single-phase transformer is connected to form a 2200/2000 V autotransformer.

(d) Draw the connection diagram with the direction of current at full load.

(e) Find the efficiency of this autotransformer at full load, unity power factor.

[**Ans.** (a) $R_{e2} = 400\ \Omega$, $X_{m2} = 229.8\ \Omega$, $X_{eq2} = 0.2 + j0.77\ \Omega$,

(b) 94.34%, (c) 6.75%]

10.19 A one-phase, 22 kVA, 2200/220 V transformer has impedance of $4.0 + j5.0$ ohm at 50 Hz (referred to HV side). Core loss at 40 Hz is 200 W and at 60 Hz it is 400 W. Assume flux density is constant.

(i) Find the hysterisis and eddy current losses at 50 Hz.

(ii) Determine the power factor for zero regulation.

(iii) What will be the loading of transformer for maximum efficiency at 50 Hz? Also find the maximum efficiency of the transformer.

[**Ans.** (i) 83.33 W, 208.33 W; (ii) 0.78 leading; (iii) 8.539 A, 96.99%]

10.20 In Figure 10.54, $v = V_{\max} \cos 100t$ volts, source resistance $R_s = 100$ ohm, $L = 10$ mH, $C = 100\ \mu$F and $R = 25$ ohm, specify the turns ration of ideal transformers *TR*-1 and *TR*-2 to provide maximum power transfer to load *R*.

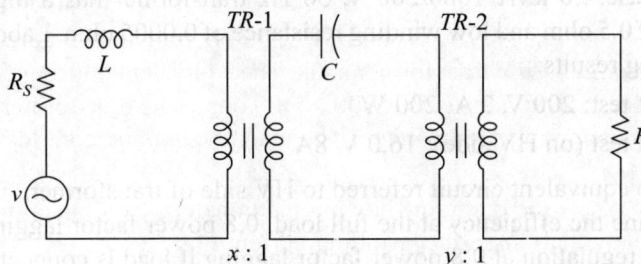

FIGURE 10.54 Problem 10.20.

[**Ans.** 1:10, 20:1]

10.21 A single phase 44 kVA, 50 Hz, 2200/220 V transformer has the following parameters: high voltage side impedance, $Z_{HV} = (0.4 + j\,0.5)$ ohm, low voltage side impedance, $Z_{LV} = (0.01 + j0.03)$ ohm, magnetizing reactance and resistance (referred to LV side) corresponding to core loss are 205 ohm and 80 ohm, respectively.

(a) Draw the exact equivalent circuit diagram (with values of impedances) referred to the HV side.

(b) Determine open-circuit test readings of wattmeter, ammeter and voltage, clearly pointing out which side the open-circuit test is conducted and which side it is open circuited.

(c) Determine the short-circuit test readings of wattmeter, ammeter and voltage, clearly pointing out which side the short-circuit test is conducted and which side it is short circuited.

(d) Find the maximum efficiency of transformer. What is load at this efficiency?

(e) Find the pu impedances ($Z_{HV,pu}$ and $Z_{LV,pu}$) of transformer at 100 kVA power base and 2000 voltage base (HV side).

(f) Above transformer is used to make a 2200/2420 V autotransformer. Draw the circuit diagram with marking of polarity, direction and magnitudes of full load currents. Find the value of kVA conductively and kVA magnetically transferred at full load.

[**Ans.** (a) $r_1 = 0.4$, $x'_1 = 0.5$, $r_2 = 1$, $x'_2 = 3$, $R_c = 8000$ and $X_m = 20500$ ohms;
(b) Open circuit test is conducted LV and HV side is open. 200 V, 2.95 A, 605 W;
(c) Short circuit test is conducted HV and LV side is shorted. 75.49 V, 20 A, 560 W;
(d) 97.42% 1.0894 timed of full load; (e) Z_{Hv}, pu = $0.01 + j0.0125$ pu, $Z_{Lv,pu}$ = 0.025 + $j0.075$ pu; (f) 440 kVA, 44 kVA]

10.22 A 50 Hz, one-phase iron core transformer with $\mu_i = 4000\ \mu_0$, area = 0.051 m^2 and nominal length of iron is 2.5 m. It has 500 turns in HV side and 200 turns in LV side. At rated voltage, the flux density (*rms*) in the iron is 1.0 Tesla.

(i) Find the rated voltage at both HV and LV sides.

(ii) Find the magnetizing current required on the HV side.

(iii) The rating of the transformer is 400 kVA, and copper loss at rated current is 3000 W. Find the equivalent circuit of transformer referred to HV side, neglecting the iron losses and leakage reactances.

[**Ans.** (i) 8011.06 kV, 3204.42 kV; (ii) 0.995 A; (iii) $X_m = 8053.6\ \Omega$, $R_{eq} = 1.2\ \Omega$]

10.23 A single-phase, 20 kVA 1000/200 V, 50 Hz transformer has a high voltage winding resistance of 0.5 ohm and low winding resistance of 0.0005 ohm. Laboratory tests showed the following results:

Open circuit test: 200 V, 2 A, 200 W

Short circuit test (on HV side): 16.0 V, 8A

(a) Find the equivalent circuit referred to HV side of transformer.

(b) Determine the efficiency at the full load, 0.8 power factor lagging.

(c) Voltage regulation at 0.8 power factor lagging if load is connected to LV side.

(d) Above single phase transformer is connected to form a 200/1200 V autotransformer. Draw the connection diagram with the direction of current at full load.

(e) Find the efficiency of this autotransformer at full load 0.8 power factor lagging.

[**Ans.** (a) $R_{c2} = 5000\ \Omega$, $X_{m2} = 2886.75\ \Omega$, $X_{eq1} = 0.5125 + j1.93$;
(b) 97.53%, (c) 3.164%]

11

Electromechanical Energy Conversion

11.1 INTRODUCTION

Electric energy is rarely used in its original form. Normally, it is converted into different forms which is used in real life, such as light, heat and rotating (mechanical) loads. Electric energy is converted to mechanical energy or vice versa through electric machines. The process of this translation is known as *electromechanical energy conversion*. The flow of energy from electrical system to mechanical system is through electric machines as shown in Figure 11.1.

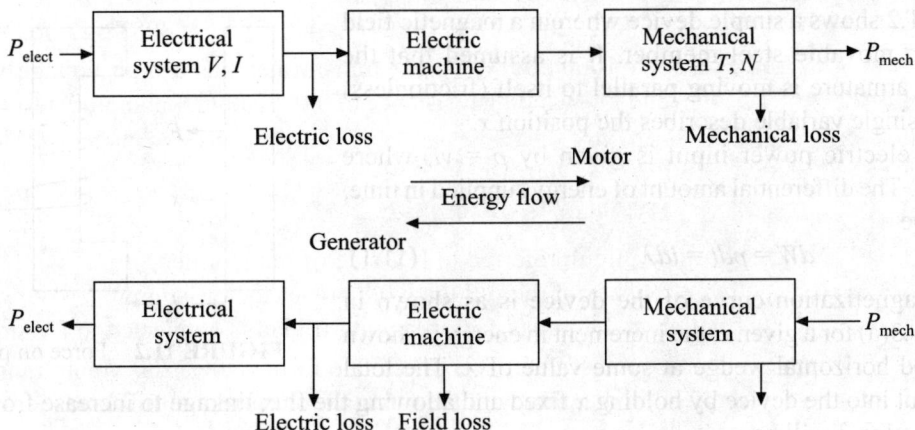

FIGURE 11.1 Electromechanical energy conversion.

An electric machine which takes power from electric system is known as *motor* (converts electric energy to mechanical energy) whereas a generator (converts mechanical energy to electrical energy) supplies power to the electric system. If an electric system is ac, the machines (generator or

279

motor) are known as *ac machines*. DC machines (generators or motors) are connected to dc electric system. AC machines (induction type and synchronous type) and dc machines (series, shunt and compound types) are widely used in electromechanical energy conversion. Three basic electromagnetic phenomenons are involved utilizing the magnetic field or electrostatic field as the coupling medium.

1. *Force on iron:*   A magnetic force is exerted on a ferromagnetic material called *armature*, tending to align it with or bring it into the position of the densest part of the magnetic field. When the magnetic field is produced by the current-carrying coil, the energy conversion process is reversible because motion of material will cause a change in the flux linking the coil and the change in the flux linkage will induce a voltage in the coil.

2. *Force on conductor:*   A current-carrying conductor experiences a mechanical force when it is placed in a magnetic field. The different current-carrying circuits also experience the forces due to the magnetic force produced by them. The energy conversion process is reversible because a voltage will be induced in a circuit undergoing motion in a magnetic field.

3. *Generation of voltage:*   A voltage is induced in a coil when there is a change in the flux linking the coil. With fixed coil position, the voltage is induced by changing the flux magnitude, is known as *transformer voltage*. Voltage can also be induced in the coil by motion of either the coil or the magnetic field relative to each other. Then it is known as *motional voltage* or *speed voltage*, or if the motion is rotary, *rotational voltage*. These voltages are given by Faraday's law. Electromechanical energy conversion takes place when the change in flux is associated with mechanical motion.

11.2   FORCE ON PLUNGER (LINEAR MOTION)

Figure 11.2 shows a simple device wherein a magnetic field attracts a movable steel member. It is assumed that the movable armature is moving parallel to itself (frictionless) where a single variable describes the position x.

The electric power input is given by $p = vi$, where $v = d\lambda/dt$. The differential amount of energy supplied in time, dt, will be

$$dW = pdt = id\lambda \qquad (11.1)$$

If the magnetization curve of the device is as shown in Figure 11.3(a) for a given x, the increment in energy is shown by shaded horizontal wedge at some value of λ. The total

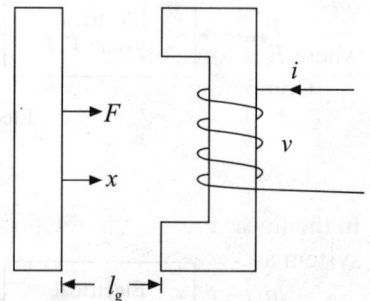

FIGURE 11.2   Force on plunger.

energy put into the device by holding x fixed and allowing the flux linkage to increase from zero to some value λ will be

$$W = \int_0^\lambda id\lambda \qquad (11.2)$$

and this energy is represented by the area to the left of the curve between zero and λ. If the limits of the integral are reversed, W is the negative of the original value. This shows that energy originally put in, is stored in the field and is completely returned to the circuit when the field is collapsed.

Magnetization curve for a new value at $(x + dx)$ is shown in Figure 11.3(b) which is different than magnetization curve for x. Thus field energy (W_F) is a function of both λ and x, and can be expressed as

$$W_F = W_F(\lambda, x) \tag{11.3}$$

and

$$dW_F = \frac{\partial W_F}{\partial \lambda}\, d\lambda + \frac{\partial W_F}{\partial x}\, dx \tag{11.4}$$

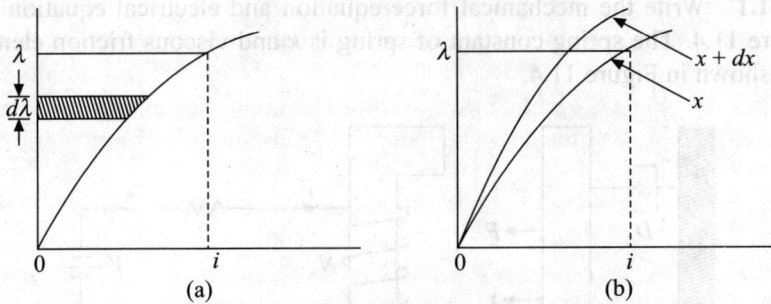

FIGURE 11.3 Magnetizing curves for device.

Due to the change in position, the energy stored is changed. Since the system is conservative, we can write equations based upon the conservation of energy as

Electrical energy input = Increase in field energy + Mechanical energy output

or

$$dW_{elect} = i\,d\lambda = dW_F + F\,dx \tag{11.5}$$

where F is force on the device.

Comparing Eq. (11.4) with Eq. (11.5), we get

$$i = \frac{\partial W_F}{\partial \lambda} \quad \text{and} \quad F = -\frac{\partial W_F}{\partial x} \tag{11.6}$$

In the linear region, the relation between i and λ can be written in terms of inductance, L, of the system as

$$i = \frac{\lambda}{L} \tag{11.7}$$

Using Eq. (11.7) and Eq. (11.6), we have

$$W_F = \frac{\lambda^2}{2L} = \frac{1}{2} L i^2 \tag{11.8}$$

where, $L = \dfrac{N^2}{\Re} = \dfrac{N^2 A \mu_0}{2(l_g - x)}$ (for the notations please refer to Chapter 9)

In Eq. (11.8), l_g stands for the length of air gap when armature is at $x = 0$. The force F can be calculated using expression (11.8) as

$$F = -\frac{\partial W_F}{\partial x} = -\frac{\partial}{\partial x} \frac{\lambda^2 (l_g - x)}{N^2 A \mu_0} = \frac{\lambda^2}{N^2 A \mu_0} \tag{11.9}$$

Using Eqs. (11.9) and (11.8), we get

$$F = \frac{N^2 A \mu_0 i^2}{4(l_g - x)^2} \tag{11.10}$$

EXAMPLE 11.1   Write the mechanical force equation and electrical equation of the system shown in Figure 11.4. The spring constant of spring is k and viscous friction element is having damping D as shown in Figure 11.4.

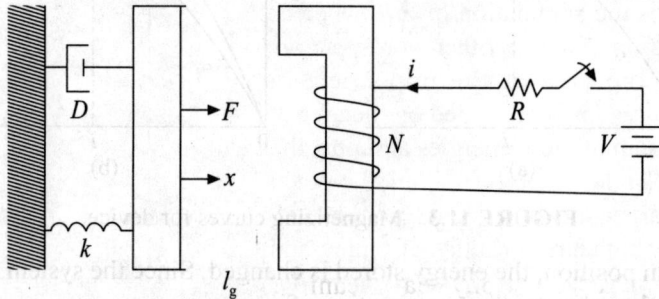

FIGURE 11.4   Example 11.1.

Solution: Let armature having mass M. The opposing forces are spring force (kx), viscous force of dash pot ($D dx/dt$) and inertia force due to mass ($M d^2x/dt^2$). Thus, the force excreted in the x-direction [using Eq. (11.10)] will be equal to the opposing forces and can be written as

$$\frac{N^2 A \mu_0 i^2}{4(l_g - x)} = M \frac{d^2 x}{dt^2} + D \frac{dx}{dt} + kx \tag{i}$$

The equation of electrical loop can be written as

$$V = Ri + \frac{d(Li)}{dt} = Ri + L \frac{di}{dt} + i \frac{dL}{dt} \tag{ii}$$

Putting the value of L from Eq. (11.8), we get

$$V = Ri + \frac{N^2 A \mu_0}{2(l_g - x)} \frac{di}{dt} + \frac{i N^2 A \mu_0}{2(l_g - x)^2} \frac{dx}{dt} \tag{iii}$$

We have two non-linear Eqs. (i) and (ii) which are to be solved for x and i subject to the initial conditions.

11.3 COENERGY

In Section 11.2, the force was derived in terms of flux linkage [Eq. (11.9)] and later was replaced with current i [Eq. (11.10)]. Keeping flux linkage to a constant value is difficult in the partial derivative. An alternate approach uses a different energy function defined as *coenergy*, W_F'.

$$W_F' = i\lambda - W_F \quad \text{or} \quad W_F' + W_F = i\lambda \tag{11.11}$$

From Eq. (11.14), we can write

$$dW_F' + dW_F = d(i\lambda) = \lambda di + id\lambda \tag{11.12}$$

or

$$W_F = \int_0^\lambda id\lambda \text{ and } W_F' = \int_0^i \lambda di \tag{11.13}$$

Thus, the *coenergy* is the summation of λdi, whereas *energy* is the summation of $id\lambda$. In other words, *energy* is represented by the area at the upper left in Figure 11.5 and *coenergy* is represented by the area at the lower right. The sum of these energies is simply the total area of the rectangle, or $i\lambda$. It is should be noted that, if λ vs i is linear, energy and coenergy are equal. Energy and coenergy both have the units of Joules, and though energy has a very real physical meaning, coenergy is just a function defined for the mathematical convenience.

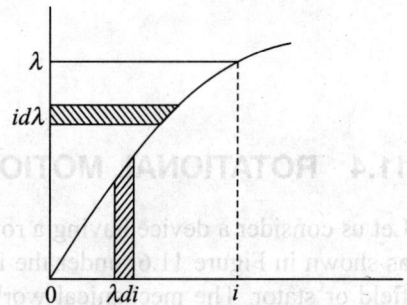

FIGURE 11.5 The relation between energy and coenergy.

Since coenergy is given by $W_F' = \int_0^i \lambda di$ for a given

x and would be different for different value of x, coenergy is a function of current i and x. Mathematically, it can be expressed as

$$W_F' = W_F'(i, x)$$

or

$$dW_F' = \frac{\partial W_F'}{\partial i} di + \frac{\partial W_F'}{\partial x} dx \tag{11.14}$$

Using Eqs. (11.5), we get

$$dW_F = id\lambda - Fdx \tag{11.15}$$

Putting this equation in Eq. (11.12), we have

$$dW_F' = \lambda di + Fdx \tag{11.16}$$

Equating Eq. (11.16) with Eq. (11.14), we get

$$\lambda = \frac{\partial W_F'}{\partial i} \text{ and } F = \frac{\partial W_F'}{\partial x} \tag{11.17}$$

The difference between the force equations using energy and coenergy can be seen below:

$$F = -\frac{\partial W_F}{\partial x} \quad \text{or} \quad F = +\frac{\partial W_F'}{\partial x}$$

Since for the linear region of λ-i curve, both energies are equal, hence

$$W_F = W_F' = \frac{1}{2} Li^2 = \frac{\lambda^2}{2L} \quad \text{and} \quad F = \frac{i^2}{2} \frac{dL}{dx}$$

That is for a linear single coil case at least where the production of mechanical force is dependent upon the variation of inductance with the mechanical displacement. And coenergy in terms of current can be written for the Figure 11.2 as

$$W_F' = \frac{N^2 A \mu_0}{4(l_g - x)} i^2 \tag{11.18}$$

$$F = \frac{N^2 A \mu_0}{4(l_g - x)^2} i^2 \tag{11.19}$$

11.4 ROTATIONAL MOTION

Let us consider a device having a rotating armature or rotor, as shown in Figure 11.6, under the influence of a stationary field or stator. The mechanical work done by the device is $Td\,\theta$ where T is torque.

Similar to force, the torque equation can be written as

$$T = \frac{\partial W_F'}{\partial \theta} \tag{11.20}$$

As torque is dependent on the variation of the inductance with angle and inductance can be approximately expressed as

$$L = L_0 + L_1 \cos 2\theta \tag{11.21}$$

Thus, the coenergy can be derived as

FIGURE 11.6 Rotational motion.

$$W_F' = \frac{1}{2} i^2 (L_0 + L_1 \cos 2\theta) \tag{11.22}$$

and torque can be

$$T = -i^2 (L_1 \sin 2\theta) \tag{11.23}$$

11.5 MOTIONAL VOLTAGE/ROTATIONAL VOLTAGE

If a conductor of length l, as shown in Figure 11.7, moves in a magnetic field B at a linear speed u, the induced voltage in the conductor will be

$$e = Blu \tag{11.24}$$

where B, l and u are perpendicular to each other.

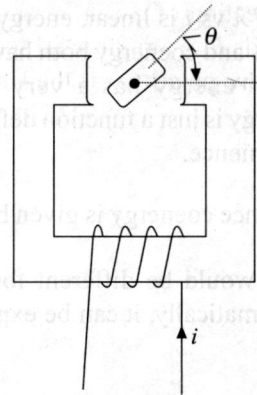

The cross (×) in Figure 11.7 shows the magnetic field perpendicular to the paper and going inside of paper (tail of an arrow). The direction of *e* as shown in Figure 11.7 is given by Fleming's right-hand rule (or right-hand screw rule). According to the Fleming's right-hand rule, if index finger of right-hand shows the direction of magnetic field and thumb shows the direction of motion, the middle finger shows the direction of induced *emf*.

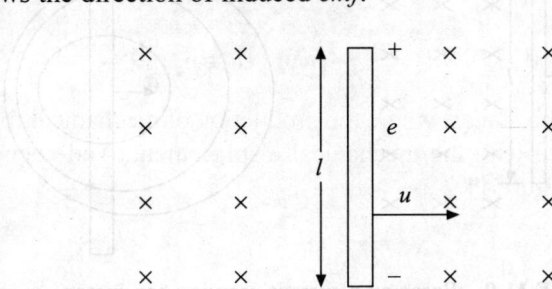

FIGURE 11.7   Induced voltage due to moving conductor.

Instead of linear motion, if the conductor is rotating and cutting the flux, voltage will also be induced. If conductor is moving with angular speed ω, the induced *emf e*, will be given by

$$e = B(\theta)\, l\, \omega\, r \tag{11.25}$$

The direction of *e* can be easily obtained by right-hand screw rule (Figure 11.8). Turning the *u* towards *B*, if a right-hand screw is in the same way, the motion of screw will show the direction of *positive* polarity. *r* is the length of conductor.

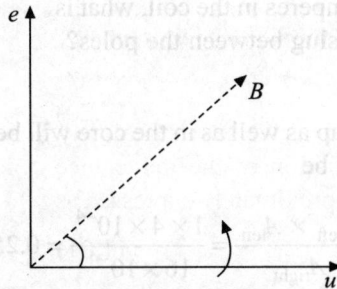

FIGURE 11.8   Induced *emf* due to rotation of conductor.

11.6 ELECTROMAGNETIC FORCE

When a current-carrying conductor is placed in a magnetic field, the force produced on the conductor (Lorenz force) is given by

$$F = B\, l\, I \tag{11.26}$$

The direction of force is governed by Fleming's left-hand rule. With the resultant flux (flux produced by the current *i* and the magnetic flux produced by another source where current-carrying conductor is placed), the direction of force can be determined. In Figure 11.9, the total flux at right-hand side of conductor will be higher than left-hand side. The motion is always from higher flux to lower flux. If B will be changing, the force will also be changing.

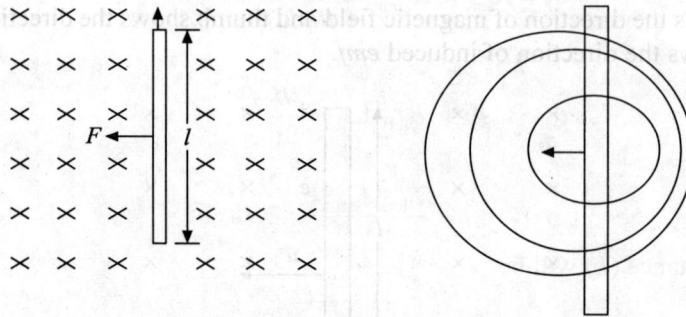

FIGURE 11.9   Force on a current-carrying conductor.

EXAMPLE 11.2   A magnetic circuit with a tapered slug of steel supported between two pole pieces as shown in Figure 11.10. The coil has 500 turns and $\mu_0 = 4\pi \times 10^{-7}$. If the permeability of the core is infinite (core consumes negligible *mmf*) and fringing in the air-gap is ignored.

(a) If B in the left hand air-gap is 1 Wb/m^2, how much current in required in the coil?
(b) What is the inductance of the device?
(c) If there is a dc current I amperes in the coil, what is the net force on the steel slug between the poles?

$A = 4 \times 10^{-4}$m^2 $A = 16 \times 10^{-4}$m^2
$l = 2 \times 10^{-3}$m $l = 1.5 \times 10^{-3}$m

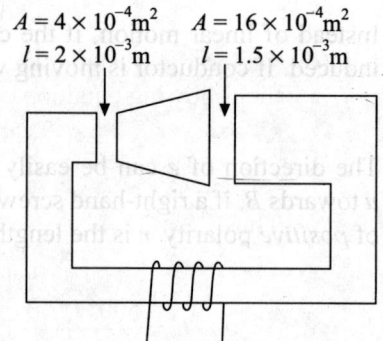

Solution: FIGURE 11.10   Example 11.2.

(a) Since the flux in the air-gap as well as in the core will be the same, the flux density in air-gap (right-hand side) will be

$$B_{\text{right}} = \frac{B_{\text{left}} \times A_{\text{left}}}{A_{\text{right}}} = \frac{1 \times 4 \times 10^{-4}}{16 \times 10^{-4}} = 0.25 \text{ Wb/m}^2$$

Total *mmf* will be equal to the *NI* (the *mmf*-drop in the core will be zero as permeability is infinite) and thus,

$$NI = H_{\text{left}} \times l_{\text{left}} + H_{\text{right}} \times l_{\text{right}} = \frac{B_{\text{left}}}{\mu_0} \times l_{\text{left}} + \frac{B_{\text{right}}}{\mu_0} \times l_{\text{right}}$$

or

$$500I = \frac{1}{4\pi \times 10^{-7}} \times 2 \times 10^{-3} + \frac{0.25}{4\pi \times 10^{-7}} \times 1.5 \times 10^{-3}$$

Hence,

$$I = 3.78 \text{ A}$$

Alternatively, let us calculate the total reluctance

$$\Re = \Re_{\text{left}} + \Re_{\text{right}} = \frac{l_{\text{left}}}{\mu_0 A_{\text{left}}} + \frac{l_{\text{right}}}{\mu_0 A_{\text{right}}}$$

and

$$\Phi = B_{\text{left}} A_{\text{left}} = \frac{NI}{\Re}$$

or

$$I = \frac{B_{\text{left}} A_{\text{left}} \Re}{N} = 3.78 \text{ A}$$

(b) The inductance (L) will be

$$L = \frac{N^2}{\Re} = \frac{500^2}{\left(\dfrac{2 \times 10^{-3}}{\mu_0 4 \times 10^{-4}} + \dfrac{1.5 \times 10^{-3}}{\mu_0 16 \times 10^{-4}} \right)} = 0.0529 \text{ H}$$

(c) Taking x to the left (slug moves in left direction), and thus, force will be in x, as positive, the following equation can be written

$$F = \frac{\partial W'_F}{dx}, \quad \text{where } W'_F = \frac{LI^2}{2} = \frac{N^2 I^2}{2\Re}$$

Thus,

$$F = \frac{500^2 I^2}{2} \frac{\partial}{\partial x} \left(\frac{1}{\dfrac{(l_{\text{left}} - x)}{\mu_0 A_{\text{left}}} + \dfrac{(l_{\text{right}} + x)}{\mu_0 A_{\text{right}}}} \right)$$

$$= \frac{500^2 I^2}{2} \times \mu_0 A_{\text{left}} A_{\text{right}} \left(\frac{-A_{\text{right}} + A_{\text{left}}}{\left((l_{\text{left}} - x) A_{\text{right}} + (l_{\text{right}} + x) A_{\text{left}} \right)^2} \right)$$

At $x = 0$, $F = -8.35 \ I^2$. Negative sign shows that slug will move in right direction.

EXAMPLE 11.3 Ten ampere current is passed in the two coils which are in series (total inductance will be $L_1 + L_2 \pm 2M$) as shown in Figure 11.11. Given that $M = 0.01/(x^2 + 3)$, $L_1 = 0.08$ H and $L_2 = 0.03$ H.

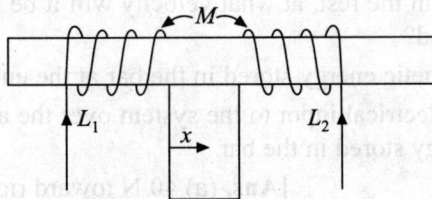

FIGURE 11.11 Example 11.3.

(a) What is the force developed between the two coils?

(b) Is force repulsive or attractive?

Solution: Looking at Figure 11.11, it is clear that fluxes produced by the coils are subtractive. Thus, the total inductance will be

$$L = L_1 + L_2 - 2M$$

(a) The coenergy will be

$$W_F' = \frac{Li^2}{2} = \frac{1}{2} \times \left(0.08 + 0.03 - \frac{2 \times 0.01}{(x^2 + 3)} \right) \times (10)^2 = 5.5 - \frac{1}{(x^2 + 3)}.$$

Therefore, the force will be

$$F = \frac{\partial W_F'}{\partial x} = 0 - (-1)(x^2 + 3)^{-2} \times 2x = \frac{2x}{(x^2 + 3)^2}$$

(b) Since the force is positive in the x direction, the force is repulsive.

PROBLEMS

11.1 Figure 11.12 shows two parallel horizontal rails 0.5 m apart with a crossbar at right angles supported by the rails but free to move in the x direction with negligible friction. A current source of 100 A passes through the crossbar because of contact with rails. A vertical magnetic field of uniform $B = 0.8$ Tesla, then reacts with the current to cause a force on the crossbar.

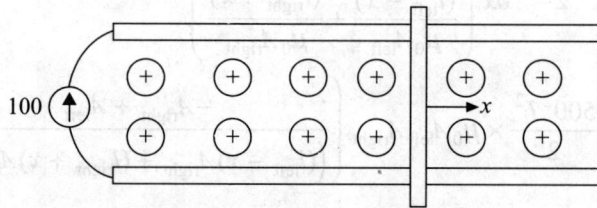

FIGURE 11.12   Problem 11.1.

(a) What is the force on the crossbar and in what direction does it act?

(b) If the bar has mass of 200 grams, what is the acceleration of the bar? (Take *f=ma*).

(c) If bars starts from the rest, at what velocity will it be moving two seconds after the current is applied?

(d) What will be kinetic energy stored in the bar at the end of two seconds?

(e) Show that the electrical input to the system over the acceleration period is equal to the kinetic energy stored in the bar.

[**Ans.** (a) 40 N toward right, (b) 200 m/s², (c) 400 m/s,
(d) 16000 joules, (e) $e = 80t$, $p = 8000t$ and $W = 16000$ joules]

11.2 A system as shown in Figure 11.13 acts on an iron rod to produce a force in the x direction when a constant current of 5 A passed. The inductance of the system is given

$$L(x) = 5 + 4x + 2(x - 1)^2$$

(a) What is the electromagnetic force developed on the iron bar as a function of x?

(b) If the restraining force of 150 N is exerted by the mass M, find the equilibrium point.

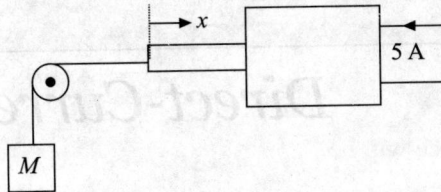

FIGURE 11.13 Problem 11.2.

[**Ans.** (a) $50x$ N, (b) $x = 3$ meter]

12

Direct-Current Machines

12.1 INTRODUCTION

The earliest electric power systems, as discussed in Chapter 1, were direct-current (dc) system having dc generation, transmission and dc utilization. When alternating-current (ac) systems become popular, the use of dc generators declined somewhat, but they were still retained for automotive and aircraft systems and for industrial drives that required the use of dc motors. The development of power electronics devices such as diodes, thyristors, transistors, has somewhat changed the scenario. It is usually more economical to obtain the dc power by rectifying the available ac supply than it is to use an ac-dc motor-generator set. Even for mobile applications, the trend is to generate ac power and rectify to direct current. There are, however, a fair number of dc generators in service and also the theory of the dc machine is important to understand the ac machine theory. That is the reason why this chapter has been written.

Every electric machine has two parts: *stator* and *rotor*. *Stator*, as its name, does not move/rotate and normally is the outer frame of the machine, whereas *rotor* is free to move and normally is inner part of the machine. Depending on the *rotor* structure, a machine can be categorized as *cylindrical* (having uniform air gap between *rotor* and *stator*) machine and *salient* (non-uniform air-gap) *pole* machine as shown in Figure 12.1.

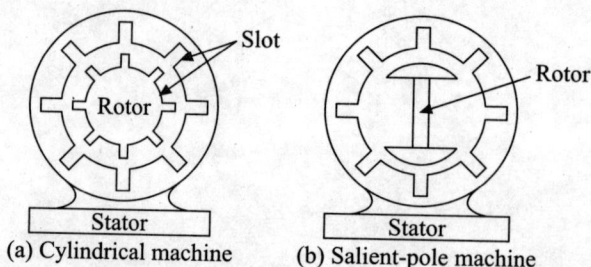

(a) Cylindrical machine   (b) Salient-pole machine

FIGURE 12.1   Cylindrical and salient pole machines.

Conductors are placed in stator and rotor slots to produce magnetic field. Therefore, stator and rotor are made of ferromagnetic materials and laminations are used to reduce the iron losses because stator and/rotor are subjected to the time varying magnetic flux. Two conductors are connected to one end by an end-connector to form a turn. A coil is formed by connecting many turns as shown in Figure 12.2. A winding is formed by connecting several winding coils. The winding in which voltage is induced is called the *armature winding*. The winding in which current is passed to produce the primary source of flux in the machine is called the *field winding*. Armature winding is a closed winding. Some machines, instead of having field winding, have permanent magnet known as *permanent magnet machine*. Others commonly used machines having field winding to produce the electric magnet are dc machines, induction machines (asynchronous), synchronous machine, etc.

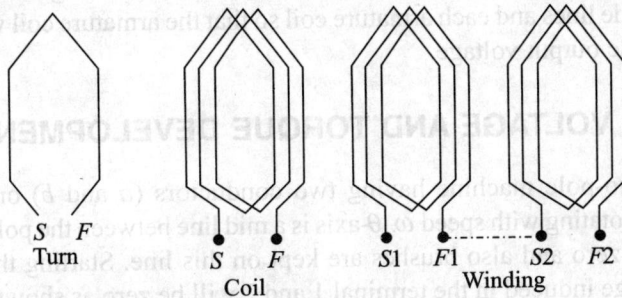

S   F
Turn   *S   F*   *S1   F1*   *S2   F2*
                Coil   Winding

FIGURE 12.2   Turns, coils and winding.

12.2 DC MACHINE CONSTRUCTION

In dc machine, the field wilding is placed on the stator (field frame) and armature winding is placed on the rotor as shown in Figure 12.3. The field frame or yoke is usually an iron or steel casting. The pole cores are usually built up of the silicon steel laminations that are stamped in the appropriate shape, riveted together, and bolted to the yoke. The armature core is basically a cylindrical stack of disc-shaped silicon steel laminations mounted on a shaft and supported between pole cores by means of bearing. The space between pole cores and the armature core is known as the air gap and usually about 3 to 5 mm. DC current is passed through the field winding to produce flux in the machine.

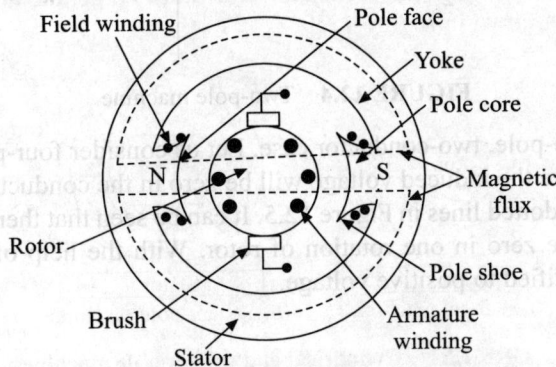

Field winding   Pole face
                            Yoke
                            Pole core
N   S   Magnetic
                            flux
Rotor
                            Pole shoe
Brush   Armature
                            winding
Stator

FIGURE 12.3   Magnetic circuit of two-pole machine.

The armature winding can be of two types: *lap* winding and *wave* winding. *Lap* winding is arranged in such a way that there should be parallel paths equal to the number of poles of machine, whereas the number of parallel paths is always two in the wave winding. Winding are connected to a switching mechanism known as the *commutator*. The commutator is comprised of axially-oriented strips of copper, insulated from each other and forming a cylindrically concentric with shaft. Electrical contacts with the rotor windings are made through graphite brushes that ride on the commutator surface to control the flow of current in the armature conductors. The brushes and commutator assembly do two things:

- They provide the electrical connection between the rotating armature coils and the stationary external circuit.
- As the armature rotates, they perform a switching action, reversing the electrical connections between outside lines and each armature coil so that the armature coil voltages add together and result in dc output voltage.

12.3 INDUCED VOLTAGE AND TORQUE DEVELOPMENT

Let us consider a two-pole machine having two conductors (*a* and *b*) on rotor as shown in Figure 12.4. Rotor is rotating with speed ω. θ-axis is a mid line between the poles where conductors induced voltages are zero and also brushes are kept on this line. Starting the rotation from this line ($\theta = 0$), the voltage induced in the terminal 1 and 2 will be zero as shown in Figure 12.4. For the movement up to 180 degree (π), the induced voltage is positive but from π to 2π, the induced voltage in coil will be negative as shown in Figure 12.4. But during that period, the position of brushes changed and again the positive voltage will appear. It can be noted that commutator-brush arrangement works like a mechanical rectifier.

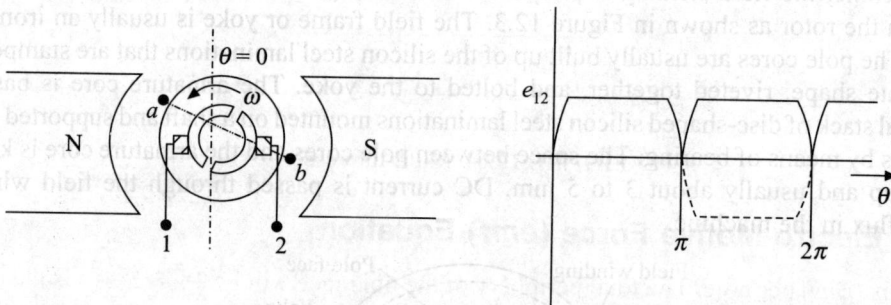

FIGURE 12.4 Two-pole machine.

Figure 12.4 shows a two-pole, two-conductor case. Let us consider four-pole and two-conductor case. The positions where the induced voltage will be zero in the conductors will be between the poles as shown with the dotted lines in Figure 12.5. It can be seen that there are four points where induced voltages will be zero in one rotation of rotor. With the help of brushes, the negative voltage generated is rectified to positive voltage.

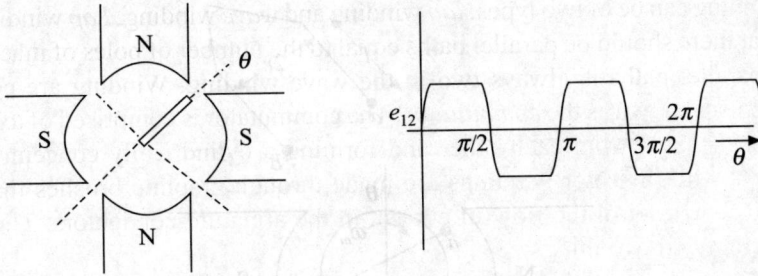

FIGURE 12.5   Four-pole machine.

It can be seen from Figure 12.5 that induced voltage makes two revolutions in one revolution of rotor. Mathematically, it can be generalized as

$$\omega_{elect} = \frac{p}{2}\,\omega_m \qquad (12.1)$$

where p is the number of poles, ω_m is the angular speed of rotor (rad/s) and ω_{elect} is the electrical angular speed of generated voltage (rad/s). Further, it can be written in terms of electrical (θ_{elect}) and mechanical (θ_m) angles as

$$\theta_{elect} = \frac{p}{2}\,\theta_m \qquad (12.2)$$

So far we discussed the two-conductor case. If there are several conductors (multi-turns) with the commutator-brush arrangements, the obtain voltage will be seen as shown in Figure 12.6.

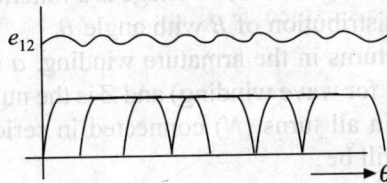

FIGURE 12.6   Multi-turn case.

12.3.1   Electro-Motive Force (*emf*) Equation

Let ϕ is flux (*rms*) per pole. The total periphery of the pole face will be $2\pi r$ where r is the distance from the rotor centre to the pole face (and approximately equal to the rotor radius as the distance between rotor and pole faces are very small). It can be assumed that space between two pole faces is negligible which is true for multi-pole machines. The area per pole (A) will be

$$A = \frac{2\pi rl}{p} \qquad (12.3)$$

where l is the length of pole face (or length of conductors).

The average induced voltage in a turn is given by the following equation [using Eq. (11.28)] which is a function of angle θ and shown in Figure 12.7.

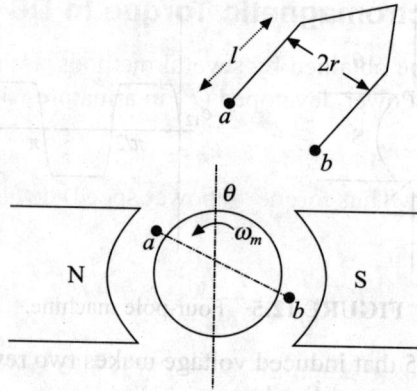

FIGURE 12.7   Induced voltage in a turn.

$$e_t = B_{av}(\theta) \times l \times \omega_m \times 2r \tag{12.4}$$

The average flux density $B_{av}(\theta)$, can be written in terms of the total flux in the air gap as

$$B_{av}(\theta) = \frac{\phi}{A} = \frac{\phi p}{2\pi r l} \tag{12.5}$$

Using Eqs. (12.4) and (12.5), we get

$$e_t = \frac{\phi p}{\pi} \omega_m \tag{12.6}$$

It is interesting to note that the average induced voltage is a function of total flux per pole, ϕ, and does not depend on the actual distribution of B with angle θ.

If N is the total number of turns in the armature winding, a is the number of parallel paths ($a = p$ for lap winding, and $a = 2$ for wave winding) and Z is the number of conductor ($= 2N$), then the total induced voltage (E_a) in all turns (N) connected in series for one parallel path across positive and negative brushes will be

$$E_a = \frac{N}{a} \cdot e_t \tag{12.7}$$

Equation (12.7) can be rewritten using Eq. (12.6) as

$$E_a = \frac{Np}{\pi a} \phi \omega_m = K_a \phi \omega_m \tag{12.8}$$

Where

$$K_a = \frac{Np}{\pi a} = \frac{Zp}{2\pi a} \tag{12.9}$$

It will be noted that K_a is a constant for a designed dc machine.

If ϕ is in Tesla (or Wb/m^2) and ω_m is in radian/second, the E_a will be in volts. The relation between E_a, ϕ and ω_m is true for motoring and generating operation. In motor operation, E_a is known as *back emf* or *counter emf* and in the case of generation operation, it is known as *generated* or *induced voltage*.

12.3.2   Developed Electromagnetic Torque in DC Machine

The expression of torque can be obtained by several methods. Using expression from previous section, it can also be derived. Power developed (P) in armature will be

$$P = E_a I_a$$

where I_a is the armature current. Thus torque (= power/speed) developed inside the machine can be written using Eq. (12.8) as

$$T = \frac{P}{\omega_m} = \frac{E_a I_a}{\omega_m} = K_a \cdot \phi I_a \tag{12.10}$$

where K_a is the same constant as given in Eq. (12.9). It is also known as *armature constant*.

Alternate method

Consider a conductor under a pole. If the total armature current I_a and the winding pattern is such that there are a parallel paths carrying current, the current in the given conductor is I_a/a. The force on a sample conductor will be, (a function of θ) using Eq. (11.29),

$$f(\theta) = B(\theta)\, l\, \frac{I_a}{a} \tag{12.11}$$

Thus, the average force using Eqs. (12.11) and (12.5) will be

$$f = \frac{p\phi}{A} l \frac{I_a}{a} = \frac{p\phi}{2\pi r l} \times l \times \frac{I_a}{a} = \frac{p\phi I_a}{2\pi r a} \tag{12.12}$$

where ϕ is the flux per pole. The force on each conductor is tangential to the surface of the rotor because of the normal direction of the B vector. If there are Z conductors, all lying at the radius r, the total torque developed is given by

$$T = Z \times r \times f = \frac{Z p \phi I_a}{2\pi a} = K_a \phi I_a \tag{12.13}$$

It can be seen that Eq. (12.13) and Eq. (12.10) are the same.

EXAMPLE 12.1   An effective length of 2-pole dc machine having armature radius of 10 cm is 50 cm. The armature winding has 210 turns. The average flux density under each pole is 0.75 T where poles cover 80% of armature periphery. Find the following for lap and wave windings of armature.

    (a) Induced armature voltage when armature rotates at 1200 rpm.
    (b) Torque developed at armature current of 20 A.
    (c) Power developed.

Solution: Given that,

$$p = 2,\ r = 10 \times 10^{-2}\,\text{m}, \quad l = 50 \times 10^{-2}\,\text{m}$$

$$N = 210 \quad \text{or} \quad Z = 420, \quad B/\text{pole} = 0.75\,\text{T}$$

Using Eq. (12.9), we have,

$$K_a = \frac{Np}{\pi a}$$

Since, for lap winding, $a = p = 2$ and for wave winding, $a = 2$. Thus $a = 2$

$$K_a = \frac{210 \times 2}{\pi \times 2} = 66.845 \text{ Volt/Wb-rad/s}$$

Area per pole, $$A = \frac{2\pi r l}{p} \times 0.8 \text{ m}^2$$

Thus,

$$\phi \text{ per pole} = B \cdot A = 0.75 \times \frac{2 \times \pi \times 10 \times 50 \times 10^{-4}}{2} \times 0.80 = 0.0884 \text{ Wb}$$

and speed in rad/second will be

$$\omega_m = \frac{2\pi}{60} \times \text{speed (in rpm)} = \frac{2\pi}{60} \times 1200 = 125.66 \text{ rad/second}$$

(a) Induced voltage, $E_a = K_a \phi \omega_m = 742.54$ Volts
(b) Developed torque, $T = K_a \phi I_a = 118.18$ N-m
(c) Power $P = E_a I_a = T \cdot \omega_m = 14850.75$ Watt.

12.4 MAGNETIZATION CURVES

Unless there is a permanent magnet device, the field flux is produced by the currents flowing in the coils wounded on the field pole structure. Opposite to ac machines and transformers where flux is alternating in nature, flux in dc machine is not alternating (i.e. constant) which is produced by dc current in field winding. The produced flux is developed into air gap, rotor path, armature and yoke as shown in Figure 12.8. Since reluctance of ferromagnetic material is very small compared to the reluctance of air gap, $\mathfrak{R}_g$, (H is much greater in the air gap than in the steel). It is assumed that the ampere-turn (mmf) is consumed by the air gap due the larger reluctance.

The flux will be

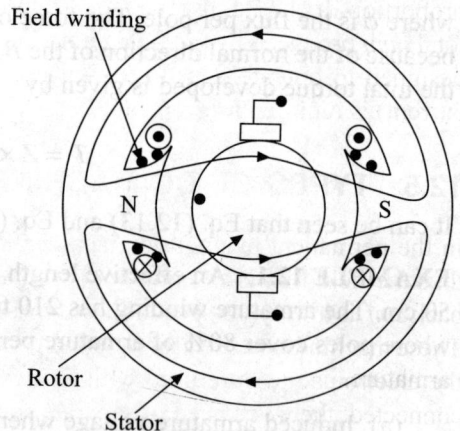

FIGURE 12.8 Magnetic flux.

$$\phi = \frac{mmf}{\mathfrak{R}_g} = \frac{N_f I_f}{\mathfrak{R}_g} \tag{12.14}$$

where N_f is the number of turns in field winding and I_f is the field current.

From equation (12.14), it can be seen that there is a linear relationship between flux and flux field current but at larger value of current, the production of flux deviates from the linear relation as shown in Figure 12.9.

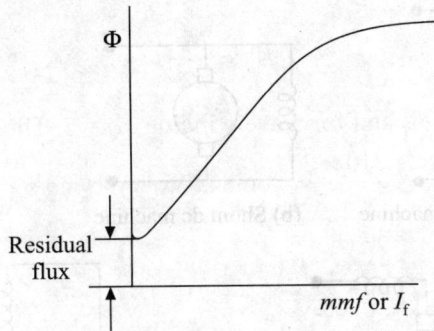

FIGURE 12.9   Flux-current curve.

With this flux, the induced voltage will be generated and will be different at different speed which can be seen from Eq. (12.8). Since the generated voltage, E_a, is proportional to the field flux per pole, ϕ, at any given speed, ω_m, a curve of E_a versus $mmf(N_f I_f)$ differs from Figure 12.9 only by constants on the axes. It should be noted that two curves are given in Figure 12.10 for two different speeds designated ω_{m1}, and ω_{m2}. Since at a given flux, the generated voltage is directly proportional to speed, any one speed could be chosen to illustrate and the value of other speeds can be found as direct proportion. It should also be noted that the mmf is the

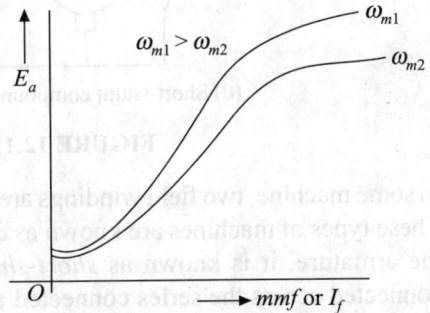

FIGURE 12.10   Magnetization curves of a dc machine.

total mmf produced by the field components. To have better utilization of material, the machine is designed to operate near to the knee-point of curve as beyond that point more mmf (or field current) is required to even small increase in the generated voltage.

12.5   TYPES OF DC MACHINES

In the permanent magnet dc machines, the pole cores are made of the permanent magnets and thus the field flux is essentially constant. The only practical way to change the output voltage or back *emf* is to vary the speed. These types of machine are known as *permanent magnet dc machines*. These machines are used for light load conditions such as for tachometer. In the large dc machines, the electromagnets are used with field windings. If field winding and armature winding are not connected, there is a need of separate source for field winding and this type of machine is known as *separately excited dc machine*. However, field winding and armature winding can be interconnected in various ways and the performance characteristics of these machines vary. If field winding is connected in series with armature, this type of machine is known as *series machine* where field current is the same as armature current. If field winding is connected across the armature, the type of machine is known as *shunt machine* where field current is a function of terminal voltage. The various types of dc machines based on the field winding connection is shown in Figure 12.11.

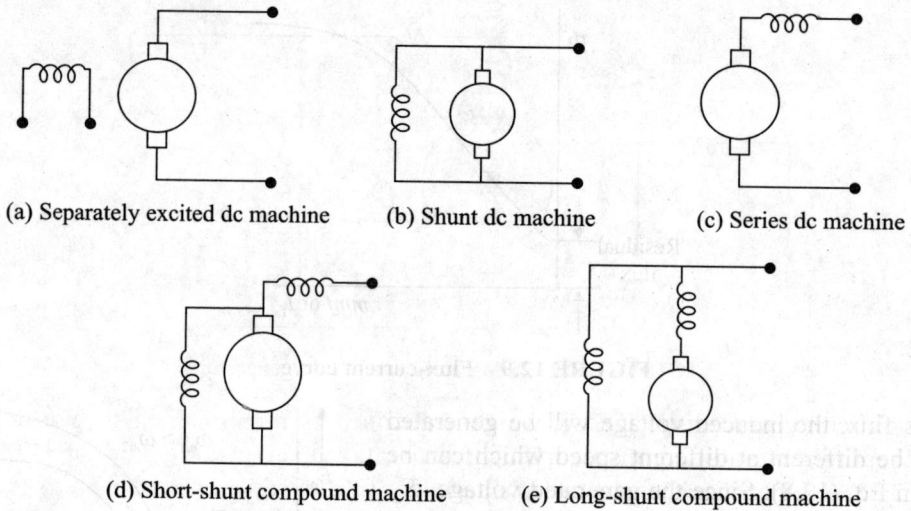

(a) Separately excited dc machine   (b) Shunt dc machine   (c) Series dc machine

(d) Short-shunt compound machine   (e) Long-shunt compound machine

FIGURE 12.11   Different types of dc machines.

In some machine, two field windings are used and are connected in series and shunt configurations. These types of machines are known as *compound machines*. If shunt winding is connected across the armature, it is known as *short-shunt machine*. In *long-shunt machine*, shunt winding is connected across the series connected armature and field winding. In compound machine series and shunt field *mmf* may odd or oppose to each other. If field *mmf*s are opposing to each other, it is known as *differential compound machine*. In cumulative compound machine, *mmf* produced by series and shunt windings is additive.

12.6   DC GENERATOR

DC generator is a device which converts mechanical power to the dc electrical power. It is rotated by a prime mover at a constant speed and armature terminal is connected to load or dc electric system. Depending on the field winding connected, dc generators are classified.

12.6.1   Separately Excited DC Generator

In a separately excited dc generator, the field winding is excited from a separate dc source as shown in Figure 12.12. The basic equations using the Kirchhoff's law can be written as

$$V_f = R_f I_f \tag{12.15}$$

$$E_a = K_a \phi \omega_m \tag{12.16}$$

$$E_a = V_t + I_a R_a + 2V_b \tag{12.17}$$

$$V_t = I_L R_L \tag{12.18}$$

$$I_a = I_L \tag{12.19}$$

where R_a is the armature resistance, I_a is the armature current, V_t is the terminal voltage, R_L is load resistance, V_b is the voltage drop per brush (in a dc machine there are two brushes).

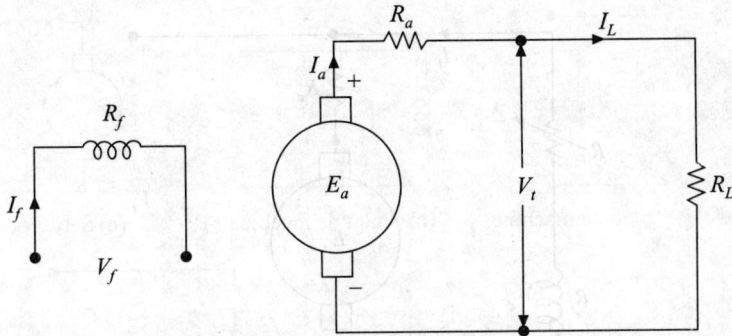

FIGURE 12.12   Separately excited dc generator.

The open circuit characteristic (OCC), which is obtained by running the generator at constant speed keeping generator terminal open, of separately excited dc generator is shown in Figure 12.13. Another characteristic called *external* or *terminal* or *load characteristic* is obtained by running the generator at rated speed and keeping the field current constant (i.e. I_f constant) but with varied load current. Due to the load current, there will be a voltage drop in the armature resistance and terminal voltage will fall. When the load current is more, a further voltage drop occurs due to armature reaction (or demagnetizing effect).

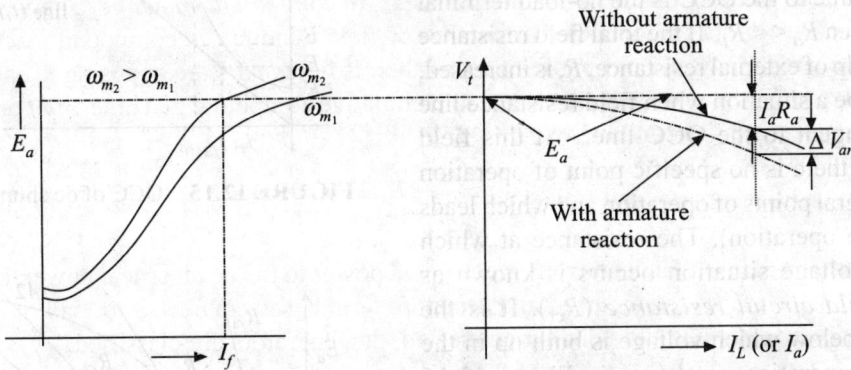

FIGURE 12.13   Open-circuit characteristics.

12.6.2   DC Shunt Generator

In dc shunt generator, field winding is connected across the armature as shown in Figure 12.14. A variable resistance is connected in the field circuit to control the field current. This machine is also known, as *self-excited*, thus some residual magnetism must exist in the magnetic circuit. The voltage build up characteristic, termed as OCC or magnetization characteristic is obtained by varying the field current while keeping the generator terminal open and running the machine at a constant speed. When field current is zero, the voltage developed E_r at generator terminal due to residual magnetism.

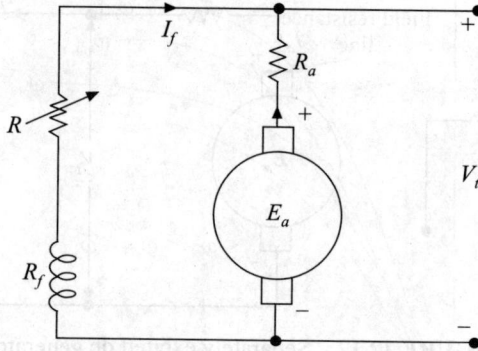

FIGURE 12.14   DC shunt generator.

Due to E_r, there will be current, I_{f1}, in field winding as shown in Figure 12.15. And thus voltage E_1 will be induced which will cause I_{f2} current in field winding and so on. This process is continued till point A. In this case, it is assumed that R_a is very-very less than R_f. The intersection of field resistance line to the OCC is the no-load terminal voltage when $R_a << R_f$. If the total field resistance with the help of external resistance, R, is increased, there will be a situation when field resistance line will be tangent to the OCC line. At this field resistance, there is no specific point of operation (rather several points of operation and which leads to unstable operation). The resistance at which unstable voltage situation occurs is known as *critical field circuit resistance* (R_{cr}). It is the resistance below which voltage is built up in the dc shunt generation as shown in Figure 12.16 where R_{f3} is the critical resistance at a given speed.

Normally, the field circuit resistance is fixed and generator is rotated to develop the induced voltage. The induced voltages at different speeds are shown in Figure 12.17. Field resistance line is tangential to the OCC for speed ω_{m3} and this speed is known as *critical speed*. Critical speed is the speed, below which there will be no voltage, built on the terminal of dc shunt generator.

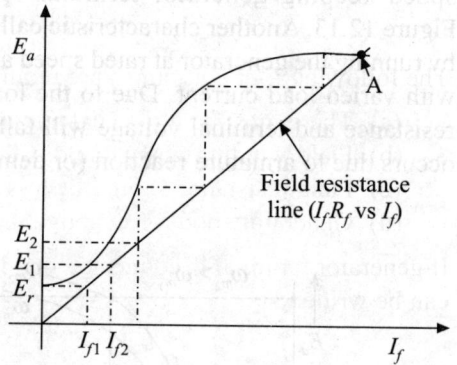

FIGURE 12.15   OCC of dc shunt generator.

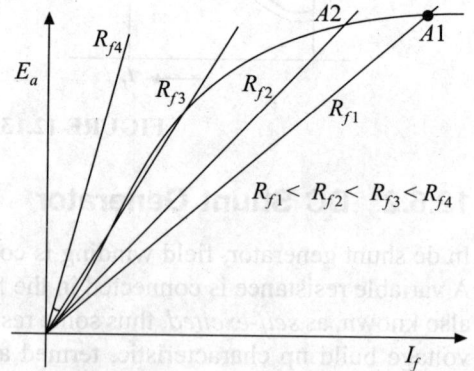

$R_{f1} < R_{f2} < R_{f3} < R_{f4}$

FIGURE 12.16   Field critical resistance.

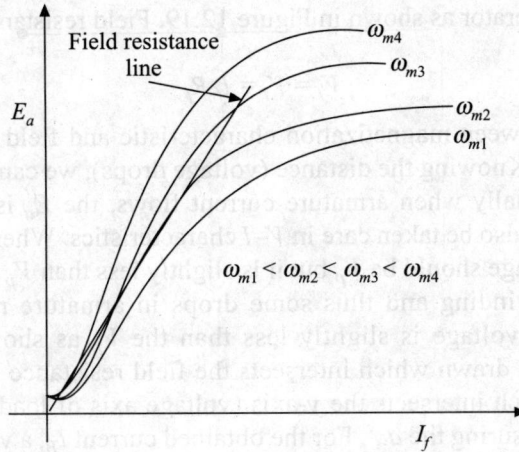

FIGURE 12.17 Critical speed.

The following four conditions are required to build up voltage in self-excited dc shunt generator.

(a) There must be a residual magnetism in the magnetic system.
(b) Field winding *mmf* should add the residual magnetism.
(c) Field resistance should be less than critical field circuit resistance.
(d) Generator should run above the critical speed.

If generator terminal is connected to a load resistance (Figure 12.18) the voltage current relation can be written as

$$E_a = K_a \phi \omega_m \tag{12.20}$$

$$E_a = V_t + I_a R_a + 2V_b \tag{12.21}$$

$$V_t = I_L R_L = I_f(R_f + R) \tag{12.22}$$

$$I_a = I_f + I_L \tag{12.23}$$

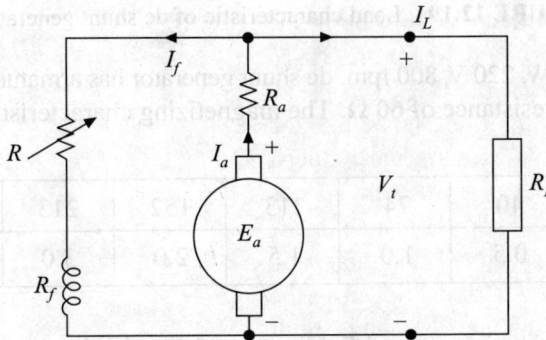

FIGURE 12.18 Circuit diagram of dc shunt generator.

The terminal voltage varies with load current, I_L, and induced voltage, E_a. The terminal characteristic or load characteristic or external characteristic can be obtained from OCC curve for self-excited dc shunt generator as shown in Figure 12.19. Field resistance line ($R = 0$) shows the terminal voltage because

$$V_f = V_t = I_f R_f$$

The vertical distance between magnetization characteristic and field resistance line (aa', bb' and cc') is the $I_a R_a$ drop. Knowing the distance (voltage drops), we can get the current I_a for the known value of R_a. Actually when armature current flows, the E_a is reduced further due to armature reaction. It can also be taken care in V–I characteristics. When there is no load current ($I_a = 0$), the terminal voltage should be V_p but it is slightly less than V_p as some currents flow in the armature for field winding and thus some drops in armature resistance remain there. Therefore, the terminal voltage is slightly less than the V_p as shown in Figure 12.19. At point a', a vertical line is drawn which intersects the field resistance line at a. The horizontal line from a is drawn which intersects the y-axis (voltage axis of load characteristic). Current will be calculated by measuring the aa'. For the obtained current I_B, a vertical line is passed and the point B will be obtained as shown in Figure 12.19. Similarly, for other load currents, we can obtain the coordinates of the load curve.

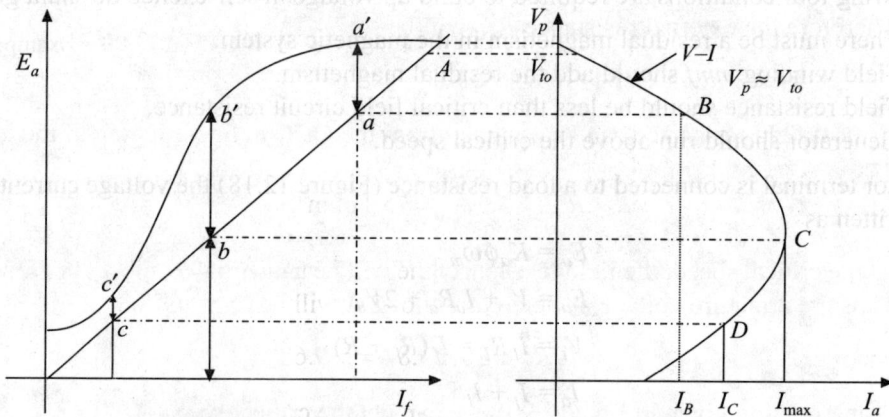

FIGURE 12.19   Load characteristic of dc shunt generator.

EXAMPLE 12.2   A 5 kW, 220 V, 800 rpm, dc shunt generator has armature resistance $R_a = 0.05\ \Omega$ and shunt field winding resistance of 60 Ω. The magnetizing characteristic of dc machine at 800 rpm is given below:

V_{occ} (V)	8	40	74	113	152	213	234	248
I_f (A)	0	0.5	1.0	1.5	2.0	3.0	3.5	4.0

Find:
   (a) The external field resistance and field current for no load voltage 220 V.
   (b) Value of critical field resistance.
   (c) Value of critical speed.
   (d) If load terminal voltage is adjusted to 220 V, field the E_a.

Solution:

(a) Field current at 220 V can be obtained by table using linear approximation between 213 V and 234 V as (see Figure 12.20) given below.

The slope of curve AB can be written as

$$m = \frac{220 - 213}{\Delta I_f} = \frac{234 - 213}{3.5 - 3.0}$$

Thus,

$$I_f = 3.0 + \left(\frac{234 - 213}{220 - 213}\right) \times (3.5 - 3.0)$$

$$= 3.15 \text{ A}$$

The total field resistance will be

$$R_f' = \frac{220}{3.15} = 69.84 \ \Omega$$

Therefore the external resistance R_{ext} will be

$$R_{fext} = R_f' - R_f = 69.84 - 60 = 9.84 \ \Omega$$

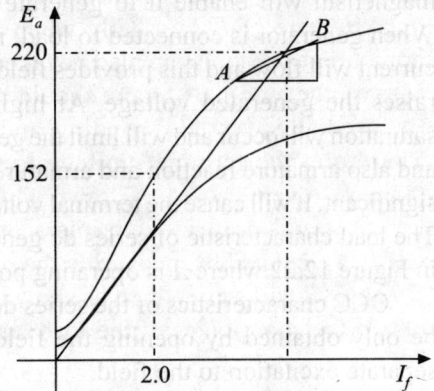

FIGURE 12.20 Example 12.2.

(b) Critical field resistance can be obtained from the OCC in linear region of the curve as

$$R_{critical} = \frac{152}{2} = 75.1 \text{ ohm}$$

(c) At the critical speed, the field resistance line will be tangential as shown in Figure 12.20. Thus the armature voltage at field current of 2 A will be

$$E_a = I_f R_f' = 2 \times 69.84 = 139.68 \text{ V}$$

Using relation given in Eq. (12.20), we can get the critical speed as

$$E_a = K_a \phi \omega_m \quad \text{or} \quad E_{a1} \propto n_1 \text{ (for the fixed value of field current)}$$

$$n_2 = \frac{E_{a2}}{E_{a1}} n_1 = \frac{139.68}{152} \times 800 = 735.16 \text{ rpm}$$

(d) The load current at the full load for 220 V terminal voltage

$$I_L = \frac{5 \times 1000}{220} = 22.73 \text{ A} \text{ and the field current will be } I_f = \frac{220}{69.84} = 3.15 \text{ A}. \text{ Thus, the}$$
armature current will be [Eq. (12.23)],

$$I_a = I_L + I_f = 25.88 \text{ A}$$

Therefore, the armature voltage [using Eq. (12.21)] assuming zero voltage drop across the brushes) will be

$$E_a = V_t + I_a R_a = 220 + 25.88 \times 0.05 = 221.29 \text{ V}.$$

12.6.3   DC Series Generator

The equivalent circuit of dc series generator is shown in Figure 12.21. Under no-load conditions, the series generator has no field excitation. However, residual magnetism will enable it to generate small voltage. When generator is connected to load, initially a small current will flow and this provides field excitation and raises the generated voltage. At high load current, saturation will occur and will limit the generated voltage and also armature reaction and armature drops become significant. It will cause the terminal voltage to decrease. The load characteristic of series dc generator is shown in Figure 12.22 where A is operating point.

   OCC characteristics of the series dc generator can

FIGURE 12.21   Series dc generator.

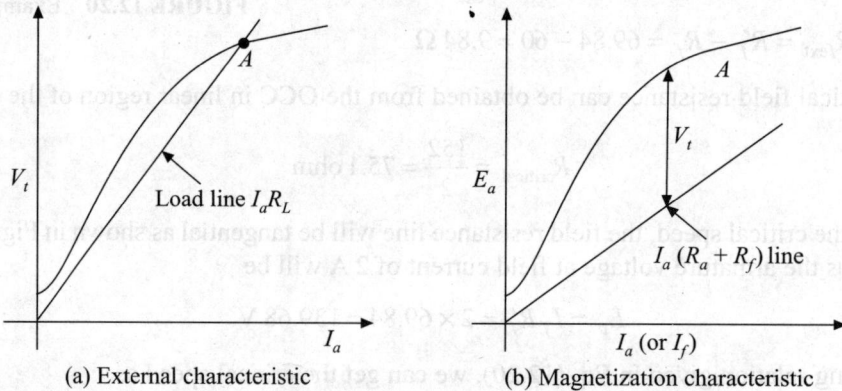

be only obtained by opening the field winding from the series connection and providing the separate excitation to the field.

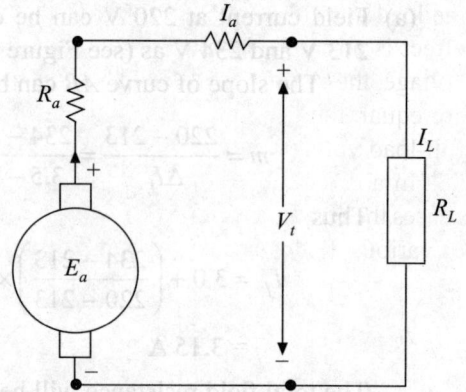

(a) External characteristic   (b) Magnetization characteristic

FIGURE 12.22   Terminal voltage characteristic of series dc generator.

Basic circuit equation for the series dc generator can be written as

$$E_a = V_t + I_a(R_a + R_f) \tag{12.24}$$
$$V_t = R_L I_L \tag{12.25}$$
$$I_a = I_L = I_f \tag{12.26}$$

12.6.4   Compound DC Generators

As shown in Figures 12.11(d) and 12.11(e), compound dc generators have both shunt and series field coils. Normally, the shunt field will have high *mmf*. Sometimes the field coils are connected such that the series field carries both the load and shunt field currents and this is known as *long shunt connection*. If the series field does not carry the shunt field current, the arrangement is known as a *short-shunt connection*. Changing from long to short shunt has little effect on the generator performance.

If the series field is connected so as to aid the shunt field (a cumulative connection), then with the load current, series field tends to offset the internal voltage drop of the machine. The overall effect is an improvement in voltage regulation. If the full-load voltage is less than the no-load voltage, the generator is said to be under-compounded. If the full load voltage and no-load voltage are equal, the generator is said to be flat-compounded. In the over-compounded generator, the full-load voltage is more than the no-load voltage as shown in Figure 12.23.

In a differential compound machine, the series field opposes the shunt field *mmf*, and this causes the terminal voltage to decrease rapidly as load is applied. Comparison of load characteristics of various dc generators is given in Figure 12.24 for equal full-load voltage.

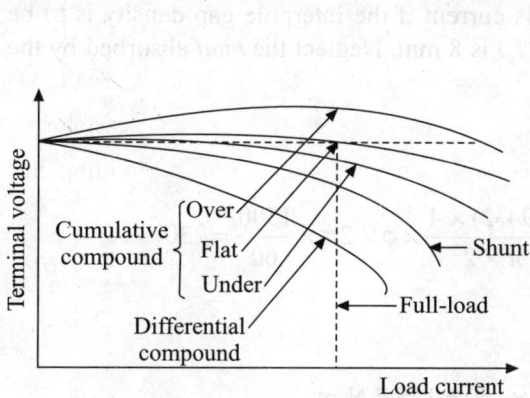

FIGURE 12.23   Terminal voltage curve with equal no-load voltage.

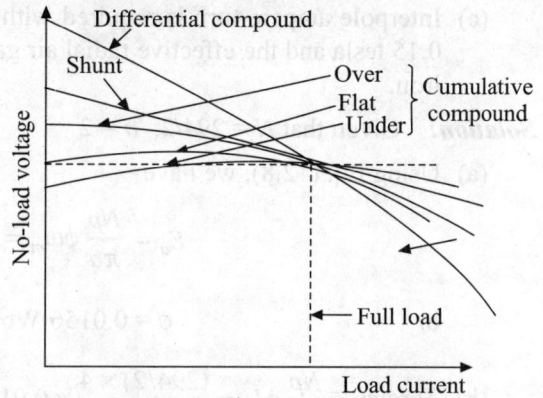

FIGURE 12.24   Terminal voltage curve with equal full-load voltage.

12.7   ARMATURE REACTION

When there is no current in the armature coil, the flux distribution due to the field coil current is uniform as shown in Figure 12.25(a). When current flows in the armature, the flux produced by armature current established as shown in Figure 12.25(b) and the resultant flux is distorted as shown in Figure 12.25(c). Armature *mmf* due to armature current distorts the flux distribution and produces demagnetizing effect, known as *armature reaction*.

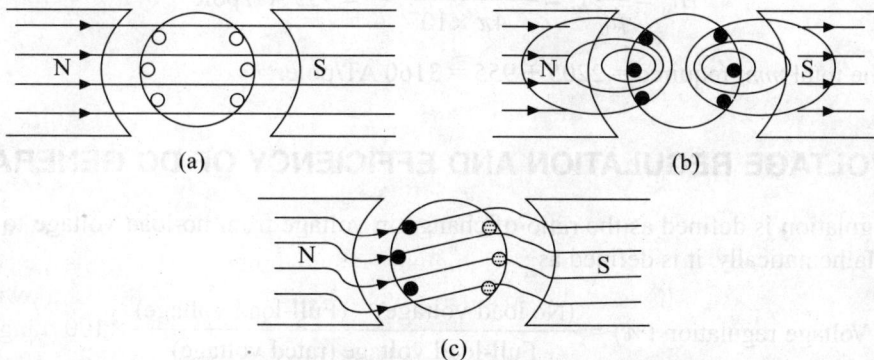

FIGURE 12.25   Armature reaction.

It can be seen that flux density under the pole is increased in one-half of the pole and decreased under the other half of the pole. Increased flux density causes magnetic saturation and thus reduction of flux per pole. To reduce the armature reaction, compensating winding is used to neutralize the armature *mmf*. Compensating winding, which is housed in the pole faces is connected in series with armature winding so that *mmf* produced is proportional the armature *mmf*.

EXAMPLE 12.3   A four-pole dc machine having wave winding has 294 conductors in armature. Find the following:

    (a) Flux per pole to generate 230 V when rotating at 1500 rpm.
    (b) Torque at this flux when rated armature current of 120 A is flowing.
    (c) Interpole ampere-turn is required with this current if the interpole gap density is to be 0.15 tesla and the effective radial air gap (l_g) is 8 mm. Neglect the *mmf* absorbed by the iron.

Solution: Given that $N = 294/2$, $a = 2$

    (a) Using Eq. (12.8), we have

$$E_a = \frac{Np}{\pi a}\phi\omega_m = \frac{(294/2)\times 4}{\pi \times 2}\times \phi \times 2\pi \times \frac{1500}{60} = 230$$

or $$\phi = 0.0156 \text{ Wb}$$

    (b) Torque $= \dfrac{Np}{\pi a}\phi I_a = \dfrac{(294/2)\times 4}{\pi \times 2}\times 0.0156 \times 120 = 175.7 \text{ N-m}$

    (c) On the interpole (quadature) axis, the maximum armature AT/pole occurs and interpole *mmf* must cancel.

$$\text{Armature AT/pole} = \frac{\text{Total AT}}{p} = \frac{\text{Ampere/conductor}\times \text{Turns } (Z/2)}{p}$$

$$= \frac{(120/2)\times (294/2)}{4} = 2205 \text{ AT/pole}$$

Hence require AT/pole in the air gap will be

$$Hl_g = \frac{B}{\mu_0}l_g = \frac{0.15 \times 8 \times 10^{-3}}{4\pi \times 10^{-7}} = 955 \text{ AT/pole}$$

The total *mmf* required $= 2205 + 955 = 3160$ AT/pole.

12.8   VOLTAGE REGULATION AND EFFICIENCY OF DC GENERATORS

Voltage regulation is defined as the ratio of change in voltage from no-load voltage to full-load voltage. Mathematically, it is defined as

$$\text{Voltage regulation (\%)} = \frac{(\text{No load voltage}) - (\text{Full-load voltage})}{\text{Full-load voltage (rated voltage)}}\times 100 \qquad (12.27)$$

or

$$\text{Voltage regulation (\%)} = \frac{V_{nl} - V_{fl}}{V_{fl}} \times 100 \qquad (12.28)$$

Efficiency of dc generator is defined as

$$\eta = \frac{\text{Output power}}{\text{Input power}} = \frac{\text{Output power}}{\text{Output power + Losses}} = \frac{P_{out}}{P_{out} + \text{Losses}} \qquad (12.29)$$

Figure 12.26 shows the various losses (energy flow) in the dc generator:

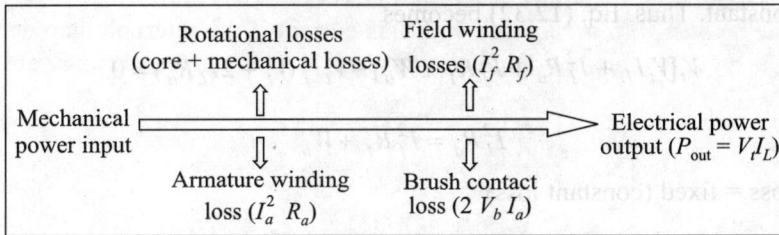

FIGURE 12.26 Losses in dc generator.

Power developed at air gap = $E_a I_a$ = Mechanical power input – rotational losses   (12.30)

12.8.1 Losses in DC Generators

1. **Rotational loss:** This includes frictional and windage loss, core loss and brush loss. Iron losses are composed of hysteresis and eddy current losses in the armature core. Raising either the armature speed or the field flux density will increase these losses.

2. **Field winding loss ($I_f^2 R_f$):** In separately excited dc generator, this loss is fixed. In shunt generator, it can be assumed as constant because terminal voltage varies in limited range. In dc series generator, this loss changes with the load current and treated as variable loss.

3. **Armature copper loss ($I^2_a R_a$):** This loss depends on the load current and hence it is a variable loss.

Thus, the efficiency of dc generator can be written as

$$\eta = \frac{V_t I_L}{V_t I_L + I_a^2 R_a + I_f^2 R_f + W_o} = \frac{\text{Power output}}{\text{Power output + Variable loss + Fixed loss}} \qquad (12.31)$$

where

V_t = Terminal voltage

I_a = Armature current

I_L = Load current

I_f = Field current

W_o = Fixed loss

The load current for maximum efficiency can be obtained by equating $\dfrac{d\eta}{dI_L}$ to zero. Differentiating Eq. (12.31) with respect to the load current, I_L, will be

$$\frac{d\eta}{dI_L} = \frac{V_t[V_t I_L + I_a^2 R_a + I_f^2 R_f + W_o] - V_t I_L\left(V_t + \dfrac{d}{dI_L}(I_a^2 R_a + I_f^2 R_f)\right)}{(V_t I_L + I_a^2 R_a + I_f^2 R_f + W_o)^2} = 0 \qquad (12.32)$$

Case A: Separately excited generator

In a separately excited generator, armature current, I_a, is equal to the load current, I_L, and field current, I_f, is constant. Thus, Eq. (12.32) becomes

$$V_t[V_t I_L + I_L^2 R_a + I_f^2 R_f + W_o] - V_t I_L(V_t + 2I_L R_a) = 0$$

or

$$I_L^2 R_a = I_f^2 R_f + W_o$$

$\Rightarrow$ Variable loss = fixed (constant loss)

Case B: Shunt generator

In this case, $I_a = I_L + I_f$ and thus Eq. (12.32) becomes

$$V_t[V_t I_L + I_a^2 R_a + I_f^2 R_f + W_o] - V_t I_L[V_t + 2(I_L + I_f)R_a] = 0$$

or

$$I_a^2 R_a + I_f^2 R_f + W - 2(I_a - I_f)I_a R_a = 0$$

i.e.

$$I_a^2 R_a = I_f^2 R_f + W_o + 2I_a I_f R_a$$

or

$$I_a^2 R_a\left[1 - 2\left(\frac{I_f}{I_a}\right)\right] = I_f^2 R_f + W_o \qquad \left(\text{Normally } \frac{I_f}{I_a} \approx 0.05\right)$$

and therefore,

$$I_a^2 R_a = I_f^2 R_f + W_o$$

Case C: Series generator

For series generator, $I_a = I_L = I_f$ and thus Eq. (12.32) becomes

$$V_t[V_t I_L + I_L^2(R_a + R_f) + W_o] - V_t I_L[V_t + 2I_L(R_a + R_f)] = 0$$

or

$$I_L^2(R_a + R_f) = W_o \quad \text{i.e. Variable loss = Fixed loss}$$

EXAMPLE 12.4   A 220 V, 8.8 kW, 1000 rpm, dc shunt generator has armature resistance of 0.5 ohm and field winding resistance of 110 ohm. Friction and windage loss is 200 W and have negligible core loss. The brush drop is 1 V per brush. Find

    (a) η at full-load and at rated speed

    (b) Voltage regulation

(c) Terminal voltage at 800 rpm

(d) Output power at 800 rpm.

Solution: From Figure 12.27,

(a) $I_L = \dfrac{8.8 \times 10^3}{220} = 40 \text{ A}, \quad I_f = \dfrac{220}{110} = 2 \text{ A}$

Thus $I_a = 42$ A

Armature resistance loss $= 42^2 \times 0.5 = 882$ W

Brush loss $= 2 \times 1 \times 42 = 84$ W

Field loss $= 2^2 \times 110 = 440$ W

Thus, efficiency will be

$$\eta = \frac{8.8 \times 10^3}{8.8 \times 10^3 + 200 + 882 + 84 + 440} \times 100 = 84.57\%$$

FIGURE 12.27   Example 12.4.

(b) The armature voltage at full-load can be calculated as

$$E_a = V_t + I_a R_a + 2 \times 1 = 220 + 42 \times 0.5 + 2 = 243 \text{ V}$$

The field current will be equal to the armature current at no-load ($I_L = 0$). Assuming the speed is constant, the armature current will be

$$\frac{E_a - \text{brush drop}}{R_f + R_a} = \frac{243 - 2}{110.5} = I_a = I_f = 2.18 \text{ A}$$

Thus no-load terminal voltage will be

$$V_t = E_a - 2 - 2.18 \times 0.5 = 239.9 \text{ V}$$

Therefore % voltage regulation is

$$VR = \frac{239.9 - 220}{220} \times 100 = 9.04\%$$

(c) Let us assume the flux ϕ is the same. Please note that there will be a change in field current which is very small. The induced voltage at 800 rpm can be calculated as

$$E_{a2} = \frac{800}{1000} E_a = 0.8 \times 243 = 194.4 \text{ V}$$

Let terminal voltage at 800 rpm is V_{t2}, thus the load current and field current will be

$$I_L = \frac{V_{t2}}{R_L} \text{ and } I_f = \frac{V_{t2}}{110}, \quad \text{where } R_L = \frac{220}{40} = 5.5 \ \Omega$$

Thus, the armature current will be

$$I_a = V_{t2} \left[\frac{1}{5.5} + \frac{1}{110} \right] = 0.191 V_{t2}$$

and

$$V_{t2} = E_{a2} - 2 - 0.191 V_{t2} \times 0.5$$

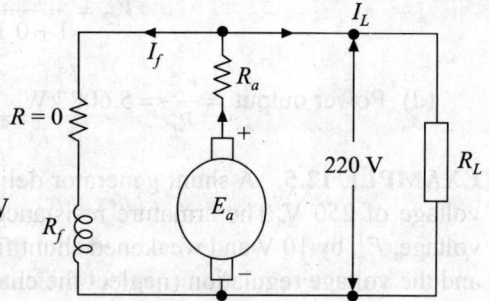

Therefore,

$$V_{t2} = \frac{194.4 - 2}{1 + 0.191 \times 0.5} = 175.63 \text{ Volt}$$

(d) Power output $= \dfrac{V_{t2}^2}{R_L} = 5.608$ kW.

EXAMPLE 12.5 ·A shunt generator delivers 30 A to load and 4 A to its own field at terminal voltage of 250 V. The armature resistance is 0.5 ohm. If armature reaction decreases induced voltage, E_a, by 10 V and weakened shunt field causes a further 9 V drop, find the no-load voltage and the voltage regulation (neglect the change in field current).

Solution: The voltage drop in the armature resistance will be $I_a R_a$ which will be equal to the (30 + 4) (0.5) = 17 V. Thus no-load terminal voltage, V_{nl}, will be

$$V_{nl} = \text{Full load terminal voltage} + \text{Armature reaction} +$$
$$\text{Weakened field voltage} + \text{Armature voltage drop}$$

or

$$V_{nl} = 250 + 10 + 9 + 17 = 286 \text{ V}$$

Thus, voltage regulation (%) will be

$$\text{Voltage regulation} = \frac{286 - 250}{250} \times 100 = 14.4\%.$$

EXAMPLE 12.6   A 125 V, 40 A, shunt generator has 500 W of mechanical losses, 300 W of iron losses, and requires 3.6 A field current. Its armature resistance is 0.19 ohm. Assuming no change in speed or field current, find

(a) The power required to run the generator at no-load and normal excitation.
(b) The power required to run the generator at no-load if generator is changed to operate with separate excitation.
(c) The driving power required with no excitation.
(d) Its full-load efficiency as a shunt generator (including the field rheostat losses).

Solution:

(a) At no-load the driving power equals the total loss. Let us compute the field winding and armature losses as

$$\text{Shunt field circuit loss} = V_t I_f = 125 \times 3.6 = 450 \text{ W}$$
$$\text{Armature circuit loss} = (I_a)^2 R_a = 3.6^2 \times 0.19 = 2.46 \text{ W}$$

$$\text{Thus the total loss} = \text{Mech. loss} + \text{iron loss} + \text{field circuit loss}$$
$$+ \text{Armature circuit loss}$$

$$= 500 + 300 + 450 + 2.46 = 1252.46 \text{ W}$$

(b) With the separate excitation, there will be no power supplied by the shaft for field loss and armature losses will be zero as armature current will be zero. Thus the driving power required will be 800 (= 500 + 330) W.

(c) With no excitation, only mechanical loss will be required by the shaft, i.e. 500 W. Since there will no flux and thus iron loss will be zero.

(d) The full-load armature current, $I_a = 40 + 3.6 = 43.6$ A

Armature circuit loss $= (I_a)^2 R_a = 43.6^2 \times 0.19 = 361.2$ W

$$\text{Total loss} = 500 + 300 + 450 + 361.2 = 1611.2 \text{ W}$$
$$\text{Output power} = 125 \times 40 = 5000 \text{ W}$$

Thus, efficiency will be

$$\eta = \frac{5000}{5000 + 1611.2} = 0.7563 = 75.63\%.$$

12.9 DC MOTORS

The need of motive power has prompted the development of various kinds of electric motors. Since early power system was direct current, the earlier motors were dc types. With the advent of ac power systems, the popularity of dc motors declined, mainly because of higher cost and need of more frequent and careful maintenance. However, it never entirely disappeared, partly because the dc motors are inherently the most suitable for smooth, efficient, wide-range speed control and partly because it was only type of machine used for automotive and aircraft applications.

The fundamental difference between motors and generators is the purpose they serve. When a dc machine is connected to the dc supply system, machine can produce mechanical power or torque. It is possible to build a machine that can operate satisfactorily as either a motor or a generator without alteration. The major differences between a motor operation and generator operation are given below:

(a) The induced voltage in the armature of motor is called back-*emf* or counter *emf* because its direction is always such as to oppose the flow of normal armature current. However, it is known as induced or generated voltage in generator.

(b) The terminal voltage is less than induced voltage in the generator ($E_a > V_t$) whereas in motor back *emf* is less than the terminal voltage ($E_a < V_t$).

(c) The reversed armature current causes the armature reaction to be reversed. Flux gets twisted in the direction opposite to the armature rotation.

(d) In generator, it is sometimes helpful to position the brushes so that commutation occurs slightly behind the neutral plane. However, in the motor due to reversed armature current, the brushes are slightly ahead of neutral plane in the direction opposite to the rotation.

(e) If a dc machine changes from generator to motor operation. the current in the field winding is reversed and so cumulative machine becomes a differential compound motor and vice versa.

DC motors are also classified in the similar ways as dc generator depending upon the connection of field and armature windings.

12.9.1 DC Shunt Motor

In shunt motor, the armature winding and field windings are connected across the supply system as shown in Figure 12.28. The field winding current is not reversed but the armature current is reversed which can be seen from Figure 12.28.

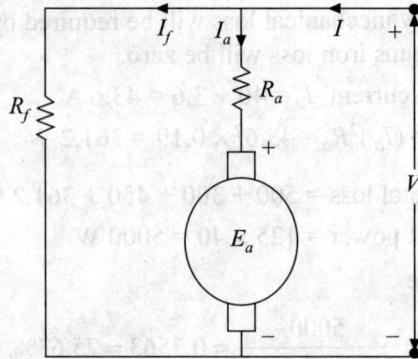

FIGURE 12.28   DC shunt motor.

The basic equations for the motor operation can be written as

$$E_a = K_a\phi\omega_m \qquad\qquad (12.33)$$
$$V_t = E_a + I_aR_a \qquad\qquad (12.34)$$
$$V_t = I_fR_f \qquad\qquad (12.35)$$
$$I = I_a + I_f \qquad\qquad (12.36)$$
$$T = K_a\phi I_a \qquad\qquad (12.37)$$

where $K_a = \dfrac{pN}{a\pi}$.

Using Eqs. (12.33) and (12.34), we get

$$K_a\phi\omega_m = (V_t - I_aR_a) \quad \text{or} \quad \omega_m = \frac{(V_t - I_aR_a)}{K_a\phi}$$

Therefore,

$$\omega_m = \frac{V_t}{K_a\phi} - \frac{R_a}{(K_a\phi)^2}T \quad \text{where } \omega_m \text{ is speed in rad/sec.} \qquad (12.38)$$

when $T = 0$,

$$\omega_m = \omega_o = \frac{V_t}{K_a\phi}$$

Thus Eq. (12.38) can be rewritten as

$$\omega_m = \omega_0 - \frac{R_a}{(K_a\phi)^2}T \qquad\qquad (12.39)$$

Since, flux ϕ is normally fixed, the speed variation of dc shunt motor with constant terminal voltage is linear as shown in Figure 12.29.

If the change in the flux is considered, the speed-torque relation will be slightly non-linear and curve will move up as reduction in field flux will cause voltage reduction with increasing torque. At the operating point, the motor torque will be equal to the load torque and can be mathematically written as

$$\text{Motor torque } (T_m) = \text{Load torque } (T_L) \qquad\qquad (12.40)$$

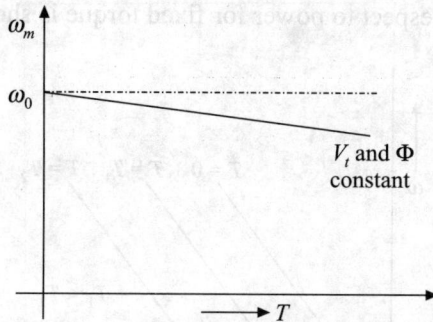

FIGURE 12.29 Torque-speed characteristic of dc shunt motor.

12.9.2 Speed Control of DC Shunt Motors

Since speed [Eq. (12.38)] is function of terminal voltage, field flux ϕ and armature resistance R_a, the speed can be controlled by

(a) Controlling terminal voltage, V_t keeping the field current constant (known as *armature voltage control*)

(b) Controlling field flux Φ by adjusting I_f and keeping armature voltage constant (termed as *field control*)

(c) Controlling armature resistance by having variable external resistance in the armature circuit (known as *armature resistance control*)

(d) Changing the number of field flux by re-arranging the field winding turns (known as *field diversion method*).

(a) Armature voltage control

If flux is kept constant, the speed varies directly with terminal voltage. This is very widely used speed-control technique for a wide range of speed (below the base case speed). The controlled armature voltage can be obtained from controlled rectified circuit receiving input power from ac source or from separately excited dc generator. Since ϕ is constant, T is function of I_a [see Eq. (12.37)] and thus for $I_{a\,max}$, T_{max} is the limiting constraint. The variation of speed with respect to terminal voltage (V_t) for different torque is plotted in Figure 12.30. It can be seen that speed control from base case speed to zero speed is obtained from armature voltage control.

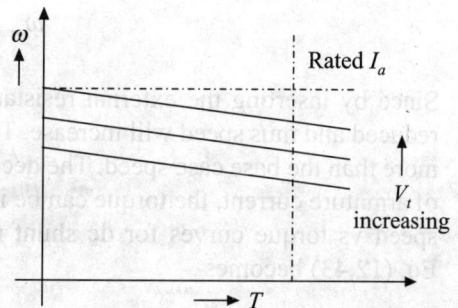

FIGURE 12.30 DC shunt motor speed-torque characteristic for different terminal voltage.

With armature voltage control, the torque is kept constant as I_a cannot exceed its rated value. Thus,

$$T = K_a \phi I_a = \text{constant}$$

or

$$P = E_a I_a = K_a \phi \omega_m I_a = K \omega_m \tag{12.41}$$

The variation of speed with respect to power for fixed torque is shown in Figure 12.31.

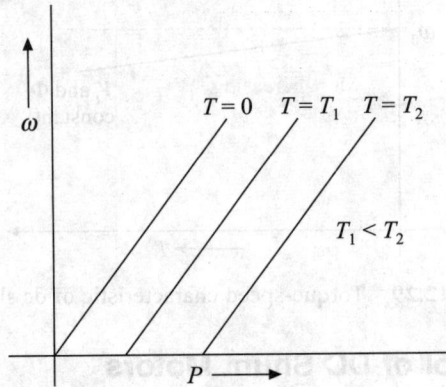

FIGURE 12.31 Speed-power characteristic of dc shunt motor.

(b) Field control

The field control method is simple in which armature voltage is kept constant and flux is varied by varying the field circuit resistance to vary the speed of motor. In this method, speed is controlled above the base speed. The power loss in the field circuit resistor is relatively small because the shunt field current is small, compared to the armature current. The maximum torque is limited by the armature current and the flux. Neglecting the saturation, the flux can be related to the field current as

$$\phi = K_f I_f \tag{12.42}$$

Using Eq. (12.38) and Eq. (12.42), we have

$$\omega_m = \frac{V_t}{K_f I_f} - \frac{R_a}{(K_f I_f)^2} T \tag{12.43}$$

Since by inserting the external resistance in the field winding, the field current (flux) will be reduced and thus speed will increase. Therefore, with this control the speed can only be achieved more than the base case speed. The decrease in the flux will decrease the torque but with the help of armature current, the torque can be increased till armature current reaches its rated value. The speed vs torque curves for dc shunt motors are shown in Figure 12.32. At no-load ($T = 0$), Eq. (12.43) becomes

$$\omega_m \simeq \frac{V_t}{K_f I_f} \tag{12.44}$$

If field circuit is opened ($I_f = 0$), the speed becomes dangerously high and will damage the motor (Figure 12.32). For any particular value of I_f, Eq. (12.43) can be written as

$$\omega_m = K_3 - K_a T \tag{12.45}$$

Since the armature current cannot exceed the rated value, the rated current should be fixed to I_a, and therefore, speed control is restricted to constant power operation.

$$P = V_t I_a \simeq E_a I_a = \text{constant}$$

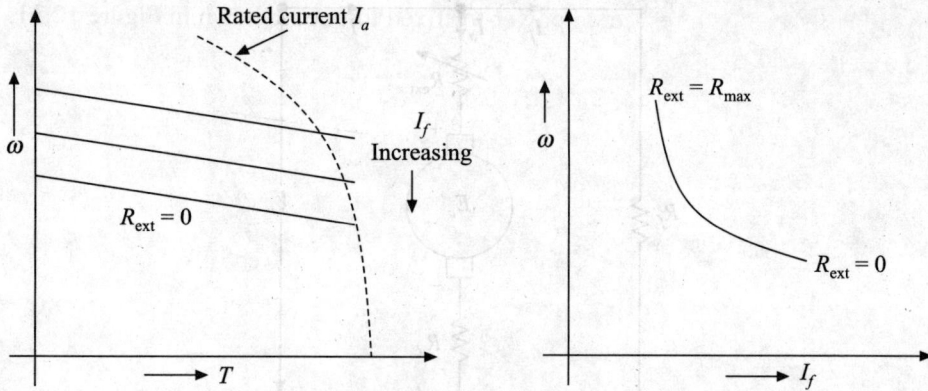

FIGURE 12.32 Field control of dc shunt motor.

Thus, torque will be inversely proportional to the speed which can be seen from the following equation.

$$T = \frac{P}{\omega_m} \tag{12.46}$$

The torque decreases with speed in the field weakening region. The torque and power variations with armature voltage and field control methods are shown in Figure 12.33.

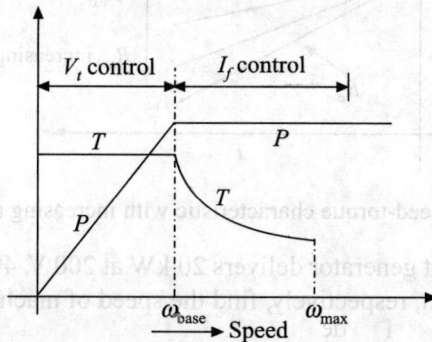

FIGURE 12.33 Power and torque w.r.t. speed for armature voltage and field controls.

(c) Armature resistance control

If terminal voltage V_t and field current I_f are kept constant, the speed variation with armature circuit resistance, as shown in Figure 12.34, can be obtained from the following expression.

$$\omega_m = \frac{V_t}{K_a \phi} - \frac{(R_a + R_{ext})}{(K_a \phi)^2} T \tag{12.47}$$

where R_{ext} is the armature circuit external resistance.

For any fixed value R_{ext}, Eq. (12.47) can be expressed as

$$\omega_m = K_5 - K_6 T \tag{12.48}$$

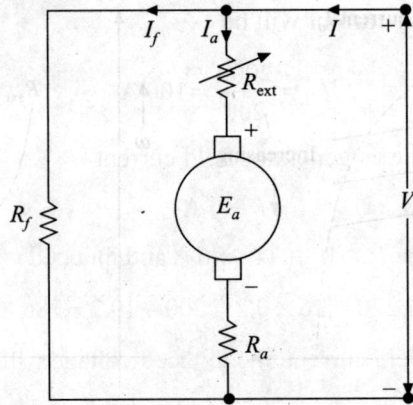

FIGURE 12.34 Armature resistance speed control method.

For increasing value of R_{ext}, the K_6 will increase. Thus, the droop will be more as shown in Figure 12.35. Armature resistance control is simple to implement but this method is less efficient due to high loss in R_{ext}.

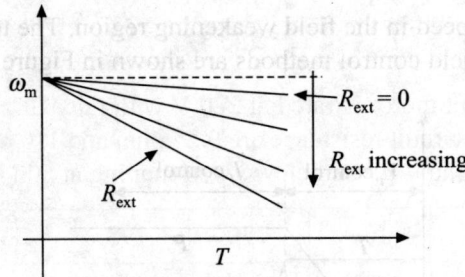

FIGURE 12.35 Speed-torque characteristic with increasing armature resistance.

EXAMPLE 12.7 A dc shunt generator delivers 20 kW at 200 V, 400 rpm. If armature and field resistances are 0.2 and 50 ohm, respectively, find the speed of machine running as a shunt motor and taking 20 kW at 200 V.

Solution: When machine is operating as shunt generator, the load current will be

$$I_L = \frac{20 \times 10^3}{200} = 100 \text{ A}$$

and field current will be

$$I_f = \frac{200}{50} = 4 \text{ A}$$

Thus, armature current will be the summation of load current and field current.

$$I_a = 100 + 4 = 104 \text{ A}$$

Therefore, induced voltage in the armature (E_{ag}) will be

$$E_{ag} = 200 + 104 \times 0.2 = 220.8 \text{ V}$$

When running as motor, the load current will be

$$I_L = \frac{2000}{200} = 100 \text{ A}$$

and field current will be the same as generator field current.

$$I_f = 4 \text{ A}$$

The armature current will be now $I_a = 100 - 4 = 96$ A and induced voltage in the armature will be

$$E_{am} = 220 - 96 \times 0.2 = 200 - 19.2 = 180.8 \text{ V}$$

Using Eq. (12.8), for the fixed field current, the induced voltage will be proportional to the speed of rotation and thus, we can have

$$\frac{E_{ag}}{E_{am}} = \frac{n_g}{n_m}$$

where n_g is the speed of rotation as generator and n_m is the speed of rotation as motor.

The speed of motor operation will be calculated as

$$n_m = \frac{n_g \times E_{am}}{E_{ag}} = \frac{400 \times 180.8}{220.8} = 327.54 \text{ rpm}.$$

EXAMPLE 12.8   A dc shunt motor is rated at 220 V with rated armature current of 40 A. The armature resistance and field circuit resistance are 0.25 ohm and 110 ohm respectively. The open circuit characteristic of machine is given below when running at 500 rpm.

emf (V)	71	133	170	195	220	232
I_f (A)	0.25	0.5	0.75	1.0	1.5	2.0

Neglecting the effect of armature reaction and brush drop and assuming $I_a = 0$ at no-load, find the range of external field required to permit speed variation from 500 rpm on no-load to 1000 rpm with the armature carrying its rated current of 40 A.

Solution: Subscript 1 denoted the quantities at no-load and subscript 2 denoted the quantities at rated armature current (40 A). Thus, induced voltages will be

$$E_{a1} = 220 - 0.25 I_{a1} = 220 \text{ V} \quad \text{and} \quad E_{a2} = 220 - 0.25 I_{a2} = 220 - 0.25 \times 40 = 210 \text{ V}$$

The field current at no-load and 500 rpm (I_{f1}) will be 1.5 A (see table). The field current at 1000 rpm will be calculated from the table as

Emf at 1000 rpm = (emf at 500 rpm) * (speed 1000 rpm)/(speed 500 rpm)

emf (V)	142	266	340	390	440	462
I_f (A)	0.25	0.5	0.75	1.0	1.5	2.0

Thus field current at 1000 rpm and 210 V will be

$$I_{f2} = \frac{(0.5 - 0.25)}{(266 - 142)} \times 210 = 0.42 \text{ A}.$$

(A linear curved is assumed. This can be accurately obtained by plotting the *emf*-field curve).
The field circuit resistances will be

$$R_{f1} = \frac{220}{I_{f1}} = \frac{220}{1.5} = 146.7 \text{ ohm} \quad \text{and} \quad R_{f2} = \frac{220}{I_{f2}} = \frac{220}{0.42} = 523.81 \text{ ohm}$$

Thus external field resistance required will be field circuit resistance minus field winding resistance (110 ohm).
Thus, $R_{ext1} = 146.7 - 110 = 36.7 \ \Omega$ and $R_{ext\,2} = 523.81 - 110 = 413.81 \ \Omega$.

12.9.3   Speed Control of Series DC Motors

Since the armature circuit is in series with field circuit, the speed control with armature resistance or field control is almost the same. The schematic diagram is shown in Figure 12.36. If the an external resistance is added as shown in Figure 12.35, the general speed relation can be written as

$$\omega_m = \frac{V_t}{K_a\phi} - \frac{(R_a + R_f + R_{ext})}{(K_a\phi)^2} T \qquad (12.49)$$

Since $\phi = K_f I_f = K_f I_a$ (assuming that the machine is operating in linear magnetization curve) and $E_a = K_a\phi\omega_m$, we can write

$$E_a = K'_a I_a \omega_m \qquad (12.50)$$

and

$$T = K_a\phi \cdot I_a = K'_a I_a^2 \qquad (12.51)$$

or

$$I_a = \sqrt{\frac{T}{K'_a}} \qquad (12.52)$$

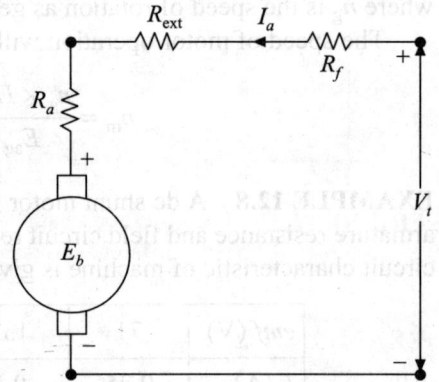

FIGURE 12.36   Speed control of dc series motor.

The induced voltage in terms of armature voltage and resistance drop can be expressed as

$$E_a = V_t - I_a (R_a + R_f + R_{ext})$$

or

$$K'_a \omega_m I_a = V_t - I_a (R_a + R_f + R_{ext})$$

Thus,

$$\omega_m = \frac{V_t}{K'_a I_a} - \frac{R_{total}}{K'_a} \qquad (12.53)$$

Putting the value of I_a from Eq. (12.52) in Eq. (12.53), we get

$$\omega_m = \frac{V_t}{\sqrt{K_a'}\sqrt{T}} - \frac{R_{\text{total}}}{K_a'} \qquad (12.54)$$

For the fixed value of terminal voltage and armature-field circuit resistance, the speed-torque characteristic is shown in Figure 12.37. When torque is zero (no load condition), speed is dangerously high. Therefore, series dc motors are always started with load. In series motor, a high torque is obtained at low speed and therefore used when large starting torque is required viz. subway car, automobile starter, cranes, hoist, etc.

The speed torque characteristics of different dc motors are given in Figure 12.38.

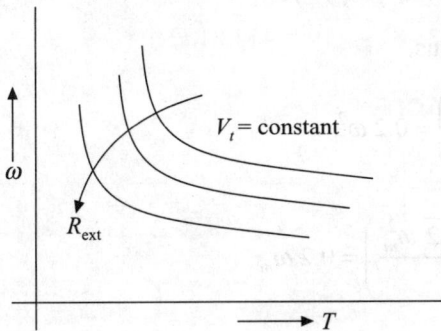

FIGURE 12.37 Speed-torque characteristics with different series external resistances.

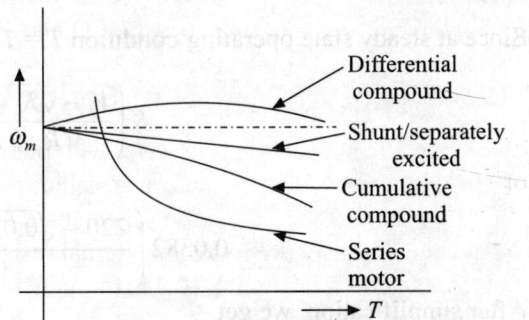

FIGURE 12.38 Speed-torque characteristics of different dc motors.

EXAMPLE 12.9 A 10 kW, 220 V, 1000 rpm dc series motor takes a full-load current of 50 A. The motor is used to drive a fan whose torque characteristic is $T_L = 0.2\,\omega_m^{\,2}$ (N-m) where ω_m is speed in rad/sec. Determine the steady state speed of motor at the load. Assume rotational loss is negligible.

Solution: Given that $P_{\text{out}} = 10 \times 10^3\,\text{W}$ and $I_a = 50$ A, the back *emf* will be

$$E_b = \frac{P_{\text{out}}}{I_a} = 200 \text{ V}$$

Since full-load speed is 1000 rpm, it will be in rad/second as

$$\omega_m = \frac{1000}{60} \times 2\pi = 104.7 \text{ rad/sec}$$

Using Eq. (12.50), we get

$$K_a' = K = \frac{200}{50 \times 104.7} = 0.0382$$

and thus, full-load torque is

$$T_{fl} = KI_a^2 = 95.51 \text{ N-m}$$

The field circuit and armature resistance will be calculated using the following equation as

$$V_t = E_b + I_a (R_a + R_f)$$

or

$$R_a + R_f = \frac{V_t - E_b}{I_a} = \frac{220 - 200}{50} = 0.4\ \Omega$$

Since $T = KI_a^2$, back *emf* can be written as $E_b = K \sqrt{\dfrac{T}{K}}\ \omega_m$

$$T = K \left(\frac{V_t - E_b}{R_a + R_f} \right)^2 = K \left(\frac{V_t - KI_a\ \omega_m}{(R_a + R_f)} \right)$$

Since at steady state operating condition $T = T_L$ and thus

$$K \left(\frac{V_t - \sqrt{K}\ \sqrt{T}\ \omega_m}{(R_a + R_f)} \right) = 0.2\ \omega_m^2$$

or

$$0.0382 \left(\frac{220 - \sqrt{0.0382 \times 0.2}\ \omega_m^2}{0.4} \right) = 0.2\ \omega_m^2$$

After simplification, we get

$$2.094\ \omega_m^2 = 220 - 0.0874\ \omega_m^2$$

or

$$\omega_m = 110.47 \text{ rad/sec} = 100 \text{ rpm.}$$

12.10 SPEED CONTROL OF DC SERIES MOTORS USING FIELD DIVERSION METHOD

In this method the ampere-turn of field is changed by diverting the currents in the field winding. If the field winding can be arranged into two coils, the ampere turns are changed. Let two coil having $N/2$ turns in each be connected in series and current I_f passes through them as shown in Figure 12.39.

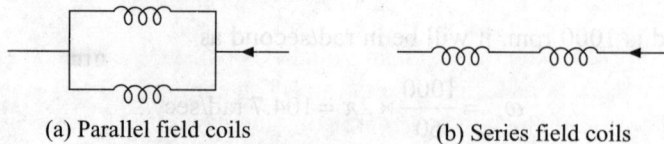

(a) Parallel field coils   (b) Series field coils

FIGURE 12.39   Series and parallel connection of field coils.

Total ampere-turn in series connection will be equal to $NI_f \left(= \dfrac{N}{2} I_f + \dfrac{N}{2} I_f \right)$. It should be noted that I_f is flowing in both the coils. In the parallel connection, the current will be divided into half if the resistances of the coil are same. If coils are assumed to be identical, the total ampere-turn in parallel connection will be

$$\frac{N}{2}\frac{I_f}{2} + \frac{N}{2}\frac{I_f}{2} = \frac{NI_f}{2}$$

Thus the AT is reduced to 50% and thus flux will also be reduced by the same. Therefore, the speed of the motor will be changed. In this method, the speed can be changed in discrete. It can be proved that the field diversion method is not applicable in dc shunt motors. Because the field current will be doubled in each coil due to reduced resistance (half) connected to the same supply voltage.

12.11 WARD LEONARD SPEED CONTROL

With Ward Leonard speed control system both armature voltage control and field control can be obtained for dc shunt or separately excited motor. It was introduced in 1890s and a motor-generator set is used to control the speed of dc drive as shown in Figure 12.40. An ac motor is used to drive the dc generator and a small shunt generator is used to supply the field voltage to the main generator and motor. By varying the dc generator excitation, the speed of dc motor can be controlled over wide range with minimum loss. This mechanism is mainly used in rolling mills in metallurgical plants.

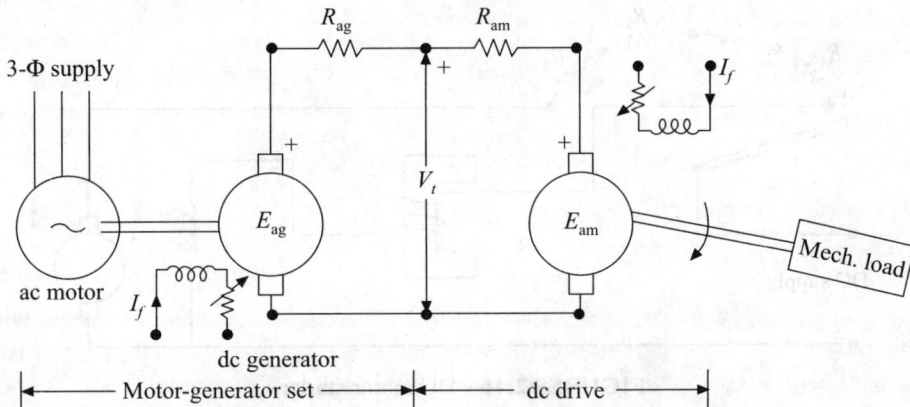

FIGURE 12.40   Ward Leonard speed control system.

In armature voltage control mode, the terminal voltage V_t of dc generator is varied with help of field excitation current I_{fg} whereas field excitation of dc motor is kept constant. The voltage appeared across the motor terminal will be generated voltage (E_{ag}) minus the drop in the armature resistances of the dc generator and dc motor. With this control, the speed variation from zero to base speed can be obtained. For the field control, terminal voltage of generator is kept constant with the help of generator excitation and dc drive (motor) field is varied for varying speed from base to high speed. In this case, armature current is also kept constant so that dc motor operates at constant power mode. Advantage of this control mechanism is the smooth control of speed from zero to above base speed. The main disadvantage of this scheme is high cost of motor-generator set. In modern age, power electronics converters are widely used.

12.12 DC MOTOR STARTERS

When a dc motor (at stand still) connected to dc supply system is switched on, there is a very high current flows in the armature. The armature current at any instant will be

$$I_a = \frac{V_t - E_a}{R_a} \qquad (12.55)$$

Since the speed is zero ($\omega_m = 0$) at starting, the back *emf* E_a ($= K_a \phi \omega_m$) will also be zero. Thus, the armature current is

$$I_a = \frac{V_t}{R_a} \qquad (12.56)$$

Since the armature resistance is very small, the current I_a is very high (sometimes more than 7-8 times of rated current). Therefore dc motor must be started through a starter. With addition of an external resistance during the starting, the armature current is limited to its rated value. The added resistance is removed when motor reaches its rated speed. A dc motor starter is shown in Figure 12.41. The armature current at any speed with the external resistance can be expressed as

$$I_a = \frac{V_t - E_a}{R_a + R_{\text{ext}}} \qquad (12.57)$$

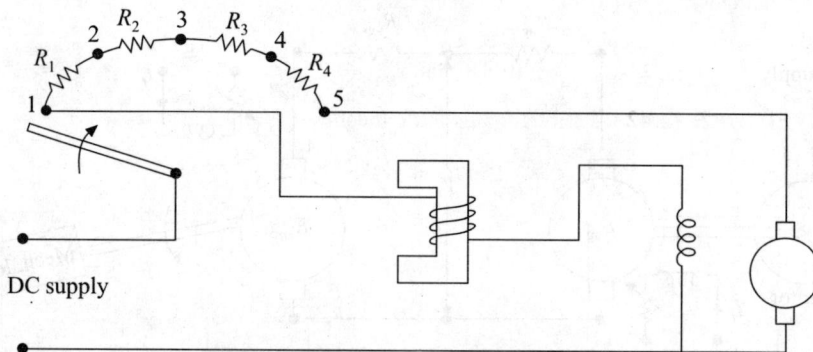

FIGURE 12.41   DC motor starter.

At point–1 (in Figure 12.41), high resistance is inserted which is the summation of all the resistances (R_1 to R_4). Once motor speeds up, the knob of starter can be moved to positions 2, 3, 4 and finally, at 5 where all the resistances (external) are out. The electromagnet attracts the handle and holds it. In recent year, the variable voltage obtained from the power electronics devices is used for starting the dc motor.

12.13 EFFICIENCY

Efficiency of any machine is defined as the ratio of output power to input power. It can also be represented in terms of losses. The various losses in a dc generator and dc motor are shown in a graphical representation of Figure 12.42(a) and 12.42(b), respectively. The notation V_a is used for voltage across the armature terminal. Please note that induced or back *emf* is not equal to the V_a. It can be easily understood by looking at Figure 12.43 for motor and generator cases. For series motor/generator, shunt field loss will be absent, whereas for shunt generator/motor, series field loss will not be present.

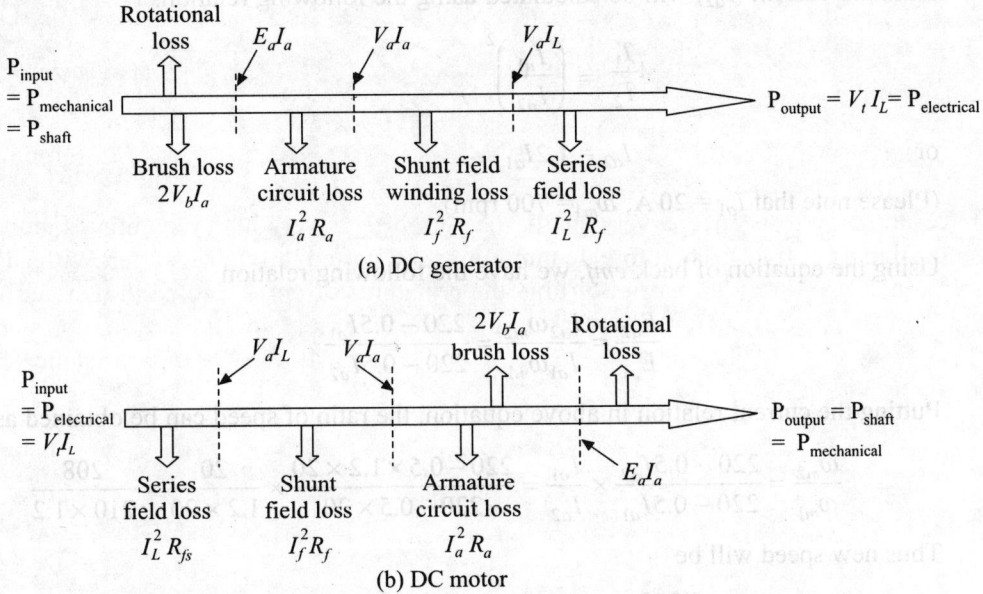

(a) DC generator

(b) DC motor

FIGURE 12.42   Various losses in dc machines (graphical representation).

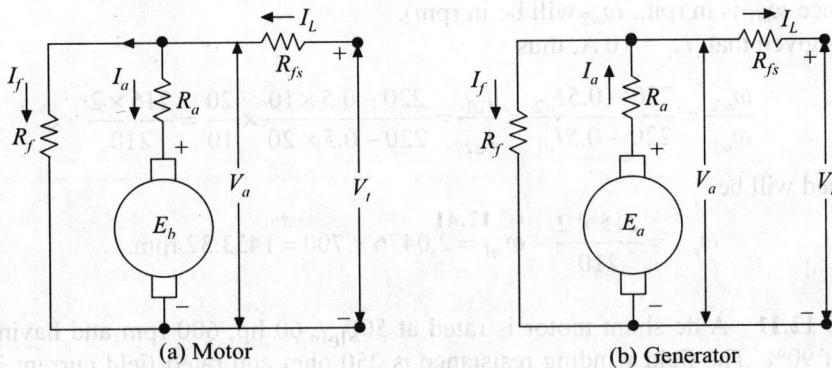

(a) Motor   (b) Generator

FIGURE 12.43   Flow of currents and voltages in dc machines.

EXAMPLE 12.10   A 220 V dc series motor runs at 700 rpm when operating at its full-load current of 20 A. The motor resistance (field and armature) is 0.5 ohm and the magnetic circuit may be assumed unsaturated (linear). What will be the speed if

(a) load torque is increased by 44 %?
(b) the motor current is 10 A?

Solution: For series motor, the following relations can be written

$$T = KI_a^2 \quad \text{and} \quad E_a = KI_a\omega_m$$

(a) For load torque increase of 44%, the new torque (T_2) will be 1.44 T_1 and thus the new armature current (I_{a2}) will be calculated using the following relation.

$$\frac{T_1}{T_2} = \left(\frac{I_{a1}}{I_{a2}}\right)^2$$

or

$$I_{a2} = 1.2I_{a1}$$

(Please note that $I_{a1} = 20$ A, $\omega_{m1} = 700$ rpm)

Using the equation of back *emf*, we have the following relation

$$\frac{E_{a2}}{E_{a1}} = \frac{I_{a2}\omega_{m2}}{I_{a1}\omega_{m1}} = \frac{220 - 0.5I_{a1}}{220 - 0.5I_{a2}}$$

Putting the current relation in above equation, the ratio of speed can be obtained as

$$\frac{\omega_{m2}}{\omega_{m1}} = \frac{220 - 0.5I_{a2}}{220 - 0.5I_{a1}} \times \frac{I_{a1}}{I_{a2}} = \frac{220 - 0.5 \times 1.2 \times 20}{220 - 0.5 \times 20} \times \frac{20}{1.2 \times 20} = \frac{208}{210 \times 1.2}$$

Thus new speed will be

$$\omega_{m2} = \frac{208}{210 \times 1.2} \times \omega_{m1} = 0.8254 \times 700 = 577.78 \text{ rpm}$$

(Since ω_{m1} is in rpm, ω_{m2} will be in rpm).

(b) It is given that $I_{a2} = 10$ A, thus

$$\frac{\omega_{m2}}{\omega_{m1}} = \frac{220 - 0.5I_{a2}}{220 - 0.5I_{a1}} \times \frac{I_{a1}}{I_{a2}} = \frac{220 - 0.5 \times 10}{220 - 0.5 \times 20} \times \frac{20}{10} = \frac{215 \times 2}{210}$$

Speed will be

$$\omega_{m2} = \frac{215 \times 2}{210} \times \omega_{m1} = 2.0476 \times 700 = 1433.32 \text{ rpm.}$$

EXAMPLE 12.11 A dc shunt motor is rated at 500 V, 60 hp, 600 rpm and having full-load efficiency of 90%. The field winding resistance is 250 ohm and rated field current is 2 A. The armature resistance is 0.2 ohm. Calculate

(a) Full-load rated current.
(b) Loss torque.
(c) The speed is to be increased up to 1000 rpm by field weakening, determine the extra resistance, over and above the field winding resistance to cover the speed range of 600–1000 rpm.

(d) Output torque and power at 1000 rpm assuming that loss torque varies proportional to the speed. The following field current and flux relation can be used for the magnetization curve.

$$\frac{I_{f1}}{I_{f2}} = \frac{0.6\left(\dfrac{\phi_2}{\phi_1}\right)}{1 - 0.4\left(\dfrac{\phi_2}{\phi_1}\right)}.$$

Solution:

(a) Full-load efficiency with given field current ($I_f = 2$ A),

$$\eta = \frac{P_{\text{output}}}{P_{\text{input}}} = \frac{60 \times 746}{500 \times (I_a + I_f)} = 0.9$$

Thus armature current, $I_a = 97.47$ A

(b) $K_a\phi_1 = \dfrac{E_{al}}{\omega_{ml}} = \dfrac{500 - 0.2 \times 97.47}{(2\pi \times 600)/60} = 7.648$ Nm/A

Thus the electrical torque developed will be

$$T = K_a\phi_1 I_a = 7.648 \times 97.47 = 745.45 \text{ Nm}$$

Shaft torque $T_{\text{mech}} = \dfrac{60 \times 746}{(2\pi \times 600)/60} = 712.38$ Nm

The loss torque $T_{\text{loss}} = T - T_{\text{mech}} = 745.45 - 712.38 = 33.07$ Nm

(c) at 1000 rpm speed, $K_a\phi_2 = \dfrac{E_{a2}}{\omega_{m2}} = \dfrac{500 - 0.2 \times 97.47}{(2\pi \times 1000)/60} = 4.589$ Nm/A

Hence flux ratio $\dfrac{\phi_2}{\phi_1} = \dfrac{4.589}{7.648} = 0.6$

Thus field current ratio $\dfrac{I_{f2}}{2} = \dfrac{0.6 \times 0.6}{1 - 0.4 \times 0.6}$ (I_{f2} is the field current required for speed

1000 rpm). Therefore $I_{f2} = 0.947$ A.

The total field circuit resistance = 500/0.947 = 528 ohm.

So external resistance = 528 − 250 = 278 ohm.

(d) The loss torque at 1000 rpm, $T_{\text{loss}} = 33.07 \times 1000/600 = 55.12$ Nm

Developed torque, $T = 4.589 \times 97.47 = 447.29$ Nm

The output torque = $T - T_{\text{loss}} = 447.29 - 55.12 = 392.17$ Nm

The power output $= \dfrac{2\pi \times 1000}{60} \times 392.17 = 41067.95$ W = 55.05 hp.

12.14 MECHANICAL LOAD

General speed torque equation of motor-load combined system can be written as

$$T_m = J \frac{d\omega_m}{dt} + D\omega_m + T_L \tag{12.58}$$

where J is the inertia of machine-load system. T_m is the output torque of machine. D is the damping constant and usually very small, and can be neglected. Thus, Eq. (12.58) becomes

$$J \frac{d\omega_m}{dt} = T_m - T_L \tag{12.59}$$

From Eq. (12.59), if $T_m > T_L$, the rate change of speed (acceleration) will be positive which shows that machine will accelerate. On the other hand, for $T_m < T_L$, machine will decelerate. At steady state, machine output torque will be equal to the load torque.

$$T_m = T_L \tag{12.60}$$

The load torque characteristic are different for different types of load and can be classified as

(a) Load torque is proportional to the square of speed

$$T_L = K\omega_m^2 \tag{12.61}$$

EXAMPLE Fans, compressor, aeroplanes, centrifugal pumps, ship propellers, etc. Figure 12.44(a) shows the torque characteristic.

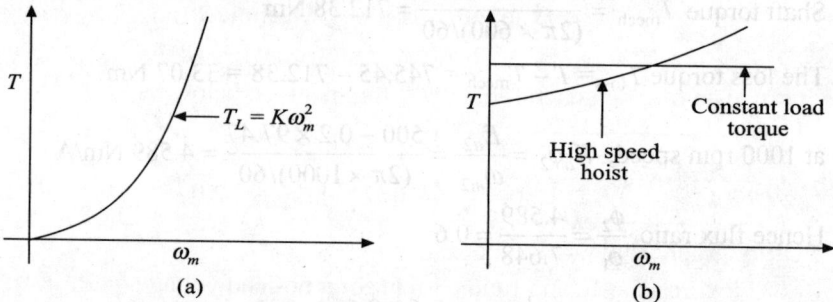

FIGURE 12.44   Different types of loads with torque characteristics.

(b) **Constant torque load:** The load torque is almost constant (or independent of speed). **Examples:** low speed hoist, lift.

(c) **High-speed hoist load:** The load torque is a combination of constant torque and windage torque as shown in Figure 12.44(b).

(d) **Locomotive or traction load:** (see Figure 12.45)

(e) **Constant power load:** $[(T_L \propto (1/\omega_m)]$ (see Figure 12.45) **Example:** Coiler, Tape record drive.

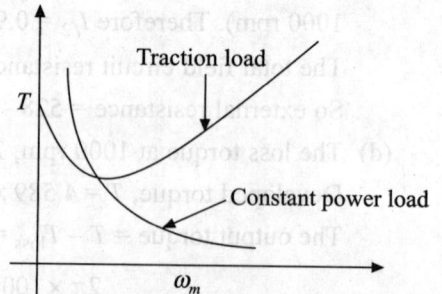

FIGURE 12.45   Traction load and constant power load characteristics.

PROBLEMS

12.1 A dc generator has two poles having 1000 turns each. The radius and length of machine are 10 cm and 15 cm, respectively. The air gap between pole face and rotor body is 2 mm and the air-gap flux density is 1.2 T. The pole face covers 80 % of the rotor circumference. The numbers of conductors are 18. If machine is running at 1000 rpm, find if armature is wave-wounded.

(a) The field current
(b) Armature voltage
(c) Developed torque if armature current is 20 A.
(d) Repeat part (b) and (c) for lap wounded armature.

[**Ans.** (a) 1.91 A, (b) 13.56 V, (c) 2.58 N-m,

(d) it will be same as part (b) and (c)]

12.2 Two 4-poles, 1200 rpm dc machines of the following rating are required.
Machine 1: 110 V;   Machine 2: 220 V
The coils are rated at 5 V, 4 A. If the same numbers of coils are to be used for both the machines, find

(a) Type of armature winding of each machine
(b) Number of coils required in each machine
(c) kW rating of each machine.

[**Ans.** (a) Machine 2 (wave wound), Machine 1 (lap wound), (b) Machine 1

(no. of coils = 88), Machine 2 (no. of coils = 88), (c) Machine 1

$(I_a = 16A , P = 1760$ watt), Machine 2 $(I_a = 8$ A, $P = 1760$ watt)]

12.3 OCC data of a dc generator at 1200 rpm is given below:

E_0 (V)	8	40	74	113	213	234	250
I_f (A)	0	0.5	1.0	1.5	2.0	3.5	4.0

Using graph, find

(a) Field resistance and field current for a no-lead voltage of 230 volt.
(b) The value of the critical field resistance.
(c) The value of the critical speed.
(d) External field resistance required to reduce the terminal voltage 200 V.

[**Ans.** (a) Field resistance = 67.65 ohm, (b) $R_{critical}$ = 76.67 ohm,

(c) $n_{critical}$ = 1100.9 rpm, (d) $R_{external}$ = 5.08 ohm]

12.4 DC generator of above problem is now run at 1000 rpm. Find

(a) The no-load voltage and field current for a field resistance of 50 ohm.
(b) Value of critical speed and critical resistance
(c) If a load resistance 5 ohm is connected cross the generator terminals with field resistance of 50 ohm, find the terminal voltage and load current. The armature resistance is 0.5 ohm.

[**Ans.** (a) $E0 = 212.5$, $I_f = 4.2$, (b) $R_{critical}$ = 63.43 ohm,

$n_{critical}$ = 882.35 rpm, (c) V = 191.44 V, I_L = 38.29 A]

12.5 The magnetization characteristic of a dc shunt generator at 1000 rpm and having negligible armature resistance is given below:

I_f (A)	0	0.2	0.3	0.4	0.5	0.6	0.7
V_{occ} (V)	7.5	93	135	165	186	202	215

If field resistance of machine is 354.5 ohm and speed is 1000 rpm,
  (i) Determine the no-load voltage (graphically)
  (ii) Determine the critical field resistance
  (iii) Determine the critical speed for given field resistance
  (iv) What additional resistance must be inserted in the field circuit to reduce the no-load voltage to 175 V?

[**Ans.** (i) 195 V, (ii) 457 Ω, (iii) 762.37 rpm, (iv) 43.23 Ω]

12.6 A 10 kW, 200 V, 1200 rpm series dc generator has armature resistance of 0.1 ohm, field winding resistance of 0.3 ohm. The frictional and windage loss of the machine is 200 W and brush contact drop is 1 V per brush. Find the efficiency of the machine. Find the load current at which this machine has maximum efficiency. Find the power output of the machine at this efficiency.

[**Ans.** $\eta = 88.5\%$, max $\eta = 90.96$, $I_L = 22.36$, % of load = 44.72, $P = 4472.14$ watt.]

12.7 A 6.6 kW, 220 V, 1000 rpm dc shunt machine is having 55 ohm field circuit resistance and 0.5 ohm armature resistance. Find the voltage regulation of machine. If the frictional and windage loss is 300 W, find the efficiency of the machine at rated output.

[**Ans.** Voltage regulation = 6.76%, $\eta = 78.96$]

12.8 A 25 kW, 250 V long-shunt compound generator has armature resistance of 0.02 ohm, series field resistance of 0.03 ohm and shunt field resistance is 125 ohm. Find the induced emf at rated load. Total brush drop is 2 V. Find the voltage regulation of the machine. Effect of series field winding is negligible.

[**Ans.** $E_a = 257.1$ volt; 2%]

12.9 A dc shunt motor runs at 1500 rpm drawing 30 A from a 250 V supply with its field halves connected in series. At what speed would it run if the field halves are reconnected in parallel? Also calculate the armature current. One half of the field has a resistance of 50 ohm. Assume load torque to be proportional to square of speed and voltage drop of armature.

[**Ans.** 750 rpm, 3.44 A]

12.10 A 50 kW, 250 V dc series motor takes a current of 210 A when running with full-load at 1000 rpm. The armature and field resistances are 0.03 ohm and 0.02 ohm respectively. Estimate the motor efficiency when the motor is drawing 55 A. What would be the maximum efficiency of the motor and the load current at which it would occur?

[**Ans.** 96.75%, 97.02%]

12.11 A dc series motor runs at 1000 rpm drawing 25 A from a 220 V supply with its field halves connected in parallel. At what speed would it run if the field halves are reconnected in series? Also calculate the armature current. Assume load torque to be proportional to the speed and voltage drop of armature and field resistances to be neglected.

[**Ans.** 794 rpm, 15.761 A]

12.12 A 80 kW, 250 V shunt motor takes 10 A when running on no-load. The resistance of the armature and field are 0.5 and 100 ohm respectively.

(a) Estimate the motor efficiency when loaded to carry 120 A.
(b) Estimate the maximum efficiency of the motor.
(c) Also estimate the maximum efficiency when it operates as generator.

[**Ans.** (a) 68.75%, (b) 72.86%, (c) 77.42%]

12.13 12 kW, 220 V dc series motor at rated armature current is running at 1500 rpm. Find the rated armature current. At light load condition, armature takes 10 A, find the speed of motor. Armature and field resistances are 0.3 and 0.2 ohms, respectively. Assume frictional and windage losses are negligible.

[**Ans.** 63.8 A, 10938.6 rpm]

12.14 A variable speed drive system uses a dc shunt motor which is supplied from a variable voltage source. The drive speed is varied from 0 to 1500 rpm (base speed) by varying the terminal voltage from 0 to 300 V with the field current maintained constant. Determine the motor armature current if the torque is held constant at 300 N-m up to base speed. Neglect the armature drop. Assume frictional and windage losses are negligible.

[**Ans.** 157.08 A]

12.15 A 15 kW, 200 V, 1500 rpm dc shunt machine has $R_a = 0.5$ ohm, $R_f = 100$ ohm and is connected to a 200 V supply.

(a) Determine the value of starting current without starting resistances.
(b) Calculate the value of starting resistance if the starting current is limited to 1.5 times that of rated armature current.

[**Ans.** (a) 100 A or 300 A, (b) 0.85]

12.16 A dc series motor has armature resistance of 0.4 ohm and series field resistance of 0.3 ohm. On certain load, the motor takes 20 A from 200 V dc supply system. Find the value of series resistance to be inserted to reduce the speed by 20%. The load torque is proportional to ω_m^3. Assume magnetic characteristic linear.

[**Ans.** 5.84]

12.17 A dc motor has the following nameplate armature information: 300 V, 20 hp, 60 A and 1500 rpm. Field nameplate information is 220 V, 2 A. The motor is to be shunt connected. Assume that the armature copper losses are equal to the rotational losses under nameplate condition.

(a) Draw the circuit diagram for the motor by putting the field rheostat resistance for nameplate condition.
(b) Determine the total efficiency of the motor under nameplate condition.
(c) Find the no-load speed assuming the rotation losses are proportional to the speed.

[**Ans.** (a) 40, (b) 80.22%, (c) 1627.17 rpm]

12.18 14 V permanent magnet field sc motor has negligible rotational losses. The stall ($\omega_m = 0$) current is 28 A. Find the maximum output power (watts) this machine can produce when supplied with 14 V.

[**Ans.** 98 watt]

12.19 A 12.6 window lift motor in car has a permanent magnet field and runs at 800 rpm raising the window with a current of 8.1 A. The stall current (motor is stopped with window is fully up) is 22 A. The torque require to lower the window is 80% of that required to lift the window because weight of window is aiding the descent. Find the current drawn lowering the window and the corresponding speed. Ignore the rotational losses in the armature.

[**Ans.** 6.474 A, 894 rpm]

12.20 A dc shunt motor has armature resistance of 0.5 ohm and field resistance of 110 ohm. On certain load armature current is 20 A when connected to 220 volt dc supply system. Find the value of field resistance to be inserted in the field circuit to increase the speed by 20%. The load torque is proportional to ω_m^2. Assume linear magnetic characteristic. With this field resistance, find the efficiency of the motor, if rotational loss is 500 W.

[**Ans.** 27.2 ohm, 81.86%]

12.21 A dc series generator has armature resistance of 0.1 ohm, field winding resistance of 0.4 ohm. It delivers 40 A (full load current) at 1000 rpm when it is connected to 5 ohm load. The frictional and windage loss of the machine is 200 W at 1000 rpm. This loss varies with square of speed.

(a) What is rating of generator? Find the efficiency of the generator.
(b) If this machine is to run as motor connected to 200 V dc supply. The load torque is such that it takes 40 A current from the source. Find the speed of motor and shaft power output. What will be the efficiency of motor?

[**Ans.** (a) 8kW, 200 V, 1000 rpm, 88.89%, (b) 818.18 rpm, 7066.12 W, 88.33%]

12.22 15 hp, 220 V, 57 A, 1500 rpm dc shunt motor has field resistance of 110 ohm and armature resistance of 0.2 ohm.

(a) Find the efficiency of motor at rating.
(b) What resistance should be added to the armature circuit to reduce the speed of motor to 1200 rpm when motor draws 40 A line current?
(c) What is efficiency and torque (in N-m) at load in (b)? Assume, rotational loss is proportional to the speed.

[**Ans.** (a) 89.23%, (b) 1.19 Ω, (c) 69.42%, 48.61 Nm]

12.23 15 hp, 250 V, 50 A, 1400 rpm dc series motor has field resistance of 0.2 ohm and armature resistance of 0.3 ohm.

(a) Find the efficiency at motor rating.
(b) What resistance should be added to the armature circuit to reduce the speed of motor to 1000 rpm when motor draws 30 A line current?
(c) What is efficiency and torque (in N-m) at load in (b)? Assume rotational loss is proportional to the speed.

[**Ans.** (a) 89.52%, (b) 4.68 Ω, (c) 37.97%, 27.20 Nm]

13

Three-Phase Induction Machines

13.1 INTRODUCTION

The induction machines are the most widely used for driving the mechanical loads. It is normally used as motor. In modern age, more than 85% of ac motors are of induction motors ranging from few tens of watts to several MW power. An induction machine can be operated as motor and generator but generator characteristic is not suitable for bulk power generator and also induction generator absorbs more reactive power than the generation. Induction generators with power electronics devices are used in wind mills. Induction machines are simple and rugged in construction. Both stator and rotor carries ac currents which are induced in the rotor winding from the stator winding field. Therefore, this machine is known as *induction machine*. Single- phase induction motor uses single-phase supply system and mainly used for lower power domestic applications such as fans, mixers, etc. Three-phase induction motors use three-phase ac supply system and are used for high power applications. Thus, this chapter is devoted to three-phase induction motors only. Single-phase induction motors are discussed in Chapter 15.

13.2 THREE-PHASE INDUCTION MOTORS

In three-phase induction motor, the stator carries distributed windings which are displaced at 120° apart in the space. The number of poles formed in rotor and stator are the same. This machine has uniform air gap unlike dc machines. Both stator and rotors are made of laminated ferromagnetic materials to reduce the core loss. Induction motors do not run at constant speed, although in usual applications, the speed variation is very small. The magnetic field is supplied from the ac source and thus, it operates at a power factor less than unity. Nevertheless, the simple and rugged construction and easy operation of these motors lead to performance and cost advantages that is why these motors are very common in actual operation. Stator winding can be wounded either in delta or star as shown in Figure 13.1. For low voltage motor, it is wounded in delta (Δ), however for high voltage machine, Y (star) is economical. The induction machines motors are classified depending on the rotor construction.

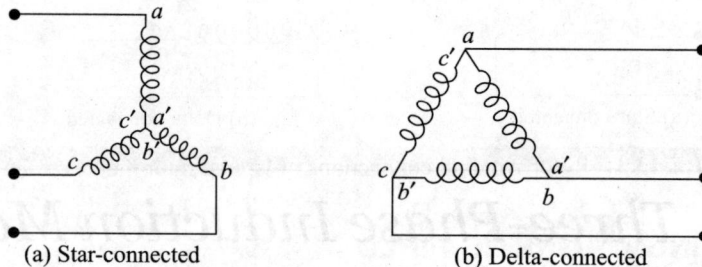

(a) Star-connected   (b) Delta-connected

FIGURE 13.1   Stator winding connections.

13.2.1   Squirrel Cage Induction Motor

In three-phase inductor motor, if stator has a three-phase winding, the rotor may not have three-phase winding but the winding should be done in such a way that the number of formed poles in stator and rotor should be the same. A two-phase rotor winding (with correct number of poles) would have two-phase balanced voltages and currents by induction, and results in a revolving field of constant magnitude. A simple and common form of construction involves a large number of straight copper/ aluminium bars laid in slots or passed through the holes in the rotor steel and connected at the end with rings to allow the current circulation. It looks like a cage and it is named as squirrel cage-type induction motor as shown in Figure 13.2. Such a construction could be regarded as a winding of many phases with only one-turn per phase.

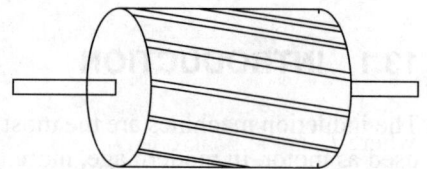

FIGURE 13.2   Squirrel cage rotor.

The squirrel cage rotors are very simple, rugged and economical to build. The squirrel cage induction motors suffer from the disadvantage that the external resistance cannot be inserted into the rotor circuit (to increase the starting torque) required in many applications.

13.2.2   Wound Rotor or Slip-Ring Induction Motor

In this type of machine, rotor is provided with 3-ϕ distributed winding similar to the stator as shown in Figure 13.3 and terminals are brought out through slip rings (three in numbers) and brushes. The rotor windings are externally short circuited. Distributed windings are used for better utilization of iron and copper and also to improve *mmf* waveform and smooth torque developed by the machine. When current flows in distributed windings placed in several slots, it produces a sinusoidal space distribution of *mmf*. Wound rotor can also have star- or delta-connected windings. Coil *aa'* presents all the distributed coils assigned to the phase *a* winding for one pair of poles.

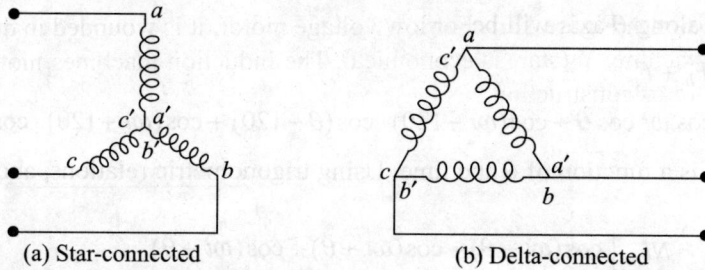

(a) Star-connected   (b) Delta-connected

FIGURE 13.3   Rotor winding connections of wound rotor induction machine.

13.3   ROTATING MAGNETIC FIELD

Let us consider a two-pole machine having windings aa', bb' and cc' on the stator as shown in Figure 13.4. Windings are displaced by 120° in space. At any instant t, the phase currents can be written as

$$i_a = I_m \cos \omega t$$

$$i_b = I_m \cos(\omega t - 120°)$$ (13.1)

$$i_c = I_m \cos(\omega t + 120°)$$

where I_m is the peak current and $\omega(= 2\pi f)$ is the supply frequency in rad/second.

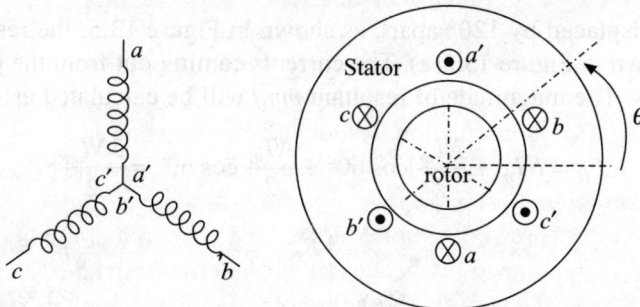

FIGURE 13.4   Two-pole induction machine.

It should be noted that at any instance, each winding produces a sinusoidal distributed *mmf* with its peak along the axis of the phase winding (perpendicular to the line aa' for phase-a) and amplitude proportional to the phase current instantaneous value. If all the phases are having N effective turns each, the *mmf* produced by these phases along θ will be given as

$$F_a = Ni_a \cos \theta$$

$$F_b = Ni_b \cos(\theta - 120°)$$ (13.2)

$$F_c = Ni_c \cos(\theta + 120°)$$

The resultant *mmf* along θ-axis will be

$$F(\theta) = F_a + F_b + F_c$$

$$= NI_m[\cos \omega t \cos \theta + \cos(\omega t - 120) \cdot \cos(\theta - 120) + \cos(\omega t + 120) \cdot \cos(\theta + 120)]$$

The resultant *mmf* is a function of θ and time. Using trigonometric relations, above equation can be simplified as

$$\left.\begin{array}{l} F(\theta, t) = \dfrac{NI_m}{2}\left[\begin{array}{l} \cos(\omega t - \theta) + \cos(\omega t + \theta) + \cos(\omega t - \theta) \\ + \cos(\omega t + \theta - 240) + \cos(\omega t - \theta) + \cos(\omega t + \theta + 240) \end{array}\right] \\[4mm] \qquad\quad = \dfrac{3NI_m}{2}\cos(\omega t - \theta) + 0 \end{array}\right\} \quad (13.3)$$

Equation (13.3) shows that resultant *mmf* wave in the air gap is rotating at the constant angular velocity $\omega (= 2\pi f)$ with constant amplitude.

> *In general, a k-phase distributed winding ($2\pi/k$ degree apart) excited by k-phase currents will produce a sinusoidal rotating magnetic field of constant amplitude. The rotating magnetic field is produced without physically rotating any magnet.*

The resultant *mmf* obtained in Eq. (13.3) can also be explained graphically.

At any instant $t = 0$, the phase currents of Eq. (13.1) can be written as

$$i_a = I_m, i_b = -\frac{I_m}{2}, i_c = -\frac{I_m}{2} \quad (13.4)$$

Since windings are displaced by 120° apart, as shown in Figure 13.5, the resultant *mmf* will be along phase-*a* as shown in Figure 13.5(a). The currents coming out from the phases are taken as positive for simplicity. The magnitude of resultant *mmf* will be calculated as

$$F_R = NI_m + \frac{NI_m}{2}\cos 60° + \frac{NI_m}{2}\cos 60° = \frac{3NI_m}{2} \quad (13.5)$$

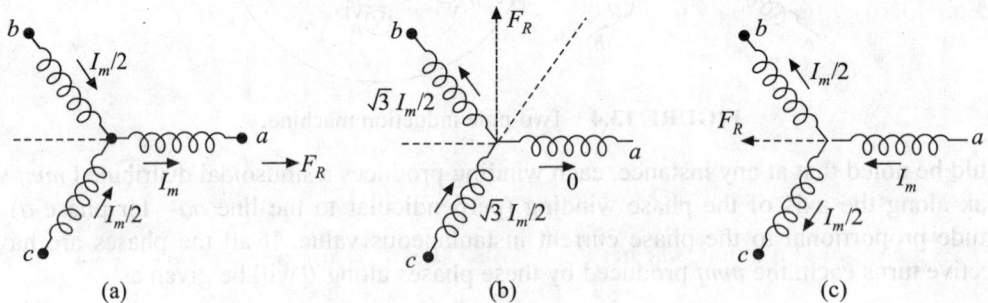

FIGURE 13.5   Graphical representation of *mmf*.

At $t = t_1 = \pi/2\omega$, the direction of currents in the phases will be as shown in Figure 13.5(b) and the magnitude will be (again using the Eq. 13.1) as

$$i_a = 0, i_b = \frac{\sqrt{3}}{2}I_m, \quad i_c = -\frac{\sqrt{3}}{2}I_m \quad (13.6)$$

The resultant *mmf* direction will be the phasor sum of the *mmf* of all the phases as shown in Figure 13.3(b). The magnitude will be obtained as

$$F_R = 0 + \frac{N\sqrt{3}I_m}{2}\cos 30° + \frac{N\sqrt{3}I_m}{2}\cos 30° = \frac{3N}{2}I_m \tag{13.7}$$

At $t = t_2 = \pi/\omega$, the direction of currents in the phases will be as shown in Figure 13.5 (iii) and the magnitude will be [again using Eq. (13.1)] as

$$i_a = -I_m, \quad i_b = \frac{I_m}{2}, \quad i_c = -\frac{I_m}{2} \tag{13.8}$$

The resultant *mmf* direction will be the phasor sum of the *mmf* of all the phases as shown in Figure 13.3(c). The magnitude will be obtained as

$$F_R = NI_m + \frac{NI_m}{2}\cos 60° + \frac{NI_m}{2}\cos 60° = \frac{3NI_m}{2} \tag{13.9}$$

From Eqs. (13.5), (13.7) and (13.9), it is clear that the magnitude of the resultant *mmf* is constant. Moreover, from the graphical representation, it is seen that resultant *mmf* is rotating.

13.4 SYNCHRONOUS SPEED

In previous chapter, we have seen that the induced *emf* frequency is related with the rotation of rotor (mechanical speed) and the number of poles of the machine and can be written as

$$\omega_{\text{elect}} = \frac{p}{2} \cdot \omega_{\text{mech}} \tag{13.10}$$

where p is the number of poles of the machines. Thus, in a p-pole machine, one cycle of variation of the current will make *mmf* wave to rotate by $2/p$ revolutions. From equation, we can also write

$$\theta_{\text{elect}} = \frac{p}{2}\theta_{\text{mech}} \tag{13.11}$$

From Eq. (13.10), if the supply frequency is f (Hz), we can write

$$2\pi f = \frac{p}{2} \times \frac{2\pi}{60} \times n_s \tag{13.12}$$

or

$$n_s = \frac{120f}{p} \tag{13.13}$$

where n_s is the speed of rotating *mmf* in rpm. It is also known as *synchronous speed of the machine*.

13.5 POLE FORMATION

Let us consider three coils aa', bb' and cc' are placed 120° apart as shown in Figure 13.6. At $\omega t = \pi/2$, the three-phase currents [as shown in Eq. (13.1)] will be

$$i_a = 0, \quad i_b = \sqrt{3}\frac{I_m}{2}, \quad i_c = -\sqrt{3}\frac{I_m}{2}$$

The coming-out current (perpendicular to the paper) is denoted by the dot (.), which is general convention, and current going inside the paper is denoted by cross (×). Also current coming-out of the paper is treated as positive and going inside the paper is considered as negative. Since i_a is zero, there will be no current in aa' winding. The current in terminal b is coming out and going to terminal b' whereas the current in terminal c is going inside the paper and coming-out terminal c' which can be seen from the figure. Using the right-hand rule, the direction of fluxes can be easily obtained and shown in Figure 13.6(a). The net flux is shown in Figure 13.6(b). It can be seen that flux is coming-out from the bottom to the top which indicates that a magnet is producing fluxes having north pole at the bottom and south pole at top.

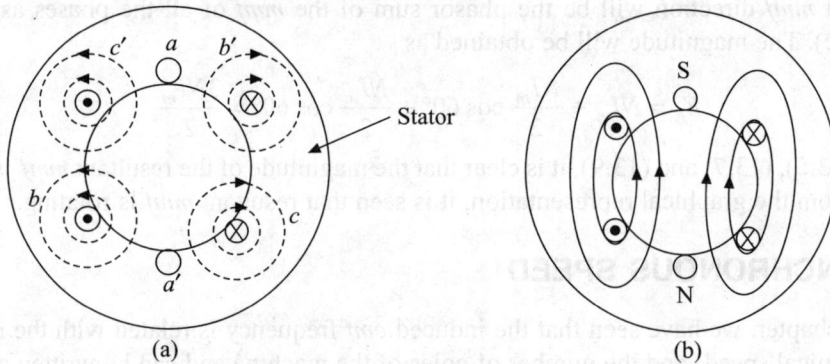

FIGURE 13.6   Pole formation.

This shows the formation of N-S poles on the stator. Similarly, for other instant pole formation can be obtained which may be at the different positions but again two poles will be formed.

13.6   INDUCED VOLTAGES

To derive the general expression for the induced voltages in the rotor and stator circuits, let us start with the case when machine is at the standstill (speed is zero).

13.6.1   Standstill Condition

The stator field (rotating field in air gap) produced by the stator current induces voltage in stator coils and rotor coils. Since stator coils are stationary and field is rotating, the flux linkage to these coils is changing sinusoidally. The rms value of induced voltage (per phase) in any phase in stator coil will be [using Eq. 10.17)]

$$E_1 = \sqrt{2}\pi f N_1 \Phi_p k_{w1} \tag{13.14}$$

where f is frequency of rotating field (supply frequency), N_1 is number of turns in stator phase winding; Φ_p is flux per pole*. For use of iron and copper and to improve the voltage waveform, each phase winding is distributed in a number of slots. For such a distributed winding, the *emf* is

---

*Please note that the ϕ used in Eq. (10.17) is the peak flux, whereas it is average flux per pole as used in Eq. (13.14).

induced in various coils placed in different slots and is not in time phase. Thus the phasor sum of *emf* in stator coils is less than the algebraic sum of *emf*. A reduction factor about (0.85 to 0.95), k_w, called *winding factor*, is used in voltage-induced equation. k_{w1} is the winding factor of stator winding.

Rotating *mmf*, which rotates at synchronous speed, n_s, induces voltage in both stator and rotor. If rotor is at standstill as stator, the induced voltage E_2 (in rms) will be

$$E_2 = \sqrt{2}\pi f N_2 \Phi_p \cdot kw_2 \qquad (13.15)$$

where N_2 is number of turns in rotor circuit and k_{w2} is winding factor of rotor winding.

If $k_{w1} = k_{w2}$, from Eqs. (13.14) and (13.15), we get

$$\frac{E_1}{E_2} \cong \frac{N_1}{N_2} = \text{Turn ratio} \qquad (13.16)$$

This shows that induction machine acts as a transformer. If rotor windings are open circuited, there will be no rotating torque developed and machine will be at standstill.

13.6.2 Running Operation

If the rotor windings are short circuited or closed, current will flow in rotor circuit. This rotor current will produce another flux in the air gap. The fluxes produced by stator current and rotor current produce a torque. If rotor is allowed to rotate, it will start rotating in the direction of rotating field so that relative speed between the rotating field and rotor decreases (Lenz's law). Rotor rotates in the direction of field to oppose its cause. But machine cannot run at synchronous speed (rotating field speed) as at this speed, there will be no induced *emf* in rotor conductors because relative speed between rotor and rotating field will be zero. Thus, no torque will be developed, Hence, a motor, at steady state, operates at slightly less than the synchronous speed. If n is the rotor speed in rpm and n_s is the synchronous speed in rpm, the slip s of machine is defined as*

$$s = \frac{n_s - n}{n_s} \qquad (13.17)$$

The relative speed between rotating field and rotor speed is called *slip rpm* $(= n_s - n = sn_s)$ because sitting on rotor one can experience the slipping of rotor behind the rotating field. The frequency of induced voltage in rotor circuit will correspond to slip *rpm* because it is relative speed between rotor and rotating field.

$$sn_s = n_s - n = \frac{120 f_2}{p} \qquad (13.18)$$

$$f_2 = \frac{p}{120}(n_s - n) = \frac{p}{120}sn_s = sf_1 \qquad (13.19)$$

---

*While calculating the slip, s, the units of n_s and n are the same. Since the slip is a dimensionless quantity, sometime, it is represented in percentage by multiplying 100 to the normal value.

where f_2 is the frequency of induced voltage in rotor and also called *slip frequency*. Hence the voltage induced voltage in the rotor circuit at any slip s will be

$$E_{2s} = \sqrt{2}\,\pi f_2 N_2 \cdot k_{w2} \cdot \Phi_p \qquad (13.20)$$

Using the Eq. (13.19), Eq. (13.20) can be rewritten as

$$E_{2s} = 4.44\,sf_1 N_2 \Phi_p k_{w2} = sE_2 \qquad (13.21)$$

where E_2 is rotor-induced *emf* at standstill. Since rotor circuit is short circuited, the rotor current of f_2 frequency will flow. This current will be produced rotating field speed n_2 (in rpm) with respect to the rotor, where

$$n_2 = \frac{120 f_2}{p} = \frac{120 sf_1}{p} = sn_s \qquad (13.22)$$

Since rotor is rotating at n rpm, the rotating field produced by rotor current with respect to stator will be

$$n_2 + n = sn_s + n = n_s \text{ rpm} \qquad (13.23)$$

Thus, both stator field and rotor field produced by the stator current and rotor currents, respectively, rotate in air gap at the same speed, i.e. synchronous speed (n_s). In other words, both fields are stationary to each other and producing unidirectional torque.

EXAMPLE 13.1   A three-phase, 440V, 50 hp, 50 Hz, induction motor delivers rated output power at 1440 rpm. Find

    (a) Number of poles of machine
    (b) Synchronous speed
    (c) Slip
    (d) Slip rpm
    (e) Rotor speed with respect to (i) rotor structure (ii) stator (iii) stator rotating *mmf*.
    (f) Rotor *emf* at operating speed, if stator-to-rotor turns ration is 1:0.5. Assume winding factors are unity.

Solution: Given that rotor speed $n_r = 1440$ rpm. Since the rotor speed is slightly less than the synchronous speed. Let us assume that rotor speed is almost equal to the synchronous speed and therefore, we can write

$$n_r \simeq n_s = \frac{120 f}{p}$$

    (a) The number of poles will be

$$p = \frac{120 \times 50}{1440} = 4.16$$

Since number of poles cannot be in fraction and odd numbers, the possible number of pole will be the nearest even number which is 4. Please note that number of poles cannot be of any value more than 4.16 as at value higher than this, the synchronous speed will be less than the rotor speed which is not possible.

(b) Synchronous speed in rpm will be $= \dfrac{120 \times 50}{4} = 1500$ rpm

(c) Slip $\dfrac{1500 - 1440}{1500} = 0.04$

(d) Slip rpm $= sn_s = 0.04 \times 1500 = 60$ rpm

(e) (i) $1440 - 1440 = 0$ rpm, (ii) $1440 - 0 = 1440$ rpm, (iii) $1500 - 1440 = 60$ rpm

(f) Using Eqs. (13.14) and (13.21), we can get

$$\frac{E_{2s}}{E_1} = s\,\frac{N_2}{N_1}$$

The voltage induced *emf* in stator is almost equal to the supply voltage (if voltage drop in the winding impedance is neglected) and will be equal to the $\dfrac{440}{\sqrt{3}}$ V/phase. It is also given that $\dfrac{N_1}{N_2} = \dfrac{1.0}{0.5}$, thus

$$E_{2s} = 0.04 \times 0.5 \times \frac{440}{\sqrt{3}} = 5.08 \text{ V/phase.}$$

or

$$E_{2s} = 5.08 \times \sqrt{3} = 8.8 \text{ V (line-to-line)}$$

13.7 EQUIVALENT CIRCUIT

The equivalent winding (star connected) diagram of three-phase wound rotor induction machine is shown in Figure 13.7. The rotor circuit, in the case of squired-cage rotor, can be represented by equivalent three-phase rotor winding. It is always convenient to represent circuit equivalent in per phase.

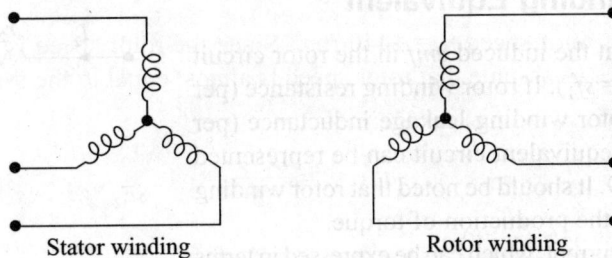

FIGURE 13.7 Stator and rotor winding.

13.7.1 Stator Winding Equivalent

The induction motor stator winding and rotor winding are equivalent to the primary winding and secondary winding, respectively, of transformer. Similar to the transformer equivalent, induction motor equivalent can be expressed. There is always a doubt that induction motor rotates but the transformer does not (even does not experience rotating) as both get three-phase balance supply. The region behind this is that the winding in the induction motor is distributed 120° in space and it produces rotating *mmf*, whereas in transformer, windings are concentric and do not produce rotating *mmf*.

If V_1 is per-phase applied voltage to stator terminal, there will an induced *emf*, E_1 (per phase). The stator winding equivalent can be represented as Figure 13.8.

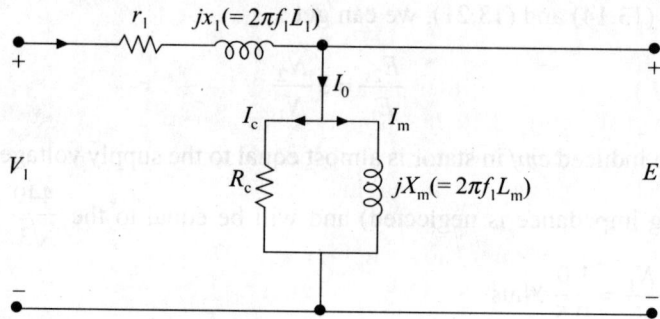

FIGURE 13.8   Stator winding equivalent circuit.

In Figure 13.8,

  r_1 = per-phase stator winding resistance

  L_1 = per-phase stator leakage inductance

  L_m = per-phase stator magnetizing inductance

  R_c = per-phase stator core loss equivalent resistance.

The value of leakage reactance in the motor is lesser than transformer due to air gap. The excitation current in the induction motor is almost 30 to 50% of rated current, whereas it is only 1 to 5% in the case of transformer.

13.7.2   Rotor Winding Equivalent

It has been showed that the induced *emf* in the rotor circuit is sE_2 at frequency $f_2 (= sf_1)$. If rotor winding resistance (per phase) is r_2 and the rotor winding leakage inductance (per phase) is L_2, the rotor equivalent circuit can be represented as shown in Figure 13.9. It should be noted that rotor winding are short-circuited for the production of torque.

If I_2 is rotor circuit current, which can be expressed in terms of rotor induced voltage and rotor winding impedance as

$$I_2 = \frac{sE_2}{r_2 + jsx_2} \qquad (13.24)$$

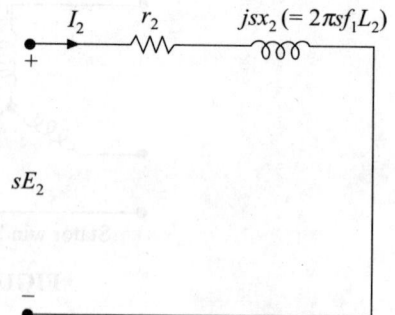

FIGURE 13.9   Rotor equivalent circuit.

where x_2 is the reactance of the rotor circuit at supply frequency (not at the rotor induced voltage frequency). E_2 is the induced voltage in the rotor circuit at standstill.

Due to the current flow in the rotor winding, there is a loss in the winding, known as *rotor copper* loss and expressed (in per phase) as

$$P_2 = I_2^2 r_2 \qquad (13.25)$$

Equation (13.24) can further be simplified as

$$I_2 = \frac{E_2}{\left(\dfrac{r_2}{s}\right) + jx_2} \qquad (13.26)$$

From Eq. (13.26), rotor circuit equivalent can be represented as shown in Figure 13.10. In this representation, the major difference is the frequency of rotor current I_2. In Eq. (13.24), I_2 is at slip frequency, f_2, whereas in Eq. (13.26), I_2 is at supply frequency f_1.

The equivalent circuit of rotor shown in Figure 13.10 is at stator supply frequency. Since equivalent circuit of stator and rotor is at the same frequency (f_1), thus stator and rotor equivalents can be joined together. The number turns in stator (N_1) and rotor (N_2), may be different, where the turns ratio is

$$a = \frac{N_1}{N_2} \qquad (13.27)$$

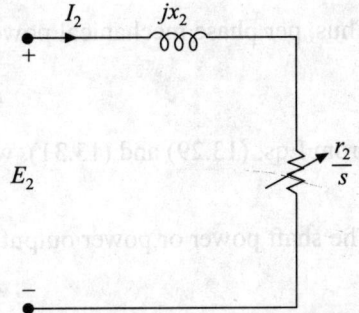

FIGURE 13.10   Equivalent circuit of rotor viewed from stator.

The equivalent circuit of induction machine referred to the stator side is shown in Figure 13.11 which is similar to the two-winding transformer.

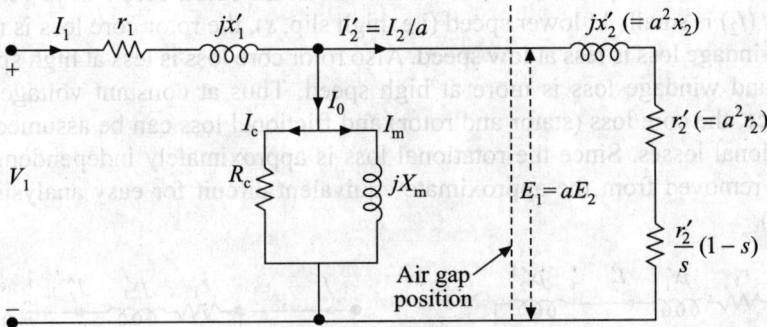

FIGURE 13.11   Exact equivalent circuit of induction motor.

The per-phase real power flow passing to the right-hand side of dotted line is the air gap power passing to the rotor circuit and can be written as

$$P_{ag} = I_2'^2 \left[r_2' + \frac{r_2'}{s}(1-s) \right] = \frac{I_2^2}{a^2}\left(\frac{r_2}{s} a^2\right) = I_2^2 \frac{r_2}{s} \qquad (13.28)$$

Equation (13.28) can be directly written from Figure 13.10 as

$$P_{ag} = I_2^2 \frac{r_2}{s} = \frac{P_2}{s} \tag{13.29}$$

Since slip of machine is normally very small (0.01 to 0.10), the value of P_{ag} is very large, which includes rotor copper loss, and mechanical power developed. Thus,

$$P_{ag} = I_2^2 \frac{r_2}{s} = I_2^2 \left[r_2 + \frac{r_2}{s}(1-s) \right] \tag{13.30}$$

$$= P_2 + P_{mech}$$

Thus, per phase mechanical power can be written as

$$P_{mech} = I_2^2 r_2 \frac{(1-s)}{s} = P_2 \frac{(1-s)}{s} \tag{13.31}$$

From Eqs. (13.29) and (13.31), we have the following relation of power and slip.

$$P_{ag} : P_2 : P_{mech} = 1 : s : (1-s) \tag{13.32}$$

The shaft power or power output is related as

$$P_{shaft} = P_{mech} - \text{Rotational loss} \tag{13.33}$$

13.7.3   Approximate Equivalent Circuit

The exact equivalent circuit (Figure 13.11) is not convenient for analyzing the motor performance. Based on shunt branch representations, there are various approximations available in the literatures.

If the voltage drop across r_1 and x_1 is small, the shunt branch can be taken near to the terminal of supply voltage, as shown in Figure 13.12(a) (because $V_1 \simeq E_1$). R_c in the equivalent circuit represents the core loss component in stator and rotor. In the rotor, core loss is very small as the rotor frequency (f_2) is small. At lower speed (i.e. high slip, s), the rotor core loss is more, whereas frictional and windage loss is less at low speed. Also rotor core loss is less at high speed (low slip) and frictional and windage loss is more at high speed. Thus at constant voltage and constant frequency source, the core loss (stator and rotor) and frictional loss can be assumed constant and termed at rotational losses. Since the rotational loss is approximately independent of the motor load, it can be removed from the approximate equivalent circuit for easy analysis as shown in Figure 13.12(b).

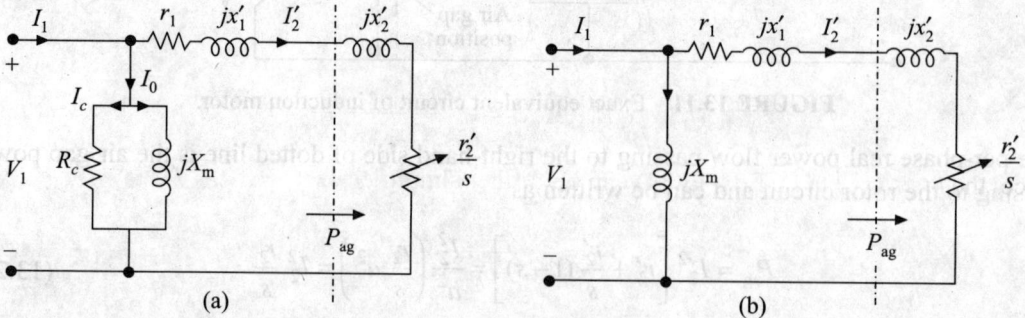

FIGURE 13.12   Approximate equivalent circuit.

13.7.4 IEEE Recommended Equivalent Circuit

Due to air gap, the exiting current is very high and also leakage inductance L_1 is more (x_1 is high). Thus, the sifting of X_m to the supply terminal size is not appropriate. Therefore, IEEE recommended equivalent circuit of induction motor is shown in Figure 13.13.

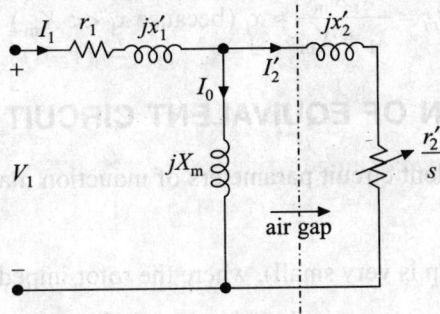

FIGURE 13.13 IEEE recommended approximate equivalent circuit.

Using Thevenin equivalent of left-hand side of dotted line, we can further simplify the equivalent circuit as shown in Figure 13.14.

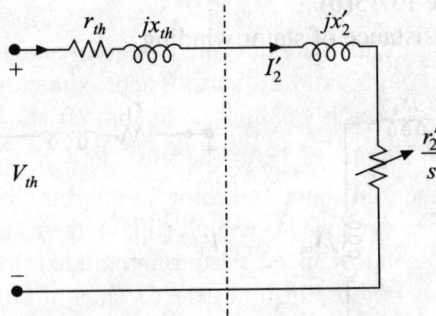

FIGURE 13.14 IEEE recommended approximate equivalent circuit.

where,

$$\overline{V}_{th} = \frac{\overline{V}_1(jX_m)}{r_1 + j(x_1 + X_m)} \tag{13.34}$$

and

$$\overline{Z}_{th} = (r_1 + jx_1) \| (jX_m) = \frac{jX_m(r_1 + jx_1)}{r_1 + j(x_1 + X_m)} = r_{th} + jx_{th} \tag{13.35}$$

Since, the $r_1 << (x_1 + X_m)$, Eq. (13.34) can be further simplified as

$$\overline{V}_{th} \simeq \frac{X_m}{(x_1 + X_m)} \overline{V}_1 \tag{13.36}$$

The Thevenin impedances can also be simplified as

$$r_{th} \simeq \left(\frac{X_m}{x_1 + X_m} \right) r_1 \tag{13.37}$$

and

$$x_{th} \simeq \frac{x_1 X_m}{x_1 + X_m} \approx x_1 \text{ (because } x_1 << X_m) \tag{13.38}$$

13.8 DETERMINATION OF EQUIVALENT CIRCUIT PARAMETERS

Like transformers, the equivalent circuit parameters of induction machine can be obtained from the results of

1. A no-load test (i.e. slip is very small), where the rotor impedance $\left(\dfrac{r_2'}{s} \right)$ is very high and can be assumed as open circuited. This gives information about exciting current and rotational losses. The equivalent circuit diagram at no-load test is shown in Figure 13.15(a).

2. A block rotor test ($s = 1$) which is equivalent to a short-circuit test of transformer and gives information about leakage impedances of rotor and stator windings. It is performed at reduced voltage and at rated current. The equivalent circuit diagram with block rotor test is shown in Figure 13.15(b).

3. Measurement of dc resistance of stator winding.

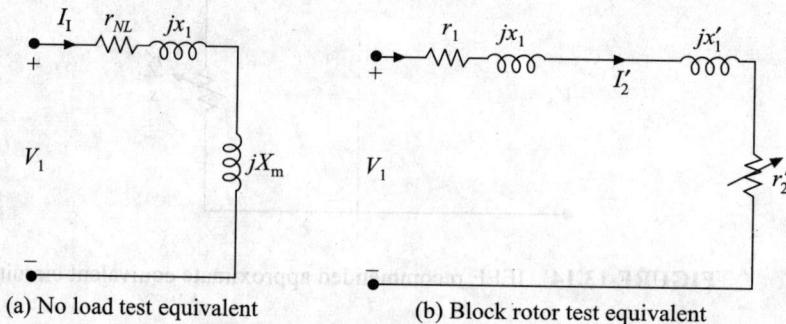

(a) No load test equivalent   (b) Block rotor test equivalent

FIGURE 13.15 Equivalent circuit of induction motor at different tests.

13.9 TORQUE CHARACTERISTIC OF INDUCTION MOTOR

The developed torque can be easily found out from the expression for mechanical power, P_{mech}, using Eq. (13.31) since torque (per phase) is given by,

$$T_{mech} = \frac{P_{mech}}{\omega_m} = \frac{I_2'^2 r_2'}{s \omega_m} (1 - s) \tag{13.39}$$

where ω_m is speed of induction machine (rotor speed) in rad/sec. If n is rotor speed in rpm, then rotor angular speed in radian/second can be obtained as

$$\omega_{\mathrm{m}} = \frac{2\pi n}{60} \tag{13.40}$$

If ω_{s} is the synchronous speed in rad/sec, ω_{m} is related as

$$\omega_{\mathrm{m}} = (1 - s)\,\omega_{\mathrm{s}} = (1 - s)\frac{2\pi n_{\mathrm{s}}}{60} \tag{13.41}$$

where n_{s} is the synchronous speed in rpm.

From Eqs. (13.39) and (13.41), we can have

$$T_{\mathrm{mech}} = \frac{I_2'^2 r_2'}{s\omega_{\mathrm{s}}}\frac{(1-s)}{(1-s)} = \frac{I_2'^2 r_2'}{s\omega_{\mathrm{s}}} \tag{13.42}$$

The per phase mechanical torque developed in induction motor (in terms of stator referred quantities) will be

$$T_{\mathrm{mech}} = \frac{1}{\omega_{\mathrm{s}}} \cdot \frac{I_2'^2 r_2'}{s} = \frac{P_{\mathrm{ag}}}{\omega_{\mathrm{s}}} \tag{13.43}$$

From the equivalent circuit as shown in Figure 13.14, the value of current can be computed and substituted in Eq. (13.43). We get,

$$T_{\mathrm{mech}} = \frac{1}{\omega_{\mathrm{s}}}\frac{V_{th}^2}{\left(r_{th} + \dfrac{r_2'}{s}\right)^2 + (x_{th} + x_2')^2} \cdot \frac{r_2'}{s} \tag{13.44}$$

Thus, the total mechanical torque will be three times per phase torque as

$$T_{\mathrm{mech}} = \frac{3}{\omega_{\mathrm{s}}}\frac{V_{th}^2}{\left(r_{th} + \dfrac{r_2'}{s}\right)^2 + (x_{th} + x_2')^2} \cdot \frac{r_2'}{s} \tag{13.45}$$

At low value of slip (i.e. high speed), $\left(r_{th} + \dfrac{r_2'}{s}\right)^2 \gg (x_{th} + x_2')^2$ and $r_{th} \ll \dfrac{r_2'}{s}$ thus, from Eq. (13.45) we can write

$$T_{\mathrm{mech}} \approx \frac{3}{\omega_{\mathrm{s}}}\frac{V_{th}^2}{r_2'} \cdot s \tag{13.46}$$

This Eq. (13.46) shows that T_{mech} (or simply T_{m}) linearly varies with slip (at higher speed range).

At high values of s (i.e. low speed), $\left(r_{th} + \dfrac{r_2'}{s}\right)^2 \ll (x_{th} + x_2')^2$, Eq. (13.45) becomes

$$T_{\mathrm{mech}} \approx \frac{3}{\omega_{\mathrm{s}}} \cdot \frac{V_{th}^2}{(x_{th} + x_2')^2} \cdot \frac{r_2'}{s} \tag{13.47}$$

Thus, torque varies almost inversely proportional to the slip near $s = 1$. The torque-speed (slip) characteristic is shown in Figure 13.16.

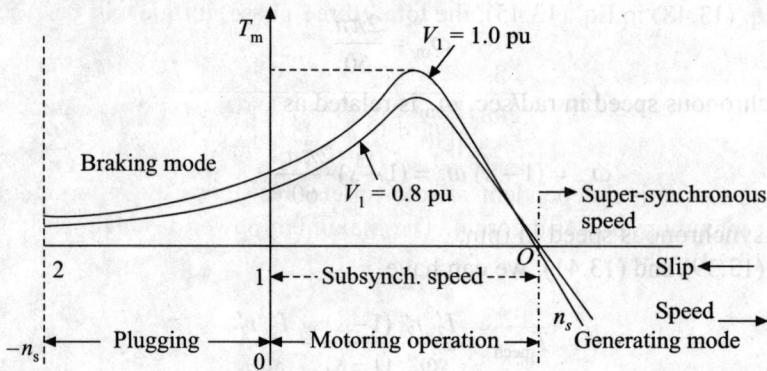

FIGURE 13.16   Rotor torque-speed characteristics of induction motor.

Induction machine can be operated in the following modes:

(a) *Motoring mode*:   ($0 < s \le 1$). Machine operates at sub-synchronous speed and runs in the direction of rotating air gap field.

(b) *Generating mode*:   ($s < 0$). Machine is rotated at super-synchronous speed in the direction of rotating air gap field. Negative s implies that mechanical output is negative. Hence machine produces electrical output from mechanical input.

(c) *Braking mode*:   ($s > 1$). In this mode machine runs in the opposite direction of rotating field. It is also known as *plugging mode*.

13.10   PERFORMANCE OF INDUCTION MOTORS

The performance of induction motor, which can be obtained from the equivalent circuit, can be described by torque-speed characteristic. The performance of induction motor is decided based on starting torque, maximum torque, pull-up torque, efficiency, power factor, current, etc. The various torques as shown in Figure 13.17 are described below.

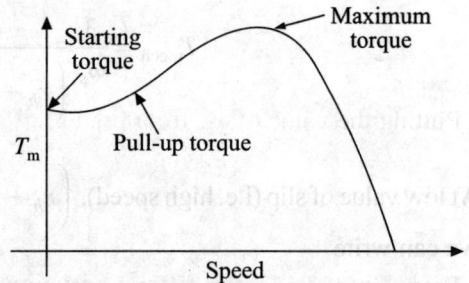

FIGURE 13.17   Various torques in induction motor operation.

(a) *Starting torque*:   It is also known as *break away torque*. This is the torque developed when motor is started from stationary position.

(b) *Maximum torque*:   It is also known as *pull-out* or *break down torque*. This is maximum value of torque, which a machine can develop. At this condition, rotor inductive reactance is equal to rotor resistance.

(c) *Pull-up torque*:   This is the torque between the starting torque and maximum torque as shown in Figure 13.17.

The slip (s_{Tmax}) corresponding to the maximum torque can be obtained from Eq. (13.45) by equating $\dfrac{dT_m}{ds} = 0$. We get the following expression after simplification,

$$\frac{r_2'}{s_{T\,max}} = [r_{th}^2 + (x_{th} + x_2')^2]^{1/2}$$

(13.48)

Substituting Eq. (13.48) in Eq. (13.45), the total (three-phase) torque will be

$$T_{max} = \frac{3}{2\omega_s} \frac{V_{th}^2}{\left\{ \left[r_{th}^2 + (x_{th} + x_2')^2 \right]^{1/2} + r_{th} \right\}} \tag{13.49}$$

Thus, maximum torque is independent of rotor circuit resistance. Equation (13.48) can also be obtained using maximum power theorem. The maximum power or maximum torque will occur (using Figure 13.14), when

$$\frac{r_2}{s_{T\,max}} = [r_{th}^2 + (x_{th} + x_2')^2]^{1/2}$$

or

$$s_{T\,max} = \frac{r_2'}{\sqrt{r_{th}^2 + (x_{th} + x_2')^2}} \tag{13.50}$$

Using Eqs. (13.45) and (13.49), we get ($T = T_{mech}$)

$$\frac{T_{max}}{T} = \frac{1}{2} \frac{\left(r_{th} + \dfrac{r_2'}{s} \right)^2 + (x_{th} + x_2')^2}{r_{th} + \sqrt{r_{th}^2 + (x_{th} + x_2')^2}} \cdot \frac{s}{r_2'} \tag{13.51}$$

Since r_{th} is very small (because r_1 is small), thus

$$\frac{T_{max}}{T} = \frac{\left(\dfrac{r_2'}{s} \right)^2 + (x_{th} + x_2')^2}{(x_{th} + x_2')} \cdot \frac{s}{2r_2'} \tag{13.52}$$

Putting the value of r_2' from Eq. (13.50), Eq. (13.52) can be written as,

$$\frac{T_{max}}{T} = \left[\left(\frac{s_{T\,max}}{s} \right)^2 + 1 \right] \cdot \frac{s}{2s_{T\,max}} = \frac{s_{T\,max}^2 + s^2}{2s \times s_{T\,max}} \tag{13.53}$$

From Eq. (13.53), the different relations with T_{max}, full-load torque (T_{full}) where slip is s_{fl} and starting torque (T_{st}) can be obtained as

$$\frac{T_{max}}{T_{fl}} = \left[\frac{s_{T\,max}^2 + s_{fl}^2}{2s_{fl} \cdot s_{T\,max}} \right] \tag{13.54}$$

$$\frac{T_{max}}{T_{st}} = \frac{s_{T\,max}^2 + 1}{2s_{T\,max}} \tag{13.55}$$

and

$$\frac{T_{st}}{T_{fl}} = \frac{(s_{T\,max}^2 + s_{fl}^2)}{(s_{T\,max}^2 + 1)} \cdot \frac{1}{s_{fl}} \tag{13.56}$$

13.11   ROTOR AND STATOR CURRENTS

From the equivalent circuit diagram, the magnitude of rotor current referred to the stator side can be expressed as

$$I'_2 = \frac{V_{th}}{\left[\left(r_{th} + \dfrac{r'_2}{s} \right)^2 + (x_{th} + x'_2)^2 \right]^{1/2}} \tag{13.57}$$

The expression for the rotor current at maximum torque, $I'_{2T\,max}$, can be derived (put $s = s_{T\,max}$ and simplified) as

$$I'_{2T\,max} = \frac{V_{th}}{\left[\left(r_{th} + \dfrac{r'_2}{s_{T\,max}} \right)^2 + (x_{th} + x'_2)^2 \right]^{1/2}} \tag{13.58}$$

If r_{th} is very small, then we can have

$$\frac{I'_{2T\,max}}{I'_2} = \left[\frac{\left(\dfrac{r'_2}{s} \right)^2 + (x_{th} + x'_2)^2}{\left(\dfrac{r'_2}{s_{T\,max}} \right)^2 + (x_{th} + x'_2)^2} \right]^{1/2} \tag{13.59}$$

Putting the value of r'_2, from Eq. (13.48), in Eq. (13.59), we get

$$\frac{I'_{2T\,max}}{I'_2} = \frac{1}{2} \left[\left(\frac{s_{T\,max}}{s} \right)^2 + 1 \right] \tag{13.60}$$

Similarly, we can get the relation between the starting current and maximum current of rotor as

$$\frac{I'_{2T\,max}}{I'_{2st}} = \frac{1}{2} [s_{T\,max}^2 + 1] \tag{13.61}$$

where I'_{2st} is the rotor current at the starting ($s = 1$).

Stator current: If input impedance is $\overline{Z}_1$ which can be obtained from the equivalent circuit, the stator current $\overline{I}_1$ will be

$$\overline{I}_1 = \frac{\overline{V}_1}{\overline{Z}_1} = \overline{I}_0 + \overline{I}'_2 \tag{13.62}$$

13.12   EFFICIENCY

To find the efficiency, various losses of the machine must be found out. The various losses in an induction machine are stator core loss, stator copper loss, rotor core loss, rotor copper loss and frictional and windage loss. The power flow in an induction motor is shown in Figure 13.18.

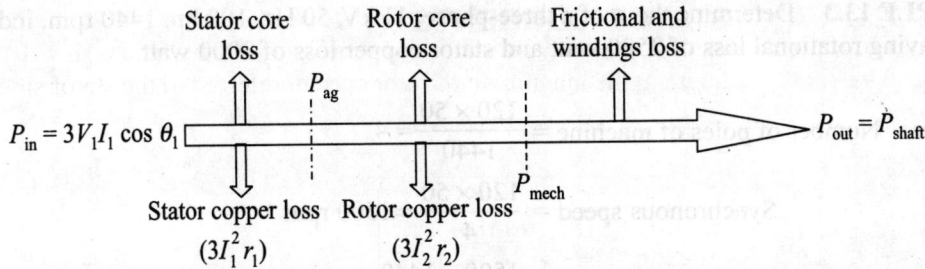

Stator core loss   Rotor core loss   Frictional and windings loss

P_{ag}

$P_{in} = 3V_1 I_1 \cos\theta_1$

$P_{out} = P_{shaft}$

P_{mech}

Stator copper loss   Rotor copper loss

$(3I_1^2 r_1)$ $(3I_2^2 r_2)$

FIGURE 13.18 Power flow in induction motor.

The efficiency of induction motor is defined as

$$\text{Efficiency } \eta = \frac{\text{Output power } (P_{out})}{\text{Input power } (P_{in})} = \frac{P_{out}}{P_{out} + \text{Total loss}} \qquad (13.63)$$

EXAMPLE 13.2 A three-phase, 415 V, 50 Hz, 960 rpm, star-connected induction motor has the following parameters per phase

$$r_1 = 0.2\ \Omega,\ x_1 = 0.6\ \Omega,\ r_2' = 0.2\ \Omega,\ x_2' = 0.7\ \Omega. \text{ Assume } X_m \text{ is very high (infinity).}$$

When motor operates at rated speed, find

  (a) The developed electromagnetic torque in N-m.
  (b) The power output (in W) of motor if frictional and windage losses are 1500 watts.
  (c) The efficiency of motor.

Solution: At rated speed, rotor speed = 960 rpm.

$$\text{Number of poles of machine} = \frac{120 \times 50}{960} \approx 6$$

$$\text{Synchronous speed} = \frac{120 \times 50}{6} = 1000 \text{ rpm.}$$

$$\text{Slip} = \frac{1000 - 960}{1000} = 0.04$$

Rotor current referred to stator

$$I_2' = \frac{415/\sqrt{3}}{\sqrt{\left(r_1 + \dfrac{r_2'}{s}\right)^2 + (x_1 + x_2')^2}} = \frac{239.6}{\sqrt{\left(0.2 + \dfrac{0.2}{0.04}\right)^2 + (0.6 + 0.7)^2}} = 44.70 \text{ A}$$

  (a) Developed torque $= T_{mech} = \dfrac{3 I_2'^2 r_2'}{s\omega_m}(1-s) = \dfrac{3 \times 44.7^2 \times 0.2 \times (1 - 0.04)}{0.04 \times 960 \times 2\pi/60} = 286.22$ Nm

  (b) Power output = air gap power – losses

$$= \omega_m T_{mech} - 1500 = 28774.02 - 1500 = 27273.97 \text{ W}$$

  (c) Stator and rotor copper loss $= (44.70)^2 \times (0.2 + 0.2) = 799.24$ W

$$\text{Efficiency, } \eta = \frac{27273.97}{27273.97 + 799.24 + 1500} \times 100 = 92.23\%.$$

EXAMPLE 13.3 Determine the η of a three-phase, 11 kV, 50 Hz, 100 hp, 1440 rpm, induction motor having rotational loss of 2500 watt and stator copper loss of 2000 watt.

Solution: Number of poles of machine $= \dfrac{120 \times 50}{1440} \approx 4$

$$\text{Synchronous speed} = \dfrac{120 \times 50}{4} = 1500 \text{ rpm.}$$

$$\text{Slip} = \dfrac{1500 - 1440}{1500} = 0.04$$

$$P_{out} = 100 \times 746 = 74600 \text{ watt}$$

$$P_{mech} = P_{out} + \text{rotational losses}$$

$$= 74600 + 2500 = 77100 \text{ watt}$$

$$= \dfrac{P_2}{s}(1 - s) \quad (P_2 = \text{rotor copper loss})$$

Thus rotor copper loss $P_2 = \dfrac{771 \times 0.04}{0.96} = 3212.5$ watt

$$\eta = \dfrac{P_{out}}{P_{out} + \text{stator copper loss} + \text{rotor coper loss} + \text{rotational losses}}$$

$$= \dfrac{74600}{74600 + 2000 + 3212.5 + 2500} \times 100 = 90.63\%.$$

EXAMPLE 13.4 The following tests are performed on a three-phase 5 hp, 440 V, 4 pole, 50 Hz, squirrel cage induction motor.

  (a) No. Load test: line voltage = 440 V, line current = 1A, input power = 200 W, supply frequency = 50Hz.

  (b) Blocked rotor test: frequency = 25 Hz, line voltage $= 60\sqrt{3}$ V, line current = 5A, input power = 400 W.

  (c) Stator dc resistance (per phase) = 2.5 Ω.

Determine the equivalent circuit parameters.

Solution: From no-load test, P = 200 W. Since at no-load slip is very small, the rotor current is also very small (Figure 13.19). Thus rotational loss will be

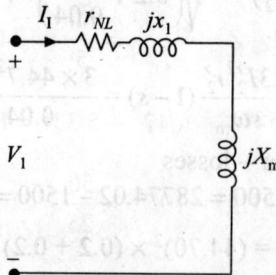

FIGURE 13.19 No-load test.

No-load rotational loss $= 200 - 3\,(1^2) \times 2.5 = 192.5$ watt

$$V_1 = \frac{440}{\sqrt{3}} = 254 \text{ V per phase}$$

$$Z_{nl} = \frac{V}{I} = \frac{254}{1} = 254 \,\Omega$$

From the equivalent circuit at no-load,

$$R_{nl} = \frac{200/3}{I_1^2} = 66.67 \,\Omega$$

Thus,

$$x_1 + X_m = \sqrt{254^2 - 66.67^2} = 245.1 \,\Omega$$

From block rotor test $(s = 1)$, the shunt impedance jx_m is parallel to the rotor circuit impedance (referred to stator side), $(r_2' + jx_2')$. Since x_m is very large compared to $\sqrt{r_2'^2 + x_2'^2}$ and thus X_m can be ignored. The equivalent circuit at blocked-rotor test is shown in Figure 13.20.

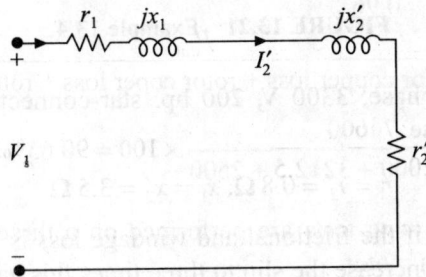

FIGURE 13.20 Block rotor equivalent.

At block rotor test is at low voltage and low frequency, stator loss is negligible. Also frictional and windage loss is zero. Thus rotational loss can be ignored.

$$P = 3I_1^2(r_1 + r_2') = 400 \text{ or } r_1 + r_2' = \frac{400}{3 \times 5^2} = 5.33 \,\Omega$$

Since r_1 is 2.5 ohm,

$$r_2' = 5.33 - 2.5 = 2.83 \,\Omega$$

As blocked rotor impedance at 25 Hz, $Z = \dfrac{V_1/\sqrt{3}}{I_1} = \dfrac{60}{5} = 12 \,\Omega$

$$x(= x_1 + x_2') \text{ at 25 Hz} = \sqrt{12^2 - 5.33^2} = 10.75 \,\Omega$$

or

$$x(= x_1 + x_2') \text{ at 50 Hz} = 10.75 \times \frac{50}{25} = 21.50 \,\Omega$$

thus,

$$x_1 = x_2' = \frac{21.50}{2} = 10.75 \ \Omega \text{ at 50 Hz (Assumed that } x_1 = x_2')$$

$$x_m = 245.1 - 10.75 = 234.35 \ \Omega$$

Since r_2' is very important in the computation of performance of the machine, r_2' can be obtained as per IEEE guidelines. Complete equivalent circuit is shown below (Figure 13.21).

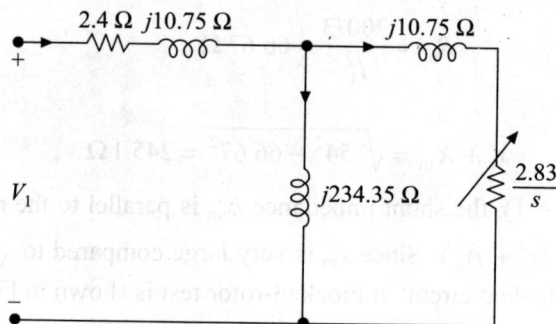

FIGURE 13.21   Example 13.4.

EXAMPLE 13.5   A three-phase, 3300 V, 200 hp, star-connected induction motor has the following parameters per phase:

$$r_1 = r_2' = 0.8 \ \Omega, \ x_1 = x_2' = 3.5 \ \Omega$$

Calculate the slip at full-load if the frictional and windage loss is 3000 watt. How much extra rotor resistance is required to increase the slip to three times this value with the full-load torque maintained? How much extra stator resistance would be necessary to achieve the same objective? Find the reduction in peak torque with external stator resistance. Magnetizing branch can be neglected.

Solution: Since, output power = $3 P_{ag}(1 - s)$ – rotational loss, thus

$$200 \times 746 = \frac{3V_1^2}{\left(r_1 + \dfrac{r_2'}{s}\right)^2 + (x_1 + x_2')^2} \times \frac{r_2'}{s}(1 - s) - 3000$$

or

$$146200 = \frac{3(3300/\sqrt{3})^2}{\left(0.8_1 + \dfrac{0.8}{s}\right)^2 + (3.5 + 3.6)^2} \times \frac{0.8}{s}(1 - s) - 3000$$

After simplifying it, we get

$$167.26 - \frac{87.445}{s} + \frac{1}{s^2} = 0$$

Solving this quadratic equation,

$$\frac{1}{s} = \frac{87.445 \pm \sqrt{87.445^2 - 4 \times 167.26}}{2} = 1.96, 85.49$$

Thus possible operating slip (lower slip) will be 0.0117 (= 1/85.49).

From the torque Eq. (13.45), it is clear that torque will be unchanged if the r_2'/s is constant. Since slip changes to three times, the total $(r_2' + R_{ext})$ will be three times of r_2' for no change in torque. Thus,

$$(r_2' + R_{ext}) = 3 \times r_2' = 3 \times 0.8 = 2.4 = (0.8 + R_{ext})$$

and $R_{ext} = 1.6\ \Omega$

For increase of stator resistance with no change in torque $\left(\dfrac{3}{\omega_s} \dfrac{V_1^2}{\left(r_1 + \dfrac{r_2'}{s} \right)^2 + (x_1 + x_2')^2} \cdot \dfrac{r_2'}{s} \right),$

$\dfrac{V_1^2}{Z^2} \times \dfrac{r_2'}{s}$ should be unchanged. This shows that $Z^2 s$ should be unaltered. Let the stator resistance with extra resistance is R_1, then we have

$$\left[\left(R_1 + \frac{0.8 \times 3}{s} \right)^2 + (3.5 + 3.5)^2 \right] \times 3s = \left[\left(0.8 + \frac{0.8}{s} \right)^2 + (3.5 + 3.5)^2 \right] \times s$$

or

$$R_1^2 + 45.584\ R_1 + 1043 = 0$$

Therefore $R_1 = 16.6\ \Omega$ and external stator resistance would be 15.8 (= 16.6 − 0.8) ohm.

Let maximum torque with stator resistance R_1 is $T_{max\,1}$ and without extra stator resistance is $T_{max\,2}$. Thus, using Eq. (13.49), we get

$$\frac{T_{max\,2}}{T_{max\,1}} = \frac{\left[r_1^2 + (x_1 + x_2')^2 \right]^{1/2}}{\left[R_1^2 + (x_1 + x_2')^2 \right]^{1/2}} = \sqrt{\frac{(0.64 + 49)}{(275.56 + 49)}} = 0.391$$

Thus, there is 60.9 % reduction in the maximum torque.

EXAMPLE 13.6 A three-phase, four-pole, 3300 V, 50 Hz, star-connected induction motor has identical primary and referred secondary impedances of value 3 + j9 ohm per phase. The turn ratio per phase is 3/1(stator/rotor) and rotor winding is connected in delta and brought out to the slip rings. Find:

(a) the full-load torque at slip of 5%,
(b) the maximum torque at nominal voltage and frequency,
(c) the supply voltage reduction which can be withstood without motor stalling,
(d) the maximum torque, if both supply and frequency fall to half of the normal value,
(e) the increase in rotor resistance which, at normal voltage and frequency, will permit the maximum torque is to be developed at starting.

Solution: Synchronous speed $= 2\pi \times 120 \times 50/(60 \times 4) = 50\pi$ rad/sec

(a) Full-load torque $= \dfrac{3}{50\pi} \times \dfrac{(3300/\sqrt{3})^2}{\left(3.0 + \dfrac{3.0}{0.05}\right)^2 + (9+9)^2} \times \dfrac{3.0}{0.05} = 969$ Nm

(b) Maximum torque $= \dfrac{3}{50\pi} \times \dfrac{(3300/\sqrt{3})^2}{2 \times \sqrt{(3.0)^2 + (9+9)^2}} = 1631$ Nm

(c) With voltage reduction, the maximum toque must not fall below 969 Nm and since $T_{mech} \propto V^2$, thus

$$\frac{969}{1631} = \left(\frac{\text{Reduced } V}{\text{Normal } V}\right)^2 \quad \text{or} \quad \text{Reduce } V = \sqrt{\frac{969}{1631}} = 0.77 = 23\% \text{ reduction.}$$

(d) The slip at the maximum torque (reduced frequency)

$$s_{T\,max} = \frac{3}{\sqrt{3^2 + \left(18 \times \dfrac{1}{2}\right)^2}} = 0.316$$

Maximum torque $= \dfrac{3}{0.5 \times 50\pi} \times \dfrac{(0.5 \times 3300/\sqrt{3})^2}{(3.0 + 3.0/0.316)^2 + (18/2)^2} \times \dfrac{3.0}{0.316} = 1388$ Nm

(e) Maximum toque occurs at slip $s_{T\,max} = \dfrac{3}{\sqrt{3^2 + (18)^2}} = 0.1644$. Now maximum torque

occurs at $s = 1$ with external resistance of rotor R_2'. Thus $s_{T\,max} = 1$, hence

$$1 = \frac{(3 + R_2')}{\sqrt{3^2 + (18)^2}} \quad \text{or} \quad R_2' = 15.25 \text{ ohm}$$

Since turn ratio is 3, the additional resistance required to be added per rotor phase $= 15.25/3^2 = 1.694$ ohm.

13.13   SPEED CONTROL OF INDUCTION MOTOR (OR SLIP CONTROL)

The speed of induction motor can be controlled by several methods, which are described below.

13.13.1   Terminal Voltage Control

The voltage applied to the induction motor is changed with the help of auto-transformer or with some static power electronics device. Since voltage across the motor cannot increase as it causes insulation problems, normally the speed is controlled by changing the voltage from rated one to lower side. Due to reduction in applied voltage, torque is reduced [Eq. (13.45)], and therefore the speed is reduced as shown in Figure 13.22. This control is used for small variation of speed as developed torque is also changed significantly.

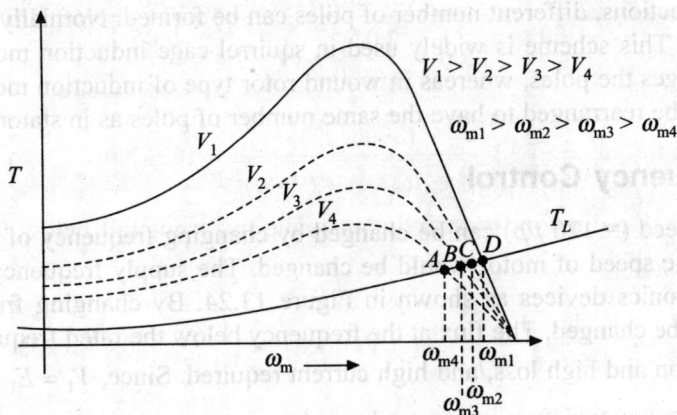

FIGURE 13.22 Torque-speed characteristics for different terminal voltage control.

From Figure 13.22, it can be seen that operating point shifts from D to C when the applied voltage is reduced from V_1 to V_2 and thus the speed of motor is reduced. The change in speed also depends on the load characteristic. The steady-state speed is the intersection of machine torque and load torque curves.

13.13.2 Rotor Resistance Control

This type of control can only be used for wound rotor induction machine because there is no provision for adding the rotor resistance in squirrel induction motor. With this control the maximum torque of machine does not change and speed is obtained below the rotor rated speed (or base speed). This control is not efficient as loss in the rotor circuit increases. Figure 13.23 shows the rotor resistance control.

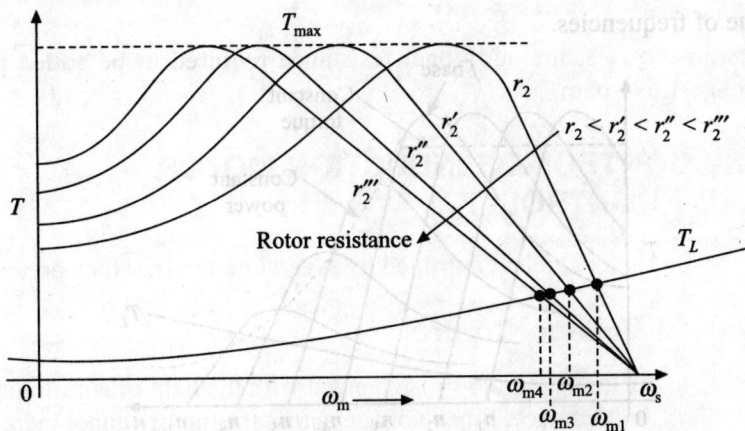

FIGURE 13.23 Effect of change in rotor resistance.

13.13.3 Pole Changing Method

This type of speed control is not smooth and also limited speed variation is possible. By making different coil connections, different number of poles can be formed. Normally pole formation is in the ratio of 2:1. This scheme is widely used in squirrel cage induction motor because rotor automatically changes the poles, whereas in wound rotor type of induction motor, the rotor coil connection should be rearranged to have the same number of poles as in stator.

13.13.4 Frequency Control

As synchronous speed (= 120 f/p) can be changed by changing frequency of supply voltage to induction motor, the speed of motor would be changed. The supply frequency can be changed using power electronics devices as shown in Figure 13.24. By changing frequency, the flux produced will also be changed. The flux at the frequency below the rated frequency will be high and causes saturation and high loss, and high current required. Since, $V_1 \simeq E_1 = K\phi_p f$, the flux

per pole $\phi_p \propto \dfrac{V}{f}$.

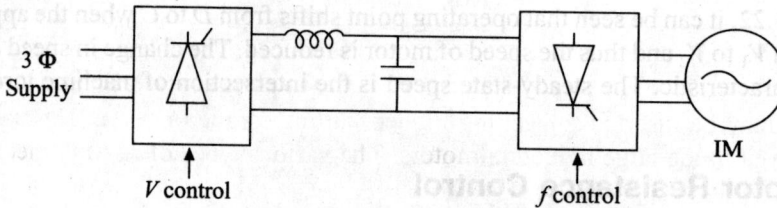

FIGURE 13.24 Frequency control scheme of induction motor.

To have constant flux operation, the ratio $\dfrac{V}{f}$ is made constant and this control is known as

constant volt per hertz $\left(\text{or } \dfrac{V}{f}\right)$ control. Figure 13.25 shows the induction motor characteristics

for different value of frequencies.

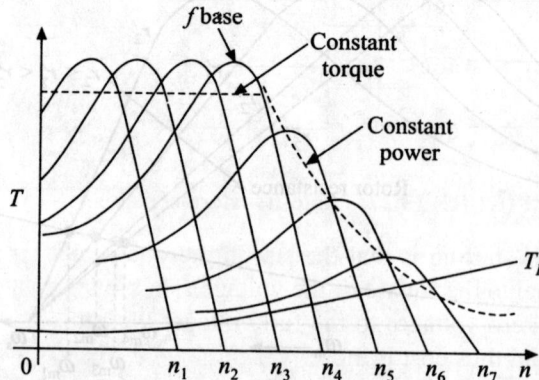

FIGURE 13.25 *V/f* characteristics of induction motor.

13.14 STARTING OF INDUCTION MOTORS

Small rating induction machine can be started by connecting directly across the supply system. In this starting method, time required in starting is very small. The large induction motors cannot be started directly by connecting to the ac supply system as the starting current is very high (5 to 10 times of full-load current) which causes excessive voltage drop and also large current produces excessive heat, which may damage the insulation of motor. It can be seen from Eq. (13.64) that the impedance at the starting (slip =1) is less compared to the impedance after the starting (slip < 1). However, the major problem is the reduction in voltage, which may cause non-starting of motor, when load torque is more than starting torque as shown in Figure 13.26. The starting current in the induction motor (neglecting the magnetizing current) is given by

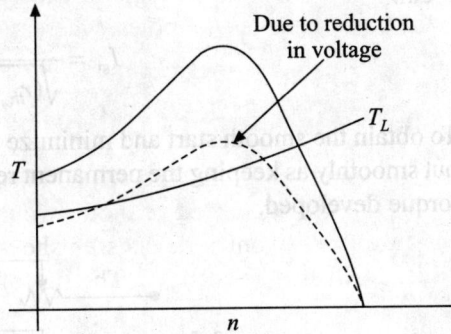

FIGURE 13.26 Direct starting of induction motor.

$$I_{st} = \frac{V_{th}}{\sqrt{(r_{th} + r_2')^2 + (x_{th} + x_2')^2}} \qquad (13.64)$$

Rotor resistance may be inserted for getting higher starting torque or reduced voltage starting should be employed in the large induction motors. The various types of starting mechanism are discussed below:

13.14.1 Auto-transformer Starters

Since the starting current is directly proportional to the supply voltage [see Eq. (13.64)], the starting current can be controlled by applying the reduced voltage to the induction motor terminal which can be obtained, using a three-phase auto-transformer as shown in Figure 13.27.

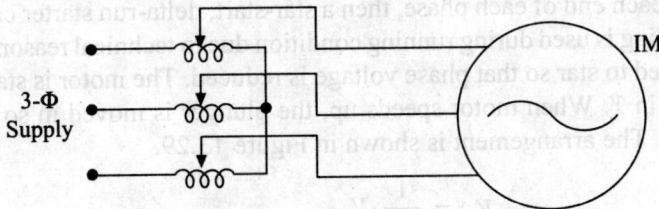

FIGURE 13.27 Auto-transformer starter.

The major problem with this starting is that the starting torque is also reduced significantly as torque is directly proportional to the square of the voltage. Thus, auto-transformer is varied from the small voltage (not with zero voltage) to the full voltage. With this starting, one three-phase autotransformer is needed and thus cost is more.

13.14.2 Primary Resistor Starters

As its name, variable resistors in all the three-phases are used between the supply voltage and induction motor as shown in Figure 13.28. The starting current with external primary resistance (R_{ext1}) will be

$$I_{st} = \frac{V_{th}}{\sqrt{(r_{th} + R_{ext1} + r_2')^2 + (x_{th} + x_2')^2}} \qquad (13.65)$$

To obtain the smooth start and minimize the voltage disturbance, the starting resistors may be cut out smoothly as keeping the permanent resistor in the circuit will increase the loss and reduces the torque developed.

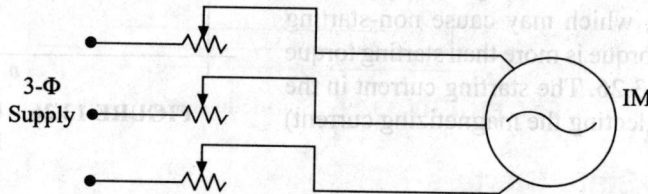

FIGURE 13.28 Primary resistor starter.

13.14.3 Primary Reactor Starters

These starters are similar to the primary resistor starters, except that reactance coils are used to limit the starting current. This is used in the large motors where the heat dissipation from the starting resistors would be a severe problem. Although the reactors reduce the starting current but they also reduce power factor on starting.

13.14.4 Star-Delta Starters

If a three-phase motor is designed to operate with the winding connected to delta and a separate lead is provided for each end of each phase, then a star-start, delta-run starter can be applied to it. Normally, delta winding is used during running condition due to technical reasons. During starting, winding are connected to star so that phase voltage is reduced. The motor is started with position where windings are in Y. When motor speeds up, the plunger is moved in so that windings are reconnected in delta. The arrangement is shown in Figure 13.29.

$$V_{ph} = \frac{1}{\sqrt{3}} \cdot V_L$$

$$I_{st(Y)} = \frac{1}{\sqrt{3}} I_{st(\Delta)} \qquad (13.66)$$

$$T_{st(Y)} = \frac{1}{3} T_{st(\Delta)}$$

3-Φ Supply

IM

Start   Run

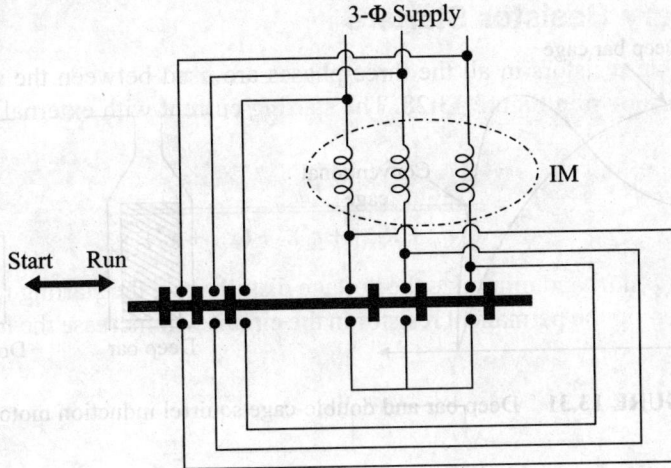

FIGURE 13.29   Star-delta starter.

13.14.5 Solid State Starters

Solid-state controllers can be used to reduce the voltage, which are more efficient than autotransformers.

13.15 INCREASING STARTING TORQUE

In slip ring induction motor, the starting torque can be increased by inserting the external rotor resistance as shown in Figure 13.30. These external resistances also reduce the starting current. These resistances are removed after starting to reduce the rotor copper loss.

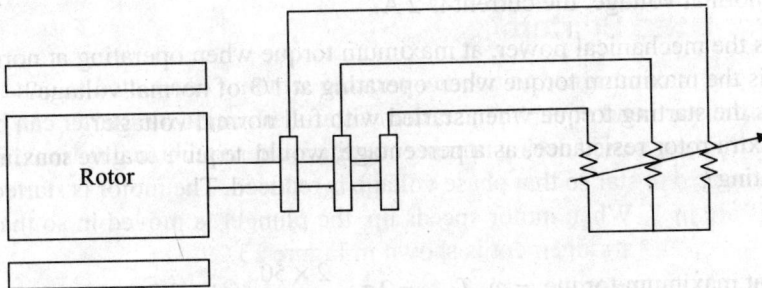

Rotor

FIGURE 13.30   Rotor resistance starting.

In squirrel cage induction motors, deep bar or bubble cage rotor is used to increase the starting torque as shown in Figure 13.31. Rotor resistance is more at high frequency than at low frequency. At starting, rotor current frequency is higher and thus more rotor resistance due to skin effect. In deep bar squirrel induction motor, leakage reaction of bottom position is higher than upper position. In double cage induction motor, the rotor current flows mainly in upper cage (at standstill) due to higher frequency. Outer cage has high resistance, as area is small. At low frequency (high speed), rotor cage shares more current. Figure 13.32 shows the equivalent circuit of double cage induction motor.

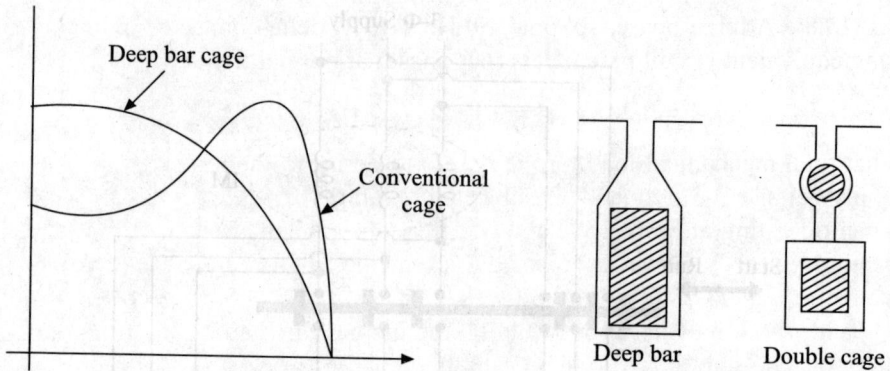

FIGURE 13.31 Deep bar and double-cage squirrel induction motor.

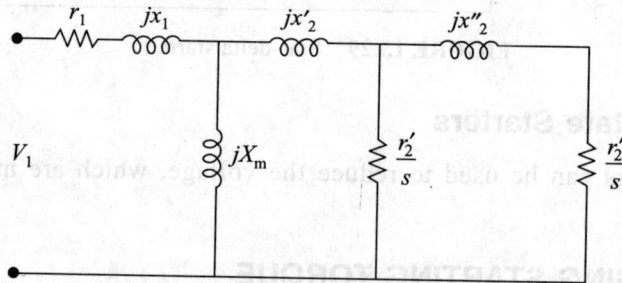

FIGURE 13.32 Equivalent circuit of double-cage induction motor.

EXAMPLE 13.7 A three-phase, six-pole, 50 Hz induction motor has a maximum torque of 6 Nm at a slip of 25% and a starting torque of 3 Nm when operating at full voltage. When started with 1/3 of the normal voltage, the current is 2 A.

(a) What is the mechanical power, at maximum torque when operating at normal voltage?
(b) What is the maximum torque when operating at 1/3 of normal voltage?
(c) What is the starting torque when started with full normal voltage?
(d) What extra rotor resistance, as a percentage, would require to give maximum torque at the strating?

Solution:

(a) Power at maximum torque $= \omega_m T_{max} = 2\pi \times \dfrac{2 \times 50}{6} \times (1 - 0.25) \times 6 = 0.471$ kW

(b) Since torque $\propto V^2$, thus maximum torque at reduced voltage $= \left(\dfrac{1}{3}\right)^2 \times 6 = 2/3$ Nm

(c) Since current $\propto V$, the starting current at full voltage $\left(\dfrac{3}{1}\right) \times 2 = 6$ A

(d) A given torque requires a particular value of $\dfrac{r_2'}{s}$. Since the slip is changed from the 0.25 to 1.0, the total rotor resistance will be changed in the same ratio, i.e. 4 times. Hence extra resistance = 300%.

EXAMPLE 13.8   A three-phase, six-pole, 50 Hz, 440 V, delta-connected induction motor has the following equivalent circuit parameters (per phase) at normal frequency.

$$r_1 = 0.2\,\Omega, \quad r_2' = 0.18\,\Omega, \quad x_1 = x_2' = 0.58\,\Omega$$

(a) What total mechanical load torque is safe to derive without stalling the motor, if both supply voltage and frequency is subjected to fall of 40%?

(b) When operating at normal voltage and frequency, calculate the speed when delivering this torque and the power developed. Also calculate the speed at which maximum torque occurs.

(c) If V and f both halved, what would be the increase in starting torque from the normal direct-on-line start at rated voltage and frequency?

Solution: $\omega_s = 2\pi \times \dfrac{2 \times f}{p} = 2\pi \times \dfrac{2 \times 50}{6} = 104.72$ rad/sec

(a) Since voltage and frequency are reduced by 40%, the maximum torque will be

$$T_{max} \cong \frac{3}{2 \times 0.6\,\omega_s} \frac{(0.6 \times 440)^2}{[0.2^2 + 0.6^2 \times (1.16)^2]^{1/2}} = 1800 \text{ Nm}$$

(b) With normal supply and frequency, the slip for load torque 1800 Nm will be

$$1800 = \frac{3}{104.72} \times \frac{400^2}{(0.2 + 0.18/s)^2 + 1.16^2} \times \frac{0.18}{s}$$

Simplifying it, we get, $\dfrac{1}{s^2} - \dfrac{214.9}{s} + 42.77 = 0$

The smaller value of slip (machine operate at smaller value of slip) = 0.0907.
The rotor speed = 1000 (1 − 0.0907) = 909 rpm

Power developed $= 2\pi \times \dfrac{909}{60} \times 1800 = 171.3$ kW = 230 hp

Slip at maximum torque $= \dfrac{0.18}{\sqrt{(0.2^2 + 1.16^2)}} = 0.1529$

Thus speed at maximum torque = 1000 (1 − 0.1529) = 847 rpm

(c) Starting torque $= T_{st} = \dfrac{1}{\omega_s} \dfrac{V_1^2}{(r_1 + r_2')^2 + (x_1 + x_2')^2} \cdot r_2'$ at normal voltage and frequency.

The starting torque at half voltage and half frequency of normal supply will be

$$T_{st-new} = \frac{3}{0.5\omega_s} \frac{(0.5\,V_1)^2}{(r_1 + r_2')^2 + 0.5^2(x_1 + x_2')^2} \cdot r_2'$$

Thus, $\dfrac{T_{st-new}}{T_{st}} = \dfrac{1}{0.5} \times \dfrac{0.25 \times (r_1 + r_2')^2 + (x_1 + x_2')^2}{(r_1 + r_2')^2 + 0.5^2 (x_1 + x_2')^2} = 0.5 \times \dfrac{(0.38^2 + 1.16^2)}{(0.38^2 + 0.25 \times 1.16^2)}$

or $\dfrac{T_{st-new}}{T_{st}} = 1.55$

Hence, the starting torque is increased by 1.55 times the normal starting current.

PROBLEMS

13.1 A three-phase, 440 V, 50 hp, 50 Hz induction motor runs at 1450 rpm when it delivers rated output power. Determine

(a) Number of poles in the machine.

(b) Speed of rotating air gap field.

(c) Frequency of the rotor current.

(d) Speed of the rotor field relative to the rotor structure, stator structure, stator rotating field.

(e) Rotor induced voltage if stator to rotor turns ratio is 1: 0.80. Assume the winding factors are the same.

> [**Ans.** (a) $P = 4$, (b) Rotating field speed = 1500 rpm, (c) $s = 0.01$, $f_z = 150$ Hz,
> (d) (i) Rotar structure = 50 rpm, (ii) Stator structure = 1500 rpm (iii) Stator
> rotating field = 0, (e) $V_1 = 254.01$ V, $E_2 = 203.25$ V, At $S = 0.08$; $E_2 = 6.1$ V]

13.2 A three-phase 30 hp, 420 V, 6 pole, 50 Hz induction motor runs at full-load with a slip of 0.05. The windage and friction loss of motor is 850 W. Determine

(a) The air gap power and mechanical power developed.

(b) Rotor copper loss.

(c) The mechanical developed torque, and output torque.

> [**Ans.** (a) Mechanical power developed = 23230 Watt, Air gap power = 24452.6 Watt
> (b) Rotor copper loss = 1222.6 Watt, (c) Mechanical torque
> developed = 233.51 N-m, O/P torque = 224.96 N-m]

13.3 The following test results are obtained from a three-phase, 50 hp, 3300 V, 8 pole, 50 Hz squirrel cage induction motor.

(a) No-load test:

    Supply frequency = 50 Hz
    Line voltage   = 3300 V
    Line current   = 6.8 A
    Input power   = 2.4 kW

(b) Blocked rotor test:

    Frequency   = 20 Hz
    Line voltage = 350 V
    Line current = 20 A
    Input power = 10.2 kW

(c) Average dc resistance per stator phase: $R_1 = 3.8$ ohm.

Determine the parameters of the equivalent circuit, and no-load rotational loss.

[**Ans.** (a) No load: rotational loss = 1872.87 W, R_{nl} = 17.30 ohm ,
Z_{nl} = 280.18 ohm, $x_1 + X_m$ = 279.65 ohm (b) From block-rotor test,
$r_2' = 4.7$ ohm, $x_1 = x_2' = 6.83$ ohm]

13.4 Using the parameters obtained in problem 13.3 determine

(a) Starting current when started direct on full voltage and starting torque.

(b) Full load current, full load power factor, full load torque at slip of 0.05.

(c) Slip at which maximum torque is developed and maximum torque developed.

(d) External resistance required in the rotor circuit so that maximum torque occurs at start.

[**Ans.** (a) T_{st} = 2457.12 N-m, (b) T_{fl} = 425.266 N-m,
(c) T_{max} = 2.03 × 10³ N-m, (d) R = 9.312]

13.5 A three-phase, 400 V, 50 Hz, four pole wound rotor induction motor drives a constant load of 120 N-m at a slip of 0.05 when rotor terminals are short circuited. An external resistance of 0.5 ohm/phase is inserted in rotor circuit, determine the speed of motor if the rotor winding resistance per phase is 0.3 ohm. Neglect rotational losses. The stator to rotor turns ratio is unity.

[**Ans.** Speed of motor = 1300 rpm]

13.6 A three-phase induction motor draws one fourth of the current as it draws at starting. If the starting torque is twice of the full-load torque,

(a) Find the full-load slip.

(b) Determine the slip at which motor develops maximum torque.

(c) Determine the maximum torque developed by motor as percent of full-load torque.

[**Ans.** (a) Full-load slip = 0.125, (b) s_{Tmax} = 0.559, (c) T_{max} = 235% T_{fl}]

13.7 If the r_{th} is small, show that

(a) $\dfrac{I_{2T\,max}'}{I_{2start}'} = \sqrt{0.5[s_{T\,max}^2 + 1]}$

(b) $\dfrac{T_{start}}{T_{fl}} = \dfrac{1}{s_{fl}}\left(\dfrac{s_{T\,max}^2 + s_{fl}^2}{s_{T\,max}^2 + 1}\right)$.

13.8 A three-phase, 11 kV, 50 Hz, four pole induction motor is taking 40 kW at 0.8 power factor lagging when running at slip of 0.03. Stator copper loss is 400 watt and rotational loss is 1.6 kW. If stator core loss is negligible. Find

(a) Rotor copper loss

(b) Output power

(c) Efficiency

(d) Shaft torque.

[**Ans.** (a) 1188 W, (b) 36.812 kW, (c) 92.03%, (d) 241.6 Nm]

13.9 A three-phase, four pole, 415 V, 50 Hz star-connected induction motor has the following parameters per phase

$r_1 = 0.1\,\Omega$, $x_1 = 0.35\,\Omega$, $r_2' = 0.125\,\Omega$, $x_2' = 0.4\,\Omega$. Assume X_m is very high (infinity).

When motor operates at a slip of 2.5%, find

(a) The developed electromagnetic torque in N-m.
(b) The power output (in hp) of motor if frictional and windage losses are 1000 watts.
(c) The efficiency of motor.

[**Ans.** (a) 206.3 Nm, (b) 41.01 hp, (c) 92.56%]

13.10 Prove that in a three-phase induction motor, the ratio of maximum torque to starting torque is $(1 + K^2)/2K$ where K is the ratio of rotor resistance to rotor reactance. Neglect the stator impedance.

13.11 A three-phase induction motor has efficiency of 0.9 when mechanical output is 50 hp. At this load, the stator copper loss and rotor copper loss each equals the iron loss. The frictional and windage losses are constant and are one-third of the iron loss. Rotor iron loss is negligible. Estimate the slip at which the motor operates.

[**Ans.** 3.19%]

14

Three-Phase Synchronous Machines

14.1 INTRODUCTION

Synchronous machine is very widely used as generators compared to the motors. As its name, synchronous machines always run at synchronous speed $(= 120 f/p$ rpm) where f is the system frequency (in Hz) to which it is connected and p is the number of poles. Since, the poles must be in even numbers, the highest possible speed, for 50 Hz supply system, will be 3000 rpm (at $p = 2$). However, for 60 Hz supply system, it is 3600 rpm. When a synchronous machine is used as generator, it is called *synchronous generator* or *alternator*. All steam, hydro, nuclear and gas power plants use synchronous generators. Synchronous machines are also used for motoring operation where constant speed is required. Large power rating synchronous motors are used in industries for compression and pumping purpose. Small size synchronous motors are used in electric clocks and other applications where constant speed is required. The important features of synchronous machines (generators and motors) are given below.

(a) It operates at a constant speed (synchronous speed).
(b) It can operate at both lagging and leading power factors.
(c) It is doubly excited (it requires both ac and dc supply system for armature and field, respectively).
(d) It has two types of rotor constructions: salient poles (non-uniform air gap) and cylindrical rotor or non-salient poles (uniform air gap).

Unlike induction machine, synchronous machine can operate at lagging, leading and unity power factors. In induction machine, magnetizing current is required to establish flux in the air gap. This magnetizing current lags the voltage and therefore, induction machine always operates at lagging power factor. On the other hand, in synchronous machine, the total air gap flux is produced by dc source and there is no use of lagging current from ac system for production of air gap flux. If dc excitation is decreased, lagging reactive power will be drawn from ac source to aid magnetization and thus machine will operate at lagging power factor. If dc excitation is more, leading current is drawn from ac source to compensate (oppose) the magnetization and the machine will operate at a leading power factor. If a motor is operating at leading power factor at no-load, it is called *synchronous condenser*, which can work as variable inductor or capacitor.

To minimize the current flow in the armature, which reduces the copper loss, the voltage rating of the alternators is increased but it is also limited due to the cost and insulation loss of the material. In India, the maximum ratings of the alternators are 21 kV, 500 MW. The voltage ratings of other Indian thermal generator units are 11 kV for 50, 100 and 110 MW, and 15.56 kV for 200/210 MW units.

14.2 CONSTRUCTIONAL DETAILS

The stator of a synchronous machine does not differ from the stator of induction machine, however, rotor is different and can be of two types: salient-pole (with pole pieces projected) and non-salient pole (cylindrical rotor). The machines having salient poles and cylindrical poles are shown in Figure 14.1. DC field current is provided in the rotor winding and stator is connected to the ac system. The electrical equivalent is shown in Figure 14.2.

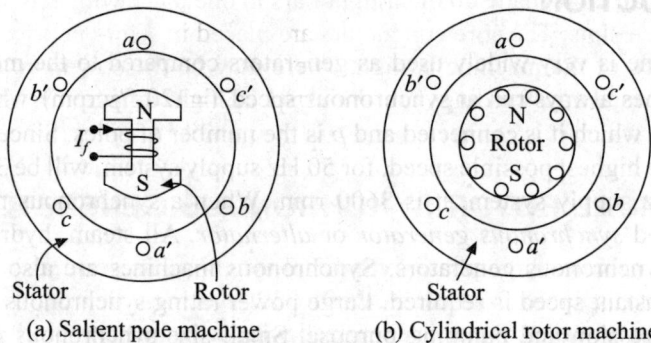

(a) Salient pole machine   (b) Cylindrical rotor machine

FIGURE 14.1   Salient and cylindrical rotor machines.

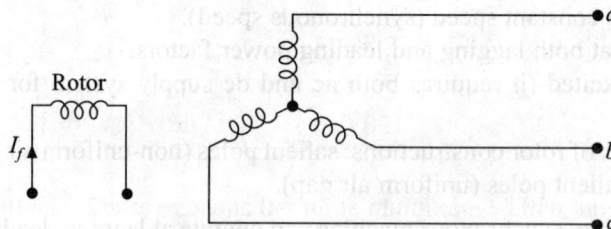

FIGURE 14.2   Electrical diagram.

Salient-pole synchronous machines are used for low speed application because of more centrifugal force on the rotor. Cylindrical synchronous machines are used for high-speed application. Steam power plants use cylindrical synchronous generators because at high speed, efficiency of steam turbines are higher. Reverse to this, hydropower plants use salient pole machine (simple in rotor construction) because hydropower turbines are more efficient at low speed.

Normally, field winding is put on the rotor and the armature winding on stator due to the following reasons:

(a) Keeping field winding on rotor requires only two brushes, whereas keeping field winding on stator will require three slip rings and brushes.

(b) The voltage and current rating of field voltage are small compared to the armature voltage and current and thus, requires less insulation.

(c) Cooling is better in stator than rotor as armature loss is more than field winding loss.

(d) The efficiency and reliability is more with field winding on rotor and armature winding on stator.

Since air gap in salient-pole machine is not uniform and thus analysis is different from cylindrical synchronous machine in which air gap is uniform. A synchronous machine is reversible, i.e. it can operate as a generator or as a motor, depending on whether its shaft is acted upon by a mechanical torque or counter-torque.

Normally concentric type or involute and distributed type windings are used for alternators. In concentric type windings, the straight bars are placed in series enclosed slots and separate end connectors are concentrically disposed. Yoke is extended at either end to protect the end windings. The both ends of each phase are brought out for differential protection. Involute and diamond windings consist of half coils made up of straight bars in one piece with two half-end connectors and are placed in open slots. The core conductors are placed in semi-enclosed slots, so designed that internal reactance of the machine is high. There may be three or more coils per slot with graded insulation. Concentric windings are preferred in high voltage machines due to several advantages.

14.3 CYLINDRICAL ROTOR SYNCHRONOUS GENERATOR

Figure 14.3 shows two-pole, three-phase cylindrical synchronous generator having three stator coils aa', bb', and cc' displaced from each other by 120 electrical degrees. Distributed winding produces sinusoidal *mmf* waves. Rotor is energized with dc source and produces an air gap flux of ϕ_f per pole and rotating at constant angular speed, $\omega (= 2\pi f)$ in electrical rad/sec. The angular speed $(= 2\pi n/60)$ in mechanical rad/sec is related as

$$\omega_m = \frac{2}{p}\omega \tag{14.1}$$

or

$$n = \frac{120f}{p} \quad \text{or} \quad f = \frac{np}{120} \tag{14.2}$$

The flux linkage in coil aa' will be maximum at $\omega t = 0$ and zero at $\pi/2$. The flux linkage with coil a (λ_a) having N turns will vary as

$$\lambda_a = N\phi_f \cos \omega t \tag{14.3}$$

and the voltage induced (e_a) in the coil aa' will be

$$e_a = -\frac{d\lambda}{dt} = \omega N\phi_f \sin \omega t = E_m \cos\left(\omega t - \frac{\pi}{2}\right) \tag{14.4}$$

where $E_m = \omega N\phi_f$

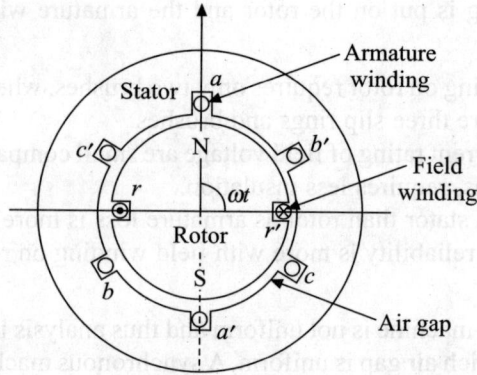

FIGURE 14.3   Three-phase cylindrical rotor synchronous machine.

If the current in phase a is lagging the e_a by an angle ψ, the instantaneous armature currents will be given by

$$i_a = I_m \sin(\omega t - \psi)$$

$$i_b = I_m \sin\left(\omega t - \psi - \frac{2\pi}{3}\right)$$

$$i_c = I_m \sin\left(\omega t - \psi + \frac{2\pi}{3}\right)$$

(14.5)

The spatial distribution of *mmf* wave of phase is shown in Figure 14.4. The *mmf* at any angle α with respect to centre of phase a, will be given by

$$F_a = Ki_a \cos \alpha$$

$$F_b = Ki_b \cos\left(a - \frac{2\pi}{3}\right)$$

$$F_c = Ki_c \cos\left(\alpha + \frac{2\pi}{3}\right)$$

(14.6)

where K is a constant.

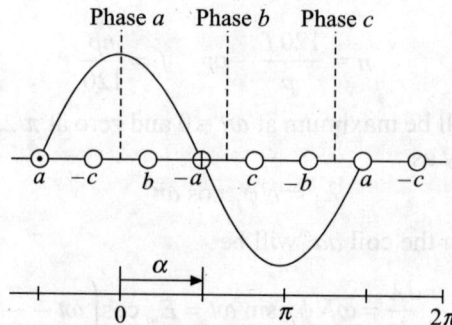

FIGURE 14.4   Spatial distribution of *mmf*.

The total *mmf* (F_T) will be

$$F_T = F_m \left(\begin{array}{l} \sin(\omega t - \psi) \times \cos\alpha + \sin\left(\omega t - \psi - \dfrac{2\pi}{3}\right) \times \cos\left(\alpha - \dfrac{2\pi}{3}\right) \\ + \sin\left(\omega t - \psi + \dfrac{2\pi}{3}\right) \times \cos\left(\alpha + \dfrac{2\pi}{3}\right) \end{array} \right) \qquad (14.7)$$

$$= \frac{3F_m}{2} \sin(\omega t - \psi - \alpha) \qquad (14.8)$$

where $F_m = KI_m$.

Equation (14.8) shows that the resultant *mmf* has constant amplitude and rotates at a constant speed and in synchronism with the field *mmf*. Thus, both the fluxes are rotating with the same speed but may have phase different and resultant flux or *mmf* can be obtained by taking the vector sum of these fluxes or *mmf*. The *rms* induced voltage, also called *excitation voltage,* is developed in stator winding having equal magnitudes and phase shift of 120° in each phase as winding are displaced by 120° in space.

Using Eq. (14.4), the root mean square (*rms*) value of voltage generated will be

$$E_f = \frac{2\pi f N \phi_f}{\sqrt{2}} = 4.44 f N \phi_f \qquad (14.9)$$

Since winding of each phase is distributed in different slots, therefore the *emf* induced in different slots is different. Their phasor sum is less, by a factor known as *winding factor* (k_w) than their algebraic sum. Thus, *rms* value of generated voltage will be

$$E_f = 4.44 \, k_w f N \phi_f \qquad (14.10)$$

or

$$E_f \propto n \times \phi_f$$

where *n* is rotor speed in rpm, ϕ_f is flux per pole due to dc field current I_f, k_w is winding factor, *N* is number of turns per phase.

It can be seen that for any rotor speed, the E_f is directly proportional to flux ϕ_f. If armature terminals are open circuited and the excitation voltage E_f is varied (vary I_f and thus ϕ_f), a characteristic, known as *open circuit characteristic (occ)* or *magnetization characteristic,* is obtained as shown in Figure 14.5.

Let three-phase balance load be connected to stator terminal, stator current will flow and will produce a rotating field at the same speed as rotor

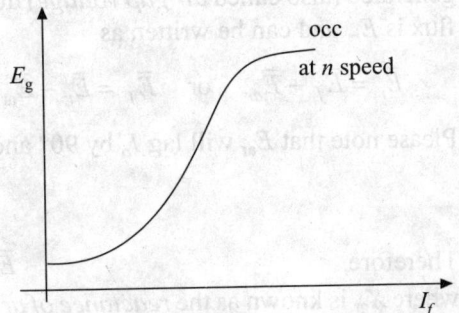

FIGURE 14.5   Magnetization characteristic.

mmf speed in the air gap. If ϕ_f is the field due to I_f and the flux produced by the armature currents I_a is ϕ_a (known as *armature flux*), thus resultant flux ϕ_r in air gap will be the phasor sum of these two fluxes as

$$\overline{\phi}_r = \overline{\phi}_f + \overline{\phi}_a \qquad (14.11)$$

The phasor diagram is shown in Figure 14.6, where θ is load power factor angle. Similarly *mmf* equation will be

$$\bar{F}_r = \bar{F}_f + \bar{F}_a \qquad (14.12)$$

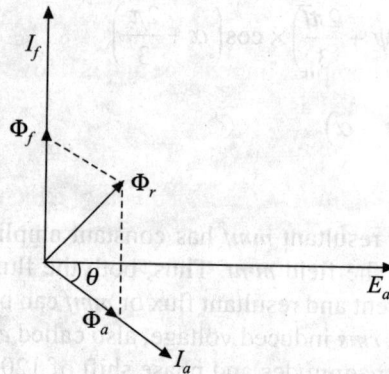

FIGURE 14.6   Phasor diagram of fluxes.

14.3.1 Equivalent Circuit

The flux, ϕ_a, produced by armature current I_a has two components: ϕ_{al} (leakage flux) and ϕ_{ar} (armature reaction) flux. The leakage flux does not link the air gap and rotor. It is confined to only stator. ϕ_{ar} is the major portion of ϕ_a that induces a voltage E_{ar} known as *armature reaction voltage* (as shown in Figure 14.7) which depends on I_a. Thus, net voltage generated (also called *air gap voltage*) due to air gap flux is E_r, and can be written as

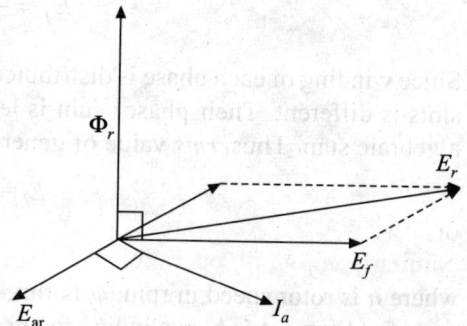

FIGURE 14.7   Phasor representation of armature reaction.

$$\bar{E}_r = \bar{E}_f + \bar{E}_{ar} \quad \text{or} \quad \bar{E}_f = \bar{E}_r - \bar{E}_{ar} \qquad (14.13)$$

Please note that E_{ar} will lag I_a by 90° and thus,

$$\bar{E}_{ar} = -j\bar{I}_a X_{ar} \qquad (14.13)$$

Therefore,

$$\bar{E}_f = j\bar{I}_a X_{ar} + \bar{E}_r \qquad (14.14)$$

where X_{ar} is known as the *reactance of armature reaction* or *magnetizing reactance*. Not all of the air gap voltage is not seen at the armature terminal. Some voltage is lost in the resistance of armature winding as $R_a\bar{I}_a$ and some more is lost in a reactance drop $jX_{al}\bar{I}_a$ where X_{al} is the leakage reactance due to ϕ_{al}. The net voltage at the terminal is then the voltage $\bar{E}_r$ minus the drops $R_a\bar{I}_a$ and $jX_{al}\bar{I}_a$. Thus,

$$\bar{V}_t = \bar{E}_r - \bar{I}_a R_a - jX_{al}\bar{I}_a \qquad (14.15)$$

From Eq. (14.14) and Eq. (14.15), we can have

$$\overline{E}_f = j\overline{I}_a X_{ar} + \overline{V}_t + \overline{I}_a R_a + jX_{al}\overline{I}_a$$

or

$$\overline{E}_f = \overline{V}_t + \overline{I}_a R_a + j\overline{I}_a X_{ar} + jX_{al}\overline{I}_a = \overline{V}_t + \overline{I}_a[R_a + j(X_{ar} + X_{al})] \qquad (14.16)$$

The coefficient of $\overline{I}_a$ is simply a complex impedance and is known as the *synchronous impedance* of the machine which can be denoted as

$$\overline{Z}_s = R_a + j(X_{ar} + X_{al}) = R_a + jX_s \qquad (14.17)$$

where X_s is known as the *synchronous reactance*, a quantity which is due to the leakage reactance and armature reaction. Thus the relation between open-circuit voltage (field excitation voltage) and terminal voltage, for generator, can be given as

$$\overline{V}_t = \overline{E}_f - \overline{I}_a(R_a + jX_s) \qquad (14.18)$$

The equivalent circuit is represented in Figure 14.8.

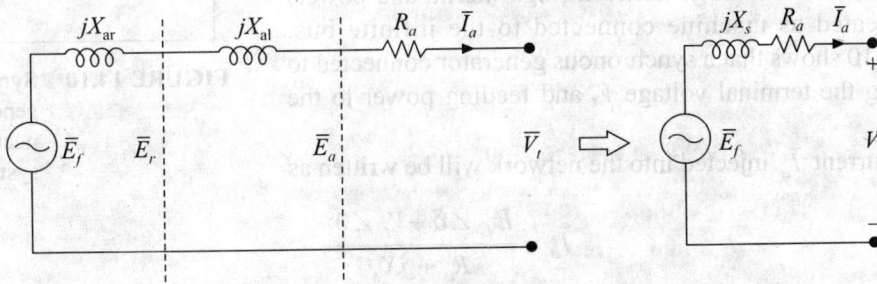

FIGURE 14.8 Synchronous generator equivalent circuit.

Similarly, for motor operation (I_a) is in reverse direction (entering into the positive terminal), the voltage equation can be written as

$$\overline{V}_t = \overline{E}_f + \overline{I}_a(R_a + jX_s) \qquad (14.19)$$

The phasor diagrams for motor and generator operations are shown in Figure 14.9. The angle between terminal voltage and generated *emf* of machine (E_f) is known as *load angle* or *torque angle* or *power angle* which is different from the power factor angle and voltage angle. Power factor angle is angle between terminal voltage and current (θ in this case). In the case of generator, the load angle is positive with reference to terminal voltage, i.e. V_t lags E_f whereas it is negative with reference to terminal voltage, i.e. V_t leads E_f, in case of motor.

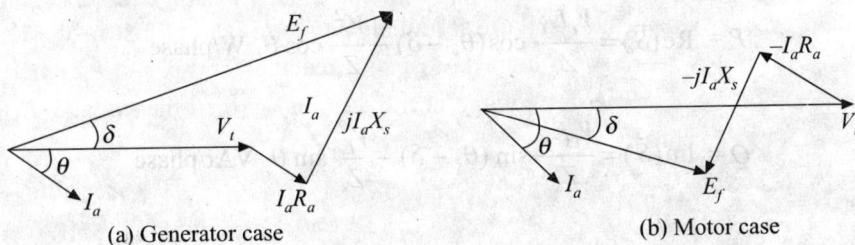

(a) Generator case   (b) Motor case

FIGURE 14.9 Phasor diagram of synchronous machine.

14.3.2 Power (Torque)-Angle Characteristic

Using the equivalent circuit, and taking terminal voltage as reference, other quantities can be written in phasor form as

$$\overline{E}_f = E_f \angle \delta$$

$$\overline{Z}_s = Z_s \angle \theta_s = R_a + jX_s$$

$$\overline{V}_t = V_t \angle 0°$$

Normally synchronous generators are connected to the bus having constant voltage and inject real and reactance power to the system at that bus, which is called *infinite bus*. This is called *infinite bus* because it can absorb or deliver (theoretically) any amount of power without any change in voltage and frequency. Ideally, there is no infinite bus in the system but a machine connected to a bus in large network (high inertia and power) can be treated as machine connected to the infinite bus. Figure 14.10 shows that a synchronous generator connected to bus having the terminal voltage V_t and feeding power to the system.

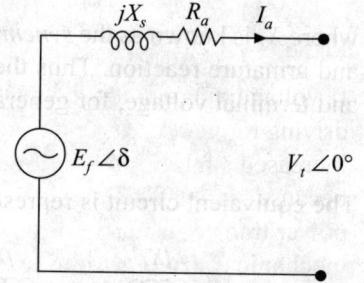

FIGURE 14.10 Synchronous generator connected to the system.

The current $\overline{I}_a$ injected into the network will be written as

$$\overline{I}_a = \frac{E_f \angle \delta - V_t \angle 0°}{R_a + jX_s} \tag{14.20}$$

The magnitude and phase angle of $\overline{I}_a$ vary widely depending upon the magnitude of $\overline{E}_f$ and the angle δ. The complex power (per phase) supplied to the system will be equal to the multiplication of phase voltage and phase current and can be written as

$$\overline{S} = P + jQ = \overline{V}_t \overline{I}_a^* = V_t \angle 0° \frac{[E_f \angle \delta - V_t \angle 0°]^*}{\overline{Z}_s^*} \tag{14.21}$$

or

$$\overline{S} = \frac{V_t E_f}{Z_s} \angle (\theta_s - \delta) - \frac{V_t^2}{Z_s} \angle \theta_s \text{ VA/phase} \tag{14.22}$$

Taking real and imaginary parts of Eq. (14.22), we get real power (P) and reactive power (Q) as

$$P = \text{Re}(\overline{S}) = \frac{V_t E_f}{Z_s} \cos(\theta_s - \delta) - \frac{V_t^2}{Z_s} \cos \theta_s \text{ W/phase} \tag{14.23}$$

and

$$Q = \text{Im}(\overline{S}) = \frac{V_t E_f}{Z_s} \sin(\theta_s - \delta) - \frac{V_t^2}{Z_s} \sin \theta_s \text{ VAr/phase} \tag{14.24}$$

For the sake of understanding, it is desirable to simplify expressions (14.23) and (14.24) by assuming that the impedance is purely inductive, i.e. $\theta_s = 90°$. This is true also as for many power system elements where impedance angle lie between 80° and 90° typically, i.e. R_a is very small. Thus, Eqs. (14.23) and (14.24) can be written as Eqs. (14.25) and (14.26).

$$P = \frac{V_t E_f}{X_s} \sin \delta = P_{max} \sin \delta \text{ W/phase} \tag{14.25}$$

and

$$Q = \frac{V_t E_f}{X_s} \cos \delta - \frac{V_t^2}{X_s} = \frac{V_t}{X_s}(E_f \cos \delta - V_t) \text{ VAr/phase} \tag{14.26}$$

If voltage magnitudes E_f and V_t are constant, the power angle can be varied by the mechanical driving torque. At $\delta = 0$, it can be seen from Eq. (14.25) that there is no power transfer. If angle is increased slightly, power flows and as δ increases power at first increases almost linearly with δ because $\sin \delta = \delta$ for small angles. At $\delta = \pi/2$, the power transfer is maximum. Increasing angle δ by mechanical driving torque will not increase the power and machine will start accelerating (generating the voltage at higher frequency than to the system frequency) which shows that machine has lost the synchronism. The power-angle curve is shown in Figure 14.11.

The plot of P vs. δ as shown in Figure 14.11 is known as *power-angle characteristic* of synchronous generator. The maximum power (P_{max}) from generator can be delivered when $\delta = 90°$. The value P_{max} is called the *steady-state stability limit* or *static stability limit*.

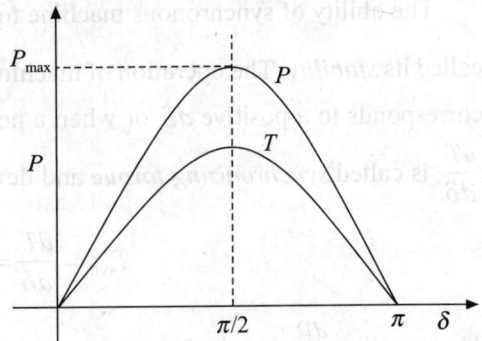

FIGURE 14.11 Power (torque)-angle curve.

If $\delta = 0$, or very small, $\cos \delta$ is nearly unity and Eq. (14.26) can be written as

$$Q \cong \frac{V_t E_f}{X_s} - \frac{V_t^2}{X_s} = \frac{V_t}{X_s}(E_f - V_t) \tag{14.27}$$

From Eq. (14.27), we can see that if $E_f = V_t$, there is no transfer of reactive power. It should be noted that even if power angle is zero, the reactive power will flow if $E_f \neq V_t$. At any power angle δ, if $E_f \cos \delta > V_t$, the reactive VAr delivered to the system is positive. This condition is said to be overexcited mode of operation of synchronous generator. The generator is called as *under-excited* ($E_f \cos \delta < V_t$) when it absorbs the reactive power from the system, i.e. the reactive VAr delivered to the system is negative. Thus, it can be summarized as

- The real power flow requires a phase angle difference between the voltages and there is a limit on the amount of power that can be transferred at a given voltage and impedance.
- The reactive power flow is dominated by the difference in voltage magnitude of two terminals. Positive Q flows from higher voltage to lower voltage end.

The three-phase real power and reactive power will be written as

$$P = 3 \frac{E_f V_t}{X_s} \sin \delta \qquad (14.28)$$

and

$$Q = 3 \frac{V_t}{X_s} (E_f \cos \delta - V_t) \qquad (14.29)$$

If the stator loss is negligible, the power developed at the terminal is also air gap power. The developed torque of machine will then be written as

$$T = \frac{P}{\omega_s} = \frac{3 V_t E_f}{\omega_s X_s} \sin \delta = T_{\max} \sin \delta \text{ Nm} \qquad (14.30)$$

where $\omega_s = \frac{2\pi n_s}{60}$ rad/sec and n_s is synchronous speed in rpm.

The ability of synchronous machine to keep in synchronism with the grid (power system) is called its *stability*. The operation of machine is stable if $\frac{dT}{d\delta} > 0$, which is true when a positive dT corresponds to a positive $d\delta$, or when a negative $d\delta$ corresponds to a negative dT. The quantity $\frac{dT}{d\delta}$ is called *synchronizing torque* and defined as

$$T_{\text{syn}} = \frac{dT}{d\delta} = \frac{3 V_t E_f}{\omega_s X_s} \cos \delta \text{ Nm/rad} \qquad (14.31)$$

The quantity $\frac{dP}{d\delta}$ is called *synchronizing power* and defined as

$$P_{\text{syn}} = \frac{dP}{d\delta} = \frac{3 V_t E_f}{X_s} \cos \delta \text{ Watt/rad} \qquad (14.32)$$

Any deviation of rotor from synchronism produces synchronizing force which tries to restore synchronous relation. This force is a kind of elastic link between machine and rest of the system. The synchronizing torque gradually decreases with increasing angle δ and reaches zero at $\delta = \pi/2$, i.e. when the electromagnetic torque of the machine attains it maximum value ($T_{\max}$). Maximum torque is the breakdown torque of the synchronous machine, beyond which stable operation is impossible. Typically the value of angle δ is kept near to 30° as beyond this value the synchronizing falls very rapidly and any small disturbance will lead to fall out of synchronism.

14.4 SALIENT-POLE SYNCHRONOUS GENERATOR

Figure 14.12 shows the schematic diagram of three-phase synchronous generator where air gap is not uniform and therefore reactance is non-uniform. The reluctance is high along the polar axis, also known as *direct axis* (d-axis) and is low along interpolar axis which is known as *quadrature axis* (q-axis). Thus reactance along the d-axis (X_d) is higher than quadrature axis reactance (X_q). If saliency is neglected, the d-axis reactance will be equal to synchronous reactance (X_s).

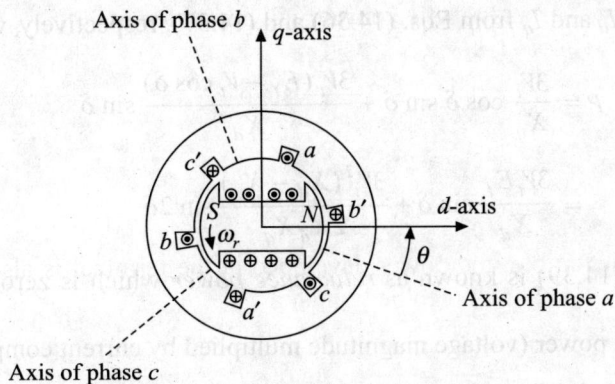

FIGURE 14.12   Three-phase salient-pole synchronous machine.

The phasor diagram with armature resistance neglected is shown in Figure 14.13.

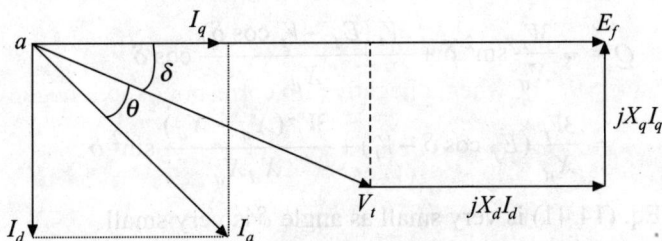

FIGURE 14.13   Phasor diagram of salient-pole synchronous generator.

From Figure 14.13, we can write the following relations

$$E_f = V_t \cos \delta + X_d I_d \tag{14.33}$$

$$X_q I_q = V_t \sin \delta \tag{14.34}$$

$$I_d = I_a \sin (\theta + \delta) \text{ and } I_q = I_a \cos (\theta + \delta) \tag{14.35}$$

From Eqs. (14.33) and (14.34), we get

$$I_d = \left(\frac{E_f - V_t \cos \delta}{X_d} \right) \tag{14.36}$$

$$I_q = \left(\frac{V_t \sin \delta}{X_q} \right) \tag{14.37}$$

From the phasor diagram, the real power generated will be given as (voltage magnitude multiplied by current component in phase with voltage),

$$P = 3V_t I_q \cos \delta + 3V_t I_d \cos(90 - \delta) = 3V_t(I_q \cos \delta + I_d \sin \delta) \tag{14.38}$$

Putting the values of I_d and I_q from Eqs. (14.36) and (14.37), respectively, we have

$$P = \frac{3V_t}{X_q} \cos \delta \sin \delta + \frac{3V_t(E_f - V_t \cos \delta)}{X_d} \sin \delta$$

$$= \frac{3V_t E_f}{X_d} \sin \delta + \frac{3V_t^2(X_d - X_q)}{2X_d X_q} \sin 2\delta \tag{14.39}$$

Second term of Eq. (14.39) is known as *reluctance power* which is zero in cylindrical rotor machine.

Similarly reactive power (voltage magnitude multiplied by current component in quadrature with voltage) will be

$$Q = -3V_t I_q \sin \delta + 3V_t I_d \sin(90 - \delta) = 3V_t(-I_q \sin \delta + I_d \cos \delta) \tag{14.40}$$

Putting the values of I_d and I_q from Eqs. (14.36) and (14.37) respectively, we have

$$Q = -\frac{3V_t}{X_q} \sin^2 \delta + \frac{3V_t(E_f - V_t \cos \delta)}{X_d} \cos \delta$$

$$= \frac{3V_t}{X_d}(E_f \cos \delta - V_t) + \frac{3V_t^2(X_q - X_d)}{X_d X_q} \sin^2 \delta \tag{14.41}$$

The second term of Eq. (14.41) is very small as angle δ is very small.

14.5   CAPABILITY CURVE OF SYNCHRONOUS GENERATOR

Synchronous generators are loaded according to operating chart which is governed by several factors. Synchronous generators are rated in terms of the maximum MVA output at a specified voltage and power factor which they can carry continuously without overheating. The continuous reactive power capability for a given rating is limited by three considerations: armature heating limit, field heating limit and end heating limit, whereas active power is limited by the prime mover capability.

14.5.1   Armature Heating Limit

Energy associated with the power loss $(I_a^2 R_a)$ in armature winding due to the flow of current must be removed to limit the rise in the temperature which can damage the insulation of the winding. Thus the generator rating is limited by maximum current that can flow in armature without exceeding the heating limitation. The rise in temperature is controlled by the cooling system of alternator. For any problem in cooling system, the maximum allowed current is reduced and thus reduced power generation. In p.u., the armature current and terminal voltage can be related with real and reactive power as,

$$P + jQ = \overline{V_t}\overline{I_a^*} = V_t I_a(\cos \theta + j \sin \theta) \tag{14.42}$$

This is an equation of circle in $P - Q$ plane with centre at the origin and radius equals to the MVA rating as shown in Figure 14.14.

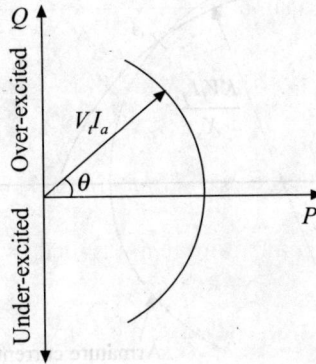

FIGURE 14.14   Armature current heating limit.

14.5.2 Field Heating Limit

Due to excitation current flow (I_f) in the field winding having resistance (R_f), $I_f^2 R_f$ power loss occurs and this causes field heating. The excitation voltage (E_f) is directly proportional to excitation current.

$$E_f = KI_f$$

Using Eqs. (14.28) and (14.29) in per unit, we get

$$P = \frac{KI_f V_t}{X_s} \sin \delta \tag{14.43}$$

$$Q = \frac{KV_t I_f}{X_s} \cos \delta - \frac{V_t^2}{X_s} \tag{14.44}$$

Equations (14.43) and (14.44) can be rearranged as follows

$$P^2 + \left(Q + \frac{V_t^2}{X_s} \right)^2 = \left(\frac{KV_t I_f}{X_s} \right)^2 \tag{14.45}$$

This is an equation of circle centered at $-\dfrac{V_t^2}{X_s}$ on Q-axis and $\dfrac{KV_t}{X_s} I_f$ as the radius. Figure 14.15 shows the maximum field heating on the capability of machine. Point A is the intersection of field current limit and armature current limit and represents the machine name plate rating MVA and power factor at the standard cooling.

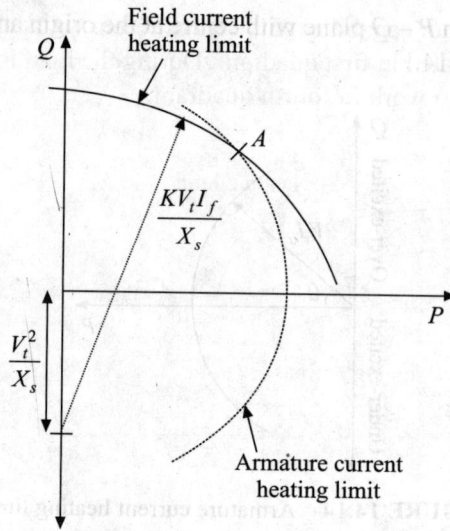

FIGURE 14.15   Field current heating limit.

14.5.3   End Region Heating Limit

Due to more end-turn leakage flux, more eddy current is produced in the stator laminations and thus it causes more localized heating in end-region. The high field current corresponding to the over-excited condition keeps the retaining ring saturated so that end leakage flux is small. In the under-excited condition, however, field current is low and the retaining ring is not saturated and this permits an increase in armature end-leakage flux. Moreover, in under-excited region, the flux produced by the armature current adds to the flux produced by the field current. Therefore, end-flux enhances the axial flux in the end region and resulting in a severe heating which severely affects the output of the generator. Figure 14.16 shows the end-region heating limit of generator.

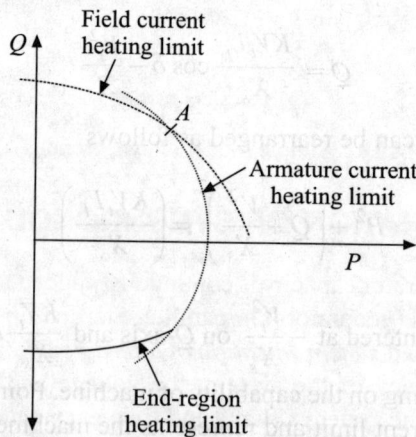

FIGURE 14.16   End-region heating limit.

Figure14.17 shows the generator capability curve for different voltages a normal cooling pressure. Normally generators are operated in first quadrant. During the light load conditions or during the black start, they are allowed to work in fourth quadrant.

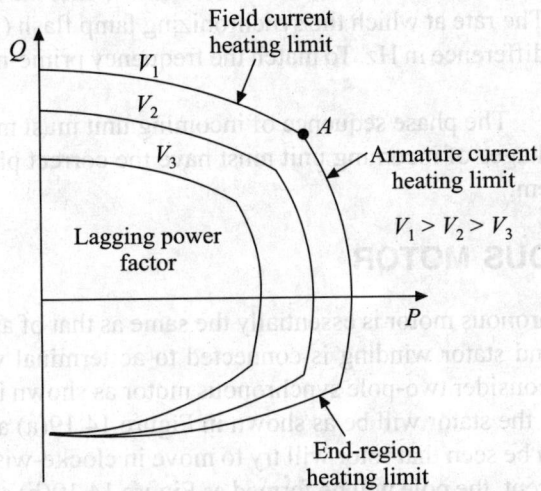

FIGURE 14.17   Generator capability curves.

For analysis purposes, synchronous machine is modelled in three timeframes: subtransient, transient and steady state. The transient and subtransient models assume constant rotor flux linkages, and the steady-state model assumes constant field current. Figure 14.18 shows the simple models of synchronous machine.

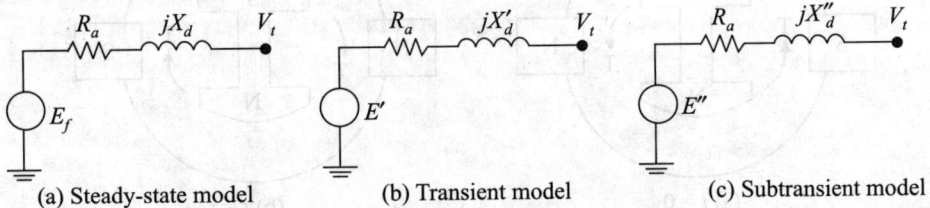

(a) Steady-state model   (b) Transient model   (c) Subtransient model

FIGURE 14.18   Simple model of synchronous machine.

14.6 PARALLEL OPERATION OF SYNCHRONOUS GENERATOR

Due to increasing demand, generator capacity is continuously added to the system. Then it becomes necessary to operate alternators (generators) in parallel. Starting up an alternator and connecting it to the existing network is called *synchronization of alternator*. Since it is impractical to change the supply voltage and frequency of the system, the voltage and frequency of synchronizing alternator is adjusted to match the system voltage and frequency. Once a unit is to be connected to the main grid, the following quantities must be the same of generator and grid.

1. **Voltage:**   The voltage of the incoming alternator must match that of the system voltage. If it is not done so, a voltage disturbance will be created on the system when switch

between alternator and grid is closed. The voltage adjustment of alternator is done using the field excitation control.

2. **Frequency:** The frequency of the incoming generator must match the system frequency (or very nearly so). To match this, a device used is known as *synchronizing lamp* (synchroscope). The rate at which the synchronizing lamp flash (flashes/second) is equal to the frequency difference in Hz. To match the frequency prime-mover control (governor control) is used.

3 **Phase sequence:** The phase sequence of incoming unit must match that of the system.

4. **Phase:** The voltages of incoming unit must have the correct phase position relative to those of the system.

14.7 SYNCHRONOUS MOTOR

The construction of synchronous motor is essentially the same as that of an alternator. Rotor field is excited by dc source and stator winding is connected to ac terminal which will produce the poles at the stator. Let us consider two-pole synchronous motor as shown in Figure 14.19. At time $t = 0$, the poles formed on the stator will be as shown in Figure 14.19(a) and will be the same till the current reverses. It can be seen that rotor will try to move in clocke-wise direction. During the negative half-cycle of current, the pole will be formed as Figure 14.19(b) as opposite to the Figure 14.19(a) and produced torque will try to move the rotor in anti-clock-wise direction. Thus, the produced torque is pulsating and the average torque will be zero. It shows that motor will not start and thus synchronous machines are not a self-starting motor.

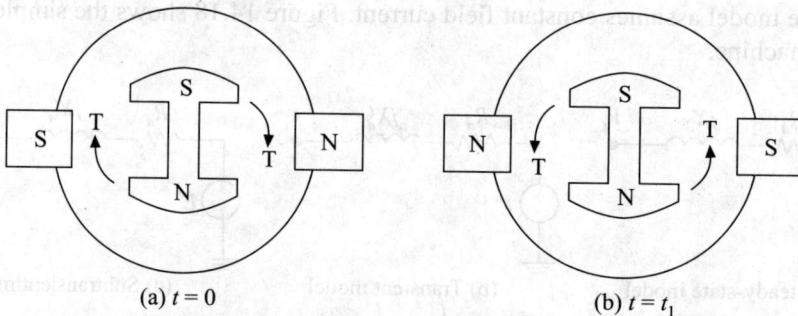

(a) $t = 0$ (b) $t = t_1$

FIGURE 14.19 Synchronous motor operation.

Thus, the most synchronous motors start with their damper windings. The rotating field induces current in the damper winding and so torque is developed on the rotor, tending to move along with the magnetic field. If the motor does not have damper windings, it is to be driven by some external means at nearly synchronous speed and then field is connected and rotor is locked with the rotating magnetic field and a steady torque is developed on the rotor.

In motor action, the armature current is reversed as shown in Figure 14.20 where excitation voltage lags behind the terminal voltage.

The current $\bar{I}_a$ drawn from the network can be written as

$$\bar{I}_a = \frac{V_t \angle 0° - E_f \angle -\delta}{R_a + jX_s} \tag{14.46}$$

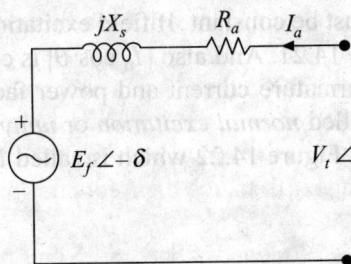

FIGURE 14.20   Synchronous motor connected to the system.

The complex power (per phase) drawn from the system will be equal to the multiplication of phase voltage and phase current and can be written as

$$\overline{S} = P + jQ = \overline{V}_t \overline{I}_a^* = V_t \angle 0° \frac{[V_t \angle 0° - E_f \angle -\delta]^*}{\overline{Z}_s^*} \tag{14.47}$$

or

$$\overline{S} = \frac{V_t^2}{Z_s} \angle \theta_s - \frac{V_t E_f}{Z_s} \angle (\theta_s + \delta) \text{ VA/phase} \tag{14.48}$$

Taking real and imaginary parts of Eq. (14.48), we get real power (P) and reactive power (Q) as

$$P = \text{Re}(\overline{S}) = \frac{V_t^2}{Z_s} \cos \theta_s - \frac{V_t E_f}{Z_s} \cos (\theta_s + \delta) \text{ W/phase} \tag{14.49}$$

and

$$Q = \text{Im}(\overline{S}) = \frac{V_t^2}{Z_s} \sin \theta_s - \frac{V_t E_f}{Z_s} \sin (\theta_s + \delta) \text{ VA/phase} \tag{14.50}$$

If $_aR$ is very small i.e. $\theta_s = 90°$. Thus, Eqs. (14.49) and (14.50) can be written as[*]

$$P = \frac{V_t E_f}{X_s} \sin \delta = P_{max} \sin \delta \text{ W/phase} \tag{14.51}$$

and

$$Q = -\frac{V_t E_f}{X_s} \cos \delta - \frac{V_t^2}{X_s} = \frac{V_t}{X_s} (V_t - E_f \cos \delta) \text{ VAr/phase} \tag{14.52}$$

If $E_f \cos \delta > V_t$ (called *overexcited motor*), the reactive VAr drawn from the system is negative, i.e. motor provides reactive power to the system. The motor is called under-excited ($E_f \cos \delta < V_t$) when it absorbs the reactive power from the system.

14.8   V-CURVE OF SYNCHRONOUS MOTOR

Let us consider a motor connected to ac supply system at terminal voltage V_t which is constant and drawing armature current $\overline{I}_a$. Thus, real power (P) taken by the motor will be

$$P = V_t I_a \cos \theta = \frac{E_f V_t}{X_s} \sin \delta \tag{14.53}$$

---

[*] Please note that P is the real power per phase. E_f and V_t are the phase voltages (line-to-neutral).

For constant power, the $E_f \sin \delta$ must be constant. If field excitation is changed, the E_f and power angle will vary as shown in Figure 14.21. And also $|I_a \cos \theta|$ is constant for constant power and thus with varying excitation, the armature current and power factor will vary. The value when armature current is minimum is called *normal excitation* or *unity power factor operation* which can be seen from Figure 14.21 and Figure 14.22 which is called *V*-curve as it looks like English letter V.

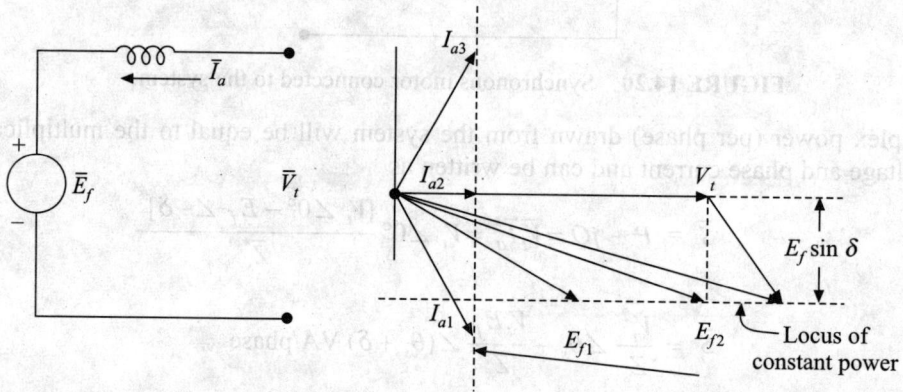

FIGURE 14.21   Locus of armature current and excitation voltage at different power factor.

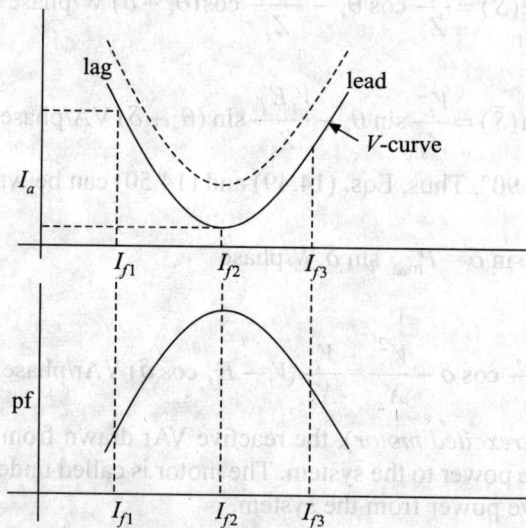

FIGURE 14.22   *V*-curve of synchronous motor.

The phasor diagrams with terminal voltage as reference for different excitation (power factors) corresponding to the lagging, unity and leading are shown in Figure 14.23.

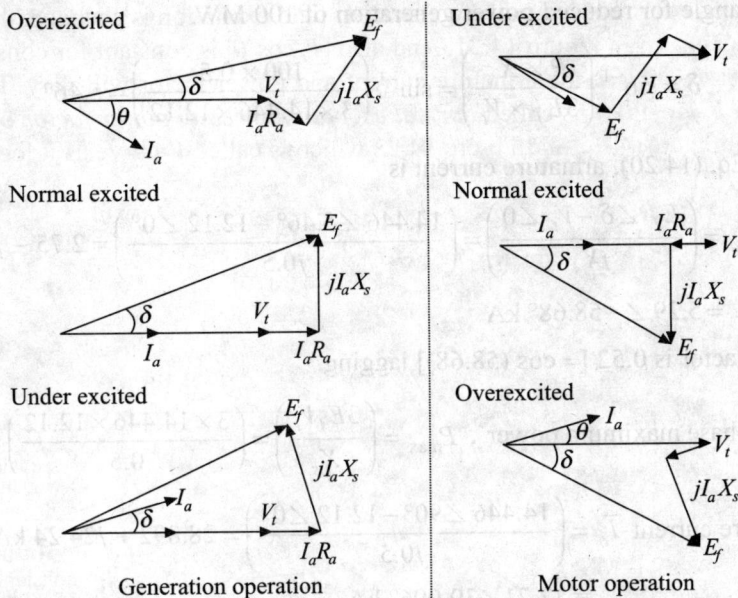

Overexcited

Normal excited

Under excited

Generation operation

Under excited

Normal excited

Overexcited

Motor operation

FIGURE 14.23 Phasor diagram of synchronous machine in generation and motor operation.

EXAMPLE 14.1 A 200 MW, 21 kV, three-phase, 50 Hz cylindrical rotor synchronous generator having synchronous reactance of 0.5 ohm/phase and negligible resistance is delivering power to an infinite bus at 0.8 power factor lagging.

(a) Find excitation voltage E_f and power angle (δ).
(b) Keeping this excitation fixed, the power generated is reduced to 100 MW, calculate the armature current and pf.
(c) Find the maximum power and armature current corresponding to E_f in part (a).

Solution: Taking terminal voltage as reference*,

Rated phase terminal voltage $\overline{V}_t = \dfrac{21}{\sqrt{3}} \angle 0° = 12.12 \angle 0°$ kV

Three-phase complex power, $\overline{S} = \dfrac{200}{0.8} \angle \cos^{-1}(0.8) = 200 + j\,150$ MVA

Rated phase current, $\overline{I}_a = \left(\dfrac{\overline{S}}{3\overline{V}_t}\right)^* = 5.498 - j4.124$ kV

(a) $\overline{E}_f = \overline{V}_t + jX_s\overline{I}_a = 12.12 + j0.5 \times (5.498 - j4.124) = 14.182 + j2.749$ kV

$= 14.446 \angle 10.97°$ kV (line-to neutral)

The excitation voltage (line-to-line) is 25.02 kV $(= \sqrt{3}\,E_f)$ and power angle is 10.97°.

---

*Calculation in per phase quantities is easier. In three-phase system, the voltage given is line-to-line voltage unless it is stated.

(b) Power angle for reduced power generation of 100 MW,

$$\delta = \sin^{-1}\left(\frac{P \times X_s}{3E_f \times V_t}\right) = \sin^{-1}\left(\frac{100 \times 0.5}{3 \times 14.446 \times 12.12}\right) = 5.46°$$

Using Eq. (14.20), armature current is

$$\bar{I}_a = \left(\frac{E_f \angle\delta - V_t \angle 0}{jX_s}\right) = \left(\frac{14.446 \angle 5.46° - 12.12 \angle 0°}{j0.5}\right) = 2.75 - j4.52 \text{ kA}$$

$$= 5.29 \angle -58.68° \text{ kA}$$

Power factor is 0.52 [= cos (58.68)] lagging.

(c) Three-phase maximum power*, $P_{max} = \left(\frac{3E_f V_t}{X_s}\right) = \left(\frac{3 \times 14.446 \times 12.12}{0.5}\right) = 1050.51 \text{ MW}$

Armature current $\bar{I}_a = \left(\frac{14.446 \angle 90° - 12.12 \angle 0°}{j0.5}\right) = 28.892 + j24.24 \text{ kA}$

$$= 37.71 \angle 39.996° \text{ kA}$$

Power factor is cos (39.996) = 0.766 leading.

14.9   VOLTAGE REGULATION AND EFFICIENCY

Voltage regulation of synchronous machine is defined as

$$\text{Voltage regulation (\%)} = \frac{\text{No load voltage} - \text{Full load volatge}}{\text{Full load voltage}} \times 100 \qquad (14.54)$$

Efficiency of synchronous machine is defined as

$$\text{Efficiency } (\eta) = \frac{\text{Output}}{\text{Input}} = \frac{\text{Output}}{\text{Output} + \text{Losses}} = \frac{\text{Input} - \text{Losses}}{\text{Input}} \qquad (14.55)$$

The various losses in the synchronous machine are stator and rotor copper losses, and iron losses which are eddy and hysteresis losses in the stator core. Iron losses are constant and independent on the load and excitation. Another loss is frictional and windage loss and is also known as *rotational loss*.

EXAMPLE 14.2   A three-phase, eight-pole, 6600 V, 50 Hz, star-connected synchronous motor has a synchronous impedance of 0.66+j 6.6 ohm/phase. When excited to give a generated *emf* of 4500 V per phase, it takes an input of 2500 kW.

(a) Calculate the electro-magnetic torque, the input current, power factor and load angle.
(b) If a motor is being operated at an input current of 180 A at unity power factor what will be the value of E_f? Under these conditions, also calculate the mechanical output and efficiency if total mechanical, excitation and iron losses is 50 kW.

*Maximum power will occur at $\delta = 90°$.

Solution: Given that $\bar{Z}_s = 0.66 + j6.6 = 6.63 \angle 84.29°$ ohm/phase, and phase voltage $\bar{V}_t = \dfrac{6600}{\sqrt{3}}$ V.

(a) Using Eq. (14.49) for electrical input of 2500 kW (three-phase power) as

$$\frac{2500000}{3} = \frac{(6600/\sqrt{3})^2}{6.63} \cos 84.29 - \frac{(6600/\sqrt{3}) \times 4500}{6.63} \cos(84.29 + \delta)$$

Solving above equation, we get $\delta = 19.46°$. Since it is a motor, the power angle will be negative with reference to the terminal voltage V_t.

$$\text{Input current, } \bar{I}_a = \frac{\dfrac{6600}{\sqrt{3}} \angle 0° - 4500 \angle -19.46}{0.66 + j6.6} = 218.7 + j87.3 = 235.5 \angle 21.8° \text{ A}$$

Power factor $= \cos(21.8) = 0.93$ leading (as angle is positive)

Stator loss $= 3I_a^2 R_a = 3 \times 235.5^2 \times 0.66 = 109811.30$ W

Air-gap power $= 2500000 - 109811.30 = 2390188.71$ W

Synchronous speed $= 2\pi \times \dfrac{2 \times 50}{8} = 25\pi$ rad/second

Hence electromagnetic torque = (Air gap power/synchronous speed)

$$= 2390188.11/25\pi = 30432.83 \text{ Nm}$$

(b) Given that $\bar{I}_a = 180 \angle 0°$ A (as power factor is unity, i.e. theta = 0)

$$\bar{E}_f = \bar{V}_t - \bar{I}_a \bar{Z}_s = \frac{6600}{\sqrt{3}} - (180 + j0) \times (0.66 + j6.6) = 3878 \angle -17.8° \text{ V}$$

Stator loss $= 3 \times 180^2 \times 0.66 = 64152$ W

Total loss $= 50000 + 64152 = 114152$ W

Input $= 3 \times \dfrac{6600}{\sqrt{3}} \times 180 \times 1 = 2057676.36$ W

Thus output = Input − Losses = 1943524.36 W

Thus, efficiency $= \dfrac{\text{Output}}{\text{Input}} \times 100 = \dfrac{1943524.36}{2057676.36} \times 100 = 94.45\%$.

EXAMPLE 14.3   If the excitation of a synchronous motor is so controlled that the power factor is unity, show that power is proportional to the tangent of power angle. Neglect all machine losses.

Solution: Let $\bar{I}_a = I_a \angle 0°, \bar{V}_t = V_t \angle 0°$ and $\bar{E}_f = E_f \angle -\delta$

The power $\bar{E}_f = \bar{V}_t - jX_s \bar{I}_a = V_t - jX_s I_a = |\bar{E}_f| \angle -\delta$

Thus, load angle or power angle $\delta = \tan^{-1}\left(\dfrac{X_s I_a}{V_t}\right)$

or $\qquad \tan \delta = \dfrac{X_s I_a}{V_t} = \dfrac{X_s P}{3 V_t^2} \qquad\qquad (\because P = 3 V_t I_a)$

Since, V_t and X_s are constant. Thus, $\tan \delta \propto P$.

EXAMPLE 14.4   An existing plant has a total load of 9000 kVA at 0.72 power factor (lagging). The management decided to add a compressor driven by a 1000 kW synchronous motor. Assuming 90% efficiency, find the new overall power factor of the plant if the motor is (a) operating at unity power factor and (b) operating at 0.8 pf leading.

Solution: Real power (P_p) of plant in kW = kVA × power factor = $9000 \times 0.72 = 6480$ kW

Reactive power (Q_p) of plant = $9000 \times \sin(\cos^{-1} 0.72) = 6246$ kVAr

Real power input to the motor $(P_m) = 1000/0.9 = 1111$ kW

(a) If motor operates at unity power factor, the reactive power (Q_m) is zero.

Thus, new real power $(P_T) = P_p + P_m = 6480 + 1111$ kW = 7591 kW and

New reactive power $(Q_T) = Q_p + Q_m = 6246 + 0 = 6246$ kVAr

The total apperant power $(|S_T|) = \sqrt{P_T^2 + Q_T^2}$ kVA

The new power factor = $P_T/|S_T| = 7591/(7591^2 + 6246^2) = 0.7722$ (lag)

(a) If motor operates at 0.8 pf leading.

Motor reactive power $(Q_m) = -1111 \times \tan(\cos^{-1} 0.8) = -833.25$ kVAr

Total reactive power $(Q_T) = 6246 - 833.25 = 5412.75$ kVAr

Thus new power factor = $7591/(7591^2 + 5412.75^2) = 0.8142$ (lag).

PROBLEMS

14.1 A three-phase, 11 kV, 250 hp, 50 Hz, Y-connected non-salient rotor synchronous motor has synchronous reactance of 10 ohm/phase. Power angle is 15° when it takes 150 kW power. Neglect the ohmic losses. Determine

(a) The excitation voltage per phase, E_f.
(b) The supply line current I_a.
(c) The supply power factor.
(d) Draw the phasor diagram.

[**Ans.** (a) 304.19 V/phase, (b) 605.75 $\angle -89.25°$ A, (c) 0.013 lagging]

14.2 In problem 14.1, if mechanical load is thrown off and all losses become negligible,

(a) Determine the new line current and supply power factor.

(b) Draw the phasor diagram.

(c) By what percent should the field current I_f be changed to minimize the line current?

[**Ans.** (a) 604.67 $\angle -90°$ A, zero (lagging), (c) 20.88% increase]

14.3 A plant load consisting of 850 kW at 0.6 lagging pf is supplied by a 1600 kVA transformer. An addition of 400 kW of power is required. If the additional load is carried by synchronous motor at 0.8 pf leading, will a new transformer be required? What will be the new plant pf?

[**Ans.** New transformer is required. pf = 0.832 (lagging)]

14.4 A 40 kVA, 220 V, three-phase cylindrical rotor generator required a field current of 2A for rated voltage at no load. With I_f= 2A and I_a = 100 $\angle -70°$ A, the terminal voltage is only 133 V. Draw the phasor diagram and evaluate the X_s.

[**Ans.** 0.5211 ohm/phase]

14.5 A three-phase, star-connected alternator has rating of 100 kVA, 440 V, 1000 rpm, 50 Hz. It has a resistance of 0.5 ohm/phase and a reactance of 2.0 ohm/phase. For the terminal voltage of 440 V and a 90 A line current, find the excitation voltage if the power factor is (a) unity, (b) 0.8 leading and (c) 0.8 lagging.

[**Ans.** (a) 349 V/phase, (b) 249.7 V/phase, (c) 414.8 V/phase]

14.6 In problem 14.5, find the real and reactive powers transferred to the system for each of the three cases. If armature resistance is neglected, find the maximum power to be transferred.

[**Ans.** (a) 132.969 kW, (b) 95.155 kW, (c) 158.088 kW]

14.7 A 3-Φ, 20 kVA, 208 V, four-pole, Y-connected synchronous machine has a synchronous reactance of X_s = 1.5 Ω/phase. The resistance of the stator winding is negligible. The machine is connected to a 3-Φ, 208 V infinite bus. Neglect rotational losses.

(a) The field current and the mechanical input power are adjusted so that the synchronous machine delivers 10 kW at 0.8 lagging power factor. Determine the excitation voltage (E_f) and the power angle (δ).

(b) The mechanical input power is kept constant but the field current is adjusted to make the power factor unity. Determine the percent change in the field current with respect to its value in part (a).

[**Ans.** (a) 156.9 V/phase, 15.4°, (b) 19%]

14.8 A 3-Φ, 2300 V, 60 Hz, 12-pole, Y-connected synchronous motor has 4.5 Ω/phase synchronous reactance and negligible stator winding resistance. The motor is connected to an infinite bus and draws 250 A at 0.8 power factor lagging. Neglect rotational losses.

(a) Determine the output power.

(b) Determine the power to which the motor can be loaded slowly without losing synchronism. Determine the torque, stator current, and supply power factor for this condition.

[**Ans.** (a) 796.743 kW, (b) 984.3 kW, 15666 Nm, 384.9 A, 0.642 lagging]

14.9 A 3-Φ, 3 MVA, 2300 V, 60 Hz, 10-pole, Y-connected cylindrical-rotor synchronous motor is connected to an infinite bus. The synchronous reactance is 0.8 pu. All the losses may be neglected. The synchronous motor delivers 3000 hp and the motor operates at 0.85 power factor leading.

(a) Determine the excitation voltage E_f.

(b) Determine the maximum power and torque and motor can deliver for the excitation current of part (a).

(c) The power output is kept constant at 3000 hp and the field current is decreased. By what factor can the field current of part (a) be reduced before synchronism is lost?

[**Ans.** (a) 1985.22 V/phase, (b) 5.61 MW, 0.0744×10^6 Nm, (c) 39.9% reduction]

14.10 A 1 MW, three-phase, 22 kV 50 Hz, star-connected synchronous motor has synchronous reactance of 250 ohm/phase. All the losses are negligible.

(i) Calculate the excitation voltage, power factor and current if power angle is 15° at rated load.

(ii) What is the minimum excitation voltage for the motor to deliver 800 kW without losing synchronism (stability)?

[**Ans.** (i) 43.91 kV, 0.486 lead, $53.96 \angle 60.89°$ A (ii) 9.09 kV]

15

Fractional kW (Horse-Power) Motors

15.1 INTRODUCTION

The majority of small motors are single-phase motors, which are extensively used for domestic applications. They provide motive power for fans, juicers, refrigerators, washing machines, etc. These motor are having low power rating, simple in construction and are economical. The most popularly used single-phase small motors are of induction type. The single-phase motor can be classified into three broad categories which are as follows:

(a) **Single-phase induction motors:**   There are several types of single-phase induction motors which are classified based on the starting methods such as
   - Resistance-start single-phase induction motor
   - Capacitor-start single-phase induction motor
   - Capacitor-run single-phase induction motor
   - Capacitor-start capacitor-run single-phase induction motor
   - Shaded pole single-phase induction motor

(b) **Single-phase series (or universal) motors:**   These motors are used for providing high starting torque and for high speed. These can be operated on dc supply or single-phase ac supply and widely used for kitchen appliances and vacuum cleaners, etc.

(c) **Single-phase synchronous motors:**   Since the applications of low power, constant speed motors are very limited, single-phase synchronous motors are suited for clocks, timer, turn-tables, etc. Since these motors neither use permanent magnets nor require dc field excitation, therefore they are simple in construction. Commonly used single-phase synchronous motors are:

   - Reluctant type
   - Hysteresis type

(d) Special motors for controlled movement: These motors are also known as *special motors* as these are used for special applications where continuous energy conversion is not required. Popularly used motors of this category are:

- Stepper motors
- Servomotors
- Synchro motors

Some of these motors are discussed in this chapter.

15.2   SINGLE-PHASE INDUCTION MOTORS

Normally, the single-phase induction motor consists of squirrel cage rotor (sometimes solid rotor) and wound stator. As the current in stator alternates, a pulsating air-gap *mmf*, hence a pulsating field is produced. This pulsating stator flux induces current in rotor circuit by transformer action. The rotor current also produces pulsating flux ϕ_r (if rotor is at stand-still) acting along the same axis as stator flux ϕ_s but opposing to each other as shown in Figure 15.1. Thus, there is no torque development at starting, hence single-phase induction motors are not self-starting and they are classified according to the methods used to start them such as resistance-start (split-phase), capacitor-start, capacitor-run, shaded-pole, etc.

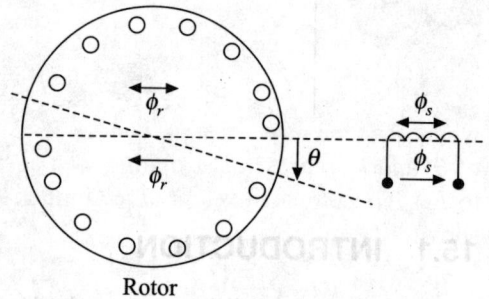

FIGURE 15.1   Single-phase induction motor fluxes.

A pulsating field (*mmf* a flux) can be considered as two revolving fields of equal magnitude but rotating in opposite direction at synchronous speed. It can be explained mathematically and graphically. Let us consider two vectors of equal magnitude **A** rotating in opposite directions as shown in Figure 15.2.

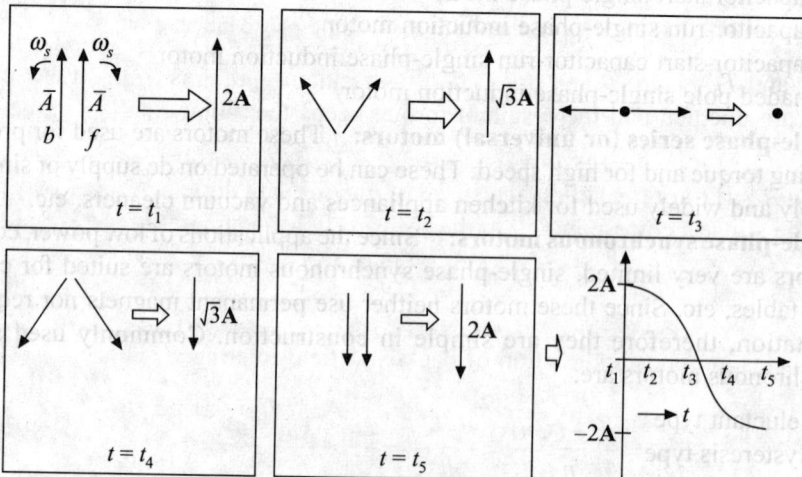

FIGURE 15.2   Pulsating and rotating field.

At time $t = t_1$, both vectors are in the same direction and thus, resultant is **2A**.

At $t = t_2$, both are forming 60° to each other and therefore, the resultant will be $\sqrt{3}$ **A**.

At $t = t_3$, both are opposing each other and resultant is zero.

Similarly, at $t = t_5$ the resultant is −2**A**. This shows that vector sum of two rotating field moving in opposite direction with the same speed is a pulsating in magnitudes between 2**A** and −2**A**. Thus, pulsating stator flux ϕ_s is equivalent to two rotating flux ϕ_b (moving backward, i.e. in the anti-clockwise) and ϕ_f (moving forward, i.e. in clockwise direction).

If N is number of turns and $i = I_m \cos \omega t$, the *mmf* along a position θ will be

$$F(\theta, t) = NI_m \cos \omega t \cdot \cos \theta$$

$$= \frac{NI_m}{2} \cos(\omega t - \theta) + \frac{NI_m}{2} \cos(\omega t + \theta)$$

$$= F_f + F_b \tag{15.1}$$

where F_f is the forward *mmf* rotating at synchronous speed in clockwise direction (in the direction of θ) and F_b represents a rotating *mmf* at synchronous speed in anticlockwise direction (opposite to F_s). These two fluxes produce induction motor torque in opposite directions and resultant is shown in Figure 15.3.

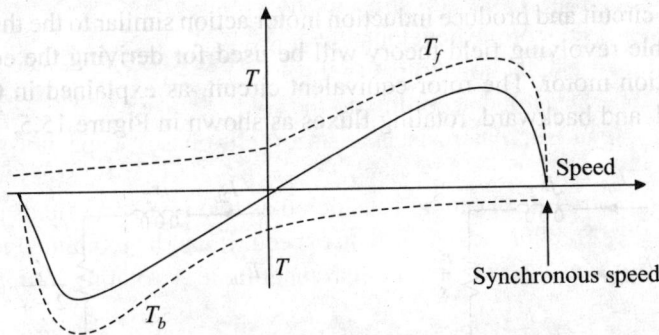

FIGURE 15.3 Resultant torque in single-phase induction motor.

At starting, T_f and T_b are equal and opposite, thus resultant torque is zero. At any other speeds, these torques are unequal and the resultant torque keeps the motor rotating in the direction of rotation.

Slip calculation: If rotor is rotating at a speed of n in forward direction (clockwise) as shown in Figure 15.4, the slip (s_f) corresponding to the forward field rotating at speed of n_s will be

$$s_f = \frac{n_s - n}{n_s} = s \tag{15.2}$$

Since rotor rotates in opposite direction of backward field rotating at speed of n_s, the backward slip (s_b) will be

$$s_b = \frac{n_s - (-n)}{n_s} = \frac{n_s + n}{n_s} = \frac{2n_s - (n_s - n)}{n_s} = 2 - s \tag{15.3}$$

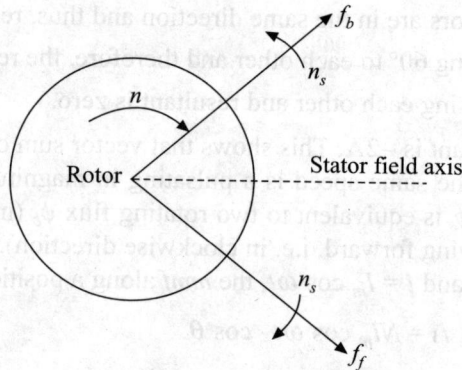

FIGURE 15.4   Rotation of forward and backward fields.

15.3   EQUIVALENT CIRCUIT OF SINGLE-PHASE INDUCTION MOTOR

When stator of a single-phase induction motor is connected to a single phase ac supply, the stator current produces a pulsating *mmf* which is equivalent to two constant amplitude *mmf* waves rotating in the opposite directions at synchronous speed as shown in Section 15.2. These two *mmfs* induce currents in the rotor circuit and produce induction motor action similar to the three-phase induction machine. This double revolving field theory will be used for deriving the equivalent circuit of single-phase induction motor. The rotor equivalent circuit, as explained in Chapter 13, can be derived for forward–and backward–rotating fluxes as shown in Figure 15.5.

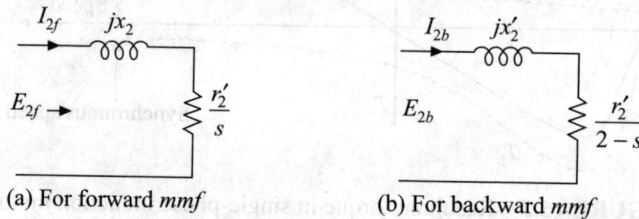

(a) For forward *mmf*   (b) For backward *mmf*

FIGURE 15.5   Rotor equivalent circuits at stator frequency.

where

r_2 = rotor winding resistance

x_2 = rotor winding leakage reactance

E_{2f} = rotor circuit induced voltage due to forward *mmf*

E_{2b} = rotor circuit induced voltage due to backward *mmf*

I_{2f} = rotor circuit current due to forward *mmf*

I_{2b} = rotor circuit current due to backward *mmf*

s = forward slip.

The equivalent circuit can be split into two-halves as per double-revolving field theory and shown in Figure 15.6.

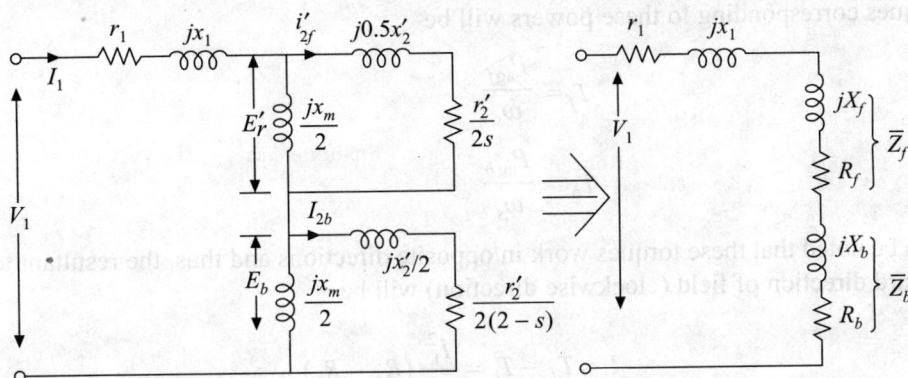

FIGURE 15.6 Equivalent circuit of single-phase induction motor.

where

r_1 = stator winding resistance

x_1 = stator winding leakage reactance

r_2' = rotor winding resistance referred to the stator

x_2' = rotor winding leakage reactance referred to the stator

E_f' = rotor circuit induced voltage due to forward *mmf* referred to the stator

E_b' = rotor circuit induced voltage due to backward *mmf* referred to the stator

I_{2f}' = rotor circuit current due to forward *mmf* referred to the stator

I_{2b}' = rotor circuit current due to backward *mmf* referred to the stator

V_1 = stator supply voltage (per phase)

I_1 = stator winding current.

The equivalent circuit can be used for analyzing the performance of induction motor by computing the various quantities such as stator current, input power, losses, developed torques, efficiency, etc. The equivalent circuit shown in Figure 15.6 is simplified and also given in Figure 15.6. The equivalent impedances, $\overline{Z}_f, \overline{Z}_b$ are calculated as

$$\overline{Z}_f = \left(j\frac{x_m}{2} \right) \left\| \left(j\frac{x_2'}{2} + \frac{r_2'}{2s} \right) \right. = R_f + jX_f \qquad (15.4)$$

$$\overline{Z}_b = \left(j\frac{x_m}{2} \right) \left\| \left(j\frac{x_2'}{2} + \frac{r_2'}{2(2-s)} \right) \right. = R_b + jX_b \qquad (15.5)$$

In running condition, $\dfrac{r_2'}{2s} > \dfrac{r_2'}{2(2-s)}$ and thus $\overline{Z}_f > \overline{Z}_b$ and $E_f > E_b$. The air gap power due to both forward and backward fields can be expressed as

$$P_{agf} = I_1^2 R_f \qquad (15.6)$$

$$P_{agb} = I_1^2 R_b \qquad (15.7)$$

The torques corresponding to these powers will be

$$T_f = \frac{P_{agf}}{\omega_s} \tag{15.8}$$

$$T_b = \frac{P_{agb}}{\omega_s} \tag{15.9}$$

It should be noted that these torques work in opposite directions and thus, the resultant torque in the forward direction of field (clockwise direction) will be

$$T = T_f - T_b = \frac{I_1^2}{\omega_S}(R_f - R_b) \tag{15.10}$$

The mechanical power developed is given by

$$P_{mech} = T \cdot \omega_m = T \cdot \omega_s(1-s) = I_1^2(R_f - R_b)(1-s) \tag{15.11}$$

$$= (P_{agf} - P_{agb})(1-s) \tag{15.12}$$

The power output from the motor will be mechanical power minus the rotational losses which include frictional and windage losses and core losses which are assumed to be constant. It can be expressed as

$$P_{out} = P_{mech} - P_{rotation} \tag{15.13}$$

The rotor current due to two revolving fields are produced at two frequencies and thus copper losses. The total rotor copper losses is the summation of the copper loss produced by the forward field and copper loss produced by the backward field. This can be written as

$$P_2 = sP_{agf} + (2-s)\,P_{agf} \tag{15.14}$$

Similarly, the total air gap power will be

$$P_{ag} = P_{agf} + P_{agb} \tag{15.15}$$

15.4 CLASSIFICATION OF SINGLE-PHASE INDUCTION MOTORS

Since single-phase induction motors are not self-starting, therefore some arrangements are made to make this motor self-starting. The most common practice is to provide some auxiliary winding on the stator in addition to the main winding to start as a two-phase motor. The different types of single-phase induction motors based on the auxiliary winding are explained below.

15.4.1 Split-Phase Induction Motors

A schematic diagram of split-phase single phase induction motor is shown in Figure 15.7. The two windings are displaced by 90° electrical degree in space. Due to resultant field, which is rotating, the torque is developed.

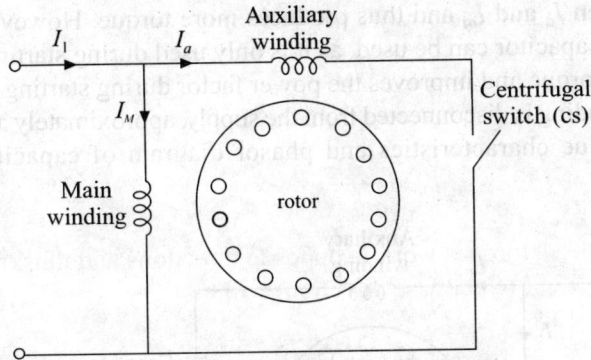

FIGURE 15.7   Split-phase single-phase induction motor.

To provide good starting torque, the resistance of auxiliary winding is kept high. Due to high resistance of auxiliary winding, it is disconnected after starting as single-phase induction motor, produces the torque while rotating. To disconnect auxiliary winding from the circuit, a centrifugal switch is provided which makes the circuit open normally at 75% of synchronous speed. This motor has low to moderate starting torque, which depends on the phase angle (θ) between main winding current (I_m) and auxiliary winding current (I_a) as shown in Figure 15.8. The speed-torque characteristic is shown Figure 15.9.

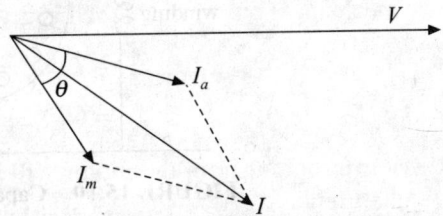

FIGURE 15.8   Phasor diagram of split-phase single-phase induction motor.

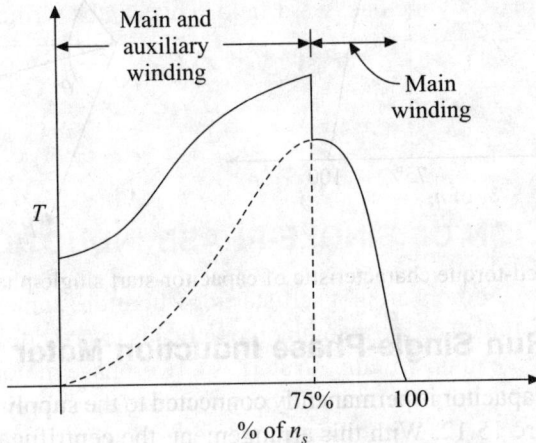

FIGURE 15.9   Speed-torque characteristic of split-phase single-phase induction motor.

15.4.2 Capacitor-Start Single-Phase Induction Motors

For various applications, high starting torque is required and it can be achieved if a capacitor is connected in series with auxiliary winding as shown in Figure 15.10. Addition of capacitor increases the phase angle between I_a and I_m and thus produces more torque. However, the capacitor adds extra cost but a single capacitor can be used, as it is only used during starting. This type of motor provides high starting torque and improves the power factor during starting period. The capacitor along with auxiliary winding is disconnected from the supply, approximately at 75% of synchronous speed. The speed-torque characteristics and phasor diagram of capacitor-start is shown in Figure 15.11.

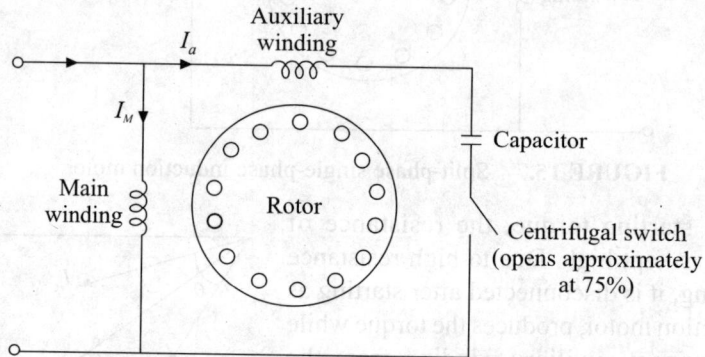

FIGURE 15.10   Capacitor-start single-phase induction motor.

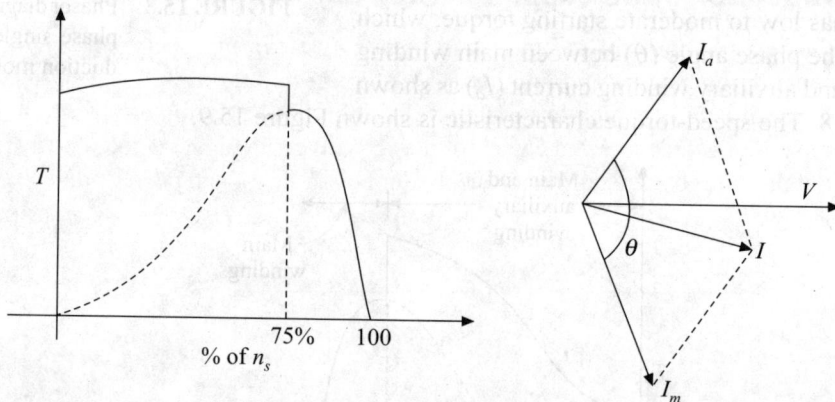

FIGURE 15.11   Speed-torque characteristic of capacitor-start single-phase induction motor.

15.4.3   Capacitor-Run Single-Phase Induction Motor

As its name suggests, the capacitor is permanently connected to the supply along with the auxiliary winding as shown in Figure 15.12. With this arrangement, the centrifugal switch is avoided and therefore the cost is reduced. Since the capacitor is permanently connected to the supply, a better quality capacitor is used. The speed-torque characteristic is shown in Figure 15.13. The main features of this type of motors are:

- Simple construction and less cost as centrifugal switch is absent.
- Performance (power factor, efficiency and pulsating torque) is improved as motor works as two-phase motor.

- Less noisy.
- High quality capacitor is used to improve the reliability.
- To compromise between starting and running values, the starting torque is reduced compared to capacitor-start and split-phase, single-phase inductor motors.

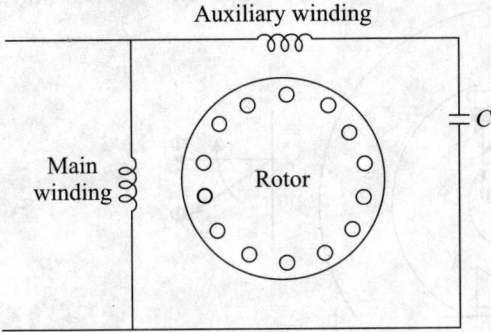

FIGURE 15.12   Capacitor-run single-phase induction motor.

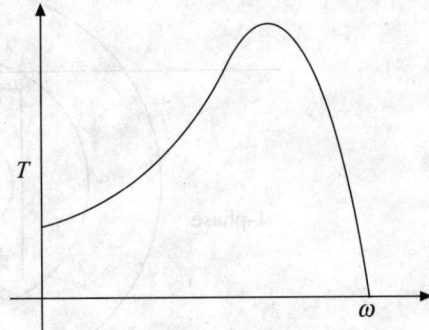

FIGURE 15.13   Speed-torque characteristic of capacitor-run single-phase induction motor.

15.4.4   Capacitor-Start Capacitor-Run Single-Phase Induction Motor

To provide the best running and starting performance, two capacitors are used out of which one (C_1) is permanently connected and another (C_2) is connected through centrifugal switch. After the starting, capacitor (C_2), which has longer value than C_1, is disconnected from the supply as shown in Figure 15.14. Due to extra capacitor along with centrifugal switch, the cost of motor is increased but it provides better starting torque, better power factor and better running performance.

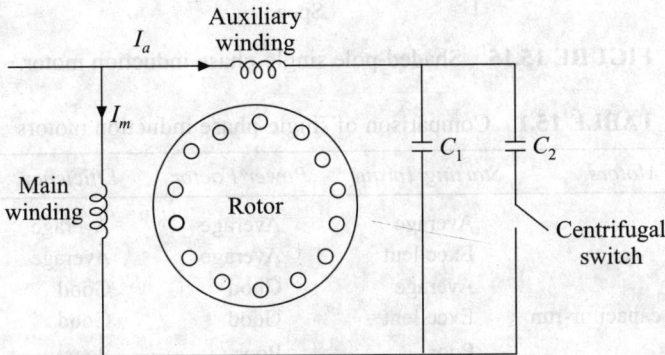

FIGURE 15.14   Capacitor-start capacitor-run single-phase induction motor.

15.4.5   Shaded Pole Single-Phase Induction Motors

These types of motors are used for very small rating (up to 0.05 hp). Stator has shaded poles as shown in Figure 15.15. The main winding is wounded on the salient poles. The current induced in shading band causes the flux in the shaded portion of pole to lag the flux produced due to the unshaded portion. The flux in shaded portion reaches its maximum value after the flux reaches its maximum value in unshaded portion of poles. This can be viewed as the rotating field moving from unshaded portion to shaded portion of poles. These types of motors are very simple and the cheapest one. Comparison of various single-phase induction motors are given in Table 15.1.

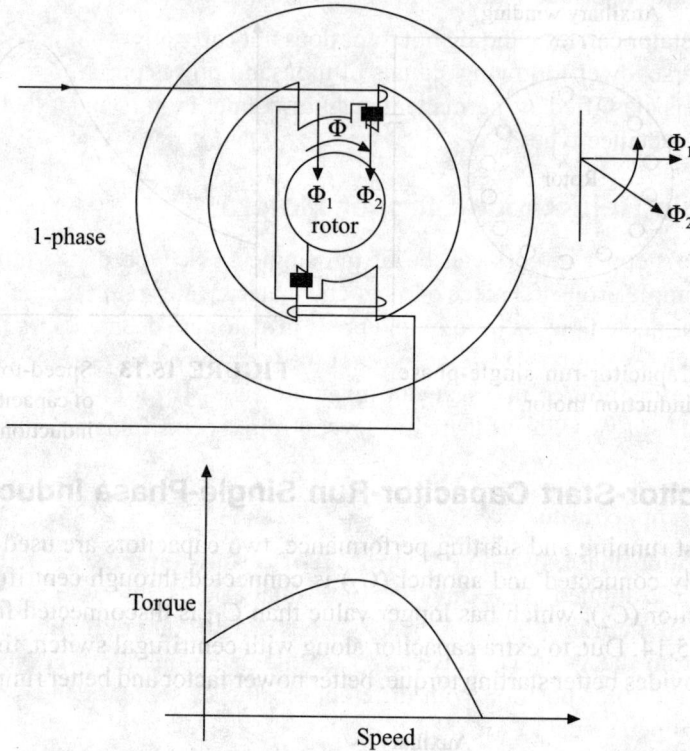

FIGURE 15.15   Shaded-pole single-phase induction motor.

TABLE 15.1   Comparison of single-phase induction motors

Types of Motors	Starting Torque	Power Factor	Efficiency	Cost
Split-phase	Average	Average	Average	Average
Capacitor-start	Excellent	Average	Average	Average-high
Capacitor-run	Average	Good	Good	High
Capacitor-start capacitor-run	Excellent	Good	Good	Very high
Shaded pole	Poor	Poor	Poor	Very cheap

15.5 STEPPER MOTORS

Stepper motor converts input information (digital) to analog output shaft motion in step sizes which can be 1°, 2° 2.5°, 5°, 7.5° and 15° for each electrical pulse. It is used in computers for driving paper feed mechanism in printers, in floppy disk/CD drives to provide precise positioning of magnetic head on disks, in plotters to drive x-y pens, in robotics, etc. It is also used in military, medical and control circuits (numerical controlled machines servo-motor, etc.). The advantages of stepper motor are as follows:

- It requires less maintenance.
- It is cheap (no position sensors or feedback mechanism).
- It offers natural interface with digital computers and circuit.

In stepper motors, stator carries winding but rotor does not carry any winding. Stepper motors can be up to several horsepower following signals of thousand pulses per seconds. Depending upon the type of rotor, it is classified as: permanent magnet stepper motors and reluctance type stepper motors (variable reluctance type).

15.5.1 Variable Reluctance Stepper Motors

Variable reluctance stepper motors can be of the single-stack type or the multiple stack type. Rotor consists of simple projected piece of magnetic material and stator wound for more than two poles. Number of projected pieces in rotor and poles in stator are decided based on the number of steps required in each revolution. The reluctance torque is produced because ferromagnetic material aligns itself in the direction of the resultant magnetic field. If on one shaft, only one rotor is provided, it is called *single-stack*. If there are several rotor stacks (magnetically isolated section) on the shaft, it is called *multi-stack*.

(a) Single-stack stepper motor

Basic circuit for a four-phase, two-pole stepper motor is shown in Figure 15.16. When switches are closed, the rotor is aligned in the direction of resultant fields. For example, when switch-A is closed, the rotor will align in yy' axis. When switches A and B are closed together, the resultant flux will be 45° from the yy' axis. This shows that 45° step operation in clockwise direction. Table 15.2 shows the rotor position with different closing of switches. 1 represents the closed switch and 0 is for open switch.

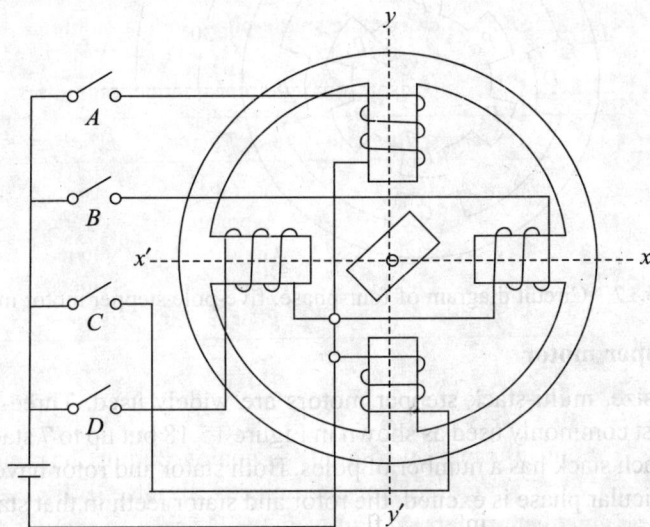

FIGURE 15.16   Circuit diagram of four-phase, two-pole stepper motor.

In general, the number of step sizes (in degree) can be calculated as

$$\text{Step size} = \frac{360°}{N \times x} \qquad (15.16)$$

where, N is number of phases and x is the number of rotor teeth.

TABLE 15.2   Rotor position with different closing of switches

Closed Switch(es)	A	B	C	D	Position from yy' Axis
A	1	0	0	0	0°
A + B	1	1	0	0	45°
B	0	1	0	0	90°
B + C	0	1	1	0	135°
C	0	0	1	0	180°
C + D	0	0	1	1	225°
D	0	0	0	1	270°
D + A	1	0	0	1	315°
A	1	0	0	0	360°, 0°

To get step size of 15°, the number of poles should be six for four-phase stepper motor as shown in Figure 15.17. When coil-A is excited, the pole P_1 will be aligned to A. When coils A and B are excited together, the magnetic axis will be 45° (between phase-A and phase-B). The nearest pole is P_2 which will move 15° in anti-clockwise direction to align the magnetic axis. Similarly, it can be understood for other coil excitements.

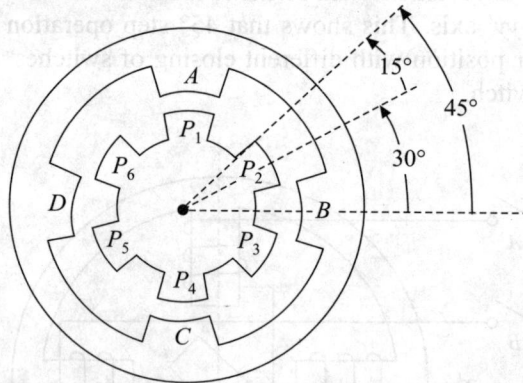

FIGURE 15.17   Circuit diagram of four-phase, five-pole stepper motor movement.

(b) Multi-stack stepper motor

To get smaller step size, multi-stack stepper motors are widely used. Three-phase, three-stack arrangements are most commonly used as shown in Figure 15.18 but up to 7 stacks and phases are available. Stator of each stack has a number of poles. Both stator and rotor have the same number of teeth. When a particular phase is excited, the rotor and stator teeth in that stack are aligned but those in other stacks are not aligned.

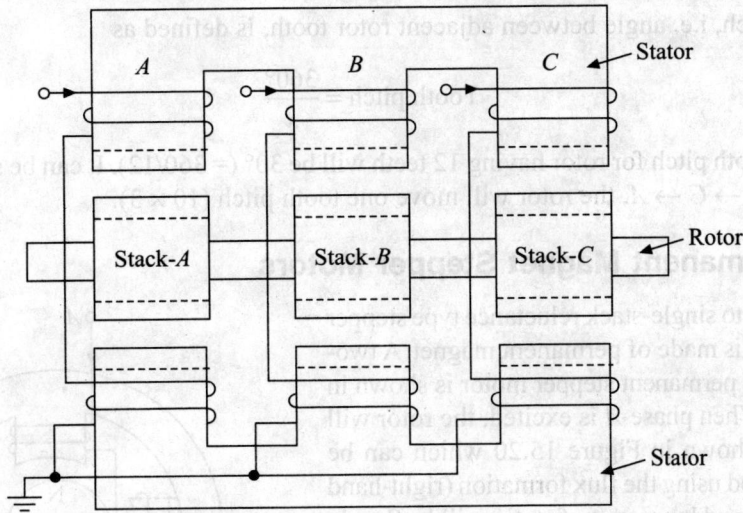

FIGURE 15.18   Multi-stack (three-stack, three-phase) stepper motor.

Figure 15.19 shows the rotor and stator teeth alignment when phase-A is excited. Since phase-A is excited, stack-A is aligned to the rotor teeth. The alignment angle between the stacks, also known as *step size*, is calculated as

$$\text{Step size} = \frac{360°}{xN} \qquad (15.17)$$

where N is the number of stacks or phases and x is the number of rotor teeth.

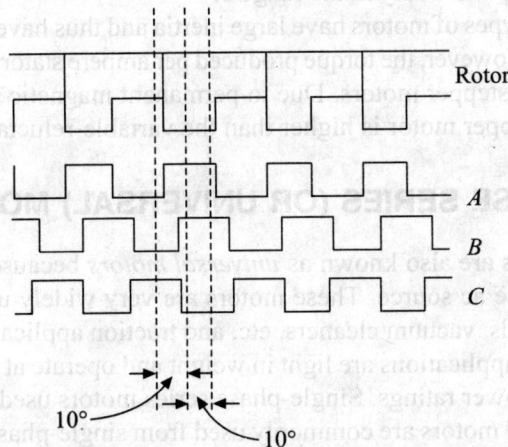

FIGURE 15.19   Rotor and stator teeth for phase-A excitation.

In Figure 15.19, there are 12 rotor teeth and thus the step size will be 10° (= 360/12 × 3). The number of steps per revolution (n) is calculated as

$$n = \frac{360}{\text{Step size}} = xN \qquad (15.18)$$

Rotor tooth pitch, i.e. angle between adjacent rotor tooth, is defined as

$$\text{Tooth pitch} = \frac{360°}{x} \qquad (15.19)$$

The value of tooth pitch for rotor having 12 teeth will be 30° (= 360/12). It can be seen that during process $A \rightarrow B \rightarrow C \rightarrow A$, the rotor will move one tooth pitch (10×3).

15.5.2   Permanent Magnet Stepper Motors

Stator is similar to single-stack reluctance type stepper motor but rotor is made of permanent magnet. A two-pole, two-phase permanent stepper motor is shown in Figure 15.20. When phase-A is excited, the rotor will be aligned as shown in Figure 15.20 which can be easily understood using the flux formation (right-hand rule) on the stator. Upper part of stator will be S-pole and bottom will be N-pole. If excitation is switched to phase-B, the rotor will move by 90° whereas the direction will be decided by the winding polarity and current direction. In this figure, it will move in the anti-clock direction, if current in phase-B flows from terminal B.

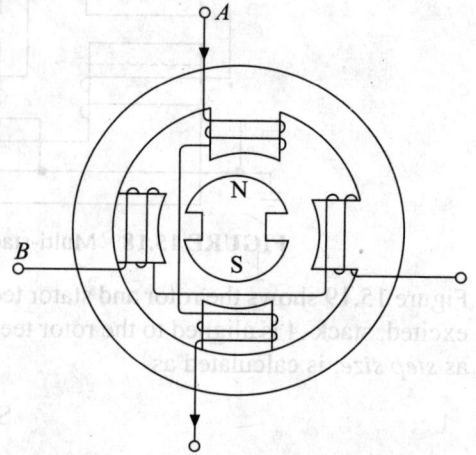

FIGURE 15.20   Two-pole permanent magnet stepper motor.

Normally permanent magnet stepper motors have less number of rotor poles due to difficulty in construction. Thus, the step sizes vary in the range of 30–90°. Moreover, these types of motors have large inertia and thus have low acceleration limiting to 300 pulses per second. However, the torque produced per ampere stator current is more, compared to the variable-reluctance stepper motors. Due to permanent magnetic rotor material, the cost of the permanent magnet stepper motor is higher than the variable-reluctance stepper motors.

15.6   SINGLE-PHASE SERIES (OR UNIVERSAL) MOTORS

Single-phase series motors are also known as *universal motors* because they can be used with a dc source or a single-phase ac source. These motors are very widely used domestically such as drills, mixers, portable tools, vacuum cleaners, etc. and traction applications. Single-phase series motors used for domestic applications are light in weight and operate at high speeds (1000 rpm to 10000 rpm) having low power ratings. Single-phase series motors used for traction purposes are of high capacity. Universal motors are commonly used from single-phase ac source as this source is easily available in domestic and commercial installations. Stator and rotor are made of laminated steel to reduce the core loss and eddy current.

Figure15.21 shows the single-phase series motor. Current i_a flowing through the field winding produces d-axis flux ϕ_d and produces q-axis flux ϕ_q due to the same current flowing in armature. ϕ_d and ϕ_q are in the same phase with i_a if the eddy current is neglected.

When this motor is connected to dc supply, it is similar to the dc series motor as discussed in Section 12.9.3. The developed torque and induced voltage can be written as

$$E_a = K_a' I_a \omega_m \qquad (15.20)$$

$$T = K_a \phi \cdot I_a = K_a' I_a^2 \qquad (15.21)$$

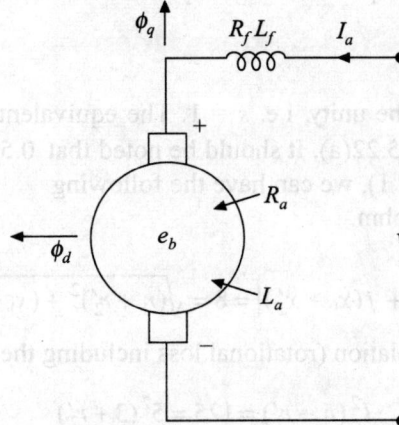

FIGURE 15.21 Single-phase series motor.

Let

$$i_a = I_{am} \cos \omega t \qquad (15.22)$$

$$\phi_d = \phi_{dm} \cos \omega t \qquad (15.23)$$

where I_{am} and ϕ_{dm} are the peak values of armature current and d-axis flux, respectively.

The instantaneous back *emf* and instantaneous torque will be given by

$$e_b = K_a \phi_d \omega_m = K_a \phi_{dm} \omega_m \cos \omega t \qquad (15.24)$$

$$T = K_a \phi_d i_a = K_a \phi_{dm} I_{am} \cos^2 \omega t$$

or

$$T = K_a \phi_{dm} I_{am} \left(1 + \cos 2\omega t\right) \qquad (15.25)$$

From Eqs. (15.22) and (15.24), it can be seen that e_a and i_a are in the same phase. The average torque will be

$$T_{\text{avg}} = K_a \phi_{dm} I_{am} = \frac{K_a \phi_d I_a}{2} \qquad (15.26)$$

where I_a is the *rms* value of current and $\Phi_d \left(= \phi_{dm}/\sqrt{2}\right)$ is the *rms* value of d-axis flux.

It should be noted that ac series motor produces pulsating torque but average torque due to ac source is positive and thus motor is self-starting. But ac excitation produces poor power factor and lowest speed due to drop in the field and armature winding reactances.

EXAMPLE 15.1 A single-phase, 0.5 kW, 220 V, 50 Hz, 1440 *rpm* induction motor has stator winding (main) resistance of 3 ohm. The following test data are obtained:

No-load test: 220 V, 3.5 A, 100 W
Block-rotor test: 40 V, 5.0 A, 125 W

Find

(a) Double revolving field equivalent circuit parameters. Assume $x_1 = x_2'$
(b) The rotational losses.

Solution:

(a) At standstill, slip will be unity, i.e. $s = 1$. The equivalent circuit will be derived from Figure 15.6 to Figure 15.22(a). It should be noted that $0.5r_2' + j0.5x_2' \ll j0.5x_m$. From the block rotor test ($s = 1$), we can have the following
Impedance = 40/5 = 8 ohm.
i.e.

$$\left| (r_1 + r_2') + j(x_1 + x_2') \right| = 8 = \sqrt{(r_1 + r_2')^2 + (x_1 + x_2')^2} \tag{i}$$

Using the power loss relation (rotational loss including the core loss is negligible). Thus

$$I^2(r_1 + r_2') = 125 = 5^2(3 + r_2') \tag{ii}$$

Solving Eqs. (15.27) and (15.28), we get

$$r_2' = 2 \text{ ohm} \quad \text{and} \quad x_1 + x_2' = 6.245 \text{ ohm}$$

Thus, $x_1 = x_2' = 3.123$ ohm

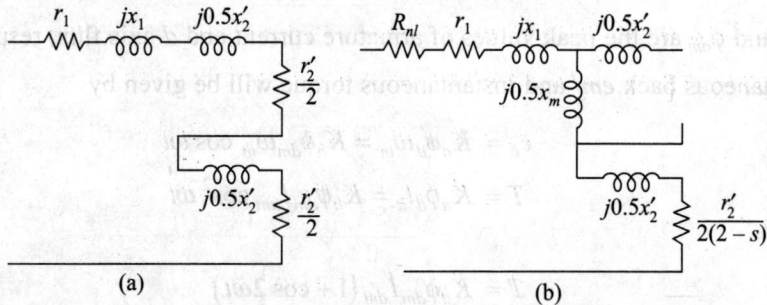

(a)   (b)

FIGURE 15.22   Equivalent circuit diagram with open circuit and block rotor tests.

At no-load, the rotational loss will be present. Let represent this rotational loss which will be in series with the stator winding resistance*. Thus, from Figure 15.22(b), we have

$$I^2\left(r_1 + \frac{r_2'}{2(2-s)} + R \right) = 100 = 3.5^2(R') \tag{iii}$$

or $R' = 8.16$ ohm

---

* $R' = r_1 + \dfrac{r_2'}{2(2-s)}$ + equivalent rotational loss resistance (R_{nl}).

Total impedance will be 62.86 (= 220/3.5) ohm and will be related as

$$\left|(R') + j(x_1 + 0.5x_2' + 0.5x_m)\right| = 62.86 = \sqrt{R'^2 + (x_1 + 0.5x_2' + 0.5x_m)^2}$$

or $$x_m = 115.29 \text{ ohm}.$$

(b) Rotational loss = Total loss at no-load – Copper loss at no-load

$$= 100 - 3.5^2(3 + 2/4)$$
$$= 57 \text{ W}.$$

EXAMPLE 15.2   A single-phase, 0.4 kW, 110 V, 50 Hz, 1425 *rpm* induction motor has the following parameters.

$$r_1 = 3.8 \ \Omega, r_2' = 4.2 \ \Omega, x_1 = j4.6\Omega, x_2' = 4.6 \ \Omega, x_m = j54 \ \Omega$$

Determine slip, input power, power factor, developed torque, output power, efficiency of motor, air gap power and rotor copper loss when the motor is running at rated speed while connected to 110 V ac supply. Rotational loss is 40 W.

Solution: The rated speed will be very close to but lower than the synchronous speed. Thus, for the 50 Hz, the nearest synchronous speed will be 1500 rpm. Therefore, the rated slip (s) will be

$$s = \frac{1500 - 1425}{1500} = 0.05$$

From the equivalent circuit shown in Figure 15.6, we have

$$\bar{Z}_f = R_f + jX_f = \frac{j0.5x_m \left[(r_2'/2s) + j0.5x_2'\right]}{\dfrac{r_2'}{2s} + j(0.5x_m + 0.5x_2')}$$

Putting the value, we get

$$\bar{Z}_f = R_f + jX_f = 11.6752 + j18.8552 \text{ ohm}$$

Similarly,

$$\bar{Z}_b = R_b + jX_b = \frac{j0.5x_m \left\{[0.5r_2'/(2 - s)] + j0.5x_2'\right\}}{\dfrac{0.5r_2'}{2 - s} + j(0.5x_m + 0.5x_2')}$$

Putting the value, we get

$$\bar{Z}_b = R_b + jX_b = 0.9133 + j2.1530 \text{ ohm}$$

Total impendence

$$\bar{Z}_{in} = r_1 + jx_1 + \bar{Z}_b + \bar{Z}_f = 16.3885 + j21.5082 \text{ ohm}$$

Thus input current at full-load will be

$$\bar{I} = \frac{110 \angle 0°}{16.3885 + j21.5082} = 1.9502 - j3.0474 = 3.618 \ \angle -57.38° \text{ A}$$

Hence, power factor = cos (57.38) = 0.539 lagging.

Input power = $VI \cos 57.38 = 214.53$ W

Synchronous speed in rad/sec = $1500 \times 2\pi/60 = 157.10$ rad/sec.

The developed torque will be

$$T = \frac{I^2}{\omega_s}(R_f - R_b) = \frac{3.618^2}{157.10}(11.6752 - 0.9133) = 0.897 \text{ N-m}$$

Mechanical power developed is

$$P_{mech} = T\omega_s(1-s) = 0.897 \times 157.10 \times (1-0.05) = 133.85 \text{ W}$$

Output power = Mechanical power developed – Rotational loss = 133.85 – 40 = 93.85 W

Thus, efficiency = Output power/input power = 93.85/214.53

$$= 43.75 \%$$

Air gap powers will be

$$P_{gf} = I^2 R_f = 3.618^2 \times 11.6752 = 152.83 \text{ W}$$

$$P_{gb} = I^2 R_b = 3.618^2 \times 0.9133 = 11.96 \text{ W}$$

The total air gap power will be 152.83 + 11.96 = 164.79 W

Rotor copper loss will be given by

$$P_2 = sP_{gf} + (2-s)P_{gb} = 0.05 \times 152.83 + (2-0.05) \times 11.96 = 46.644 \text{ W.}$$

EXAMPLE 15.3 A single-stack, six-phase stepper motor has four rotor teeth. These phases are excited one at time. Find

(a) Step size
(b) Step per revolution.

Solution:

(a) Step size = $360°/(6 \times 4) = 15°$

(b) Step per revolution = $360°/15° = 24$.

PROBLEMS

15.1 A single phase, 0.7 kW, 200 V, 1440 rpm, 50 Hz, induction motor has the following equivalent circuit parameters.

$$r_1 = 4.2 \ \Omega, r_2' = 4.0 \ \Omega, x_1 = j5 \ \Omega, x_2' = 5 \ \Omega, x_m = j80 \ \Omega$$

Determine slip, input power, power factor, developed torque, output power, efficiency of motor and air gap power when the motor is running at rated speed while connected to 220 V ac supply. Rotational loss is 40 W.

[**Ans.** 0.04, 668.66 W, 0.6 lagging, 3.49 N-m, 486.41 W, 72.74%, 604.4 W]

15.2 A three-stack rotor is used to produce a step size of 3°. Find the steps per revolution and number of rotor teeth.

[**Ans.** 120 steps per revolution, 40 teeth]

15.3 A single-stack, six-phase stepper motor produces 15° step motion. Determine the number of rotor poles.

[**Ans.** 4]

16

Electric Power Generation Technologies

16.1 INTRODUCTION

The electric power requirement, all over the world, has been increased manyfolds and it will continue in future also, due to its several advantageous features. To meet the required electric demand, power generation capacity is also expanding. Electric energy sources can broadly be classified into two categories: conventional energy sources and non-conventional (or alternate) energy sources. Conventional energy sources include coal, diesel, hydro (large), gas and nuclear. On the other hand, non-conventional energy sources are wind, solar, fuel cells, tidal, biogas, hydro (small), etc., which are having minimal operating cost and pollution, and inexhaustible. These are also called *renewable and alternate energy sources*. Limited fossil fuels resources and concern of global warming, the exploration of electric power generation from renewable and alternate energy sources are the present and future needs.

16.2 CONVENTIONAL ELECTRIC POWER PLANTS

16.2.1 Thermal Power Plants

In thermal power plants, steam is generated by heating the water and is used to rotate the turbines which are coupled with the synchronous generators (also known as *alternators*) to produce electricity. Steam can be generated from coal, gas or nuclear as main fuel. Coal, which is main fuel used in thermal power plants, is fired in the boiler to generate heat for producing steam. The thermal efficiency of a steam power plant mainly depends on the choice of steam cycle and varies from 28% to 35%. The principal equipment of steam power plants are the boiler, super-heater, feed water pump, steam reheater, condenser, turbine, and generator. The major components of steam generating plants are shown in Figure 16.1. In large thermal power plant unit, several stages of turbines such as high pressure, intermediate pressure and low pressure

408

are used to extract more power from the steam and thus, to increase the efficiency of the machine with minimum cost.

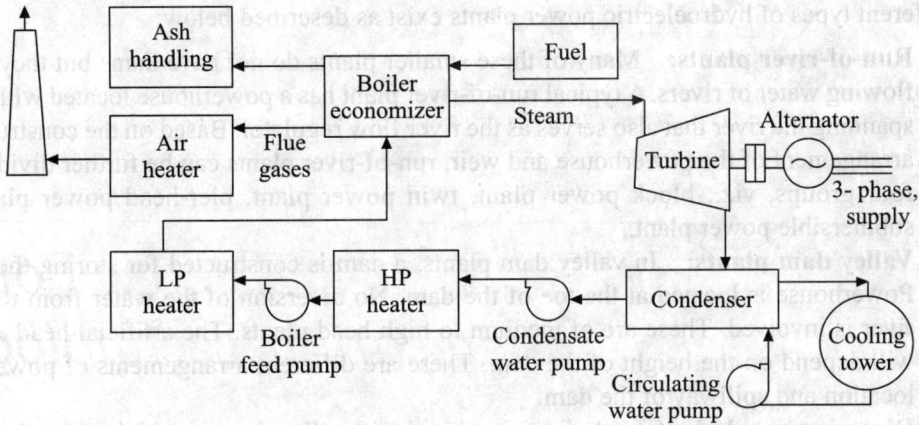

FIGURE 16.1   Layout of a typical steam power plant.

When water is heated, its enthalpy and physical state change. As heating takes place, the temperature of water rises and generally its density decreases. The steam is a gaseous state but does not entirely follow the laws of a perfect gas. The Carnot cycle cannot be applied to the steam turbines as compression phase does not exist in steam plants. A steam power plant basically works on the Rankine cycle with small deviations from ideal Rankine cycle.

Air is required for complete combustion of the fuel, which is supplied through the forced draught (FD) fans and/or induced draught (ID) fans. In all large thermal power stations both fans are used and normally they are in pairs to balance the boiler air. The air, which is fed to the boiler, is passed through the air pre-heater to extract some energy of flue gases coming out from the boiler after the burning the coal. It also helps in the proper burning of the coal. The flue gases consist of several gases and ash, which are passed through the precipitator (or dust collector) and then finally go to atmosphere through chimney.

For generating the electricity using steam turbine, high-speed synchronous generators are used because the efficiency of steam turbines is high at high speed. Since the speed of turbo-alternator is high, the diameters of the machine is kept minimum so that the centrifugal force acting on the rotor is minimized. To keep the same electrical loading (which is proportional to area × length), the length of the turbo-alternators is increased.

16.2.2 Hydropower Plants

Hydropower is one of the cheapest and cleanest sources of energy, although, with big dams, there are many environmental and social problems. For the electricity generation, ample quantity of water is used where water-flow from higher level to a lower level is passed through a turbine. The energy of water utilized for electric power generation may be either kinetic or potential. When water is in motion, it has kinetic energy and is passed through the turbine for production of electricity. Potential energy, available in form of the level difference of water (called *head*), is converted into kinetic energy that is used for rotating the generator to produce electricity. Depending on the capacity, hydroelectric plants are categorized into microhydro (≤ 100 kW), minihydro (101 kW to 1 MW), small hydro (1 MW to few MW, i.e. normally 6 MW) and hydroelectric plants (more than few MW).

Different types of hydroelectric power plants exist as described below:

   (a) **Run-of-river plants:**   Many of these smaller plants do not have dams but they run by flowing water of rivers. A typical run-of-river plant has a powerhouse located with a weir spanning the river that also serves as the river flow regulator. Based on the constructional arrangement of the powerhouse and weir, run-of-river plants can be further divided into four groups, viz., block power plant, twin power plant, pier-head power plant and submersible power plant.

   (b) **Valley dam plants:**   In valley dam plants, a dam is constructed for storing the water. Powerhouse is located at the toe of the dam. No diversion of the water from the main river is involved. These are of medium to high head plants. The artificial head created, will depend on the height of the dam. There are different arrangements of powerhouse location and spillway of the dam.

   (c) **Diversion canal plants:**   A diversion canal with a flat slope in which water flow from the river is diverted through a canal to powerhouse. The water from powerhouse is drained back into the original river at downstream point. A weir is constructed at the end of the canal to create a small pool of water, called the *forebay*. The water from forebay is fed by means of penstocks to the powerhouse situated in the lower reach of the river. In high head diversion plants, water is diverted through a system of channels and tunnels.

   (d) **Pumped storage power plants:**   These power plants are used for meeting the peak demand. Such plants utilize the concept of recycling the same water by pumping the water back to storage during off-peak hours.

Hydroelectric generators are low speed machines compared to the steam turbine-driven generators because the performance of hydro-turbine is better at low speed. As speed of hydro generator is low, for the same operating frequency, the number of poles is high. It is economical to have salient pole type synchronous machine. To accommodate the large number of poles, the diameter of the rotor has to be large enough. However, the axial length of the poles may be comparatively low.

16.2.3   Gas Power Plants

When natural gas is used to produce steam for running the turbine to generate the electricity, it is known as *thermal power plant*. But normally, natural gas is used as fuel for running the power plants without utilizing the steam as energy transfer media and it is called *gas power plant*. It is very simple as it does not have the boiler or steam supply system, condenser and waste heat disposal system. Moreover, it has ability to start and take load quickly as its start up time is only 2–8 minutes. Gas turbines are widely used in aircraft due to several advantages over the reciprocating engines such as (1) use of kerosene or distillate, (2) less fire hazard, (3) no unbalance forces, (4) simple cooling system, (5) fewer moving parts, (6) easy installation, (7) small frontal area, (8) not being limited by propeller characteristics and (9) less weight per horsepower.

Two types of gas power plants are available: open cycle and combined cycle. Figure 16.2 shows the typical open cycle gas power plant. A simple gas turbine plant consists of a combustor, an air compressor, a turbine and several auxiliaries for lubrication, speed control, fuel supply and starting. There are several practical limitations such as blade temperature, blade speed, compressor efficiency, turbine efficiency and heat transfer, are encountered in gas turbine operation.

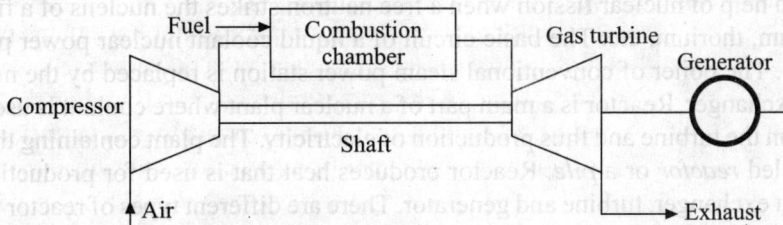

FIGURE 16.2 Gas power plant.

In gas power plants, compressed air is delivered to the combustor where fuel is supplied in steady flow and a continuous flame is maintained. A spark is used to make the original ignition. The heated air in the combustor expands through nozzles and develops a high velocity. Some of the kinetic energy of air stream is delivered to the blades of the turbine. Part of this energy is used to drive the compressor and the remainder is available for shaft work. A combustion chamber connects the compressor air outlet and the turbine gas inlet. When unit runs, air enters the compressor inlet from the atmosphere. Air pressure and temperature rise as the compressor forces the air through its outlet. The compressed air then enters the combustion chamber or furnace, where fuel oil or gas also enters and burns in the compressed air, raising its temperature. The heated air and the product of combustion, comprising a mixture of gases, then enter the turbine and produce the shaft work. The gas pressure drops as it flows through the turbine and the gas finally exhausts at atmospheric pressure into the surrounding air. Gas turbines (GTs) work on Brayton's cycle.

The Brayton cycle has high source temperature and rejects heat at a temperature that is conveniently used as the energy source for the Rankine cycle for thermal power generation. The most commonly used working fluids for combined cycles are air and steam. Combined cycle power plants have efficiency well over 50%. Combined-cycle systems that utilize steam and air-working fluids have achieved widespread commercial application due to:

- High thermal efficiency through application of two complementary thermodynamic cycles.
- Heat rejection from the Brayton cycle (gas turbine) at a high temperature that can be utilized in a simple and efficient manner.
- Working fluids (water and air) that are readily available, inexpensive, and nontoxic.

The main features of the combined-cycle power plants are:

- High thermal efficiency and low installed cost
- Fuel flexibility
- Flexible duty cycle and short-installation cycle
- High reliability/availability and low operation and maintenance costs
- High efficiency in small capacity increments.

16.2.4   Nuclear Power Plants

A nuclear power plant is almost similar to the coal-fired power plant except the steam generation part where nuclear energy is used for the steam production. In nuclear power plants, the heat is generated with help of nuclear fission when a free neutron strikes the nucleus of a fissile material such as uranium, thorium, etc. The basic circuit of a liquid coolant nuclear power plant is shown in Figure 16.3. The boiler of conventional steam power station is replaced by the nuclear reactor and the heat exchanger. Reactor is a main part of a nuclear plant where coolant is used to generate the steam to run the turbine and thus production of electricity. The plant containing the radioactive material is called *reactor* or a *pile*. Reactor produces heat that is used for production of electric energy via heat exchanger, turbine and generator. There are different types of reactor with different coolants and moderators. A moderator is a substance that causes neutrons to slow down, hence increasing their probability of interacting with fassile nuclei.

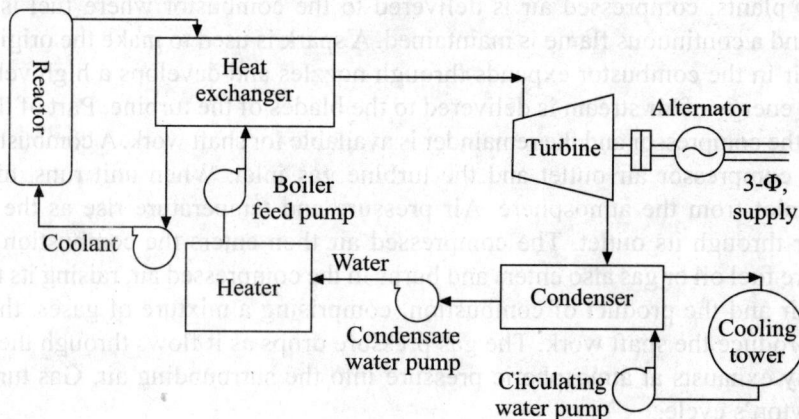

FIGURE 16.3   Nuclear power plant using heat exchanger.

A mass of fissile material such as uranium is brought together in the form of fuel rods and inserted into the core of the reactor. The energy from fission heats the core. To cool the core, and remove the heat so that it can be used to generate power, a coolant is run through the system and taken to a heat exchanger for removing the heat to a separate system. It is then used to generate steam, which runs the turbines to drive the electric generators. The main components of the nuclear reactor are given below.

(a) **Fuel rods:**   A fuel rod is a zircaloy tube, filled with pellets of uranium. These fuel assemblies can be lifted into and out of the reactor mechanically, allowing fuel replenishment while the reactor is in operation.

(b) **Shielding:**   Shielding is used to give the protection against the alpha, beta and gamma radiations during the process of fission in reactor.

(c) **Moderator:**   A moderator is used to slow down the neutrons released during fission. Graphite, heavy water or beryllium can be used as a moderator for natural uranium however, ordinary water is used with enriched uranium.

(d) **Control rods:**   To prevent the melting of fuel rods, disintegration of coolant, destruction of reactor due to excessive energy release and to control the chain reaction at steady-state value during the operation of reactor, it is necessary to have controls. Chain reaction can be controlled either by removing the fuel rods or by inserting neutron-absorbing material. These rods are made of boron carbide, cadmium or hafnium and are inserted from the bottom of the core to absorb neutrons, which controls the rate of fission. Additional main control rods are inserted from the top down to provide automatic, manual or emergency control.

(e) **Coolant:**   Coolants are used for transferring heat produced inside the reactor to a heat exchanger for utilization of power generation. The commonly used coolants are gas (carbon dioxide, air, hydrogen, helium), ordinary water, heavy water, liquid metals (sodium or potassium) and some other organic liquids. Coolant flows through and around the reactor. There is also an emergency core cooling system, which will come into operation if either coolant circuit is interrupted.

(f) **Steam separator:**   Steam from the heated coolant is fed to turbines to produce electricity in the generator. Ninety-five percent of the heat from fission is transferred through the coolant. The steam is then condensed and fed back into the circulating coolant.

(g) **Containment:**   The reactor core is located in a concrete lined cavity that acts as a radiation shield. The upper shield or pile cap above the core, is made of steel and supports the fuel assemblies.

16.2.5   Diesel Engines

Diesel engines are used to drive the prime mover of electric generators for producing the electricity. Diesel engines are used for two purposes. First, as a stand-by set for start up of auxiliaries in steam, hydro and gas turbine power plants and for emergency supply to hospitals, hotels, factories and in other commercial/domestic applications. Second, as continuous power generation, diesel engines have several advantages such as

- high operating efficiency;
- no stand-by loss;
- need very little water for cooling;
- quick start and stop is possible;
- easier handling of fuel.

However, it has disadvantage of high noise and air pollution.

16.3   RENEWABLE AND ALTERNATE ENERGY SOURCES-BASED POWER GENERATION

With increased concern of global warming, efforts are ongoing to reduce the emission level by substituting the energy production from renewable energy sources. The development of renewable and alternative sources of energy is being widely promoted in several countries by providing government subsidy, incentives, etc.

16.3.1   Wind Power Generation

The use of wind power is in place from long time for different applications such as in ships, in agriculture, etc. The exploitation of wind power, at present, is increasing due to environmental concerns, high price of oil and a better knowledge of possible depleting of fossil fuels. Wind energy is plentiful, inexhaustible and pollution-free but great drawback on the utilization arises from both the intermittence and unreliability. There are several initiatives taken by various countries for speedy and technical developments in the area of renewable energy sources including the wind. In terms of the installed capacity of wind power generation, Germany, USA, Spain, China and India are the leaders however Denmark is the first country to use wind for the generation of electricity.

In wind power generation, wind energy is converted into electrical power with the help of wind turbines. Selection of windmill sites depends on several factors such as windy area, scattered population, cheaper to grid electricity cost due to transmission, etc. It is more suitable near to coastal and remote areas. Now-a-days, both on-shore and off-shore wind farms provide electricity to the grid at various voltage levels as per grid code of the respective country. The greatest advantage of windmills is that they may be installed in any locality where the topographical and meteorological conditions are suitable and require no outside supplies for its operation except for a stand-by battery and lubricating oil.

Typical wind-turbine power generation systems range from 30 kW to 6 MW of an individual unit. Hub heights are around 80 meters and rotor diameters are up to 65 meters. Most of wind turbines designed for production of electricity are consisted of a two- or three bladed propeller rotating around a horizontal axis. Rotor construction either is variable blade angle or non-variable and conversion from mechanical to electrical energy is via either synchronous or induction generators. Synchronous generators are equipped with pulse-width-modulated converters and controls for these converters are essential for regulating the behaviour of windmill in the electric grid. Windmills are often installed in groups, or wind farms, and are seldom used in isolation. A disadvantage of wind power is its irregularity and this further complicates the connection to power grids. Doubly-fed induction generators are mostly used for large wind power generation units. Typical costs are around 1200 \$/kW and electrical efficiencies are around 25%. Footprint size is in the order of 0.01 kW/m^2 and its operating costs are between 4 to 12 cents-USD/kWh.

16.3.2 Solar Power Generation

When the sun passes through the atmosphere, part of the sun's radiation is reflected, scattered and absorbed by heating air, dust and by evaporating water. The solar power, which falls upon the whole earth, is of the magnitude of 1.77×10^{14} kW. The rate at which solar energy reaches the earth's atmosphere (approximately 15 km from sea level) is known as of *solar constant* which is equal to 1.39 kWh/ (m^2-h).

Most of the solar energy used today is harnessed as heat or electricity. The solar power densities at the surface of earth depend upon the sun's position and upon the clarity and humidity of the atmosphere. Solar power generation describes a number of methods of producing electrical energy from the light of the sun. It is already in widespread use where other supplies of power are absent such as in remote locations and in space. The main problem in solar power energy is its availability. Solar power generation may be classified as direct and indirect. Direct solar power generation involves only one transformation into a usable form, such as sunlight hits a photovoltaic cell to create electricity. However, indirect solar power generation involves more than one transformation to reach a usable form such as solar thermal power generation. Solar power can also be classified as passive or active.

Passive solar systems do not involve the input of any other forms of energy apart from the incoming sunlight, whereas *active solar systems* use additional mechanisms such as circulation pumps, air blowers, or automatic systems to collect solar energy. The effective use of solar radiation often requires the radiation (light) to be focused to give a higher intensity beam (such as parabolic dish, parabolic trough, etc. are used to concentrate sun light at a point or a line). At the focus, high-concentration photovoltaic cells (solar cells) or a thermal energy "receiver" may be placed. Solar design aims the use of architectural features to replace the use of grid electricity and fossil fuels with the use of solar energy and decrease the energy needed in a home or building.

Solar energy can also be used to meet our electricity requirements. Solar cells, also known as *photovoltaic cells*, use the photovoltaic (PV) effect of semiconductors to generate dc electricity directly from the sunlight. PV cells are made primarily of silicon, the second most abundant element in the earth's crust. When the silicon is combined with one or more other materials, it exhibits unique electrical properties in the presence of sunlight. Electrons are excited by the light and move through the silicon. This is known as the *photovoltaic effect.* Other than electric power generation, there are several distinct applications of solar power as

(a) Space and water heating in domestic and commercial buildings; e.g. photovoltaic collectors as shown in Figure 16.4;
(b) The chemical and biological conversion of organic materials to liquid, solid and gaseous fuel;
(c) water pumping;
(d) desalination of salty water.

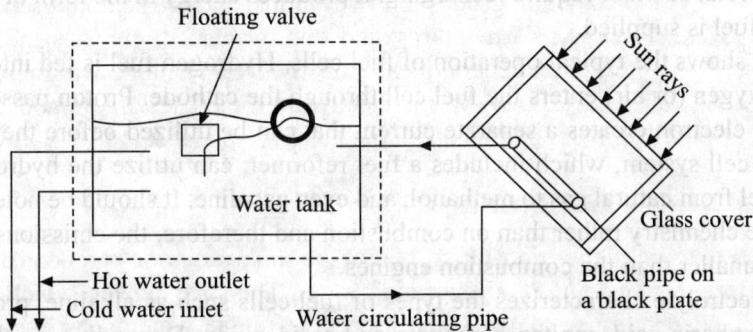

FIGURE 16.4   Solar water heating arrangement.

India is one of the few countries with long days (around 300 days) and plenty of sunshine, especially in the Thar Desert region. India supports development of both solar thermal and solar photovoltaics (PV) power generation. To demonstrate and commercialize solar thermal technology in India, Ministry of New and Renewable Energy (MNRE) is promoting megawatt scale projects such as the proposed 35 MW solar thermal plant in Rajasthan and encouraging private sector projects by providing financial assistance. Solar thermal energy is being used in India for heating water for both industrial and domestic purposes. A 140 MW integrated solar power plant is to be set up in Jodhpur but the initial expense incurred is still very high. Figure 16.5 shows schematic diagram of a solar power plant.

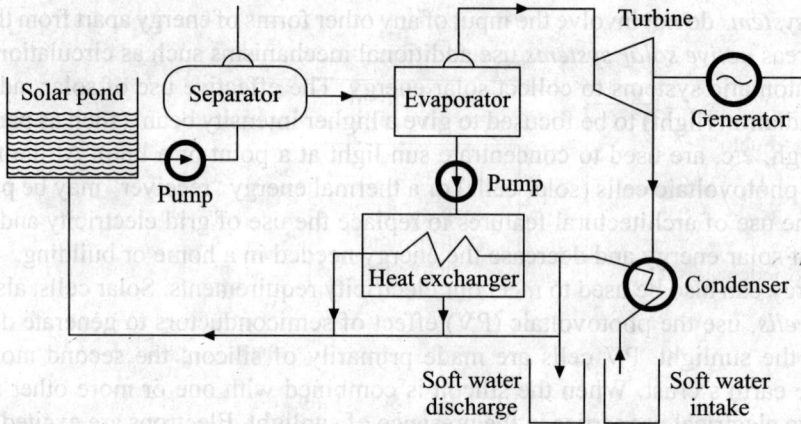

FIGURE 16.5   Schematic diagram of solar power plant.

16.3.3   Fuel Cells

A fuel cell, which consists of an electrolyte sandwitched between two electrodes, is electrochemical device that converts the chemical energy of a fuel directly and very efficiently into dc electricity and heat without any combustion. The most suitable fuel for these cells is hydrogen or a mixture of compounds containing hydrogen. Oxygen passes over one electrode and hydrogen over the other, and they react electrochemically to generate electricity, water and heat. Unlike a battery, a fuel cell does not run down or require recharging. It produces energy in the form of electricity and heat as long as fuel is supplied.

Figure 16.6 shows the typical operation of fuel cells. Hydrogen fuel is fed into the anode of the fuel cell. Oxygen (or air) enters the fuel cell through the cathode. Proton passes through the electrolyte. The electron creates a separate current that can be utilized before they return to the cathode. A fuel cell system, which includes a fuel reformer, can utilize the hydrogen from any hydrocarbon fuel from natural gas to methanol, and even gasoline. It should be noted that the fuel cell relies on the chemistry rather than on combustion and therefore, the emissions from the fuel cells are much smaller than the combustion engines.

The used electrolyte characterizes the types of fuel cells such as alkaline, proton exchange membrane, phosphoric acid, molten carbonate and solid oxide. Depending on the electrolyte, ignoring produced heat, the fuel cell efficiency, operated between 80 and 1000°C, can range between 35–65%. Utilizing the produced heat can raise the efficiency to over 80%. Target capital costs (assuming large volume manufacturing) ranges from 800–1300 \$/kW and footprint size ranges from 1–3 kW/m². Operating cost is estimated at between 8–10 cent-\$/kWh and emissions of NO_x gasses are extremely low at 0.003–0.03 lb/BTU. Fuel cells are typically aimed at single installation site that requires between 50 and 1000 kW, e.g. high rise office buildings, hospitals, schools, hotels, etc. However, new small fuel cells are available for residential purposes at about 5–10 kW.

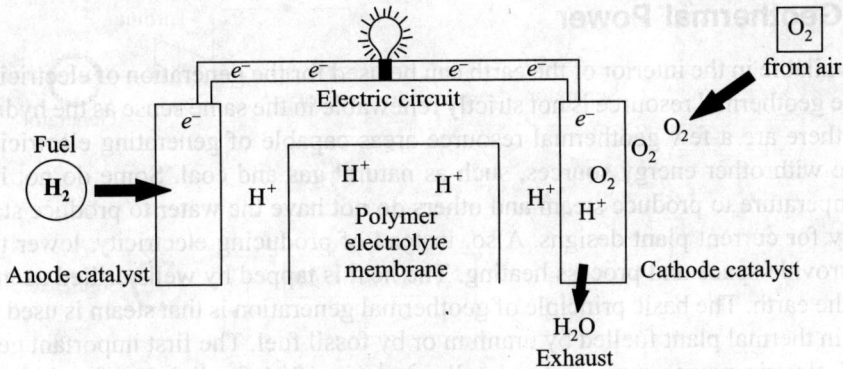

FIGURE 16.6 Fuel cell.

16.3.4 Tidal Power

The tides are generated due to changing gravitation force of the orbiting and rotating earth, moon and sun. The electricity can be generated from the kinetic energy of currents due to the tides and potential energy from the difference in height (or *head*) between high and low tides. The extraction of potential energy involves building a barrage. The barrage traps a water level inside a basin. Head is created when the water-level outside the basin changes relative to the water level inside. The head is used to drive turbines. Till date there has been little development of tidal projects despite the fact that these schemes are quite attractive. There are at least three ways in which tidal energy might conceivably be harnessed.

The first way of harnessing the tidal energy is simply to place a water wheel in a tidal stream as shown in Figure 16.7(a). It is analogous to using a water wheel in the river and is not suitable for electric generation due to variability of the tidal stream flow. This scheme can be used for pumping water or milling grain. In the second scheme, a large floating object such as a barge is raised by an incoming tide; it is constrained by pilings as shown in Figure 16.7(b). It can be then held and dropped later or allowed to fall with the tide to drive the electric generator. Third choice, as shown in Figure 16.7(c), is to build a low dam across the mouth of bay or tidal estuary. As the tide comes in, gates are opened and water flows into the bay or estuary. After the high tide, the gates are closed. During the off tide the stored water can be used for electricity generation.

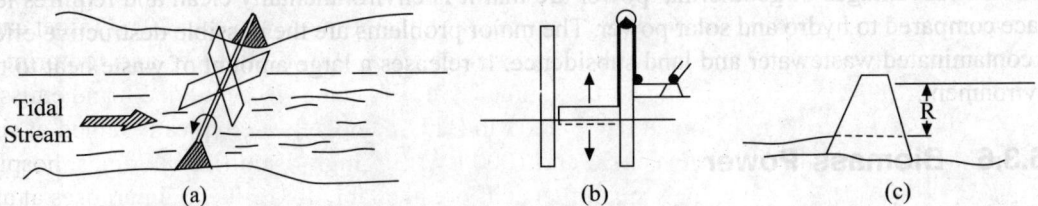

FIGURE 16.7 Tidal power.

16.3.5 Geothermal Power

The heat available in the interior of the earth can be used for the generation of electricity. It can be said that the geothermal resource is not strictly renewable in the same sense as the hydro resource. Currently, there are a few geothermal resource areas capable of generating electricity at a cost competitive with other energy sources, such as natural gas and coal. Some do not have a high enough temperature to produce steam and others do not have the water to produce steam, which is necessary for current plant designs. Also, instead of producing electricity, lower temperature areas can provide space and process heating. The heat is tapped by wells drilled as much as two miles into the earth. The basic principle of geothermal generation is that steam is used to drive the turbine, as in thermal plant fuelled by uranium or by fossil fuel. The first important generation of geothermal electric power came in Larderello, Italy in 1904 for lighting few bulbs. Italy and Geysers area of San Francisco, USA are the largest geothermal producing region in the world. Figure 16.8 shows the typical geothermal power plant at the Geysers.

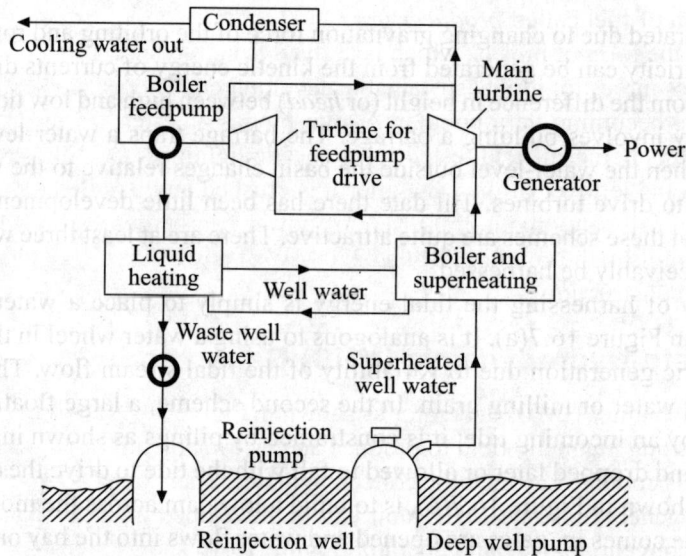

FIGURE 16.8   Typical geothermal power plant.

The main advantages of geothermal power are that it is environmentally clean and requires less space compared to hydro and solar power. The major problems are the possible destructive effect of contaminated wastewater and land subsidence. It releases a large amount of waste heat to the environment.

16.3.6 Biomass Power

Biomass is an important source of energy and the most important fuel worldwide after coal, oil and natural gas. This energy is being used for cooking, mechanical applications/pumping, power generation and transportation. Biomass is a renewable energy resource derived from the carbonaceous waste of various human and natural activities such as the timber industry, agricultural crops, raw material from the forest, major parts of household waste and wood. Biomass does not add carbon dioxide to the atmosphere as it absorbs the same amount of carbon when consumed as a fuel. Its advantage is that it can be used to generate electricity with the same equipment or power plants that are now burning fossil fuels.

Traditional use of biomass is more than its use in modern application. Biomass energy is gaining significance as a source of clean heat for domestic and community heating applications. In fact, in the countries like Finland, USA and Sweden, the per capita biomass energy used is higher than in India, China or in Asia. Biomass fuels used in India account for about one third of the total fuel used in the country, being the most important fuel used in over 90% of the rural households and about 15% of the urban households. Instead of burning the loose biomass fuel directly, it is more practical to compress it into briquettes (compressing them through a process to form blocks of different shapes) and thereby improve its utility and convenience of use. Such biomass in the dense briquetted form can either be used directly as fuel instead of coal in the traditional *chulhas* and furnaces or in the gasifier. Gasifier converts solid fuel into a more convenient to use gaseous form of fuel called *producer gas*.

Bio-energy, in the form of biogas, which is derived from biomass, is expected to become one of the key energy resources for global sustainable development. Biogas plants have been set up in many areas and are becoming very popular. A mini biogas digester has recently been designed and developed, and is being in-field tested for domestic lighting. Indian sugar mills are rapidly turning to bagasse, the leftover of the cane after it is crushed and its juice extracted, to generate electricity. From an estimate, about 3500 MW of power can be generated from bagasse in the existing 430 sugar mills in the country. Around 270 MW of power has already been commissioned and more is under construction.

16.3.7   Magneto Hydrodynamic (MHD) Generation

In magneto hydrodynamic (HMD) generator as shown in Figure 16.9, an ionized gas or plasma is passed through a strong magnetic field to produce an electric potential. The ionized gas (positive and negative ions) is collected by metallic collecting plates. There are a number of possible forms of working MHD generators. The MHD unit can be developed alone or combined with a gas turbine, or with a conventional steam generator. Figure 16.10 shows the MHD cycle with steam cycle. Fuel is introduced to the burner along with the seed material such as potassium which increases the conductivity of the gas to permit the practical operation of the device. The magnet deflects some of the ions to the plate, which become charged, producing a dc electric potential. The exhaust gas passes first through an air heater that heats the outside air that has been compresses by the compressor attached to the steam turbine. The heated air is then used in the burner. The hot exhaust gas from MHD generator is then passed into the steam generator to drive the turbine. The steam turbine drives both the compressor and an electric generator. The exhaust gases continue to flow through a seed recovery stage where the seed is captured and fedback to burner. Since recovery is not perfect, some make-up seed must be added. Next the exhaust gas passes through a nitrogen and sulphur removal stage before being released by the stack.

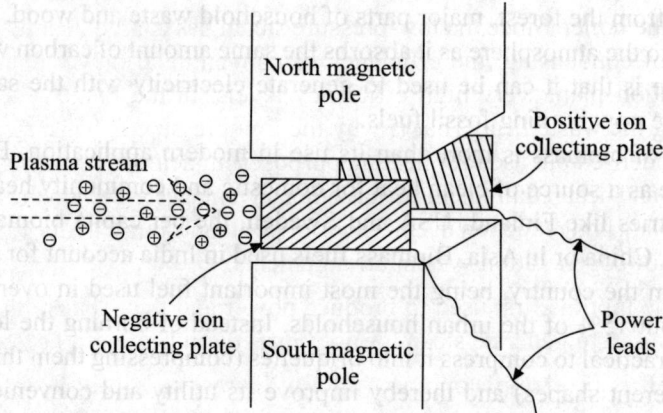

FIGURE 16.9   MHD power generation.

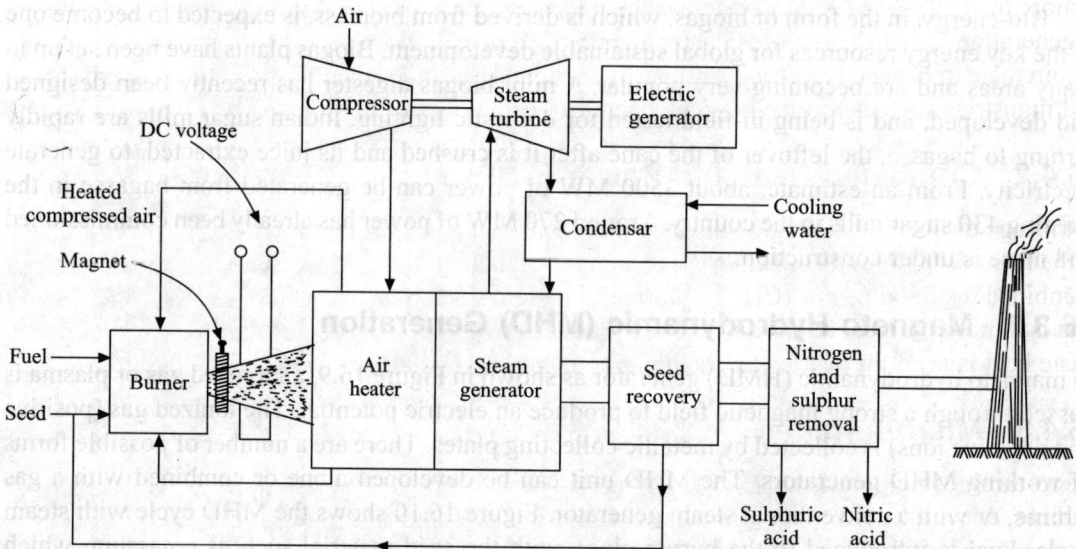

FIGURE 16.10   MHD cycle with steam.

16.4   SOME OTHER ENERGY SOURCES

16.4.1   Cogeneration

Cogeneration is the concept of producing two forms of energy from one fuel. One of the forms of energy must always be heat and the other may be electricity or mechanical energy. In cogeneration process, waste heat from manufacturing, industrial process or heating or cooling systems is captured to generate electric power that can be used to run one's own equipment or can be sold to the electric utility company. In a conventional power plant, fuel is burnt in a boiler to generate high-pressure steam. This steam is used to drive a turbine, which in turn drives an alternator through a steam turbine to produce electric power. The exhaust steam is generally condensed to water, which goes back to the boiler. Since the low-pressure steam has a large quantum of heat, which is lost in the process of condensing, the efficiency of conventional power plants is only around 35%. In a cogeneration plant, very high efficiency levels, in the range of 75%–90%, can be achieved by utilizing the waste heat. That is why the low-pressure exhaust steam coming out of the turbine is not condensed but used for heating purposes in factories or houses. As cogeneration can meet both power and heat requirements, it has other advantages as well in the form of significant cost savings for the plant and reduction in emissions of pollutants due to reduced fuel consumption. There are two types of cogeneration system arrangements.

Topping cycle: The input is first used to produce power and the exhaust heat from the power producing system is used to generate steam or used directly in heating process.

Bottoming cycle: It is reverse of topping cycle, utilizing waste heat from a heating process such as industrial furnace, to produce electricity. Commonly waste heat is used to produce steam in heat recovery boiler and used to run turbine for producing electricity.

Even at conservative estimates, the potential of power generation from cogeneration in India is more than 20,000 MW. Since India is the largest producer of sugar in the world, bagasse-based cogeneration is being promoted. The potential for cogeneration thus lies in facilities with joint requirement of heat and electricity, primarily sugar and rice mills, distilleries, petrochemical sector and industries such as fertilizers, steel, chemical, cement, pulp and paper, and aluminum.

16.4.2   Combined Heat and Power (CHP)

In this age of liberalization, globalization and climate change, the optimum use of energy is need of the hour. A flexible and effective use of energy will help to reduce the greenhouse gas emissions. Combined heat and power (CHP) is normally more common in cold countries where the use of heat is more prominent in buildings. They produce power (from any source using heat cycle) and waste heat is used for heating up the building by air or by water.

16.4.3   Micro-turbines

Micro-turbines ranging from 25–500 kW units operate on the same principles as traditional gas turbines. Air is drawn into the compressor where it is pressurized and forced into the cold side of the recuperator. Here, exhaust heat is used to preheat the air before it enters the combustion chamber. The combustion chamber then mixes the heated air with fuel and burns it. This mixture expands through the turbine, which drives the compressor and generator (typically at high speeds such as 70,000 to 120,000 rpm). Since the generator is mounted on the same shaft as the turbine, it rotates at the same speed. The combusted air is then exhausted through the recuperator before being discharged at the exhaust outlet. The generator thus produces high frequency ac power that is converted to 50/60 Hz by power electronic interfaces. Typical capital costs are in the 500–1000 $/kW range and electrical efficiency ranges from 27–32%. Utilizing the exhaust heat can improve the overall efficiency up to 80%. Typically, micro-turbines use natural gas as fuel, but usage of other fuels such as diesel, propane and kerosene are possible.

Appendix
A

Matrix Algebra

A.1 MATRIX REPRESENTATION

A set of simultaneous linear equations can be written in a matrix form and matrix methods can be used to find the solution. The following set of m equations having n unknowns (also known as *variables*) is written in vector-matrix form as shown in equation (A1.1).

$$a_{11}x_1 + a_{12}x_2 \cdots + a_{1n}x_n = b_1$$
$$a_{21}x_1 + a_{22}x_2 \cdots + a_{2n}x_n = b_2$$
$$\vdots \qquad \vdots \qquad \vdots \qquad \vdots \qquad \text{(A1.1)}$$
$$a_{m1}x_1 + a_{m2}x_2 \cdots + a_{mn}x_n = b_m$$

In compact form, we can write as

$$\mathbf{AX} = \mathbf{B} \qquad \text{(A1.2)}$$

where $\mathbf{X}$ is referred to as unknown vector, $\mathbf{B}$ is referred to as known vector and $\mathbf{A}$ is connection matrix. These can be represented as

$$\mathbf{A} = \begin{bmatrix} a_{11} & a_{12} & \cdots & a_{1n} \\ a_{21} & a_{22} & \cdots & a_{2n} \\ \vdots & \vdots & \cdots & \vdots \\ a_{m1} & a_{m2} & \cdots & a_{mn} \end{bmatrix}; \quad \mathbf{B} = \begin{bmatrix} b_1 \\ b_2 \\ \vdots \\ b_m \end{bmatrix} \quad \text{and} \quad \mathbf{X} = \begin{bmatrix} x_1 \\ x_2 \\ \vdots \\ x_n \end{bmatrix}$$

If in any equation, any variable is not seen, that coefficient of x will be taken as zero.

EXAMPLE A.1   Represent following three equations having three unknowns into vector-matrix form.

$$3x_1 + 4x_2 + x_3 = 0$$
$$-x_1 + 4x_2 = 2 \qquad\qquad \text{(A1.3)}$$
$$2x_2 + x_3 = -1$$

Solution: From Eq. (A1.3), it can be written as

$$\begin{bmatrix} 3 & 4 & 1 \\ -1 & 4 & 0 \\ 0 & 2 & 1 \end{bmatrix} \begin{bmatrix} x_1 \\ x_2 \\ x_3 \end{bmatrix} = \begin{bmatrix} 0 \\ 2 \\ -1 \end{bmatrix} \qquad\qquad \text{(A1.4)}$$

A.2   MATRIX ADDITION AND MULTIPLICATION

Let us consider a matrix $\mathbf{A}$ having dimension m_A rows by n_A columns with ordered array of values $a_{ij}, i = 1, ..., m_A; j = 1,..., n_A$. A matrix $\mathbf{C}$ of dimension m_C rows by n_C columns is another ordered array of values $c_{ij}, i = 1,..., m_C; j = 1, ..., n_C$ as shown below.

$$\mathbf{A} = \begin{bmatrix} a_{11} & a_{12} & \cdots & a_{1n_A} \\ a_{21} & a_{22} & \cdots & a_{2n_A} \\ \vdots & \vdots & \cdots & \vdots \\ a_{m_A1} & a_{m_A2} & \cdots & a_{m_An_A} \end{bmatrix}; \quad \mathbf{C} = \begin{bmatrix} c_{11} & c_{12} & \cdots & c_{1n_C} \\ c_{21} & c_{22} & \cdots & c_{2n_C} \\ \vdots & \vdots & \cdots & \vdots \\ c_{m_C1} & c_{m_C2} & \cdots & c_{m_Cn_C} \end{bmatrix}$$

Transpose $\mathbf{A}^T$ of an $m_A \times n_A$ matrix $\mathbf{A}$ is defined as an $n_A \times m_A$ matrix whose (i, j) element is the (j, i) element of $\mathbf{A}$ as shown below.

$$\mathbf{A}^T = \begin{bmatrix} a_{11} & a_{21} & \cdots & a_{m_A1} \\ a_{12} & a_{22} & \cdots & a_{m_A2} \\ \vdots & \vdots & \cdots & \vdots \\ a_{1n_A} & a_{2n_A} & \cdots & a_{n_Am_A} \end{bmatrix}$$

$\mathbf{A}^T$ is the mirror image of $\mathbf{A}$ flipped around the main diagonal.

(a) **Matrix Addition/Subtraction:**   Two matrices can be added or subtracted only when the order of matrices is the same, i.e. they should have the same number of rows and columns. To add matrices $\mathbf{A}$ and $\mathbf{C}$, we should have

$$m_A = m_C \quad \text{and} \quad n_A = n_C \qquad\qquad \text{(A1.5)}$$

The additive matrix will have the same dimension and can be written as

$$\mathbf{A} + \mathbf{C} = \begin{bmatrix} a_{11}+c_{11} & a_{12}+c_{12} & \cdots & a_{1n_A}+c_{1n_C} \\ a_{21}+c_{21} & a_{22}+c_{22} & \cdots & a_{2n_A}+c_{2n_C} \\ \vdots & \vdots & & \vdots \\ a_{m_A1}+c_{m_C1} & a_{m_A2}+c_{m_C2} & \cdots & a_{m_An_A}+c_{m_Cn_C} \end{bmatrix} \qquad\qquad \text{(A1.6)}$$

Similarly, difference of these matrices can be written as

$$\mathbf{A} - \mathbf{C} = \begin{bmatrix} a_{11} - c_{11} & a_{12} - c_{12} & \cdots & a_{1n_A} - c_{1n_C} \\ a_{21} - c_{21} & a_{22} - c_{22} & \cdots & a_{2n_A} - c_{2n_C} \\ \vdots & \vdots & \cdots & \vdots \\ a_{m_A 1} - c_{m_C 1} & a_{m_A 2} - c_{m_C 2} & \cdots & a_{m_A n_A} - c_{m_C n_C} \end{bmatrix} \tag{A1.7}$$

(b) Scalar Multiplication: The operation of multiplication of a matrix **A** by a scalar k, is defined as element-wise multiplication of A by k as

$$kA = \begin{bmatrix} ka_{11} & ka_{12} & \cdots & ka_{1n_A} \\ ka_{21} & ka_{22} & \cdots & ka_{2n_A} \\ \vdots & \vdots & \cdots & \vdots \\ ka_{m_A 1} & ka_{m_A 2} & \cdots & ka_{m_A n_A} \end{bmatrix} \tag{A1.8}$$

(b) Matrix Multiplication: Two matrices can be multiplied as $\mathbf{A} \times \mathbf{C}$ if the column dimension of the left matrix **A** equals the row dimension of the right matrix **C** in which case matrices are called *conformable*. Thus, $n_A = m_C$.

The matrix product $\mathbf{D} = \mathbf{A} \times \mathbf{C}$, which is having dimension m_A rows by n_C columns with ordered array of values $d_{ij}, i = 1,\ldots, m_A; j = 1,\ldots, n_C$, is defined as

$$d_{ij} = \sum_{k=1}^{n_A} a_{ik} c_{kj} \tag{A1.9}$$

EXAMPLE A.2   Find the product matrix $\mathbf{D} = \mathbf{A} \times \mathbf{C}$ where

$$\mathbf{A} = \begin{bmatrix} 3 & 4 & 1 \\ -1 & 4 & 0 \\ 0 & 2 & 1 \end{bmatrix}; \quad \mathbf{C} = \begin{bmatrix} 1 & 3 \\ 0 & 2 \\ 4 & -3 \end{bmatrix}.$$

Solution: Matrix $\mathbf{D} = \begin{bmatrix} d_{11} & d_{12} \\ d_{21} & d_{22} \\ d_{31} & d_{32} \end{bmatrix}$ can be computed using Eq. (A1.9) as

$$d_{11} = \sum_{k=1}^{3} a_{1k} c_{k1} = a_{11} c_{11} + a_{12} c_{21} + a_{13} c_{31} = 3 + 0 + 4 = 7$$

$$d_{12} = \sum_{k=1}^{3} a_{1k} c_{k2} = a_{11} c_{12} + a_{12} c_{22} + a_{13} c_{32} = 9 + 8 - 3 = 14$$

$$d_{21} = \sum_{k=1}^{3} a_{2k} c_{k1} = a_{21} c_{11} + a_{22} c_{21} + a_{23} c_{31} = -1 + 0 + 0 = -1$$

$$d_{22} = \sum_{k=1}^{3} a_{2k}c_{k2} = a_{21}c_{12} + a_{22}c_{22} + a_{23}c_{32} = -3 + 8 + 0 = 5$$

$$d_{31} = \sum_{k=1}^{3} a_{3k}c_{k1} = a_{31}c_{11} + a_{32}c_{21} + a_{33}c_{31} = 0 + 0 + 4 = 4$$

$$d_{32} = \sum_{k=1}^{3} a_{3k}c_{k2} = a_{31}c_{12} + a_{32}c_{22} + a_{33}c_{32} = 0 + 4 - 3 = 1$$

Thus, multiplication matrix will be

$$\mathbf{D} = \begin{bmatrix} 7 & 4 \\ -1 & 5 \\ 4 & 1 \end{bmatrix}.$$

A.3 MATRIX DETERMINANT

The determinant of a square matrix $\mathbf{A}$ (denoted as $|\mathbf{A}|$ or Δ) is a scalar function of $\mathbf{A}$ computed using co-factors and minors. The minor A_{ij} of the element a_{ij} is the determinant of matrix left after the *i*th row and *j*th column are removed. The cofactor C_{ij} of element a_{ij} is given by

$$C_{ij} = (-1)^{i+j} A_{ij} \qquad (A1.10)$$

The value of determinant of matrix $\mathbf{A}$ will be the sum of products of the elements in any row or column and their cofactors as

$$\Delta = |\mathbf{A}| = \sum_{k=1}^{n} a_{ik}c_{ik} \ \forall i \in n \qquad (A1.11)$$

For simple matrix up to second order, determinant can be obtained without knowing the minors and cofactors and given below.
For $n = 1$,

$$\Delta = a_{11}$$

For $n = 2$,

$$\Delta = \begin{vmatrix} a_{11} & a_{12} \\ a_{21} & a_{22} \end{vmatrix} = a_{11}a_{22} - a_{12}a_{21}$$

It can be stated that it is difference of the product of diagonal terms.
For $n = 3$, it is better to use cofactors and minors as

$$\Delta = \begin{vmatrix} a_{11} & a_{12} & a_{13} \\ a_{21} & a_{22} & a_{23} \\ a_{31} & a_{32} & a_{33} \end{vmatrix} = a_{11}C_{11} + a_{12}C_{12} + a_{13}C_{13} \qquad (A1.12)$$

$$= a_{11}(-1)^{1+1}A_{11} + a_{12}(-1)^{1+2}A_{12} + a_{13}(-1)^{1+2}A_{13} = a_{11}A_{11} - a_{12}A_{12} + a_{13}A_{13}$$

Putting the values of minors in Eq. (A12), we get

$$\Delta = a_{11}\begin{vmatrix} a_{22} & a_{23} \\ a_{32} & a_{33} \end{vmatrix} - a_{12}\begin{vmatrix} a_{21} & a_{23} \\ a_{31} & a_{33} \end{vmatrix} + a_{13}\begin{vmatrix} a_{21} & a_{22} \\ a_{31} & a_{32} \end{vmatrix}$$ (A1.13)

$$= a_{11}(a_{22}a_{33} - a_{23}a_{32}) - a_{12}(a_{21}a_{33} - a_{23}a_{31}) + a_{13}(a_{21}a_{32} - a_{22}a_{31}).$$

A.4 MATRIX INVERSION

A square matrix **A** of dimension $n \times n$ will have its inverse, if the determinant of the matrix is not zero (It can also be stated that the rank of the matrix is equal to its order, n and it is also known as *full matrix*). Inverse of matrix **A** is denoted as $\mathbf{A}^{-1}$ and is related as

$$\mathbf{A}^{-1}\mathbf{A} = \mathbf{I}$$ (A1.14)

where **I** is an identity matrix of order $n \times n$ in which diagonal elements (i, i) are unity and other elements (i, j) are zero for i not equal to j. Inverse of matrix **A** can be defined as

$$\mathbf{A}^{-1} = \frac{\text{adj } \mathbf{A}}{\Delta}$$ (A1.15)

where adj **A** is a transpose a matrix of dimension $n \times n$ whose (i, j)th elements is the cofactor C_{ij} defined in Eq. (A1.10).

EXAMPLE A.3 Find the inverse of a matrix **A** given as

$$\mathbf{A} = \begin{bmatrix} a_{11} & a_{12} \\ a_{21} & a_{22} \end{bmatrix}.$$

Solution: Each cofactor C_{ij} is found by striking out the ith row and jth column of matrix **A**, computing the determinant of remaining matrix part and multiplied by $(-1)^{i+j}$. Thus,

$$C_{11} = a_{22} (-1)^{1+1} = a_{22}, \quad C_{12} = a_{21} (-1)^{1+2} = -a_{21},$$
$$C_{21} = a_{12} (-1)^{2+1} = -a_{12}, \quad C_{22} = a_{11} (-1)^{2+2} = a_{11}$$

Hence, adj $\mathbf{A} = \begin{bmatrix} C_{11} & C_{12} \\ C_{21} & C_{22} \end{bmatrix}^T = \begin{bmatrix} a_{22} & -a_{21} \\ -a_{12} & a_{11} \end{bmatrix}^T = \begin{bmatrix} a_{22} & -a_{12} \\ -a_{21} & a_{11} \end{bmatrix}$

It can be seen that the adj **A** *matrix of order two is a matrix having exchanged the diagonal elements and changed sign of off-diagonal elements.*
Thus,

$$\mathbf{A}^{-1} = \frac{1}{\Delta}\begin{bmatrix} a_{22} & -a_{12} \\ -a_{21} & a_{11} \end{bmatrix}$$

where $\Delta = a_{11}a_{22} - a_{12}a_{21}$.

A.5 CRAMER'S RULE

The solution of linear simultaneous equations as shown in Eqs. (A1.1) and (A1.1) can be obtained using Cramer's rule. To get the solution, the number of unknowns should be equal to the number of equations. In other words, the following condition should hold for Eq. (A1.1).

   (a) The matrix **A** should be square matrix and
   (b) The determinant of matrix **A** should not be zero[*].

Cramer's rule states that the ith component x_i, $i = 1, 2, \ldots, n$ of the Eq. (A1.2) can be written as

$$x_i = \frac{\Delta_i}{\Delta} \tag{A1.16}$$

where Δ_i is the determinant of the matrix formed by replacing ith column of matrix **A** by the known vector-matrix **B**. However, Δ is the determinant of matrix **A**.

A.6 GAUSS ELIMINATION METHOD

It is a very common direct method to solve linear equations. In this method, the first equation is used to eliminate the first unknown from the remaining equations. This process is repeated sequentially for second unknown, the third unknown, etc. until the elimination process is completed.

Gaussian elimination is the process by which matrix **A** is augmented as

$$[\mathbf{A} : \mathbf{B}]$$

and is converted to the matrix $n \times (n + 1)$ matrix

$$[\mathbf{I} : \mathbf{B}']$$

through a series of elementary row operations.

If a series of elementary row operations exits that can transform the matrix **A** into identity matrix **I**, then the application of the same set of elementary row operations will also transform the vector-matrix **B** into the solution vectors **X**. The elementary row operations consist of one of the three possible actions:

   • multiply any row by a constant.
   • interchange any two rows of the matrix
   • take a linear combination of rows and add it to another row.

If the elementary row operations are performed in such a manner that matrix **A** is transformed into an upper triangular matrix that has ones on the diagonals and zeros in the lower off-diagonal positions (known as *upper triangular matrix*). The solution obtained by this is known as *forward elimination method*, whereas the solution obtained by lower triangularization matrix is known as *backward elimination process*.

---

[*]If the determinant of the matrix **A** of dimension $n \times n$ is zero, this shows that the rank of **A** is less than n. It also indicates that the some rows or columns are dependent on one another.

$$\begin{bmatrix} 1 & x & x & . & . & x \\ 0 & 1 & x & . & . & x \\ 0 & 0 & 1 & . & . & x \\ . & . & . & & & . \\ . & . & . & & & . \\ 0 & 0 & 0 & . & . & 1 \end{bmatrix} \qquad \begin{bmatrix} 1 & 0 & 0 & . & . & 0 \\ x & 1 & 0 & . & . & 0 \\ x & x & 1 & . & . & 0 \\ . & . & . & & & . \\ . & . & . & & & . \\ x & x & x & . & . & 1 \end{bmatrix}$$

Upper triangular matrix   Lower triangular matrix

where, x = non-zero/zero elements.

EXAMPLE A.4 Find the solution of the following three equations using Cramer's rule.

$$3x_1 + 4x_2 + x_3 = 0$$
$$-x_1 + 4x_2 = 2$$
$$2x_2 + x_3 = -1.$$

Solution: These equations can be written in matrix-vector form as

$$\begin{bmatrix} 3 & 4 & 1 \\ -1 & 4 & 0 \\ 0 & 2 & 1 \end{bmatrix} \begin{bmatrix} x_1 \\ x_2 \\ x_3 \end{bmatrix} = \begin{bmatrix} 0 \\ 2 \\ -1 \end{bmatrix}$$

Determinant of **A** will be calculated using Eq. (A1.12) as

$$\Delta = 3(4-0) - 4(-1-0) + 1(-2-0) = 14$$

Now,

$$\Delta_1 = \begin{vmatrix} 0 & 4 & 1 \\ 2 & 4 & 0 \\ -1 & 2 & 1 \end{vmatrix}; \quad \Delta_2 = \begin{vmatrix} 3 & 0 & 1 \\ -1 & 2 & 0 \\ 0 & -1 & 1 \end{vmatrix}; \quad \Delta_3 = \begin{vmatrix} 3 & 4 & 0 \\ -1 & 4 & 2 \\ 0 & 2 & -1 \end{vmatrix}$$

The values of Δ_i will be $\Delta_1 = 0$, $\Delta_2 = 7$, $\Delta_3 = -28$. Thus,

$$x_1 = \frac{0}{14} = 0, \quad x_2 = \frac{7}{14} = 0.5, \quad x_3 = \frac{-28}{14} = -2.$$

Appendix

B

Trigonometric Identities and Functions

B.1 SINE AND COSINE VALUES

The values of sine and cosine for different angles can be obtained using the following circle diagram. The values on the periphery of the circle are presented as (cos θ, sin θ). For a given angle θ (written inside the circle), the first value in the small bracket is cosine and second one for sine.

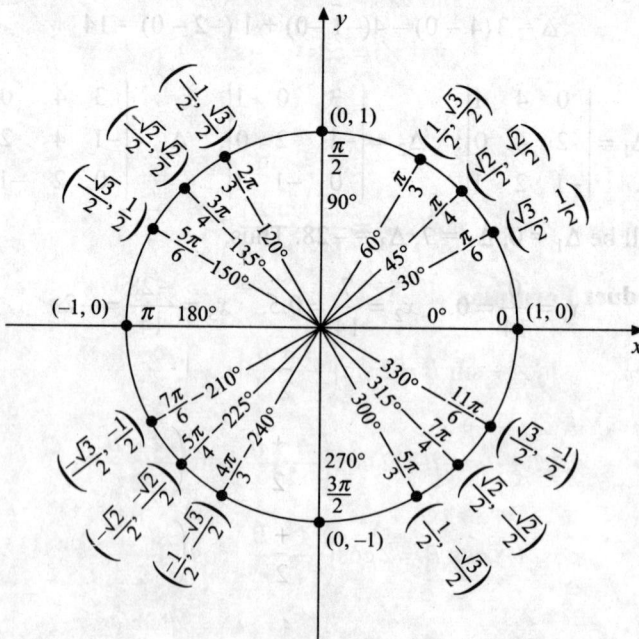

The sine and cosine values are related with each other. It is given below.

	$\sin(\pi/2 - \theta) = \cos\theta$
	$\cos(\pi/2 - \theta) = \sin\theta$
$\sin(-\theta) = -\sin\theta$	$\sin(\pi/2 + \theta) = \cos\theta$
$\cos(-\theta) = \cos\theta$	$\cos(\pi/2 + \theta) = -\sin\theta$
$\sin(\pi - \theta) = \sin\theta$	$\sin(2\pi - \theta) = -\sin\theta$
$\cos(\pi - \theta) = -\cos\theta$	$\cos(2\pi - \theta) = \cos\theta$
$\sin(\pi + \theta) = -\sin\theta$	$\sin(2\pi + \theta) = \sin\theta$
$\cos(\pi + \theta) = -\cos\theta$	$\cos(2\pi + \theta) = \cos\theta$

B.2 TRIGONOMETRIC IDENTITIES

The useful trigonometric identities are given below.

(a) Sum-Difference Formulas

$$\sin(A \pm B) = \sin A \cos B \pm \cos A \sin B$$

$$\cos(A \pm B) = \cos A \cos B \mp \sin A \sin B$$

(b) Double-Angle Formulas

$$\sin(2A) = 2\sin A \cos A$$

$$\cos(2A) = \cos^2 A - \sin^2 A$$

$$= 2\cos^2 A - 1$$

$$= 1 - 2\sin^2 A$$

$$\tan(2A) = \frac{2\tan A}{1 - \tan^2 A}$$

(c) Sum-to-Product Formulas

$$\sin A + \sin B = 2\sin\left(\frac{A+B}{2}\right)\cos\left(\frac{A-B}{2}\right)$$

$$\sin A - \sin B = 2\cos\left(\frac{A+B}{2}\right)\sin\left(\frac{A-B}{2}\right)$$

$$\cos A + \cos B = 2\cos\left(\frac{A+B}{2}\right)\cos\left(\frac{A-B}{2}\right)$$

$$\cos A - \cos B = -2\sin\left(\frac{A+B}{2}\right)\sin\left(\frac{A-B}{2}\right)$$

(d) Product-to-Sum Formulas

$$\sin A \sin B = \frac{1}{2}\left[\cos(A-B) - \cos(A+B)\right]$$

$$\sin A \cos B = \frac{1}{2}\left[\sin(A+B) + \sin(A-B)\right]$$

$$\cos A \cos B = \frac{1}{2}\left[\cos(A-B) + \cos(A+B)\right]$$

$$\cos A \sin B = \frac{1}{2}\left[\sin(A+B) - \sin(A-B)\right]$$

(e) Pythagorean Identities

$$\sin^2 A + \cos^2 A = 1$$
$$\tan^2 A + 1 = \sec^2 A$$
$$\cot^2 A + 1 = \operatorname{cosec}^2 A.$$

Appendix
C

Useful Laplace Transform Pairs

S.No.	Function Name	Function, $f(t)$	Laplace Transform, $F(s)$
1.	Unit impulse	$\delta(t)$	1
2.	Unit step	$u(t)$	$1/s$
3.	Unit ramp	$r(t) = tu(t)$	$1/s^2$
4.	Unit parabola	$p(t) = \dfrac{1}{2}t^2 u(t)$	$1/s^3$
5.	Unit doublet	$\delta'(t)$	s
6.	Constant	1	$1/s$
7.	t	t	$1/s^2$
8.	Power of $t(n \geq 2)$	$t^{n-1}/(n-1)!$	$1/s^n$
9.	Exponential	e^{-at}	$1/(s+a)$
10.	sine	$\sin \omega t$	$\omega/(s^2 + \omega^2)$
11.	cosine	$\cos \omega t$	$s/(s^2 + \omega^2)$
12.	Rectified sine	$\lvert \sin \omega t \rvert$	$\dfrac{w}{(s^2 + \omega^2)} \coth \dfrac{\pi s}{2\omega}$
13.	Function multiplies exponential*	$f(t)e^{-at}$	$F(s+a)$
14.	Differential of function	$\dfrac{d}{dt}[f(t)]$	$sF(s)$
15.	Integral of function	$\int f(t)$	$F(s)/s$

---

* Laplace transform of $f(t)$ is $F(s)$.

Bibliography

Cogdell, J.R., *Foundation of Electric Power*, Prentice Hall of India.

Husaain, Ashfaq, *Fundamentals of Electrical Engineering*, Dhanpat Rai Publications.

Nagrath, J. and D.P. Kothari, *Basic Electrical Engineering*, Tata McGraw Hill, New Delhi.

Prasad, R., *Fundamentals of Electrical Engineering*, Prentice Hall of India.

Rizzoni, Giorgio, *Principles and Applications of Electrical Engineering*, McGraw Hill, UK.

Sen, P.C., *Principles of Electrical Machine and Power Electronics*, 2nd ed., John Wiley & Sons, USA.

Smith, Ralph J., *Circuits, Devices and Systems*, John Wiley & Sons, New York.

Toro, V. Del, *Electrical Engineering Fundamentals*, Prentice Hall of India.

Valkenburg, Van, *Network Analysis*, Prentice Hall of India, New Delhi.

Wadhawa, C.L., *Network Theory*, New Age International, India.

Index

Absolute instrument, 143
AC instruments, 143
Active solar systems, 415
Admittance
   definition of, 84
Air friction damping, 148
Air gap voltage, 370
Air-core transformer, 228
All day efficiency, 253
Alternative sources of energy, 413
Alternator, 365
Ampere's Law, 204
Analog instrument, 142, 143
Angular frequency, 70
Apparent power, 103
Approximate equivalent circuit, 242, 342
Armature
   copper loss, 307
   flux, 369
   heating limit, 376
   reaction, 305, 371
   reaction voltage, 370
   resistance control, 313, 315
   voltage control, 313
   winding, 291
Attraction type, 151
Autotransformer, 254
   starters, 357
Average power, 72, 101
Average value, 73

Backward elimination process, 428
Bandwidth, 194, 195, 199
Basic requirements of shunt, 154
Bilateral network, 24
Biomass power, 418
Braking mode, 346
Bulk supply system, 5

Capacitor, 15, 18
Capacitor-run single-phase induction motor, 396
Capacitor-start capacitor-run single-phase induction
   motor, 389
Capacitor-start single-phase induction motors,
   389, 395
Circuit approach, 9, 28
   branch, 29
Coefficient of coupling, 213
Coenergy, 283
Coercive force, 236
Cogeneration, 420
Coil, 149, 291
Coloured bands, 17
Combined heat and power, 421
Commutator, 292
Comparison instruments, 144
Complementary solution, 120
Complex power, 372
Compound DC generators, 304
Compound machines, 298

Conductance
  Siemens, 16
Conductors, 291
Connected balanced load, 171
Conservation of complex power, theorem of, 104
Control and data acquisition, 7
Controlled sources 13
Controlling mechanism, 144, 146
Conventional electric energy sources, 5
Conventional energy sources, 408
Copper loss, 240, 394
Core loss, 239, 252
Core type, 229
Cosine wave, 72
Coupling
  coefficient of, 213
Coupled coils, 15, 212
Cramer's rule, 428
Critical field circuit resistance, 300
Critical speed, 300
Current coil, 157
Current density, 207
Current dependent voltage source, 13
Current source, 12, 14
  current dependent, 13, 14
Cut-off frequencies, 194, 199
Cylindrical rotor, 366

Damper windings, 380
Damping mechanism, 145, 147
DC
  generator, 12, 298
  instruments, 143
  junction, 28
  motor starters, 322
  motors, 290, 311
  series generator, 304
  shunt generator, 299
  shunt motor, 311, 313
  system, 2, 290
Decay transient, 123
Deflecting or operating mechanism, 144
Deflecting torque, 149
Deflection instruments, 143
Deflection torque, 152, 153
Delta configuration, 168
Delta-connected balanced load, 173
Delta-connection, 170
Delta-star transformation, 36, 87, 88

Dependent sources, 13
Diamagnetic materials, 205
Dielectric loss, 252
Diesel engines, 413
Differential compound machine, 298, 305
Digital instrument, 142, 143
Direct axis, 374
Direct measuring instruments, 144
Discharge transients, 122
Distributed network, 24
Distributed winding, 367
Distribution systems, 6
Distribution transformer, 227
Diversion canal, 410
Dot convention, 215
Double-cage induction motor, 359
Double-revolving field theory, 392

Eddy current loss, 238, 239
Effective (RMS) value, 72
Efficiency of an instrument, 162
Electric circuit, 11
Electric energy, 279
Electric power, 5
Electric power system, 2
Electricity, 1, 2
Electro-motive force (emf)
  equation, 11, 21, 211, 293
Electrodynamometer instruments, 153
Electrodynamometer type instrument, 157
Electromagnetic damping, 148
Electromagnetic effect, 146
Electromagnetic force, 285
Electromagnetic induction, 21
Electromechanical energy conversion, 279
Electrons current, 11
Electrostatic effect, 145
End-heating limit, 376
End-region heating limit, 378
  steady-state, 379
  subtransient, 379
  transient, 379
Energy, 18, 20
  alternative sources, 413
  dissipation, 193
  efficiency, 253
  management centre, 8
  management system, 7
  meter, 159

source, 12, 408
  stored in capacitor, 20, 120
  stored in inductor, 117
Equivalent circuit, 370, 372
  rotor, 341
Euler's formula, 74
Exponential decay, 117
Extra high voltage, 3, 5

Farad, 19
Faraday's law, 21
Ferromagnetic materials, 205
Field
  approach, 9, 28
  coil, 157
  control, 313
  control method, 314
  diversion method, 313
  heating, 377
  heating limit, 376
  winding, 291
  winding loss, 307
First-order circuit, 113
First-order differential, 120
Flat-compounded, 305
Fleming's left-hand rule, 285
Flexible AC transmission systems (FACTS), 3
Flux, 21
  density, 203
  line of, 203
  linkage, 211
  magnetic, 203
Force on conductor, 280
Force on iron, 280
Forcing function, 121
Forward elimination method, 428
Four-terminal, 15
Frequency control, 356
Friction damping, 148
Frictional and windage loss, 307
Fuel cell, 416

Galvanometers, 142
Gas power plant, 410
Gaussian elimination, 428
Generating mode, 346
Generating transformer, 227
Generation of voltage, 280

Generator, 279
Geothermal power, 418
Gravity control, 146
Green power, 5

Hall effect, 146
High-voltage direct current, 3
Homogeneous solution, 120
Hydropower, 409
Hysterisis, 237
  loop, 207, 237
  loss, 237, 239
  type, 389

Ideal inductor, 20
  flux, 21
  reluctance, 21
  self-inductance, 21, 211, 212
Ideal passive element
  bilateral, 23
  linear, 23
Ideal source, 12
Ideal transformer, 230
Ideal voltage, 12, 14
IEEE recommended equivalent circuit, 343
Impedance
  definition of, 84
Impedance diagram, 267
Impedance transformation, 233
Independent loop equations, 42
Independent node, 28
Independent source, 12
Independent voltage source, 12
Indicating instrument, 143
  performance of, 160
Inducatnce, 219
Induced emf, 232
Induced voltage, 292, 294, 336
Induced voltage in rotor, 337
Inductance, 21, 211
Induction effect, 146
Induction machines, 331
Induction motor
  equivalent, 339
  starting of, 357
Inductor, 15
Infinite bus, 372
Initial and steady-state values, 136

Instantaneous current, 101
Instantaneous power, 102, 169, 232
    three-phase ac circuit, 169
Instantaneous voltage, 101
Instrumentation systems, 142
Instruments, 142
    efficiency of, 162
Integrating instruments, 144
Iron-cored transformer, 228

Kirchhoff's law, 29
    current, 29
    voltage, 30

Lagging, 102
    leading concept, 79
    reactive power, 365
Lap winding, 292
Laplace transform, 133
Leading power, 102
Leakage flux, 230
Leakage reactance, 371
Lenz's law, 231
Line of flux, 203
Linear motion, 280
Linearity, 51
Load angle, 371
Load curve, 302
Loading effect, 161
Long shunt connection, 304
Long-shunt machine, 298
Loop, 29
Loop current, 44

Magnet, 149
Magnetic damping, 148
Magnetic effect, 145
Magnetic energy, 22
Magnetic field, 203
Magnetic field intensity, 204, 236
Magnetic flux, 203
Magnetization
    characteristic, 369
    current, 239
    curve, 237, 281, 296
    inductance, 239
    reactance, 370

Magneto hydrodynamic (hmd) generator, 419
Magneto-motive force (mmf), 28, 204, 205
Matrix
    addition/subtraction, 424
    full, 427
    inversion, 427
    methods, 423
    multiplication, 425
Maximum power transfer, 61, 62, 98
    theorem, 98
Maximum torque, 346
Mechanical load, 326
Mechanical power, 344
Mechanical torque, 345
Mesh, 29
    current, 44
    current (loop current) method, 42, 90
    equation, 43, 44
Meters, 142
Micro-turbines, 421
Millman's theorem, 64
Motional voltage, 284
Motor, 279
Motoring mode, 346
Moving coil instruments, 150
    advantages, 150, 152, 153
    disadvantages, 150, 152, 153
Moving iron (MI) instruments
    advantages, 152
    attraction type, 151
    disadvantages, 152
    repulsion type, 151
Multi-stack, 399
    stepper motors, 400
Multipliers, 155
Mutual flux, 230
Mutual inductance, 212, 219

Natural response, 120
Natural solution, 120
Network analysis approaches, 9
Neutral point, 166
Node, 28
Node pair voltage, 47
Node voltage, 47, 49
    method, 47, 91
Non-linear network, 24
Non-salient pole, 366
Normal excitation, 382

Norton's theorem, 53, 58, 96
Nuclear power plant, 412
Nuclear power station, 5
Nuclear reactor
    containment, 413
    control rods, 412
    coolant, 413
    fuel rods, 412
    moderator, 412
    shielding, 412
    steam separator, 413
Null-type instruments, 143

Ohm's law, 15
Open-circuit characteristic (OCC), 299, 369
Open-circuit test, 244
Over-compounded generator, 305
Overexcited mode, 373

Parallel capacitors, 83
Parallel case, 127
Parallel inductors, 82
Parallel resistances, 34
Parallel RLC resonance, 196
Paramagnetic materials, 205
Parameters of transformers, 243
Passive elements, 15
Passive solar systems, 415
Per unit representation, 267
    value, 264
Performance of an indicating instrument, 160
Periodic signals, 71
Permanent magnet
    dc machines, 297
    machine, 291
    moving coil, 149
    stepper motors, 402
Permeability, 205
Phase angle, 70
Phase sequence, 380
Phasor diagram, 75, 382
Phasor representation, 74, 75
Photovoltaic (pv) effect, 415
Photovoltaic cells, 415
Pole, 290
Pole-changing method, 356
Potential difference, 11
Potential coil, 157

Power angle, 371
Power calculation, 100
Power factor, 102, 104, 365
Power factor angle, 100
Power factor determination, 178
Power in three-phase circuit, 169
Power losses, 232
Power meter, 157
Power transformer, 227
Power triangle, 103
Power-angle characteristic, 373
Practical transformer, 236, 240
Practical voltage, 14
Pressure coil, 157
Primary distribution system, 6
Primary reactor starters, 358
Primary resistor starters, 358
Primary winding, 228
Pull-up torque, 346
Pulsating field, 390

Quadrature axis, 374
Quality factor, 193, 198

Rating of transformer, 236
Reactive power, 102, 193, 372
Reactive power capability, 376
Real power, 101, 175, 372
Reciprocity theorem, 63
Recording instruments, 144
Relative permeability, 205
Relative permittivity, 19
Reluctance, 21, 206
    power, 376
    type, 389
Renewable energy sources, 408
Repulsion type, 151
Residual flux density, 236
Residual magnetism, 301
Resistance-start single-phase induction motor, 389
Resistor, 15
Resonant frequency, 195
Resultant mmf, 369
Resultant torque, 394
Retentivity, 207
Right-hand rule, 203
RLC resonance, 191

Root-mean-square, 72
Rotating magnetic field, 333
Rotational motion, 284
Rotational voltage, 284
Rotor, 290
Rotor copper loss, 341
Rotor current at maximum torque, 348
Rotor equivalent circuit, 340
Rotor resistance control, 355
Run-of-river plant, 410

Salient pole machines, 290
Salient-pole, 366
Salient-pole synchronous generator, 374
Salient-pole synchronous machines, 366
Scalar multiplication, 425
Second-order circuits, 126
Secondary instruments, 143
Secondary transmission, 7
Secondary winding, 228
Self-starting motor, 380
Separately excited dc generator, 298
    external characteristics, 299
    load characteristics, 299
    terminal characteristics, 299
Separately excited dc machine, 297
Series and parallel combinations, 33
Series capacitors, 83
Series case, 127
Series inductors, 82
Series machine, 297
Series resistances, 34
Series RL circuit, 85
Series RLC resonance, 191
Shaded pole single-phase induction motors, 389, 397
Shell type, 229
Short-circuit test, 245
Short-shunt connection, 304
Short-shunt machine, 298
Shunt
    basic requirements, 154
    machine, 297
Sign of mutually induced voltage, 215
Simultaneous linear equations, 423
Sine wave, 72
Single-phase
    circuit, 164
    equivalent, 261
    load, 171

motors, 389
    series (or universal) motors, 389, 402
    synchronous motors, 389
    wattmeters, 159
Single-stack, 399
Single-stack reluctance type, 402
Sinusoidal signals, 70
Slip, 337
    control, 354
    ring induction motor, 332
Solar constant, 414
Solar power generation, 414
Solid state starters, 359
Source transformation, 24
Speed control, 313
    dc series motors using field dive, 320
    induction motor, 354
    series dc motors, 318
Split-phase single phase induction motor, 394
Spring control, 147
Square matrix
    determinant of, 426
Squirrel cage induction motors, 332
Star (or wye) connection, 166
Star-connected unbalanced load, 181
Star-connection, 170
Star-delta, 87
    starters, 358
    transformation, 36, 38, 89
Starting of induction motors, 357
Starting torque, 346,347
Static stability limit, 373
Static VAV compensator, 3
Stator, 290
Steady-state condition, 115, 119
Steady-state response, 121
Steady-state stability limit, 373
Step-down transformer, 228
Step-up transformer, 228
Stepper motors, 390, 398
Stray loss, 252
Subtransmission systems, 6
Superposition, 51, 107
    theorem, 51
Supervisory control and data acquisition, 7
Synchronization
    alternator, 379
    condenser, 366
    force, 374
    generator, 365

impedance, 371
machine, 365
motor, 380
reactance, 371
speed, 335
torque, 374

Tank circuit, 196
Taut ribbon, 149
Tertiary winding, 228
Thermal effect, 145
Thermal power plants, 408, 410
Thevenin equivalent, 54
Thevenin impedances, 344
Thevenin resistance, 55
Thevenin voltage, 54
Thevenin's equivalent, 92
Thevenin's theorem, 53, 92
Three-phase, 164
    AC generator, 165
    induction motor, 331
    power, 175
    supply system, 171
    system, 265
    transformer, 260, 263
    wattmeter, 159
Time constant, 115, 118
Time period, 71
Torque, 149
    angle, 371
    developed, 295
    (per phase), 344
    (= power/speed), 295
Total air gap, 394
Transformation ratio, 231
Transformer, 227
    parameters of, 243
    rating of, 236
Transient current, 118
Transient in RL circuit, 114
Transient in series RC circuit, 117
Transient period, 113
Transient response, 115, 121
Transients in RC circuit, 120
Transients response in series RLC circuit, 127
    critically-damped, 128

over-damped, 128
under-damped, 128
Transmission loss, 2
Transmission system, 6
Trigonometric identities, 431
Turns ratio, 231
Turn ratio of 3-Φ transformer, 260
Two-terminal, 15, 64
Two-wattmeter method, 177

Unbalanced delta-connected load, 181
Unilateral network, 24
Unit of power, 22
Unity power factor operation, 382
Universal motors, 402
Upper triangular matrix, 428

V-curve, 382
Valley dam plants, 410
Variable reluctance stepper motors, 399
Volt-ampere, 103
    auto transformer, 257
    two-winding transformer, 256
Voltage coil, 157
Voltage and current dividers, 31
Voltage regulation, 248, 306, 384
Voltage source, 12
    current dependent, 13
Voltage-dependent voltage source, 13, 14
Voltmeter design, 155

Ward leonard speed control system, 321
Wattmeter, 157
    method, 175
Wave winding, 292
Wind power, 413
Wind power generation, 414
Winding, 291
Winding factor, 369
Wound rotor, 332

Zero-voltage regulation, 250

impedance, 391
machine, 363
motor, 380
reactance, 374
speed, 335
torque, 374

link circuit, 196
Tent ribbon, 140
Tertiary winding, 258
Thermal effect, 145
Thermal power plants, 408, 410
Thevenin equivalent, 54
Thevenin impedances, 341
Thevenin resistance, 55
Thevenin voltage, 54
Thevenin's equivalent, 92
Thevenin's theorem, 53, 92
Three-phase, 164
AC generator, 168
induction motor, 354
power, 175
supply system, 171
system, 265
transformer, 260, 263
wattmeter, 156
Time constant, 115, 118
Time period, 77
Torque, 349
angle, 371
developed, 295
(two-phase), 341
(= power speed), 295
Total air gap, 394
Transformation ratio, 251
transformer, 237
parameters of, 243
rating of, 237
Time constant, 118
Transient in LC circuit, 114
Transient in series RC circuit, 112
Transient period, 117
Transfer functions, 116, 121
Transients in RC circuit, 110
Transient response in series RLC circuit, 126
critically damped, 128

over-damped, 128
under-damped, 128
transmission loss, 2
Transmission system, 6
Trigonometric identities, 431
Turns ratio, 251
Tata radio 2-φ transformer, 260
Two-terminal, 13, 64
Two-wattmeter method, 177

Unbalanced delta-connected load, 181
Unbalanced network, 24
Unit of power, 22
Unity power factor operation, 382
Universal motors, 402
Upper triangular matrix, 128

V-curve, 382
Valley dam plants, 410
Variable reluctance stepper motors, 394
Volt-ampere, 103
auto-transformer, 252
two-winding transformer, 256
Voltage built, 337
Voltage and current dividers, 31
voltage regulation, 248, 300, 384
Voltage source, 12
current dependent, 13
Voltage-dependent voltage source, 13, 14
Voltmeter design, 155

Ward leonard speed control system, 321
Wattmeter, 157
method, 175
Wave winding, 362
Wind power, 413
Wind power generation, 414
Winding, 297
Winding factor, 360
Wound rotor, 332

Zero-voltage regulation, 250